Lecture Notes in Control and Information Sciences

Edited by A. V. Balakrishnan and M. Thoma

Vol. 1: Distributed Parameter Systems: Modelling and Identification
Proceedings of the IFIP Working Conference, Rome, Italy, June 21–26, 1976
Edited by A. Ruberti
V, 458 pages. 1978

Vol. 2: New Trends in Systems Analysis
International Symposium, Versailles, December 13–17, 1976
Edited by A. Bensoussan and J. L. Lions
VII, 759 pages. 1977

Vol. 3: Differential Games and Applications
Proceedings of a Workshop, Enschede, Netherlands, March 16–25, 1977
Edited by P. Hagedorn, H. W. Knobloch, and G. J. Olsder
XII, 236 pages. 1977

Vol. 4: M. A. Crane, A. J. Lemoine
An Introduction to the Regenerative Method for Simulation Analysis
VII, 111 pages. 1977

Vol. 5: David J. Clements, Brian D. O. Anderson
Singular Optimal Control: The Linear Quadratic Problem
V, 93 pages. 1978

Vol. 6: Optimization Techniques
Proceedings of the 8th IFIP Conference on Optimization Techniques, Würzburg, September 5–9, 1977
Part 1
Edited by J. Stoer
XIII, 528 pages. 1978

Vol. 7: Optimization Techniques
Proceedings of the 8th IFIP Conference on Optimization Techniques, Würzburg, September 5–9, 1977
Part 2
Edited by J. Stoer
XIII, 512 pages. 1978

Vol. 8: R. F. Curtain, A. J. Pritchard
Infinite Dimensional Linear Systems Theory
VII, 298 pages. 1978

Vol. 9: Y. M. El-Fattah, C. Foulard
Learning Systems:
Decision, Simulation, and Control
VII, 119 pages. 1978

Vol. 10: J. M. Maciejowski
The Modelling of Systems with Small Observation Sets
VII, 241 pages. 1978

Vol. 11: Y. Sewaragi, T. Soeda, S. Omatu
Modelling, Estimation, and Their Applications for Distributed Parameter Systems
VI, 269 pages. 1978

Vol. 12: I. Postlethwaite, A. G. J. McFarlane
A Complex Variable Approach to the Analysis of Linear Multivariable Feedback Systems
IV, 177 pages. 1979

Vol. 13: E. D. Sontag
Polynomial Response Maps
VIII, 168 pages. 1979

Vol. 14: International Symposium on Systems Optimization and Analysis
Rocquentcourt, December 11–13, 1978;
IRIA LABORIA
Edited by A. Bensoussan and J. Lions
VIII, 332 pages. 1979

Vol. 15: Semi-Infinite Programming
Proceedings of a Workshop, Bad Honnef, August 30 – September 1, 1978
V, 180 pages. 1979

Vol. 16: Stochastic Control Theory and Stochastic Differential Systems
Proceedings of a Workshop of the „Sonderforschungsbereich 72 der Deutschen Forschungsgemeinschaft an der Universität Bonn"
which took place in January 1979 at Bad Honnef
VIII, 615 pages. 1979

Vol. 17: O. I. Franksen, P. Falster, F. J. Evans
Qualitative Aspects of Large Scale Systems
Developing Design Rules Using APL
XII, 119 pages. 1979

Vol. 18: Modelling and Optimization of Complex Systems
Proceedings of the IFIP-TC 7 Working Conference
Novosibirsk, USSR, 3–9 July, 1978
Edited by G. I. Marchuk
VI, 293 pages. 1979

Vol. 19: Global and Large Scale System Models
Proceedings of the Center for Advanced Studies (CAS)
International Summer Seminar
Dubrovnik, Yugoslavia, August 21–26, 1978
Edited by B. Lazarević
VIII, 232 pages. 1979

Vol. 20: B. Egardt
Stability of Adaptive Controllers
V, 158 pages. 1979

Vol. 21: Martin B. Zarrop
Optimal Experiment Design for Dynamic System Identification
X, 197 pages. 1979

Lecture Notes in Control and Information Sciences

Edited by A.V. Balakrishnan and M.Thoma
Series: IFIP TC7 Optimization Conferences

23

Optimization Techniques

Proceedings of the
9th IFIP Conference on Optimization Techniques
Warsaw, September 4–8, 1979

Part 2

Edited by
K. Iracki, K. Malanowski, S. Walukiewicz

Springer-Verlag
Berlin Heidelberg GmbH 1980

Editors
K. Iracki, K. Malanowski, S. Walukiewicz
Systems Research Institute
Polish Academy of Sciences
ul. Newelska 6, 01-447 Warszawa

ISBN 978-3-540-10081-2 ISBN 978-3-540-38253-9 (eBook)
DOI 10.1007/978-3-540-38253-9

Originally published by Springer-Verlag Berlin Heidelberg New York in 1980.

2061/3020-543210

PREFACE

These Proceedings contain most of the papers presented at the 9th IFIP Conference on Optimization Techniques held in Warsaw, Poland on September 4-8, 1979.

The Conference was sponsored by the IFIP Technical Committee on System Modelling and Optimization /TC7/.

It was organized by the Systems Research Institute of the Polish Academy of Sciences with cooperation of the Mathematical Institute of the Polish Academy of Sciences and the Institute of Automatic Control of the Technical University of Warsaw.

The European Research Office /ERO/ in London has contributed to the Conference with a grant appropriated for partial covering of travel expenses of invited speakers.

The Conference was attended by 284 scientists from 29 countries. The Conference Program comprised 8 plenary lectures, 131 contributed papers classified into 15 sections, and a Round Table Session on Systems Techniques in Economics.

The program offered a broad view of recent developments in theory and computational methods of optimization and their applications in various fields of science and technology. The emphasis was on advances in optimal control and mathematical programming techniques as well as their applications to modelling and control in particular in economics, and environmental and energy systems.

The Proceedings are composed of two volumes. The first volume contains plenary lectures, pannel addresses of the Round Table Session and the contributed papers dealing essentially with optimal control. In the second volume there are collected papers devoted essentially to mathematical programming and various applications.

The International Program Committee of the Conference consisted of:

A.V. Balakrishnan /USA/ - Chairman ,
R. Kluge /GDR/ ,
R. Kulikowski /Poland/ ,
J.L. Lions /France/ ,
G.I. Marchuk /USSR/ ,
C. Olech /Poland/ ,
L.S. Pontryagin /USSR/ ,
A.Ruberti /Italy/ ,
J. Stoer /FRG/ ,
J. Westcott /UK/ .

TABLE OF CONTENTS

MATHEMATICAL PROGRAMMING: THEORY

MATHEMATICAL PROGRAMMING: ALGORITHMS

SOFTWARE PROBLEMS

GRAPHS AND NETWORKS

SCHEDULING AND ALLOCATION PROBLEMS

APPLICATIONS: ECONOMICS AND ECONOMETRICS

APPLICATIONS: INDUSTRIAL PROCESSES

PART 1

(published as Lecture Notes in Control and Information Sciences, Vol.22)

TABLE OF CONTENTS

OPTIMAL CONTROL: PARTIAL DIFFERENTIAL EQUATIONS

OPTIMALITY CONDITIONS FOR SOME NONCONVEX PROBLEMS

K.-H. Elster and R. Nehse
Technische Hochschule Ilmenau
Sektion Mathematik, Rechentechnik
und ökonomische Kybernetik

DDR-63 Ilmenau, Am Ehrenberg

1. Introduction

The recent development in theory of nonlinear programming is characterized by the consideration of nonconvex problems. In connection with investigations about operator optimization some interesting results on separation of convex and, moreover, nonconvex sets could be obtained (cf. [3], [4], [5], [7], [10]). A survey on different directions of generalisations for separation theorems is given in [1].

As an application of these mentioned results for problems involving an objective functional we give in the present paper optimality conditions for nonconvex optimization problems. It is wellknown that the research in local and global optimality conditions is closely connected with studies about separation of sets. Using the obtained results it is possible to answer the question for a weakest condition on convexity under which the existence of a saddle-point is a necessary and sufficient optimality condition for rather general optimization problems. A suitable class of functions which is to take into consideration, is that of convex-like functions. Finally, some remarks are made about corresponding results of Ioffe/Tichomirov and about control theory, too.

Throughout the paper we consider real vector spaces.

If E and F are vector spaces, then $\mathcal{L}(E,F)$ denotes the real vector space of all linear operators $L : E \to F$. If $F = R$ (the space of the reals), we write $E^* := \mathcal{L}(E,R)$. For a non-empty subset $A \subset E$ we define

${}^iA$ - the algebraical relative interior of A, i. e.

${}^iA := \{x \in A / \forall y \in {}^1A \;\exists\, \varepsilon := \varepsilon(x,y) > 0 \;\; \forall \lambda \in (-\varepsilon,\varepsilon) : x + \lambda(y-x) \in A\}$,

where ${}^{l}A$ means the affine manifold spanned by A;
${}^{c}A$ - the convex hull of A;
K(A) - the cone generated by A, i. e. $K(A) := \bigcup_{\lambda \in R_+} \lambda A$,

where R_+ means the non-negative real line.
For a non-empty subset A of a product space $E \times F$ we define the projection $P_E(A)$ of A on E according to

$$P_E(A) := \{x \in E / \exists y \in F : (x,y) \in A\}.$$

The vector space F is called quasiordered if the order-relation "$\lesssim$" in F is reflexive, transitive and compatible with the structure of the vector space F, i. e. the order cone $K_F := \{y \in F / y \gtrsim 0\}$ is a convex cone such that $0 \in K_F$.

2. Some Remarks on Separation of Sets

In [1] the assertions are restricted to convex sets (separation of finite families; separation in product spaces, in projective spaces, in convexity spaces). But in the following we shall consider nonconvex sets separated by affine manifolds or hyperplanes, respectively.

Theorem 2.1 (cf. [11]):

Let A and B be subsets of a vector space $\bar{E}$ being a product of vector spaces $E \times F$ such that the following conditions are satisfied:

(a) F is quasiordered and conditionally complete,

(b) $(0,y) \in {}^{c}K(A-B) \Longrightarrow y \gtrsim 0$,

(c) ${}^{l}P_E\,[{}^{c}K(A-B)] \subset P_E\,[{}^{c}K(A-B)]$.

Then there are $L \in \mathcal{L}(E,F)$ and $y^0 \in F$ such that

(2.1) $L(x^1) - y^1 \lesssim y^0 \lesssim L(x^2) - y^2 \quad \forall (x^1,y^1) \in A, \ \forall (x^2,y^2) \in B.$

A proof of this theorem is given in [7], [8] in an extensive form. Using Theorem 2.1 we may obtain new results in operator-optimization (see [3], [9]), but it is possible to extend results for usual optimization problems, too. For this the following separation theorems (as special cases of Theorem 2.1) are more convenient.

Corollary 2.2 (cf. [5]):

Let A and B be subsets of a topological vector space $\bar{E}$, let $\operatorname{int} {}^{c}K(A-B) \neq \emptyset$. A and B can be separated properly by a closed hyperplane if and only if $0 \notin \operatorname{int} {}^{c}K(A-B)$.

Corollary 2.3 (cf. [5]):

Let A and B be subsets of a vector space $\bar{E}$, let ${}^{ic}K(A-B) \neq \emptyset$. A and B can be separated properly by a hyperplane if and only if $0 \notin {}^{ic}K(A-B)$.

It was proved in [4], [10] that the conditional completeness of F and condition (b) in Theorem 2.1 are necessary conditions for (2.1), while (c) is a more "technical" assumption which may be weakened but can't be dropped.

Since, by construction, $(0,0) \in {}^{c}K(A-B)$ and, therefore, $0 \in P_E[{}^{c}K(A-B)]$, condition (c) in Theorem 2.1 is equivalent to

(2.2) $\quad 0 \in {}^{i}P_E[{}^{c}K(A-B)]$.

This follows from

Lemma 2.4:

Let E be a vector space, and let $C \subset E$ be a convex cone. Then ${}^{l}C \subset C$ and $0 \in C$ if and only if $0 \in {}^{i}C$.

Proof: If ${}^{l}C \subset C$, then ${}^{l}C = C$. ${}^{l}C$ is a vector space, since $0 \in C$. Hence ${}^{l}C = {}^{il}C = {}^{i}C$ and $0 \in {}^{i}C$. Conversely, let $0 \in {}^{i}C$. C is a cone and, therefore, ${}^{i}C$ is a cone, too. $C = -C$ follows easily by $0 \in {}^{i}C$. Thus we obtain ${}^{l}C = C - C = C + C \subset C$.

3. A Global Optimality Condition

In several papers of the last few years generalizations of local optimality conditions using certain differentiability assumptions have been studied. Simultaneously, many authors have used weakened assumptions on convexity. Nevertheless, in the field of global necessary optimality conditions there are no assertions including weakened convexity for sets and functions. In this section we shall give such a theorem. Moreover, we characterize some classes of nonconvex mappings for which a global necessary optimality condition holds.

We consider the problem

(P) $\quad f(x) \longrightarrow \min$

$\quad x \in G,\ G := \{x \in P_o \,/\, -g(x) \in K_W\}$

where E is a vector space, W is a quasiordered vector space with the order-cone K_W, D(g) and D(f) are subsets of E;

$P_o := D(g) \cap D(f) \neq \emptyset$, $f : D(f) \rightarrow R$, $g : D(g) \rightarrow W$.

Let $x^o \in D(f)$ be fixed. Let $A \subset W \times R$ be a set defined by

(3.1) $\quad A := \{(g(x) + k,\ f(x) - f(x^o) + \lambda)/x \in P_o,\ k \in K_W, \lambda \in R_+\}$.

We need the following conditions (cf. Theorem 2.1 and (2.2))

(K) $(0,\mu) \in {}^{c}K(A) \Longrightarrow \mu \geq 0$,

(R) $0 \in {}^{i}P_W [{}^{c}K(A)]$.

Clearly, if $x^0 \in G$ and (K) is fulfilled, then x^0 is a solution of (P). Namely, if $\bar{x} \in G$, then $-g(\bar{x}) \in K_W$ holds and, for a suitable $\bar{k} \in K_W$, we get $g(\bar{x}) + \bar{k} = 0$. Hence $f(\bar{x}) - f(x^0) + \lambda \geq 0$ for all $\lambda \in R_+$, that means $f(\bar{x}) \geq f(x^0)$ for all $\bar{x} \in G$.

In order to get another equivalent condition for (R) let us prove (cf. [8])

<u>Lemma 3.1:</u>

Let C be a convex subset of a vector space W, let ${}^{i}C \neq \emptyset$. Then $0 \in {}^{i}C$ if and only if $0 \in {}^{i}K(C)$.

Proof: Let $0 \in {}^{i}K(C)$. Assuming $0 \notin {}^{i}C$, then, by Corollary 2.3, there exists $u \in W^*$, $u \neq 0$, such that $\langle u,x\rangle \geq 0$ for all $x \in C$. Therefore $\langle u,x\rangle \geq 0$ for all $x \in K(C)$.

From $0 \in {}^{i}K(C)$ we obtain $\langle u,x\rangle = 0$ for all $x \in K(C)$.

This is a contradiction to the proper separation of the sets $\{0\}$ and C. The converse implication is proved in [2], Satz 2.

Using the definition of the set A and Lemma 3.1 it is easy to see that each of the following conditions is equivalent to (R):

$$0 \in {}^{ic}K\,[g(P_0) + K_W],$$

$$0 \in {}^{ic}\,[g(P_0) + K_W].$$

Now, we are enabled to give global optimality condition using the Lagrangian L of (P) defined by

$$L(x,v) := f(x) + \langle v,g(x)\rangle\,, \quad x \in P_0,\ \ v \in K_W^*\,,$$

where $v \in K_W^*$ means that $\langle v,x\rangle \geq 0$ holds for all $x \in K_W$.

<u>Theorem 3.2</u> (cf. [3], [7], [11]):

(1) If $x^0 \in G$, (K) and (R) are fulfilled, then there exists a $v^0 \in K_W^*$ such that

(3.2) $L(x^0,v) \leq L(x^0,v^0) \leq L(x,v^0) \quad \forall x \in P_0,\ \forall v \in K_W^*$

and $\langle v^0,g(x^0)\rangle = 0$, that means (x^0,v^0) is a saddle-point of the Lagrangian with respect to $P_0 \times K_W^*$.

(2) Let K_W be algebraically closed. Let ${}^{i}K_W \neq \emptyset$ or let exist a countable algebraical basis of W.

If (x^0,v^0) is a saddle-point of the Lagrangian, then x^0 is a solution of (P).

Our convexity-condition (K) is a weakest condition under which (3.2) holds. In order to show that we assume (3.2) is true. Then because of $v^0 \in K_W^*$

$$0 = \langle v^0, g(x^0)\rangle \leqq f(x) - f(x^0) + \lambda + \langle v^0, g(x) + k\rangle$$
$$\forall x \in P_0, \forall \lambda \in R_+, \forall k \in K_W.$$

Therefore, the set A defined above is contained in the halfspace $H^+ := \{(y,\mu) / \mu + \langle v^0, y\rangle \geqq 0\}$.
Then ${}^cK(A) \subset H^+$ holds, too. Hence $0 \leqq \mu + \langle v^0, y\rangle$ for all $(y,\mu) \in {}^cK(A)$ and $(0,\mu) \in {}^cK(A)$ implies $\mu \geqq 0$.

Therefore, Theorem 3.2 is the most universal global optimality condition (with respect to convexity).
As we will show in the next assertions, Theorem 3.2 fills out a gap in the theory of necessary optimality conditions given by Ioffe/Tichomirov [6], Chapter 1.
A first result in this direction is contained in

Lemma 3.3:

Let A be given by (3.1).

(1) If K(A) is convex, then the following condition is equivalent to (K):

(3.3) $$(0, \mu_1) \in A \Rightarrow \mu_1 \geqq 0.$$

(2) Let K(A) be convex and let $x^0 \in G$.
x^0 is a solution of (P) if and only if (3.3) is fulfilled.

(3) If the set A is convex and (3.3) holds, then (K) is satisfied.

Proof: (1) It is clear that (K) implies (3.3). Now, let (3.3) be true and $K(A) = {}^cK(A) = K({}^cA)$. Let $(0,\mu) \in K(A)$. If $\mu = 0$, then $\mu \geqq 0$ holds trivially. If $\mu \neq 0$, then there is a $\nu > 0$ such that $\nu(0,\mu) \in A$. Using (3.3) we get $\mu = \nu^{-1}\mu_1 \geqq 0$.
(2) and (3) are clear by use of the remark given above.
Useful results are included in

Lemma 3.4 (cf. [7]):

Let A be given by (3.1) and let $x^0 \in G$ be a solution of (P). (K) is fulfilled if for arbitrary $x^1, x^2 \in P_0$, $k^1, k^2 \in K_W$, λ_1, λ_2, μ_1, μ_2 of R_+ and $\lambda \in (0,1)$ there are $x^3 \in P_0$, $k^3 \in K_W$, λ_3, $\mu_3 \in R_+$ such that

$$\lambda\mu_1(g(x^1)+k^1) + (1-\lambda)\,\mu_2(g(x^2)+k^2) = \mu_3(g(x^3)+k^3),$$
$$\lambda\mu_1(f(x^1)+\lambda_1-f_0)+(1-\lambda)\,\mu_2(f(x^2)+\lambda_2-f_0) = \mu_3(f(x^3)+\lambda_3-f_0)$$

where $f_o := f(x^o)$.
Clearly, the condition named in Lemma 3.4 ensures the convexity of K(A). The proof is easy by Lemma 3.3.

Now, let us give an essential condition which is (in connection with Lemma 3.3) sufficient for (K).

Lemma 3.5 (cf. [12]):

Let A by given by (3.1). A is convex if and only if the pair $(g,f):P_o \to W \times R$ is convex-like on P_o, that means: for each $x^1, x^2 \in P_o$ and any $\lambda \in (0,1)$ there is $x^3 := x^3(x^1,x^2,\lambda) \in P_o$ such that

$$\lambda g(x^1) + (1-\lambda)g(x^2) - g(x^3) \in K_W,$$
$$\lambda f(x^1) + (1-\lambda)f(x^2) - f(x^3) \geq 0.$$

As a condition of convexity given by Ioffe/Tichomirov [6], S. 74, the convex-likeness of (g,f) is used. While in [6] the convex-likeness is utilized in order to develop only local necessary optimality conditions, it is easy to see (by Lemma 3.5, Lemma 3.3, Theorem 3.2) that this condition for (g,f) in connection with (R) is also sufficient in order to derive global necessary optimality conditions. Furthermore, the convex-likeness of (g,f) on P_o is not necessary. This is shown by the following example (see [3], [7]).
We consider

$$W := R^2,\ K_W := R^2_+,\ E := R^2,$$

$$f(x_1,x_2) := |\sin^2 x_1 + \cos^2 x_2 - 1|,\quad g(x_1,x_2) := \begin{pmatrix} -x_1^2 - x_2^2 + \frac{1}{4} \\ x_1 + x_2 - \frac{3}{2} \end{pmatrix},$$

$$P_o := \{(x_1,x_2) \in R^2 \,/\, 0 \leq x_1 \leq 1,\ 0 \leq x_2 \leq 1\}.$$

f is not convex, not quasiconvex, not concave, and not differentiable; the feasible region G is not convex; A defined by (3.1) is not convex (for $x^1 = (1,\frac{1}{2})$, $x^2 = (1,0)$, $\lambda \in (0,1)$ and $k^1 = k^2 = (0,0)$ there are no $x^3 \in P_o$ and $k^3 \in R^2_+$ such that $g(x^3) + k^3 = \lambda g(1,\frac{1}{2}) + (1-\lambda)g(1,0)$), and, by Lemma 3.4, the pair (g,f) is not convex-like on P_o. Nevertheless, $K(A) = R^2 \times R_+$ is convex and, therefore, Theorem 3.2 is available for our problem:

$x^o = (\frac{1}{4}\sqrt{2}, \frac{1}{4}\sqrt{2})$ is a solution of that problem and $(x^o, v^o) = (\frac{1}{4}\sqrt{2}, \frac{1}{4}\sqrt{2}, 0, 0)$ is a saddle-point of the Lagrangian.

Since also in [6] conditions for the convex-likeness of (g,f) are omitted, it is useful to give some sufficient conditions for the convex-likeness of (g,f) in the following theorem.

Theorem 3.6 (cf. [11]):

(1) Let E be a vector space, let $D(f) \subset E$ with $D(f) \neq \emptyset$, let $\tilde{f} : D(\tilde{f}) \to W$ be isoton and convex on a convex set $D(\tilde{f})$, where $D(\tilde{f}) \supset f(D(f))$.
Then the mapping (g,f) is convex-like with $g(x) := \tilde{f}(f(x))$.
An analogous result we obtain by changing f and g.

(2) Let E be a vector space, let $P_o \subset E$ be convex, and let (g,f) be strongly v-pseudoconvex, that means: there exist a $p : P_o \times P_o \to R_+ \setminus \{0\}$ and, for any $x^2 \in P_o$, an $L_g \in \mathcal{L}(E,W)$ and a $u_f \in E^*$ such that
$$p(x^1,x^2)[g(x^1) - g(x^2)] - L_g(x^1-x^2) \in K_W \quad \forall x^1 \in P_o,$$
$$p(x^1,x^2)[f(x^1) - f(x^2)] - \langle u_f, x^1 - x^2\rangle \geqq 0 \quad \forall x^1 \in P_o.$$
If the equation
$$[(1-\lambda)p(x^1,x) + \lambda p(x^2,x)]\, x = \lambda p(x^2,x)x^1 + (1-\lambda)p(x^1,x)x^2$$
has a solution $\bar{x} := \bar{x}(x^1,x^2,\lambda)$ for any $x^1,x^2 \in P_o$ and any $\lambda \in (0,1)$, then (g,f) is convex-like on P_o.

(3) Let E be a topological vector space, let $P_o \subset E$ be convex, and let (g,f) be strongly v-pseudoconvex on P_o. If $p(x,\cdot)$ is continuous on P_o for each fixed $x \in P_o$, then (g,f) is convex-like on P_o.

Since every functional in convex-like on any part of the real line the first property named in Theorem 3.6 is convenient to obtain simple examples.

Let $E := R^2$, $W = R$, $K_W = R_+$,
$$g(x_1,x_2) = x_1^2 - x_2^3,\quad f(x_1,x_2) = e^{g(x_1,x_2)} + g(x_1,x_2) = e^{x_1^2 - x_2^3} + x_1^2 - x_2^3,$$
$$P_o := \{(x_1,x_2) \in R^2 / |x_1| \leqq 1,\ x_2 \leqq \tfrac{1}{2}x_1^2 + \tfrac{1}{2}\}.$$
The associate optimization problem (P) is very strong nonconvex, but Theorem 3.2 is available (cf. Theorem 3.6, Lemma 3.5, Lemma 3.3).

A geometric consideration says that only one of the points

$$x_{o1} = (0,\tfrac{1}{2}),\ x_{o2} = \left(\sqrt{-1+\sqrt{\tfrac{8}{3}}},\ \tfrac{1}{2}\sqrt{\tfrac{8}{3}}\right),\ x_{o3} = \left(-\sqrt{-1+\sqrt{\tfrac{8}{3}}},\ \tfrac{1}{2}\sqrt{\tfrac{8}{3}}\right)$$

can be a solution of (P). But it is possible to show that there is no $l_o \geq 0$ such that (x_{o2}, l_o) and (x_{o3}, l_o), respectively, is a saddle-point of the Lagrangian. Since $(0, \frac{1}{2}, 0)$ is a saddle-point, x_{o1} is the unique solution of (P).

As it was shown in [6], convex-like mappings (g,f) may be used fruitfully in problems of optimal control, too. Therefore, Theorem 3.6 is also applicable for such problems.

References

[1] Deumlich, R.; Elster, K.-H.; Nehse, R.: Recent results on separation of convex sets. Math. Operationsforsch. u. Statist., Ser. Optimization 9(1978), 273 - 296.

[2] Elster, K.-H.; Nehse, R.: Ein Bipolarensatz. Math. Nachr. 62(1974), 111 - 119.

[3] Elster, K.-H.; Nehse, R.: Nichtkonvexe Optimierungsprobleme. 21. Internat. Wiss. Koll. TH Ilmenau 1976, Vortragsauszüge Reihe B_1, 69 - 72.

[4] Elster, K.-H.; Nehse, R.: Necessary and sufficient conditions for the order-completeness of partially ordered vector spaces. Math. Nachr. 81(1978), 301 - 311.

[5] Elster, K.-H.; Nehse, R.: Separation of certain non-convex sets. Proc. of the Conf. MMÖ in Smolenice (CSSR) 1979 (to appear).

[6] Ioffe, A. D.; Tichomirov, V. M.: Theorie der Extremalaufgaben. Deutscher Verlag d. Wiss., Berlin 1979.

[7] Nehse, R.: Some general separation theorems. Math. Nachr. 84(1978), 319 - 327.

[8] Nehse, R.: Ordnungstheoretische Untersuchungen zu nichtkonvexen Optimierungsproblemen. Diss. B, TH Ilmenau 1978.

[9] Nehse, R.: Methoden der konvexen Optimierung für nichtkonvexe Probleme. Internat. wiss. Tagung "Math. Optimierung - Theorie und Anwendungen" Eisenach 1978, Vortragsauszüge, 59 - 62.

[10] Nehse, R.: Separation of two sets in a product space. Math. Nachr. (to appear).

[11] Nehse, R.: Problems in connection with the Hahn-Banach-Theorem. Proc. of the VIIth Summer School "Nonlinear Analysis" Berlin 1979 (to appear).

[12] Yu, P. L.: Cone convexity, cone extreme points, and nondominated solutions in decision problems with multiobjectives. J. Optimiz. Theory Appl. 14(1974), 319 - 377.

A GENERAL PERTURBATION THEORY FOR OPTIMIZATION PROBLEMS

B. Gollan
Mathematisches Institut der Universität
87 Würzburg, West Germany

1. Introduction

The aim of this paper is to report some new results in perturbation theory. Because of space limitations we can only sketch the ideas and have to refer readers interested in a complete presentation to Refs. [3 , 4].

We deal with optimization problems of the following form:

$$\begin{aligned} \text{minimize} \quad & f(x) \quad \text{subject to} \\ & x \in A \\ & g(x) \in B + p_1 \\ & h(x) = p_2 \end{aligned} \qquad (P)_p$$

where $f: X \to \mathbb{R}$, $g: X \to Y$, $h: X \to Z$, with locally convex linear spaces X , Y and Z. A is an arbitrary subset of X and $B \subseteq Y$ a convex cone with nonempty relative interior. $p = (p_1, p_2) \in Y \times Z$ plays the role of perturbations and for obvious reasons $(P)_o$ is called the unperturbed problem.

We limit ourselves to these standard perturbations. If $F := (f,g,h)$ is assumed to be Gateaux differentiable, then more general perturbations $f(x,p)$, $g(x,p) \in B$, $h(x,p) = 0$ can be considered as well. In [3 , 4] it is shown, how these can be reduced to the case of standard perturbations.

If $x_o(p)$ denotes an optimal solution of problem $(P)_p$, then $V(p) := f(x_o(p))$ defines the optimal value function. We are interested in the local behaviour of V at $p = 0$, i.e. in the effect of small perturbations in the constraints to the optimal value. To be more precise, we want to get estimates for

$$v(p) := \liminf_{\substack{\varepsilon \to 0+ \\ r(\varepsilon) = o(\varepsilon)}} (V(\varepsilon p + r(\varepsilon)) - V(0)) / \varepsilon \qquad (1)$$

taken over all functions r of type $o(\varepsilon)$, i.e. $\lim_{\varepsilon \to o} r(\varepsilon) / \varepsilon = 0$. If even the limit exists, this is the well known directional Hadamard derivative of V at 0 in direction p, denoted by $V'(0)p$.

The estimates for $v(p)$ will be determined by multipliers belonging to an optimal solution x_o of the unperturbed problem. Hence before entering the topic of perturbation theory we first make some comments on multipliers.

2. Multiplier rules and multiplier sets

Depending on the structure of an optimization problem and on the particular regularity assumptions being made, there is a lot of multiplier rules in literature, all of them resulting in the existence of multipliers. Here we want to dispens from particular regularity assumptions and only look at some crucial common properties of all these multiplier rules.

One starts with an optimal solution x_o of the unperturbed problem. Roughly speaking first one needs a convex cone K, that approximates the set F(A). In optimal control problems Pontryagin's cone of attainibility is an example of a possible choice of K. Then there is another convex cone T representing the set of points that are out of reach because of the optimality of x_o.

A <u>multiplier</u> λ is an element of the algebraic dual of $\mathbb{R} \times Y \times Z$, that defines a hyperplane separating the cones K and T (if such a hyperplane exists). In general λ is only a linear functional, but not necessarily continuous.

Of course λ need not be unique. Obviously the better (i.e. larger) the approximation K is, the smaller is the set of multipliers, a fact, which will be important later.

In the sequel the following notations are useful:

$$\begin{aligned}
\Lambda(K) &= \{ \lambda = (\lambda_o,\lambda_1,\lambda_2) \in \mathbb{R} \times Y' \times Z' \mid \lambda k \geq 0 \text{ for all } k \in K , \\
&\qquad\qquad \lambda t \leq 0 \text{ for all } t \in T \} \\
\Lambda_o(K) &:= \{ \lambda \in \Lambda(K) \mid \lambda_o = 0 \} \qquad\qquad (2) \\
\Lambda_1(K) &:= \{ \lambda \in \Lambda(K) \mid \lambda_o = 1 \}
\end{aligned}$$

If K is clear from the context, we simply write Λ, Λ_o, Λ_1. We note, that Λ_o and Λ_1 characterize Λ completely. If $\Lambda_o = \emptyset$, then the underlying optimal solution x_o is called <u>normal</u>. The <u>abnormal</u> case $\Lambda_o \neq \emptyset$ is a degenerate situation and in optimization most people try to avoid it by assuming regularity conditions. In this approach it can be considered as well. But it is useful to distinguish further between the cases (i) $\Lambda_o \neq \emptyset$ and $\Lambda_1 \neq \emptyset$

and (ii) $\Lambda_o \neq \emptyset$ and $\Lambda_1 = \emptyset$.

What (i) and (ii) mean with respect to perturbation theory, is carried out in chapter 4 of this paper.

3. Perturbation theory

In the last years quite some work has been done in this field, and we want to comment first on those results, that are closely related to ours. For different types of optimization problems and under different assumptions perturbation results were obtained by Levitin (see [9,10]), Gauvin and Tolle, [2], Lempio and Maurer, [7, 8], and Auslender, [1]. The common characteristics of these papers are about the following:

3.1 Depending on the respective optimization problem suitable constraint qualifications are chosen. Sometimes additional regularity assumptions are made.

3.2 A particular approximation K of F(A) is chosen. In case of differentiable constraints and if A = X, $K = \{ F'(x_o)x \mid x \in X \}$.

3.3 A multiplier rule holds, that guarantees a compact (resp. weak-*-compact) set $\Lambda_1 \neq \emptyset$ of continuous multipliers. $\Lambda_o = \emptyset$ as a consequence of the regularity conditions, i.e. only the case of normal solutions is treated.

3.4 The result is of the following form:

$$\liminf_{\epsilon \to 0+} (V(\epsilon p) - V(0))/\epsilon \leq \sup_{\Lambda_1} (-\lambda_1 p_1 - \lambda_2 p_2)$$

3.5 The main tool of the proof is modern duality theory.

Our approach differs significantly from the above. It can be characterized as follows:

3.6 No regularity assumptions are made.

3.7 Some convex cone K contained in the tangent cone to the set F(A) at $F(x_o)$ is chosen. It will come out below, that K should be chosen as large as possible to obtain good results.

3.8 $\Lambda(K)$, $\Lambda_o(K)$ and $\Lambda_1(K)$ are defined as in (2). Here $\Lambda_1(K)$ may also be empty or unbounded.

3.9 We obtain the following perturbation results, with v(p) defined as in (1) :

a) If $\Lambda_1(K) \neq \emptyset$, then $v(p) \leq \sup_{\Lambda_1(K)} (-\lambda_1 p_1 - \lambda_2 p_2)$.

b) If $\Lambda_1(K) = \emptyset$ and $\sup_{\Lambda_o(K)} (-\lambda_1 p_1 - \lambda_2 p_2) < \infty$, then $v(p) = -\infty$.

3.10 The proof relies only on a general separation theorem (see [7], Theorem 2.1).

Remarks.

We want to point out several ways, how known perturbation results can be improved by the theory described by 3.6 through 3.10.

a. Since no continuity, differentiability or regularity assumptions are made about the constraints, the general theory can be applied relatively easily to quite different types of optimization problems. For differentiable and nondifferentiable problems and several types of optimal control problems this has been worked out in detail in [3].

b. The linear space K defined in 3.2 may be only a poor approximation of the set F(A). As a consequence it may happen, that the upper bound in 3.4 is only a rough estimate and does not say too much about the value function. [2] presents a very simple example of this effect. In such situations convex approximations K defined via derivatives of F of higher than first order may lead to much better results: a larger set K may decrease the set $\Lambda(K)$ (see (2)) and thus yield a better upper bound (or even the exact value) in 3.9 . In case of finite dimensional spaces Y and Z such a theory is presented in [5]. In the example of [2] mentioned above second order approximations suffice to obtain even the exact values of v(p).

c. We obtain results also in case of abnormality. It is true that abnormal solutions are rather the exception than the rule, but they show up at points of special interest. This topic is treated in the following chapter.

4. Abnormality and perturbed problems

The main purpose of this chapter is to create some intuitive feeling for the meaning of abnormality. Nevertheless, the considerations are far from being heuristical and can be made precise in the following way (see [3 , 4]) : Under few additional assumptions any of the situations 4.1 through 4.3 listed below is sufficient for $(P)_o$ to have only abnormal solutions.

A solution x_o of $(P)_o$ is usually called abnormal, if $\Lambda_o \neq \emptyset$. In this case either $\Lambda_1 = \emptyset$ or $\Lambda_1 + \Lambda_o \subseteq \Lambda_1$, and since $\Lambda_o \cup \{0\}$ is a convex cone, Λ_1 is either empty or unbounded. Hence, if $\Lambda_1 \neq \emptyset$, there are directions p, such that $s_1(p) := \sup \{ -\lambda_1 p_1 - \lambda_2 p_2 \mid \lambda \in \Lambda_1 \}$ is infinite for at least one p. If the problem is not totally pathological, there will be other directions p, such that $s_1(p)$ is finite.

If $s_1(\bar{p}) = \infty$ for some $\bar{p}$, this may correspond to one of the following situations:

4.1 At least for small positive ε problem $(P)_{\varepsilon p}$ has no feasible solutions.

4.2 V is continuous at $p=0$, but has a vertical tangent in direction $\bar{p}$ with $V'(0)\bar{p} = \infty$ (i.e. V increases rapidly in direction $\bar{p}$).

4.3 V is discontinuous at $p = 0$ with a positive jump in direction $\bar{p}$.

If $\Lambda_1 = \emptyset$, then $s_o(p) := \sup \{ -\lambda_1 p_1 - \lambda_2 p_2 \mid \lambda \in \Lambda_o \}$ is either infinite or zero, since $\Lambda_o \cup \{0\}$ is a cone. If $s_o(\bar{p}) = 0$ for some $\bar{p}$, this may correspond to one of the following situations:

4.4 V is continuous at $p = 0$, but has a vertical tangent in direction $\bar{p}$ with $V'(0)\bar{p} = -\infty$ (i.e. V decreases rapidly in direction $\bar{p}$).

4.5 V is discontinuous at $p = 0$ with a negative jump in direction $\bar{p}$.

There seems to be little hope to make these correspondences strong. For instance there are examples, where the first order necessary conditions admit both multipliers $\lambda \in \Lambda_o$ and $\bar{\lambda} \in \Lambda_1$, while higher order conditions reduce the multiplier set such that $\Lambda_o = \emptyset$. This shows, that some more work has to be done to clarify the concept of a normality.

Concluding we give some simple examples, where different types of abnormality occur. In each example $X = \mathbb{R}^2$ and the elements are denoted by (x,y).

<u>Example 1</u>

Minimize y subject to $y - x^2 = p_1$ and $y^3 = p_2$.

$x_o = (0,0)$ is the optimal solution of the unperturbed problem and K can be chosen as $K = \{ (u,u,0) \mid u \in \mathbb{R} \}$. Then $\Lambda_o = \{ (0,0,\lambda_2) \mid \lambda_2 \neq 0 \}$, and $\Lambda_1 = (1,-1,0) + \Lambda_o$. $s_1(p) = +\infty$, if $p_2 \neq 0$, and $s_1(p) = p_1$, if $p_2 = 0$. From the constraints one can check directly: If $p_2 < 0$, then case 4.1 occurs, and if $p_2 > 0$, then case 4.2. If $p_2 = 0$ and $p_1 > 0$, $s_1(p) = p_1$ seems to contradict the fact, that $(P)_{(\varepsilon p_1,0)}$ has no feasible solutions for $\varepsilon > 0$. But $(P)_{(\varepsilon p_1,\varepsilon^3 p_1{}^3)}$ has a feasible solution and since $p = (p_1,0)$ is tangent to the curve $(\varepsilon p_1,\varepsilon^3 p_1{}^3)$ at $\varepsilon = 0$, this curve is a candidate in the definition of $v(p)$. $v(p) = s_1(p) = p_1$ holds here, too.

<u>Example 2</u>

Minimize y^2 subject to $\sin x - y^3 = p_1$ and $x - y^2 = p_2$.

$x_o = (0,0)$ is the optimal solution of the unperturbed problem. First order approximations of F result in $K_1 = \{ (0,u,u) \mid u \in \mathbb{R} \}$, which is only a poor approximation of the image set. Using second order derivatives (see [3, 5]) one obtains $K_2 = K_1 + \{ (a,0,-a) \mid a \geq 0 \}$ as a better approximation. Hence $\Lambda_o(K_2) = \{ (0,b,-b) \mid b > 0 \}$ and $\Lambda_1(K_2) = \{ (1,b,-b) \mid b \geq -1 \}$.

If $\bar{p} = (0,p_2)$ with $p_2 > 0$, then $s_1(\bar{p}) = \infty$. Here case 4.3 occurs, since $V(0,0) = 0$, but $\lim_{\varepsilon \to 0+} V(r(\varepsilon),\varepsilon p_2) = d > 0$, where r is any function of type $o(\varepsilon)$ and d is the unique solution of $\sin(y^2) - y^3 = 0$.

Example 3   Minimize y subject to $y^2 - x \le p_1$ and $x \le p_2$.
$x_o = (0,0)$ is optimal for $p = 0$ and we may choose $K = \{ (w,-u,u) \mid u,w \in \mathbb{R} \}$. Then $\Lambda = \Lambda_o = \{ (0,\lambda_1,\lambda_2) \mid \lambda_1 = \lambda_2 > 0 \}$. If $p_1 + p_2 \ge 0$, then $s_o(p) = 0$ and hence $v(p) = -\infty$. This corresponds to a situation of type 4.4, as one can check directly (e.g. $V(\varepsilon p_1,0) = -(p_1)^{1/2}$).

References

[1] Auslender,A.,Differentiable Stability in Non Convex and Non Differentiable Programming, Mathematical Programming Study 10, Edited by P.Huard, North Holland, Amsterdam, 1979.

[2] Gauvin,J.,and Tolle,J.W.,Differential Stability in Nonlinear Programming,SIAM J. on Control and Optimization,Vol.15,No.2,1977.

[3] Gollan,B.,Störungstheorie für abstrakte Optimierungsprobleme mit Anwendungen auf die Theorie optimaler Steuerungen,Dissertation, Universität Würzburg,1979.

[4] Gollan,B.,Perturbation Theory for Abstract Optimization Problems, Preprint No.47,Mathematisches Institut,Universität Würzburg,1979.

[5] Gollan,B.,Higher Order Necessary Conditions for an Abstract Optimization Problem,Preprint No.48,Mathem.Institut,Univ.Würzburg,1979.

[6] Klee,V.,Separation and Support Properties of Convex Sets - aSurvey, Control Theory and the Calculus of Variations, Edited by A.V. Balakrishnan, Academic Press, New York, 1969.

[7] Lempio,F.,and Maurer,H.,Differentiable Perturbations of Infinite Optimization Problems, Lecture Notes in Economics and Mathematical Systems,Vol.157, Edited by R.Henn,B.Korte and W.Oettli, Springer, Berlin - Heidelberg - New York, 1978.

[8] Lempio,F.,and Maurer,H.,Differential Stability in Infinite Dimensional Nonlinear Programming,to appear in Applied Mathematics and Optimization.

[9] Levitin,E.S.,On the Local Perturbation Theory of Mathematical Programming in a Banach Space,Soviet Mathematics Dokl.,Vol.16,No.5,1975

[10] Levitin,E.S.,On the Perturbation Theory of Nonsmooth Extremal Problems with Constraints,Soviet Mathematics Doklady,Vol.16,No.5, 1975.

On the theoretical basis for methods for parametric optimization problems

J. Guddat, M. Lips

Humboldt-Universität, Sektion Mathematik

DDR-1086 Berlin, Unter den Linden 6

We consider the following parametric optimization problem

$$P(\lambda) \min \{ f(x,\lambda) \mid x \in M(\lambda) \}, \lambda \in E_k, \qquad (1)$$

where

$$M(\lambda) = \{ x \in E_n \mid g_i(x,\lambda) \leq 0, i=1,\ldots,m; \ h_j(x,\lambda)=0, j=1,\ldots,p \}$$

and f, $g_i (i=1,\ldots,m)$, $h_j (j=1,\ldots,p)$ are maps from $E_n \times E_k$ into R.

We introduce the following notions

$$M: \lambda \longrightarrow M(\lambda) \text{ (constraint set map)} \qquad (2)$$

$$\varphi: \lambda \longrightarrow \varphi(\lambda) \underset{df}{=} \inf \{ f(x,\lambda) \mid x \in M(\lambda) \} \text{ (extreme value function)}, \qquad (3)$$

$$\psi: \lambda \longrightarrow \psi(\lambda) \underset{df}{=} \{ x \in M(\lambda) \mid f(x,\lambda) = \varphi(\lambda) \}$$

$$\text{(optimal set map)} \qquad (4)$$

$$\mathfrak{A} = \{ \lambda \in E_k \mid \psi(\lambda) \neq \emptyset \} \text{ (solvability set)} \qquad (5)$$

For many applications of parametric optimization it is important to find at least one optimal point of the problem $P(\lambda)$ for all $\lambda \in \Lambda$ (where Λ is a given subset of $\mathfrak{A}$) (see [9]). Especially the possibilities of applications in vector optimization ([5]) and stochastic optimization ([12], [13]) are a sufficient motivation for the development of effective algorithms for large problems too. In this paper the situation for different special classes of $P(\lambda)$ is discussed from the theoretical point of views.

Firstly we consider the following classes of parametric optimization problems

P (c,b): $\min \{ c^T x \mid Ax = b,\ x \geq 0 \},\ c \in E_n,\ b \in E_m,$ (6)

P (p,b): $\min \{ x^T C x + p^T x \mid Ax = b,\ x \geq 0 \},\ p \in E_n,\ b \in E_m,$ (7)

where C is positive semidefinite,

P (d): find an element of

$$\Upsilon(d) = \{ (x,y)^T \in E_{2n} \mid -Kx + y = d,\ x^T y = 0,\ x \geq 0,\ y \geq 0 \},\ d \in E_m, \quad (8)$$

P (C,p,b) $\min \{ x^T C x + p^T x \mid Ax = b,\ x \geq 0 \},\ C \in K,\ p \in E_n,\ b \in E_m$ (9)

where

$$K = \{ C \mid C \text{ positive semidefinite} \}.$$

P (c,b) is investigated in [11], P (p,b) is analysed in [2] and [4], P (d) is studied in [2] and [15], P (C,p,b) is analysed in [2] and [7].

Decomposition and partition theorems are a theoretical basis for the development of algorithms.

Let $\mathfrak{A}, \mathfrak{A}^1, \ldots, \mathfrak{A}^N$ be convex sets. Then $\mathfrak{A}^1, \ldots, \mathfrak{A}^N$ is called a partition of $\mathfrak{A}$ if the following properties hold:

1° $\mathfrak{A} = \bigcup_{j=1}^{N} \mathfrak{A}^j,$

2° $\text{ri}\, \mathfrak{A}^{j_1} \cap \text{ri}\, \mathfrak{A}^{j_2} = \emptyset,\ j_1, j_2 \in \{1, \ldots, N\},\ j_1 \neq j_2$ (ri=relative interior)

3° $\dim \mathfrak{A} = \dim \mathfrak{A}^j,\ j = 1, \ldots, N.$

Let $\mathfrak{A}, \mathfrak{A}^1, \ldots, \mathfrak{A}^N$ be arbitrary subsets of E_k, then $\mathfrak{A}^1, \ldots, \mathfrak{A}^N$ is called a decomposition of $\mathfrak{A}$ if

1° $\mathfrak{A} = \bigcup_{j=1}^{N} \mathfrak{A}^j,$

2° $\mathfrak{A}^{j_1} \cap \mathfrak{A}^{j_2} = \emptyset,\ j_1, j_2 \in \{1, \ldots, N\},\ j_1 \neq j_2$

For the problem P (d), (8), we have the following partition theorem.

Theorem 1. ([2],[15]): There exists a unique partition $\mathfrak{A}^1, \ldots, \mathfrak{A}^N$ of the solvability set $\mathfrak{A}$, (5), with the following properties

1° $\mathfrak{A}$, $\mathfrak{A}^j$, j=1, ..., N are convex polyhedral cones,

2° there exists a linear selection function $(x^j(\lambda), y^j(\lambda))^T \in \psi(\lambda)$ on $\mathfrak{A}^j$ (j=1, ..., N).

Since P (c,b), (6) and P (p,b),(7) are special cases of P (d), theorem 1 also holds for these classes of parametric optimization problems. More details are contained in the given references [2], [4], [11], [15]. From the theorem 1 we expect good possibilities to compute the partition and corresponding linear selection functions or at least such a piecewise linear selection function on $\mathfrak{A}$ or on a subset Λ of $\mathfrak{A}$ for practical applications having an interesting size of P (d), -P (c,b) and P (p,b). We note that we can use the simplex technique ([12],[15]) in all these cases. In most practical applications it is sufficient to solve -instead of the multiparametric optimization problem P (d), P (p,b) and P (c,b) - a sequence of one-parametric optimization problems which we obtain when we substitute $d = d^o + t(d^1 - d^o)$, $t \in R$ for given several pairs of parameter points d^o, $d^1 \in \mathfrak{A}$. Indeed we have algorithms for solving the problems P (c,b), P (p,b), P (d) and any special classes of them in the above sense in an effective way by a simplex technique for large problems too. Especially the algorithms for the linear and quadratic problem, (6), (7),can be used to generate efficient points for a vector optimization problem ([5]) or a stochastic optimization problem ([13], [14]). We note that in both cases the computing of the partition is not necessary.

Now we consider the parametric optimization problem P (C,p,b), (5), which is analysed in [2] and [6]. The most important results for our discussion are contained in the following theorem.

<u>Theorem 2 ([2], [6]):</u>

Let $W = \{(C,p,b) \in K \times E_n \times E_m / \psi(C,p,b) \neq \emptyset\} \neq \emptyset$.

Then there exists a unique decomposition $W^1, \ldots, W^N$ of W with the following properties:

1° W is convex, $W^1, \ldots, W^N$ are connected.

2° the restriction $\psi/_{W^j}$ of ψ on W^j is closed and lower semicontinuous on W^j (j = 1, ..., N)
(i.e. at each $(C^o, p^o, b^o) \in W^j$ there exists a continuous selection function $x^j(C,p,b) \in \psi(C,p,b)$.

3° $\varphi/_{W^j}$ is continuous on W^j (j=1, ..., N).

In difference to theorem 1 we note that we have here only a decomposition of $\mathfrak{A}$ in connected not necessariely convex sets. From this point we can't expect to find this decomposition and the corresponding continuous selection functions $x^j(C,p,c)$ (j=1, ..., N) in an effective way. Indeed there are only theoretical algorithms. The implementation on the computer shows that we can solve only small problems. Therefore P (d), (8), is the largest class in which we can solve this problem with a practical interesting size in the above sense. For complicated classes of parametric optimization it is necessary to take other ways. Here we follow Levitin, [8], and Fiacco [1], [13]. We do not require the calculation of a selection function, but only the computing of an approximation of a selection function. We introduce the following notions for P (λ), (1):

<u>Definition 1:</u> Let $D \subseteq \mathfrak{A}$, $\lambda^o \in D$. The map $x:D \to E_n$ is called an approximation of a selection function in the sense of Fiacco with respect to λ^o and D (shortly F-function with respect to λ^o and D) if there exists a selection function $x:D \to E_n$, $\hat{x}(\lambda) \in \psi(\lambda)$ with

$$\| x(\lambda) - \hat{x}(\lambda) \| = o(\| \lambda - \lambda^o \|).$$

<u>Definition 2:</u> Let $D \subseteq \mathfrak{A}$, $\lambda^o \in D$. The map $x:D \to E_n$ is called an approximation of a selection function in the sense of Levitin with respect to λ^o and D (shortly L-function with respect to λ^o and D) if it is holds that

1° $d(x(\lambda), M(\lambda)) = o(\| \lambda - \lambda^o \|)$,

2° $|\varphi(\lambda) - f(x(\lambda), \lambda)| = o(\| \lambda - \lambda^o \|)$.

For a general class of parametric twice continuously differentiable nonlinear optimization problems of the form P (λ) Fiacco obtained a theoretical basis for locally characterizing the differentiable properties of a local solution $x(\lambda)$ and the associated Lagrange multipliers $u(\lambda)$, $w(\lambda)$, with respect to a variation of λ, and he established the use of a penalty function method to estimate the sensitivity information, i.e., first derivatives of the Kuhn-Tucker-triple. Computational experience was reported by Armacost and Fiacco ([1]). This approach is used in [10] to compute a F-function with respect to λ^o and D.

We refer to [2] for the relations between an F-function and an L-function.

We introduce the following linear multi-parametric optimization problem as a special case of P (c, b), (6).

$P^{\mathcal{L}}(\lambda)$: $\min \{ \nabla_x f(x^0, \lambda^0)^T y \mid y \in M^{\mathcal{L}}(\lambda) \}, \quad \lambda \in E_k,$

where

$$M^{\mathcal{L}}(\lambda) = \{ y \in E_n \mid \nabla_x g_i(x^0, \lambda^0)^T y \leq - \nabla_\lambda g_i(x^0, \lambda^0)^T (\lambda - \lambda^0),\ i \in I_0,$$
$$\nabla_x h_j(x^0, \lambda^0)^T y = - \nabla_\lambda h_j(x^0, \lambda^0)^T (\lambda - \lambda^0),\ j = 1, \ldots, p \},$$

$$I_0 = \{ i \in \{1, \ldots, m\} \mid g_i(x^0, \lambda^0) = 0 \},$$

$(x^0, \lambda^0)^T \in E_n \times E_k$ is a given parameterpoint.

Here we assume that

(A1) f, g_i (i=1, ..., m), h_j (j= 1,...,p) are continuous differentiable on $E_n \times E_k$.

By $\psi^{\mathcal{L}}(\lambda)$ we denote the optimal set map for $P^{\mathcal{L}}(\lambda)$.

The relation between solutions of $P^{\mathcal{L}}(\lambda)$ and an L-function is discussed by Levitin [8] and extended in [2] and [10] especially for the case that x^0 is not the unique solution of $P(\lambda^0)$.

Some of the important results are contained in the following theorems, where we give sufficient conditions so that a selection function of $P^{\mathcal{L}}(\lambda)$ will be an L-function.

<u>Theorem 3:</u> Let $D \subseteq \mathcal{O}\!\mathcal{l}$, $\lambda^0 \in D$. Further it holds that

(A1) f, g_i (i=1,...,m), h_j (j=1,...,p) are continuous differentiable on $E_n \times E_k$.

(A2) $x^0 \in \psi(\lambda^0)$,

(A3) there exists a vector $a^0 \in E_n$ with

$$\nabla_x g_i(x^0, \lambda^0)^T a^0 < 0 \ (i \in I_0),\ \nabla_x h_j(x^0, \lambda^0)^T a^0 = 0 \ (j = 1, \ldots, p)$$

the vectors $\nabla_x h_j(x^0, \lambda^0)$ $(j = 1, \ldots, p)$ are linear independent,

(A4) there exists a selection function $y^{\mathcal{L}}(\lambda) \in \psi^{\mathcal{L}}(\lambda)$ defined on D and being locally Lipschitzian at λ^{0};

(A5) there exists a selection function $x(\lambda) \in M(\lambda)$ defined on D and being locally Lipschitzian at λ^{0} with the properties

a) $f(x(\lambda), \lambda)$

b) $x(\lambda^{0}) = x^{0}$.

Then $x^{0} + y^{\mathcal{L}}: D \rightarrow E_n$ is an L-function with respect to λ^{0} and D.

<u>Theorem 4:</u> Let $D \subseteq \mathfrak{A}$, $\lambda^{0} \in D$. Further (A1), (A2), (A3), (A4) hold and

(A5) there exists a selection function $x(\lambda) \in M(\lambda)$ defined on D and being continuous at λ^{0} with the properties

a) $f(x(\lambda), \lambda) \leq \varphi(\lambda) + o(\|\lambda - \lambda^{0}\|)$

b) $x(\lambda^{0}) = x^{0}$.

(A6) $f(\cdot, \lambda^{0})$, $g_i(\cdot, \lambda^{0})$, $i \in I_0$ are convex, $h_j(\cdot, \lambda^{0})$ $(j=1,\dots,p)$ are affine-linear.

Then $x^{0} + y^{\mathcal{L}}$ is an L-function with respect to λ^{0} and D.

<u>Theorem 5:</u> Let $D \subseteq \mathfrak{A}$, $\lambda^{0} \in D$. Further (A1), (A2), (A3), (A4) hold and

(A5) a) there are functions $f_1(x)$ and $f_2(\lambda)$ such that $f(x, \lambda)$ has the description

$$f(x, \lambda) = f_1(x) + f_2(\lambda),$$

b) there are functions $g_{i1}(x)$, $g_{i2}(\lambda)$ $(i \in I_0)$ such that

$$g_i(x, \lambda) = g_{i1}(x) + g_{i2}(\lambda) \quad (i \in I_0),$$

c) there are functions $h_{j1}(x)$, $h_{j2}(\lambda)$ $(j=1,\dots,p)$ such that

$$h_j(x,\lambda)=h_{j1}(x)+h_{j2}(\lambda) \quad (j=1,\dots,p).$$

(A6) $f(\cdot,\lambda^0)$, $g_i(\cdot,\lambda^0)$, $i\in I_0$ are convex, $h_j(\cdot,\lambda^0)$, $j=1,\dots,p$, are affine-linear.

Then $x^0 + y^{\mathcal{L}}$ is an L-function with respect to λ^0 and D.

In [2], [10] the assumptions of theorems 3, 4, 5 are discussed, especially the existence of a continuous selection function of the constraint set maps M or the optimal set map Ψ. Moreover is shown that a locally Lipschitzian selection function $y^{\mathcal{L}}(\lambda)\in\Psi^{\mathcal{L}}(\lambda)$ can be calculated by a simplex technique.

References

[1] R. L. Armacost, A. V. Fiacco, Second-order parametric sensitivity analysis in NLP and estimates by penalty function methods; Technical paper, serial T-324, Georg Washington University, Washington 1975

[2] B. Bank, J. Guddat, D. Klatte, B. Kummer, K. Tammer, Nichtlineare parametrische Optimierungsaufgaben, Seminarbericht Humboldt-Universität (to appear 1980)

[3] A. V. Fiacco, Sensitivity analysis for nonlinear programming using penalty methods; Mathematical Programming 10(1976), North-Holland Publishing Company, 287-311

[4] J. Guddat, Stability in convex quadratic programming, Math. Operationsforschung Statist. 7(1976), 223-245

[5] J. Guddat, Parametrische Optimierung und Vektoroptimierung, in: K. Lommatzsch (Hrg.) [9]

[6] D. Klatte, Untersuchungen zur lokalen Stabilität konvexer parametrischer Optimierungsprobleme, Dissertation A, Humboldt-Universität, Berlin 1977

[7] B. Kummer, Global Stability of optimization problems, Math. Operationsforschung Statist., 8(1977), 367-383

[8] E. C. Levitin, On linear correction of the solution of nonlinear programming with incomplete information (Russian), Methods of optimization and their application (All-Union-Summer-Seminar, Baikal-lake) 1972

[9] K. Lommatzsch (Hrg.), Anwendungen der linearen parametrischen Optimierung, Berlin 1979

[10] M. Lips, Näherungsweise Bestimmung von Auswahlfunktionen für nichtlineare parametrische Optimierungsprobleme, Diplomarbeit, Humboldt-Universität, Berlin 1979

[11] F. Nožička, J. Guddat, H. Hollatz, B. Bank, Theorie der linearen parametrischen Optimierung, Berlin 1974

[12] Panne, C., van de, Methods for linear and quadratic Programming, Amsterdam, Oxford, New York 1975

[13] K. Tammer, Relations between stochastic and parametric programming, Math. Operationsforschung Statist., Series Optimization 9(1978), 4, 535-547

[14] K. Tammer, Über den Zusammenhang von parametrischer Optimierung und Entscheidungsproblemen der stochastischen Optimierung, in: K. Lommatzsch (Hrg.) [9]

[15] K. Tammer, Beiträge zur Theorie der parametrischen Optimierung, zu den math. Grundlagen ihrer Anwendung und zu Lösungsverfahren, Dissertation B, Humboldt-Universität, Berlin 1979

BASIC SOLUTIONS AND A 'SIMPLEX' METHOD FOR A CLASS OF CONTINUOUS LINEAR PROGRAMS

E.J. Anderson
Engineering Department
University of Cambridge
Cambridge U.K.

Abstract

A continuous linear program is an optimal control problem which can be posed as a linear program on a space of functions. In this paper a special class of such problems is discussed. A member of this class is called a separated continuous linear program (SCLP). Such problems occur in continuous models of production scheduling/inventory control.

There is a natural generalisation of the idea of a basic solution for the SCLP. Furthermore, a simple characterisation of the basic solutions can be given. It can be shown that if there is an optimal solution to an SCLP, then there is a basic optimal solution. This is an essential prerequisite for a simplex-like algorithm. Such an algorithm is developed by finding analogues to the reduced costs and pivot operation of the finite simplex method. The algorithm is illustrated by an application to a model of a production/inventory problem.

1. INTRODUCTION

A number of papers have appeared on the subject of continuous linear programs (e.g. those by Tyndall [7] and Grinold [5]. Such problems were first considered by Bellman [4] who introduced them in an economic context (and called them bottleneck problems). In this paper a particular class of such problems is discussed. A member of this class is called a Separated Continuous Linear Program (or SCLP). Such problems arise in continuous models of production scheduling/inventory control.

As motivation for what follows consider the production/inventory problem shown in figure 1.

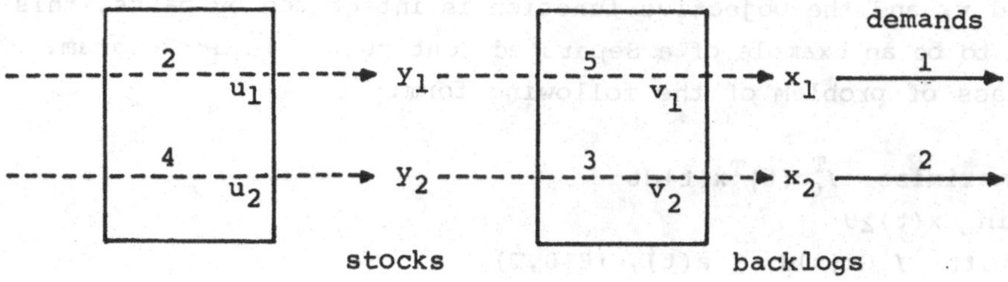

u and v are the efforts on the two machines

Figure 1.

There are two machines processing two kinds of product. Each machine can share its available effort between the two products, which are then processed at a rate proportional to the effort applied. Different maximum rates are given for each combination of machine and product. The stocks of each product between the machines must remain non-negative. Production is aimed at meeting some constant demands for the two products. Initially there are some given backlogs. The aim is to minimise the integral of these backlogs as they vary over time. The backlogs are reduced to zero but do not become negative (i.e. over-production is not allowed).

The particular problem of figure 1 can be formulated mathematically as follows:

$$\begin{aligned}
&\text{minimise} \quad \int_0^{10} (x_1(t) + x_2(t))\, dt \\
&\text{in} \quad x(t)\geq 0,\ y(t)\geq 0, \\
&\qquad u(t)\geq 0,\ v(t)\geq 0, \\
&\qquad\quad u_1(t) + u_2(t) \leq 1,\ v_1(t) + v_2(t) \leq 1, \quad t\in(0,10), \\
&\text{with} \quad \dot{y}_1(t) = 2u_1(t) - 5v_1(t), \\
&\qquad\quad \dot{y}_2(t) = 4u_2(t) - 3v_2(t), \\
&\qquad\quad \dot{x}_1(t) = 1 - 5v_1(t), \\
&\qquad\quad \dot{x}_2(t) = 2 - 3v_2(t), \quad t\in(0,10), \\
&\text{and initial conditions.}
\end{aligned}$$

Here 10 is an arbitrary time horizon. If x and y are expressed in terms of u and v, and the objective function is integrated by parts, this can be seen to be an example of a Separated Continuous Linear Program. This is a class of problem of the following form:

$$\text{maximise} \quad \int_0^T c(t)^T x(t)dt$$
$$\text{in} \quad x(t) \geq 0$$
$$\text{with} \quad \int_0^t Gx(s)ds \leq a(t), \; t\in(0,T), \qquad (1)$$
$$Hx(t) \leq b(t), \; t\in(0,T). \qquad (2)$$

Here x and c are bounded measurable functions of dimension n_1, a is a continuous function of dimension n_2, b is a bounded measurable function of dimension n_3, G is an $n_2 \times n_1$ matrix and H is an $n_3 \times n_1$ matrix. The description "separated" is applied because the constraints are in two sets; the integral constraints, (1), and the instantaneous constraints, (2).

The SCLP is actually a specialisation of the general continuous linear program which has the following form:

$$\text{maximise} \quad \langle x, c^* \rangle$$
$$\text{in} \quad x \geq 0$$
$$\text{with} \quad Ax \leq b$$

where x is in $L^n_\infty(0,T)$, c* is in $(L^*_\infty(0,T))^n$, b is in $L^m_\infty(0,T)$ and A is a bounded linear operator mapping $L^n_\infty(0,T)$ into $L^m_\infty(0,T)$.

In this paper an outline account is given of the way that a simplex like algorithm can be developed for the SCLP (proofs for the assertions that are made can be found in the references) . In order to do this analogues are developed to the three essential elements of the ordinary simplex algorithm: basic solutions, reduced costs and the pivot step. Some of the work described here is similar to that of Perold [6] . However, because he considers a more general problem, the results he gives are generally less well developed than those given here.

2. BASIC SOLUTIONS FOR SCLP

First we need to make a boundedness assumption as follows:

A: (i) a(t) is everywhere differentiable with bounded derivative
(ii) x(t) is bounded by constraints (1) and (2).

It is easy to show that the set, F, of feasible solutions to an SCLP, is convex and closed in the weak* topology [2]. Under assumption A this set is weak* compact. So from the Krein-Milman theorem, F is the closed convex hull of its extreme points (in the weak* topology). This is sufficient to ensure that there is an extreme point of the set of feasible solutions where the objective function is maximised. Thus, in looking for a solution to a SCLP, we need only consider the weak* extreme points of the set of feasible solutions.

In fact these extreme points have a simple characterisation. Some more notation is needed before giving this. First slack variables y and z are included in the formulation of the SCLP as follows:

SCLP: maximise $\int_0^T c^T(t)x(t)dt$

in $x(t),y(t),z(t)\geq 0 \quad t\in(0,T)$,

with $\int_0^t Gx(s)ds + y(t) = a(t), \; t\in(0,T)$, (3)

$Hx(t) + z(t) = b(t), \; t\in(0,T)$. (4)

It is convenient to make the constraints of SCLP apply only almost everywhere on (0,T). In what follows all the statements made about the interval (0,T) will apply only almost everywhere on that interval.

Let $n=(n_1+n_2+n_3)$ and $m=(n_2+n_3)$. For any x in $L^n_\infty(0,T)$ define the <u>support</u> of x, S_x to be the set valued function on (0,T) whose value at time t is the set of non-zero components of x at that time. That is

$$S_x(t) = \{k:x_k(t)\neq 0\}, \quad t\in(0,T).$$

Write $\underline{x}(t)$ for the n-vector given by

$$\underline{x}(t)=\begin{pmatrix} x(t)\\ y(t)\\ z(t)\end{pmatrix}, t\in(0,T).$$

Note that if $\underline{x}$ is feasible for SCLP then it is determined from x alone.

Define the mxn matrix K by

$$K=\begin{pmatrix} G & I & 0\\ H & 0 & I\end{pmatrix}$$

and let

$$\bar{x}(t)=\begin{pmatrix} x(t)\\ \dot{y}(t)\\ z(t)\end{pmatrix}, \quad t\in(0,T).$$

Thus equations (3) and (4) are equivalent to

$$K\bar{x}(t) = \begin{pmatrix} \dot{a}(t) \\ b(t) \end{pmatrix}, \quad t \in (0,T),$$

$$y(0) = a(0).$$

If $\underline{x}$ is feasible for SCLP and the columns of K indexed by the support of $\underline{x}$, $S_{\underline{x}}(t)$, are linearly independent almost everywhere in (0,T), then $\underline{x}$ is called basic. Thus if $\underline{x}$ is basic, $S_{\underline{x}}(t)$ contains no more than m elements for almost all t in (0,T). This definition is analogous to the definition of basic feasible solutions in ordinary LP. It turns out that the basic solutions are exactly those amongst which we must look for an optimal solution. In fact the following theorem can be proved (see [3]).

Theorem

Suppose $\underline{x}$ is feasible for SCLP and assumption A holds. Then $\underline{x}$ is basic if and only if $\underline{x}$ is an extreme point of F in the weak* topology.

3. REDUCED COSTS

In ordinary linear programming the reduced costs for a particular primal solution can be viewed as a solution to the dual problem, which is complementary slack to the primal solution, but may not be dual feasible (i.e. positive). If the reduced costs are all positive then they are an optimal solution to the dual problem and the primal solution is also optimal. In this section an analogue to the reduced costs of ordinary LP is found for SCLP.

First the dual problem SCLP* is formulated as follows:

SCLP*: minimise $\int_0^T \{u(t)^T a(t) + v(t)^T b(t)\}dt$

in $u(t), v(t), w(t) \geq 0, \quad t \in (0,T)$,

with $\int_t^T G^T u(s)ds + H^T v(t) - w(t) = c(t), \; t \in (0,T)$.

The problem SCLP* plays much the same role here as the dual problem plays in ordinary linear programming.

Let $\underline{u}(t)$ be defined as

$$\underline{u}(t) = \begin{pmatrix} w(t) \\ u(t) \\ v(t) \end{pmatrix}, \quad t \in (0,T).$$

In the same way as for ordinary LP, it can be shown that if $\underline{x}$ is primal feasible, $\underline{u}$ is dual feasible and

$$\int_0^T \underline{u}(t)^T \underline{x}(t)dt = 0, \tag{5}$$

then $\underline{x}$ is optimal for SCLP and $\underline{u}$ is optimal for SCLP*. Now define

$$\underline{c}(t) = \begin{pmatrix} c(t) \\ 0 \\ 0 \end{pmatrix}, \quad t\in(0,T),$$

$$\bar{u}(t) = \begin{pmatrix} -\hat{w}(t) \\ u(t) \\ -\hat{v}(t) \end{pmatrix}, \quad t\in(0,T).$$

Write $K_B(t)$ for the columns of K indexed by $S_{\underline{x}}(t)$ and $\underline{c}_B(t)$ for the elements of $\underline{c}(t)$ indexed by $S_{\underline{x}}(t)$. For convenience it is assumed that $K_B(t)$ is of full rank for almost all $t\in(0,T)$. Then define the <u>matching dual solution</u>, $\underline{u}(t)$, for any basic feasible solution $\underline{x}$ from

$$\bar{u}(t) = \underline{c}(t) - K^T(K_B^{-1}(t))^T \underline{c}_B(t), \quad t\in(0,T),$$

and the boundary conditions:

$$H^T v(T) - w(T) = c(T),$$
$$\underline{u}_i(T) = 0, \quad i \in S_{\underline{x}}(T).$$

This turns out to be nearly complementary slack with $\underline{x}$. In fact though (5) may not hold, it is true that:

$$\int_0^T \bar{u}(t)^T \underline{x}(t)dt = 0 .$$

The matching dual solution fills the same role for SCLP as the reduced costs do for the ordinary linear program.

4. THE PIVOT STEP

Before proceeding any further it is necessary to put some restrictions on the functions a, b and c. The following conditions will be assumed,

$$\left.\begin{aligned} a(t) &= a_1 + ta_2, \\ b(t) &= b, \\ c(t) &= c_1 + tc_2. \end{aligned}\right\} \tag{6}$$

These conditions are certainly stronger than is necessary for most of what follows. However, they do include the production/inventory problem given in the introduction. An SCLP which satisfies (6) will be called a linear SCLP. It can be shown [3] that for a linear SCLP the optimal solution x(t) is piecewise constant. In fact, it consists of a finite number of intervals (called regions), in each of which the choice of support is constant.

As in ordinary LP the pivot step is a means of moving from one basic feasible solution to another while improving the payoff. This is done by introducing a new variable into the support of $\underline{x}$ at some time t_0. Essentially the choice of variable and time which will give an improvement is made by choosing a negative part of the matching dual solution, and then selecting the primal variable with which it corresponds. Unfortunately the pivot procedure is quite complicated and difficult to describe. Only an outline account will be given here. A full description can be found in [1].

First, if a w or v variable from the matching dual solution is negative at t_0, say, then the corresponding x or z variable is increased from zero on an interval $(t_0, t_0+\varepsilon)$. Furthermore, all the region lengths between t_0 and the first time that this variable appears in the support of $\underline{x}$ are changed, in order to retain feasibility.

Just as in ordinary LP, having decided the variable to introduce into the support, "basicness" of the solution is retained by increasing that variable till some other variable becomes zero. The interval on which the change happens, $(t_0, t_0+\varepsilon)$, is made as large as possible, to obtain the maximum improvement in the value of the objective function.

If a u variable from the matching dual solution is negative on some region, then the corresponding $\dot{y}$ variable is increased from zero on the first part of the region and decreased from zero on the second part. The ratio of the lengths of these parts is chosen so as to retain feasibility.

The convergence properties of the algorithm are obviously of great interest. For all the examples which have been tried the algorithm has reached the optimal solution in a finite number of steps, just as the simplex algorithm does in ordinary LP. Whether or not this is the case for every linear SCLP is still an open question.

5. AN EXAMPLE

To clarify some of the ideas given above, consider a very simple example:

$$\text{maximise} \quad \int_0^2 (2-t)(x_1(t)+x_2(t))dt$$
$$\text{in} \quad x(t)\geq 0, \quad t\in(0,2),$$
$$\text{with} \quad \int_0^t x_1(s)ds \leq 4+t, \quad t\in(0,2),$$
$$\int_0^t x_2(s)ds \leq 3+2t, \quad t\in(0,2),$$
$$x_1(t) + 2x_2(t) \leq 10, \quad t\in(0,2),$$

For this problem:

$$K = \begin{bmatrix} 1 & 0 & 1 & 0 & 0 \\ 0 & 1 & 0 & 1 & 0 \\ 1 & 2 & 0 & 0 & 1 \end{bmatrix},$$

$$\underset{\sim}{c}(t) = (-1,-1,0,0,0)^T.$$

An initial basic feasible solution is given by

$$\bar{x}(t) = \bar{x}^{(1)} = (0,5,1,-3,0)^T, \quad t\in(0,1),$$
$$= \bar{x}^{(2)} = (6,2,-5,0,0)^T, \quad t\in(1,2).$$

It is easy to see that the matching dual solution is given by

$$\bar{u}(t) = \bar{u}^{(1)} = (-1/2,0,0,0,1/2)^T, \ t\in(0,1),$$
$$= \bar{u}^{(2)} = (0,0,0,-1,1)^T, \ t\in(1,2),$$

with

$$\underset{\sim}{u}(T) = (0,0,0,-1,0)^T.$$

Now $\underset{\sim}{u}$ is not dual feasible. As w_1 is negative over (0,1) an improvement can be obtained by increasing x_1 on some interval $(0,\varepsilon)$. The new basic feasible solution is as follows:

$$\bar{x}(t) = \bar{x}^{(0)} = (10,0,-9,2,0)^T, \quad t\in(0,4/9),$$
$$= \bar{x}^{(1)}, \quad t\in(4/9,47/27),$$
$$= \bar{x}^{(2)}, \quad t\in(47/27,2).$$

The corresponding matching dual solution is still not dual feasible, and a further step of the algorithm is needed. The solution which is then reached is given by

$$\bar{x}(t) = \bar{x}^{(0)}, \quad t\in(0,4/9),$$
$$= (1,9/2,0,-5/2,0)^T, \quad t\in(4/9,2).$$

Figure 2 shows this solution together with its matching dual solution. It is easy to see that there is primal and dual feasibility with complementary slackness. Hence this solution is optimal.

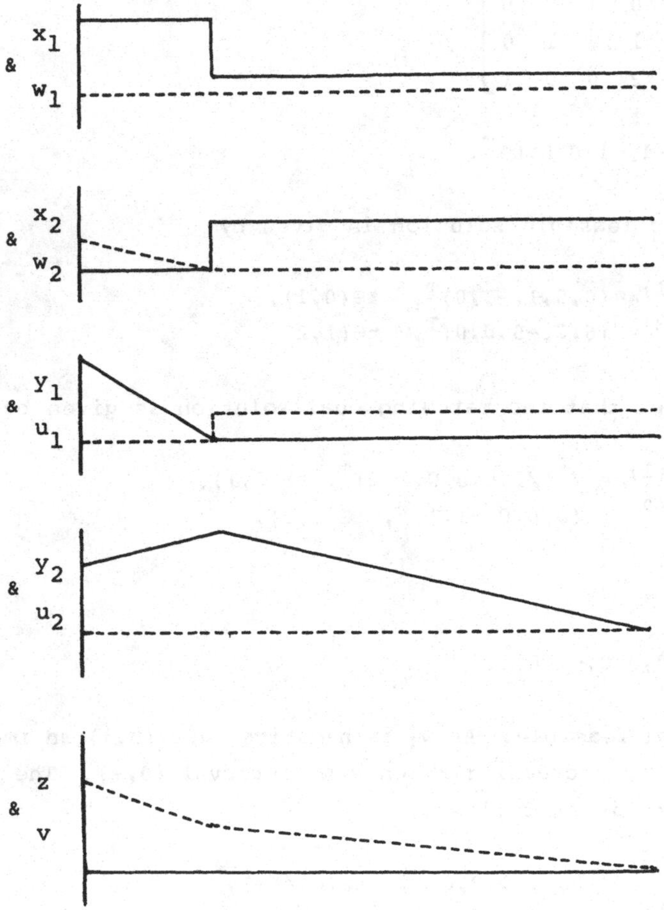

Solid lines are the primal solution; x,y,z,
dashed lines are the matching dual solution; w,u,v.

Figure 2.

References

1. Anderson, E.J. "A continuous model for job-shop scheduling", unpublished Ph.D. thesis, University of Cambridge, 1978.
2. Anderson, E.J. and Nash, P. "Continuous linear programming: Duality and elements of a simplex procedure", internal technical report CUED/F-CAMS/TR176, University of Cambridge, 1978.
3. Anderson, E.J. and Nash, P. "Proper solutions in a class of continuous linear programs", internal technical report CUED/F-CAMS/TR 177, University of Cambridge, 1978.
4. Bellman, R. "Dynamic Programming", Princeton University Press, 1957.
5. Grinold, R. "Symmetric duality for a class of continuous linear programming problems", SIAM J. Appl. Math., 18, pp. 84-97, 1970.
6. Perold, A.F. "Fundamentals of a continuous time simplex method", Stanford University Technical Report, SOL 78-26, 1978.
7. Tyndall, W.F. "An extended duality theory for continuous linear programming problems", SIAM J. Appl. Maths., 15 pp. 1294-1298, 1967.

A PROBABILISTIC ALGORITHM FOR GLOBAL OPTIMIZATION PROBLEMS WITH A DIMENSIONALITY REDUCTION TECHNIQUE.

Francesco Archetti

Istituto di Matematica - Università di Milano

Probabilistic methods have been proposed since the earliest studies in global optimization and have been gaining recently increasing attention as a suitable numerical tool for global optimization problems [1].

In this paper we shall be concerned with a class of methods based on a stochastic model of the objective function, which were first suggested in [2] and [3], and are now being increasingly investigated [4], [5], [6].

In these methods, the objective function $f(x)$ is considered as a realization -a sample path- of a stochastic process $f(x,w)$ -the model of $f(x)$- where w belongs to some probability space.

In the simpler 1-dimensional problem:

find f^* such that $f^* = f(x^*) = \min_{x \in K} f(x)$, where $K=[a,b]$,

the Wiener process is assumed as a stochastic model:

i) $f(a)=\mu$ ii) $f(x)-f(y) \sim N(0,\sigma^2|y-x|)$ $x,y \in K$.

If $f(x)$ is considered as a sample path of the Wiener process in $[a,b]$, its expected value and variance, conditioned by the values of $f(x)$ already observed $z_n=(f(x_i) = f_i;\ i=1,\dots,n)$ are given by the following formulas:

$$\mu(x) = E(f(x)|z_n) = f_i \frac{x_{i+1}-x}{x_{i+1}-x_i} + f_{i+1} \frac{x-x_i}{x_{i+1}-x_i}$$

(1)

$$\sigma^2(x) = \mathrm{var}(f(x)|z_n) = \sigma^2 \frac{(x-x_i)(x_{i+1}-x)}{x_{i+1}-x_i}$$

for $x \in \Delta_i=[x_i,x_{i+1}]$ $a \le x_i \le x_{i+1} \le b$

For $i=n$, $\Delta_i=[x_n,b]$ and we have, for $x \in \Delta_n$:

$$\mu(x) = E(f(x)|z_n)= f(x_n)$$

(2)

$$\sigma^2(x) = \mathrm{var}(f(x)|z_n) = \sigma^2(x-x_n)$$

If $f(x)$ is considered as a sample path of the Wiener process in $[a,b]$, the following theorem can be shown to hold, about the distribution of its global essential infimun.

Let $F(z) = \text{Prob}\{\inf_{a \le x \le b} f(x) \le z \mid f(b) = f_b\}$.

It turns out:

$$(3) \quad F(z) = \begin{cases} 1 & z \ge \min(f_a, f_b) \\ \exp\left(-2 \dfrac{(f_a - z)(f_b - z)}{\sigma^2 (b-a)}\right) & \text{otherwise.} \end{cases}$$

A basic advantage of the Wiener process as a stochastic model is that the behaviour of the 1-dimensional objective function, conditioned by the observations already performed, can be characterized, in probabilistic terms, rather easily by formulas (1), (2) and (3).

The Wiener process is also a "good model" in a statistical sense of the global behaviour of the objective function: its goodness of fit can be easily checked by statistical tests.

Thus it is possible to design an effective 1-dimensional global optimization algorithm with a probabilistic control of the error in the approximation to the global minimum and a sensible strategy for the selection of the points where f(x) is to be evaluated.

Unfortunately the actual use of optimization algorithms based on stochastic models is still severely restrained in multidimensional problems both by the theoretical difficulties of handling stochastic processes with a multidimensional parameter and the relevant overhead computations connected to their numerical implementation.

In order to overcome this basic difficulty some simplified stochastic models of a multidimensional objective function have been tentatively proposed which can be used either to choose the points for evaluating f(x) [7] [8] or to control in a probabilistic sense, the progress of the optimization algorithm. [9].

Another possibility is reducing the dimensionality of the problem so that the 1-dimensional algorithm can be used.

It is well known that N continuous functions exist, $X(t) = (x_1(t), x_2(t), \ldots, x_N(t))$ mapping the unit interval onto the unit hypercube of R^N. [10] [11]. For any point $\bar{X} = (\bar{x}_1, \bar{x}_2, \ldots, \bar{x}_N) \in \prod_{i=1}^{N} [0,1]$, a value $\bar{t} \in [0,1]$ existe such that $X(\bar{t}) = \bar{X}$.

Thus, at least in principle, a problem in any finite number of variables can be reduced to an 1-dimensional problem.

A wide experience already exists, both in the computer implementation [12] [13] and in the application of these space filling curves (and of its reciprocal i.e. the "scanning" which gives a 1-dimensional image of a set of points in $\prod_{i=1}^{N}[0,1]$) to data transmission [14], pattern recognition [15], image processing [16] and numerical methods for optimization problems [17] [18] [19].

In the case of optimization we may write:

$$(4)\qquad \min_{x\in\prod_{i=1}^{N}[0,1]} f(x) = \min_{t\in[0,1]} f(X(t)) = \min_{t\in[0,1]} \phi(t).$$

A somewhat disturbing feature of $\phi(t)$, apart from the trivial case $f(x)$=const., is its non differentiability due to the non differentiability of $X(t)$.

For this reason the algorithm we are considering, in which the objective function is assumed to be a sample path of the Wiener process, and therefore non differentiable, is a natural condidate for being connected with a dimensionality reduction technique based on space filling curves.

Sect. 2 The structure of the optimization algorithm.

A theoretical framework of this algorithm, including the proofs of convergence is given in [6], where details are also given about its implementation.

We assume that n function evaluations have been already performed in the search domain $K = [a,b]$ at $x_1, x_2, \ldots, x_n$ - rearranged so that $x_1 \leq x_2 \leq \ldots \leq x_n$ and $\Delta_i = [x_i, x_{i+1}]$.

The next point x_{n+1} of the algorithm is chosen according to the following rule:

1) The interval Δ_j is chosen such that:

$$\text{Prob}\{\min_{x\in\Delta_j} f(x) < f_n^* - \gamma \mid z_n\} = \max_i \text{Prob}\{\min_{x\in\Delta_i} f(x) < f_n^* - \gamma \mid z_n\}$$

where $z_n = (f(x_1)=f_1, f(x_2)=f_2, \ldots, f(x_n)=f_n)$,
$f_n^* = \min(\mu, f_1, f_2, \ldots, f_n)$ and γ is some positive value.

By the Markov property:

$$\text{Prob}\{\min_{x\in\Delta_j} f(x) < f_n^* - \gamma \mid z_n\} = \text{Prob}\{\min_{x\in\Delta_j} f(x) < f_n^* - \gamma \mid f(x_j)=f_j, f(x_{j+1})=f_{j+1}\}.$$

This probability can be computed by the formula (3), as:

$$\exp -2 \frac{(f^*_n-\gamma-f_{i+1})(f^*_n-\gamma-f_i)}{\sigma^2(x_{i+1}-x_i)}$$

2) X_{n+1} is now chosen in Δ_j such that:

$$\mathrm{Prob}\{f(x_{n+1})< f^*_n-\gamma \mid z_n\}=\max_{x\varepsilon\Delta_j} \mathrm{Prob}\{f(x)< f^*_n-\gamma \mid z_n\}=\max_{x\varepsilon\Delta_j} P_j(x)$$

where, for $x \varepsilon \Delta_j$, we have:

$$P_j(x) = \int_{-\infty}^{f^*_n-\gamma} \frac{1}{\sqrt{2\Pi}\,\sigma(x)} \exp\left(-\frac{1}{2}\frac{(z-\mu(x))^2}{\sigma^2(x)}\right) dz$$

$P_j(x)$ attains its maximum in Δ_j at the point:

$$x_{n+1} = x_j + \frac{(f^*_n-\gamma-f_j)(x_{j+1}-x_j)}{2(f^*_n-\gamma)-f_j-f_{j+1}} .$$

As far as the parameter γ is concerned, an adaptive control of it can be accomplished, according to the progress of the algorithm by the following rule.

Let $C_\gamma = \mathrm{Prob}\{\min_{x\varepsilon K} f(x) > f^*_n-\gamma \mid z_n\}$

γ is kept constant as long as $C_\gamma > P_T$, a prefixed probability level; when $C_\gamma < P_T$, γ is halved and the process is restarted comparing the new C_γ with P_T.

Higher values of P_T imply smaller values of γ: the algorithm gets faster but less reliable.

The 1-dimensional optimization algorithm can be completed with a procedure for the identification of those subintervals of $K = [a,b]$ where a quadratic model would be more appropriate than the Wiener process. These regions are therefore to be excluded from the progress of the global search and handled by a local optimization routine [20].

This procedure is to be omitted when the algorithm is applied to the minimization of $\phi(t)$, defined by (4), which is not locally quadratic in the neighborhood of its minima. Moreover there are many "parasitic" minima, whose identification gives no information about the local minima of $f(x)$.

Indeed, if $f(x)$ has a local minimum in $X=X^*$, any point t^* such that $X(t^*)=X^*$ is, by the continuity of $X(t)$, a local minimum of $\phi(t)$, while the converse is not true.

Therefore the 1-dimensional probabilistic optimization algorithm is used mainly to identify those points in R^1, and hence in R^N, where the probability is maximum of improving by more than γ over tha values of $f(x)$ already computed.

The identification of the regions which can be associated to the minima of $f(x)$ and the final approximation to these minima are performed in R^N respectively by a statistical technique known as clustering or cluster analysis [21], and by the use of local optimization routines.

Sect. 3 Numerical results.

Two 2-dimensional test functions have been considered from the standard set of global optimization problems given in [1].

1) $B(x_1,x_2) = (x_2-bx_1^2+cx_1-d)^2+e(1-f)\cos x_1+e$

where $b=\frac{5.1}{(4\Pi)^2}$, $c=\frac{5}{\Pi}$, $d=6$, $c=10$ and $F=\frac{1}{8\Pi}$.

The search domain is $K = \{-5\leq x_1\leq 10,\ 0\leq x_2\leq 15\}$.

3 global minima exists in K:

$x_1^*=(\Pi,2.275)$, $x_2^*=(-\Pi,12.27)$ and $x_3^*=(0.425,2.475)$

whose value is $f^* = 0.398$.

The 1-dimensional optimization algorithm, applied to the resulting $\phi(t)$, has been terminated after 90 function evaluation, with a value t_{90}^*, whose image in R^2 is (-3.1415,12.23), and $f_{90}^*=0.4042$.

The cluster analysis and the use of local routines, subsequently performed in R^2 identified the exact value and position of the global minima.

2) $GP(x_1,x_2) = [1+(x_1+x_2+1)^2](19-14x_1^2-14x_2^2+6x_1x_2)$.

The search domain is $K = \{-2\leq x_1\leq 2,\ -2\leq x_2\leq 2\}$.

The global minimum is $x^*=(0,1)$ whose value is $f^*=3.0$.

The 1-dimensional optimization algorithm, applied to the resulting $\phi(t)$, has been terminated after 150 function evaluations, with a value t_{150}^* whose image in R^2 is (0.0033,-1.004) and $f_{150}^*=3.006$.

The cluster analysis and the use of local routines, subsequently performed in R^2, gave the exact value and position of the global minimum and two local minima.

REFERENCES

1) Towards global optimization, vol. II , L. Dixon & G.P. Szego eds., North-Holland 1978.

2) H.J. Kushner "A new method for locating the maximum point of an arbitrary multipeak curve in presence of noise", Journal of Basic Engineering, pp. 97-106 (1964).

3) J. Mockus "On a method of distribution of random experiments in the solution of multi-extremal problems" USSR Computational Mathematics & Mathematical Physics, vol. 4, 1964.

4) J. Mockus-V; Tiesis-A. Zilinskas "The application of Bayesian methods for seeking the extremum" in 'Towards Global Optimization', vol. II, L. Dixon & G.P. Szego eds., 1978, North-Holland.

5) F. Archetti-B. Betrò "Stochastic models and optimization" to appear in Bollettino U.M.I. 16-A (1979).

6) F. Archetti "Stochastic strategies in global optimization problems" in 'Numerical Techniques and Stochastic Systems' M. Cugiani and F. Archetti eds., 1980, North-Holland.

7) A. Zilinskas "On statistical models for multimodal optimization" Math. Operations Forsch. - Stat. Ser. Statistic, vol. 9 (1978) n.2.

8) J. Mockus "The simple Bayesian algorithm for the multidimensional global optimization" in 'Numerical Techniques and Stochastic Systems' M. Cugiani and F. Archetti eds., 1980, North-Holland.

9) F. Archetti-B. Betrò "A stopping criterion for global optimization algorithms" (1979) Research Report A61 of the Dept. of Operations Research & Statistics, University of Pisa.

10) G. Peano "Sur une courbe, qui remplit toute une aire plane" Mathematische Annalen 36, 1980.

11) B. Mandelbrot "Fractals, forms, chance and dimension" 1977, Freeman & C.

12) J. Quinqueton "Un algorithm adaptif de balayage de Peano" Rapport de Recerche IRIA n. 344 (1979).

13) A.R. Buts "Alternative algorithm for Hilbert's space filling curve" IEEE Computers, April 1971.

14) T. Bially "Space filling curves: their generation and their application to Boundwidth reduction" IEEE Ing. Theory, vol. II 15, n. 6.

15) L. Kanal-T. Harley-K. Abend "Classification of binary random patterns" IEEE Ing. Theory, vol. II 11, n. 4.

16) V. Alexandrov-A. Polyekov-V. Tochtinov "Synthese et application d'analogue multidimensionnel d'une courbe de Peano" Congres AFCET-IRIA, Reconnoissance des Formes e Traitment des Images, Paris, (1979).

17) A.R. Buts "Space filling curves and mathematical programming" Information & Control 12, (1968), pp. 314-330.

18) R.G. Strongin "On the convergence of an algorithm for finding a global extremum" Engineering Cybernetics (1973).

19) F. Archetti-B. Betrò "Some remarks on dimensionality reduction techniques in global optimization problems" 1977, Research Report A49, Dept. of Operations Research and Statistics, Univ. di Pisa.

20) F. Archetti-B. Betrò "A probabilistic algorithm for global optimization" to appear in 'Calcolo'.

21) A. Torn "A search clustering approach to the global optimization problem" in 'Towards Global Optimization' vol. II, L. Dixon & G.P. Szego eds., (1978), North-Holland.

THE METHOD OF FEASIBLE DIRECTIONS FOR OPTIMIZATION PROBLEMS WITH SUBDIFFERENTIABLE OBJECTIVE FUNCTION

K. Beer
Technische Hochschule Karl-Marx-Stadt
DDR-9010 Karl-Marx-Stadt, Postfach 964

1. The algorithm FDM

Let us consider the problem

$$\inf \{ f(x) + g(y) : h(x) + k(y) \leq b \,,\; x \in X \,,\; y \in Y \} \qquad (1)$$

In so-called right-hand-side decomposition methods (see Geoffrion/5/) the problem (1) leads to

$$\varphi^{*} = \inf \{ \varphi(u) : u \in U \} \,, \qquad (2)$$

where

$$\varphi(u) = \varphi_1(u) + \varphi_2(u)$$

$$\varphi_1(u) = \inf \{ f(x) : x \in D(u) \} \qquad (3)$$

$$\varphi_2(u) = \inf \{ g(y) : y \in G(u) \} \qquad (4)$$

$$D(u) = \{ x \in R^n : h(x) \leq u \,,\; x \in X \}$$

$$G(u) = \{ y \in R^m : k(y) \leq b - u \,,\; y \in Y \}$$

$$U = \{ u \in R^p : D(u) \neq \emptyset \,,\; G(u) \neq \emptyset \}$$

$$X = \{ x \in R^n : v(x) \leq \mathbb{O} \}$$

$$Y = \{ y \in R^m : w(y) \leq \mathbb{O} \} \,,$$

i.e., φ and also U are given only implicitly.

Let us suppose that

(V1) $f(x), h_i(x), i=1(1)p$, are convex and continuous differentiable functions at a neighbourhood of $X, v_j(x), j=1(1)s$, are convex and continuously differentiable over R^n .

Analogous assumptions we made for $g(y), k_i(y), i=1(1)p, w_j(y), j=1(1)t$.

(V2) $\varphi^* > -\infty$,

$N(\varphi^*) = \{ u \in U : \varphi(u) = \varphi^* \} \neq \emptyset$ and compact .

(V3) The sets X and Y satisfy the Slater-condition .

From (V1) - (V3) concludes that $\varphi(u)$ is a proper,convex,closed,subdifferentiable function over U .

We shall solve the Problem (1) by solving the equivalent problem (2). Algorithms for such a subdifferentiable problem are considered by Clarke /4/,Lemarechal /7/,Mifflin /8/,Nurminski /9/ and others.They described subgradient or conjugate subgradient methods,which work with a-priori determined step-length and result generally in a non-monotonical decreasing of $\varphi(u)$.

We shall solve the problem (2) by the method of feasible directions, i.e.,a method of the form :

ALGORITHM FDM

Step 0 : Given is a point $u^0 \in U$ and a point-set-mapping P(u), defined for each $u \in U$. We put $k := 0$.

Step 1: Determine some direction $r^k \in R^p$ such that $\|r^k\| \leq 1$ and

$$\delta_k = \sup \{ \langle u^*, r^k \rangle : u^* \in P(u^k) \} = \inf_{\|r\| \leq 1} \sup \{ \langle u^*, r \rangle : u^* \in P(u^k) \} . \quad (5)$$

If $\delta_k < 0$,we go to Step 2.Otherwise we go to Ende .

Step 2 : Determine γ_k such that

$$\varphi(u^k + \gamma_k r^k) = \inf \{ \varphi(u^k + \gamma r^k) : \gamma \geq 0 \}.$$

Step 3 : $u^{k+1} := u^k + \gamma_k r^k$, $k := k+1$, go to Step 1 .

Remarks:

a) If $u^k \in U$ and $\delta_k < 0$, then from (V1) - (V3) follows, that γ_k exists and $0 < \gamma_k < +\infty$.

b) We temporaryly assume, that we may determine δ_k and γ_k exactly. In practice we desire only

$$\delta_k \leq \inf_{\|r\| \leq 1} \sup \{\langle u^*, r\rangle : u^* \in P(u^k)\} + \alpha_k ,$$

$$\varphi(u^{k+1}) \leq \inf \{\varphi(u^k + \gamma r^k) : \gamma \geq 0\} + \beta_k ,$$

without difficulties for convergence, if the series $\{\alpha_k\}$ and $\{\beta_k\}$ are monotonically decreasing to zero.

2. Realization of the algorithm: How to choose P(u) ?

When we shall apply the algorithm FDM to problem (2), then the basic question is: How take the extension P(u) of the subdifferential $\partial\varphi(u)$. The following theorem is well-known:

Theorem 1

Let a fixed $\varepsilon > 0$ be given. Let the assumptions (V1), (V2) be satisfied and let us additionally assume

$$P(u) \supset \partial_\varepsilon \varphi(u) := \{u^* : \varphi(v) - \varphi(u) \geq \langle u^*, v-u\rangle - \varepsilon , \ \forall v \in U\} \ \forall u \in U .$$

Then

a) There exists k_o : $\emptyset \in P(u^{k_o})$, i.e. , the algorithm terminates in a finite number of steps .

b) $\varphi(u^{k+1}) \leq \varphi(u^k) - \varepsilon$, $k = 0,1,2,\ldots,k_o-1$.

We are interested to select such a mapping P(u) that from $\mathbb{O} \in P(u)$ follows that u is near the optimum of problem (2). We cannot put $P(u) = \partial_\varepsilon \varphi(u)$ because it is unknown explicitly in practice, so that we cannot realize the Step 1 in the algorithm FDM .

Therefore J.Käschel and B.Schwartz /6/ suggest to put

$$P(u) = \Big\{ u^* = \big[\overset{*}{u}^{(1)} + \bar{\mu}\,,\ \overset{*}{u}^{(2)} + \bar{\mu}\big] :$$
$$\operatorname{grad} f(\bar{x}) - \sum_{i=1}^{p} \overset{*}{u}_i^{(1)} \operatorname{grad} h_i(\bar{x}) + \sum_{j=1}^{s} \lambda_j \operatorname{grad} v_j(\bar{x}) = \mathbb{O},$$
$$\operatorname{grad} g(\bar{y}) - \sum_{i=1}^{p} \overset{*}{u}_i^{(2)} \operatorname{grad} k_i(\bar{y}) + \sum_{j=1}^{t} \mu_j \operatorname{grad} w_j(\bar{y}) = \mathbb{O},$$
$$\langle -\overset{*}{u}^{(1)}, u \rangle \geq \varphi_1(u) - \varepsilon\,,\quad \overset{*}{u}^{(1)} \leq \mathbb{O}$$
$$\langle -\overset{*}{u}^{(2)}, b-u \rangle \geq \varphi_2(u) - \varepsilon\,,\quad \overset{*}{u}^{(2)} \leq \mathbb{O}$$
$$\bar{\mu} \in R^p,\quad \lambda \in R^s,\quad \mu \in R^t \text{ are arbitrary vectors} \Big\}, \quad (6)$$

where $\bar{x}$ (respectively $\bar{y}$) is an arbitrary optimal solution of the problem (3) (resp. (4)), i.e. , $\varphi_1(u) = f(\bar{x}),\ \bar{x} \in D(u),$

$$\varphi_2(u) = g(\bar{y})\,,\ \bar{y} \in G(u)\,.$$

We denote, that P(u) is a convex, polyhedral set.
If we take P(u) according to (6) then valid is the

<u>Theorem 2</u> (Käschel, Schwartz /6/)

Let a fixed $\varepsilon > 0$ be given. Let the assumptions (V1)-(V2) be satisfied, where we assume that the functions h(x), v(x), k(y), w(y) are affin-linear.
Then the statements a), b) from theorem 1 holds and from $\mathbb{O} \in P(u^k{}_o)$ follows, that

$$\varphi(u^k{}_o) - \varphi^* \leq 2 \cdot \varepsilon\,,$$

i.e. , the algorithm FDM leads to a near-optimal solution of problem (2) .

An another way to get a suitable set P(u) is desired below.

There exists a formula for the subdifferential $\partial\varphi(u)$.

Theorem 3

From (V1) - (V3) follows that $\forall u \in U$

a) $\partial\varphi := \{ u^*: \varphi(v) - \varphi(u) \geq \langle u^*, v - u\rangle, \forall v \in U\} \neq \emptyset$

b) $\partial\varphi(u) = \{ u^* = [\overset{*}{u}^{(1)} + \bar{\mu}, \overset{*}{u}^{(2)} + \bar{\mu}]$:

$$\mathbb{O} = \text{grad } f(\bar{x}) - \sum_{i=1}^{\ell} \overset{*}{u}_i^{(1)} \text{ grad } h_i(\bar{x}) + \sum_{j=1}^{s} \lambda_j \text{ grad } v_j(\bar{x}),$$

$$\mathbb{O} = \text{grad } g(\bar{y}) - \sum_{i=1}^{\ell} \overset{*}{u}_i^{(2)} \text{ grad } k_i(\bar{y}) + \sum_{j=1}^{t} \mu_j \text{ grad } w_j(\bar{y}),$$

$$\overset{*}{u}^{(1)} \leq \mathbb{O}, \quad \overset{*}{u}^{(2)} \leq \mathbb{O}, \quad \lambda \geq \mathbb{O}, \quad \mu \geq \mathbb{O},$$

$$\overset{*}{u}_i^{(1)} = 0, \forall i \bar{\in} I_1(\bar{x}), \quad \overset{*}{u}_i^{(2)} = 0, \forall i \bar{\in} I_2(\bar{y}),$$

$$\lambda_j = 0, \forall j \bar{\in} J_1(\bar{x}), \quad \mu_j = 0, \forall j \bar{\in} J_2(\bar{y}),$$

$\bar{\mu} \in R^p$ an arbitrary vector $\}$, (7)

where $\bar{x}$ (respectively $\bar{y}$) is an arbitrary, but fixed, optimal solution of the problem (3) (resp. (4)),

$$I_1(\bar{x}) = \{ i : h_i(\bar{x}) = u_i \}, \quad I_2(\bar{y}) = \{ i : k_i(\bar{y}) = b_i - u_i \},$$

$$J_1(\bar{x}) = \{ j : v_j(\bar{x}) = 0 \}, \quad J_2(\bar{y}) = \{ j : w_j(\bar{y}) = 0 \}.$$

We define now for a given $\varepsilon > 0$ the extended Index sets

$$I_1^{\varepsilon}(\bar{x}) = \{ i: u_i - \varepsilon \leq h_i(\bar{x}) \leq u_i \}, \quad I_2^{\varepsilon}(\bar{y}) = \{ i: b_i - u_i - \varepsilon \leq k_i(\bar{y}) \leq b_i - u_i \},$$

$$J_1^{\varepsilon}(\bar{x}) = \{ j: -\varepsilon \leq v_j(\bar{x}) \leq 0, \quad J_2^{\varepsilon}(\bar{y}) = \{ j: -\varepsilon \leq w_j(\bar{y}) \leq 0 \}.$$

Let us denote by P(u) now the set what we recieve from formula (7) if we replace there $I_1(\bar{x})$, $I_2(\bar{y})$, $J_1(\bar{x})$, $J_2(\bar{y})$ by $I_1^{\varepsilon}(\bar{x})$, $I_2^{\varepsilon}(\bar{y})$, $J_1^{\varepsilon}(\bar{x})$, $J_2^{\varepsilon}(\bar{y})$ respectively. P(u) is a convex, polyhedral set.

Obviously we recieve

$$\partial_{2\varepsilon}\varphi(u) \supset P(u) \supset \partial\varphi(u), \forall u \in U,$$

so that $u \in U$, $\varnothing \in P(u)$ implys $\varphi(u) - \varphi^* \leq 2 \cdot \varepsilon$. Moreover, there valids (see Schwartz /10/)

Theorem 4

Let us assume that (V1) - (V3) be satisfied and that P(u) is a continuously point-set-mapping $\forall u \in U$. Then the algorithm FDM generates a sequence $\{u^k\}$ with the properties

a) $\varphi(u^{k+1}) < \varphi(u^k)$, $k = 0, 1, 2, \ldots$

b) If for some $k = k_o$ the Ende is reached, then $\varnothing \in P(u^{k_o})$.

c) If the sequence $\{u^k\}$ is not finite, then it is bounded an we recieve $\varnothing \in \underline{\lim}\, P(u^k)$.

We remarke: If the functions h(x),k(y),v(x),w(y) are affin-linear, then P(u) is continuous.

3. Realization of the algorithm : How to solve the problem (5) ?

Let $\|\ldots\|_P$ denote a norm in an Euclidean space and $\|\ldots\|_D$ the connected dual norm, i.e. ,

$$\|u^*\|_D = \sup \{ |\langle u^*, u\rangle| : \|u\|_P \leq 1 \}.$$

Above in the algorithm FDM we determined the direction r by the way of solving the problem (5). In this connection we put

$$\delta = \inf_{\|r\|_P \leq 1} \sup \{ \langle u^*, r\rangle : u^* \in P(u) \} , \tag{8}$$

$$\Delta = \inf \{ \|u^*\|_D : u^* \in P(u) \} . \tag{9}$$

The dual problems (8) , (9) are equivalent to a linear, respectively quadratic problem, when we apply the sum or maximum respectively Euclidean norms and when P(u) is a convex polyhedral set. This problems may be solved by column generation as we discussed in Beer /1/, Beer/2/, Beer, Käschel /3/ .

We obtain

Theorem 5

Suppose: $u \in U$, (V1)-(V3) are satisfied, P(u) we put as defined in theorems 2 or 4 .

Then: a) $\Delta = -\delta^{\sim}$

b) There exists an $\bar{u}^* \in P(u)$ such that $\| \bar{u}^* \|_D = \Delta$.
There exists an $\bar{r} \in R^p$ with $\| \bar{r} \|_P \leq 1$ such that $\delta^{\sim} = \langle \bar{u}^*, \bar{r} \rangle$.

c) If $\delta^{\sim} \geq 0$, then $\emptyset \in P(u)$.

d) If $\delta^{\sim} < 0$, then $\bar{r}$ is a suitable feasible direction in problem (2) at the point u and we obtain $\varphi'(u, \bar{r}) \leq \delta^{\sim} < 0$.

4. Numerical experiments

Applied to linear programming problems with block-structure we have compared the algorithm FDM with the simplex method.

Example I (J.Käschel): The example has 4 diagonal blocks , each of these having 6 rows and 16 columns. The coupled row-block has 2 rows and 64 columns. The density was 32 % . The CPU-time for solving an optimal solution by the algorithm FDM (with P(u) respsectively theorem 2) was 6.08 sec. at the ES 1040 computer and 6.07 sec. by the simplex algorithm at the same computer.

Example II (W.Remke): The example has 8 diagonal blocks, each of these having 20 rows and 49 columns. The coupled column-block has 15 columns and 160 rows. The density was 11.07 % . The CPU-time at the ES 1040 computer was 282 sec. for solving an optimal solution by the algorithm FDM (with P(u) respectively theorem 4) and 390 sec. by the simplex method.

algorithm FDM		simplex method	
number of Iteration	value of the objective function	number of iteration	value of the objective function
0	7 452	38	9 231
1	13 773	46	13 627
2	14 240	47	14 287
3	15 189	50	14 796
4	15 698	51	15 668
5	15 720.9	52	15 721
6	15 720.9		

Example I

algorithm FDM		simplex method	
number of iteration	value of the objective function	number of iteration	value of the objective function
0	667	420	0
1	722	430	367
2	745	440	608
3	751	450	697
4	756	470	761
5	813,3	490	820
6	813,4	500	840
		505	843

Example II

5. References

/1/ Beer,K. : Über zulässige,brauchbare Fortschreitungsrichtungen in Optimierungsaufgaben mit SALK-Zielfunktion, Wissenschaftliche Zeitschriftt der THK,XIX (1977), Nr.4,465-469

/2/ Beer,K. : Lösung großer linearer Optimierungsaufgaben,VEB Deutscher Verlag der Wissenschaften,Berlin 1977

/3/ Beer,K.,J.Käschel : Column generation in quadratic programming,Mathem.Operationsforsch.Statistik,Series Optimization,10(1979),Nr.2,179-184

/4/ Clarke F.H. : Generalized gradients and applications,Transactions Amer. Math.Soc. 205(1975),247-262

/5/ Geoffrion A.M. : Primal resource-directive approaches for optimizing nonlinear decomposable systems,Operat. Research,18(1970),Nr.3,375-403

/6/ Käschel,J.,B.Schwartz : Finding directions in decomposition algorithms ,Seminar "Nichtkonventionelle Optimierungsprobleme",Mogilany,VR Polen,October 1978

/7/ Lemarechal,C. : Nonsmooth optimization and descent methods, IIASA,RR-78-4

/8/ Mifflin,R. : An algorithm for constrained optimization with semismooth functions,Mathematics of Operat.Res., 2(1977),Nr.2,191-207

/9/ Nurminski,E.A.:Numerical methods for solving deterministic and stochastic minimax problems (in russian),Kiev, Naukova dumka 1979

/10/ Schwartz,B. : Zur Minimierung konvexer,subdifferenzierbarer Funktionen über dem R^n,Math.Operationsforsch. Statistik,Series Optimization,9(1978),Nr.4,545-557

FACTORIZED VARIABLE METRIC ALGORITHMS FOR UNCONSTRAINED OPTIMIZATION

L. GRANDINETTI

Dipartimento di Sistemi - Università della Calabria

87036 Arcavacata (Cosenza) - Italy

ABSTRACT

In recent years several complex alternatives to the straightforward BFS variable metric algorithm have been proposed. In this paper a numerical comparison has been done between the LDL^T decomposition of the BFS algorithm versus its unfactorized version. The special feature of this comparison consists in the choice of test problems most of them possessing severe ill-conditioning characteristics.
Numerical results show with sufficient evidence that, in view of performance, there is no superiority of the factorized versus unfactorized implementation of BFS algorithm.

Combining these results with further results concerning other complex strategies incorporating factorization procedures and recently documented [11], it seems reasonable to infer that a straightforward BFS implementation is both more efficient and more robust then any other known variable metric algorithm.

1. INTRODUCTION

A number of quite complex strategies has been recently proposed in order to improve numerical performances of variable metric algorithms for unconstrained minimization of smooth objective functions.

One of the most interesting among these strategies is based on the hypothesis that updating, at any iteration k, an approximation J_k to the Hessian matrix of the objective function rather than the approximation H_k to the inverse Hessian, could improve numerical stability and thus numerical efficiency if J_k were updated in the factorized LDL^T form, where L is unit-lower triangular and D is diagonal.
In this case, effective techniques which guarantee that J_{k+1} remains positive definite can be used for updating the factors L_k and D_k. The most important consequence is that the downhill property of the algorithm, which is an imperative requisite, is satisfied even in difficult numerical situations in which the effects of rounding error could be expected to cause loss of positive definiteness of the update.

In this paper the LDL^T decomposition of the BFS algorithm has been compared with the unfactorized straightforward BFS implementation using a set of test functions possessing severe ill-conditioning characteristics.

In spite of appealing theoretical features embodied in the factorized version of the algorithm, numerical results show with sufficient evidence that does not exist, in view of performances, any motivation in using the LDL^T decomposition.

The same considerations apply when more sophisticated strategies, still incorpora ting a factorization procedure, are used making reasonable the conclusion that the straightforward BFS implementation is more efficient than any other known variable metric algorithm.

2. THE LDL^T DECOMPOSITION OF THE BFS VARIABLE METRIC ALGORITHM

The BFS algorithm for unconstrained nonlinear minimization, due to Broyden [1] Fletcher [2] and Shanno [3], calculates the least value of the objective function $F(x)$, $x \in E^n$, by generating iteratively a sequence of points x_k such that:

$$x_{k+1} = x_k - \alpha_k H_k g_k$$

where $g_k = \nabla F(x_k)$ and α_k is an appropriately chosen scalar: it is usually computed as the minimum of the one-variable function $F(x_k - \alpha H_k g_k)$ ("exact line search") or in such a way to ensure a sufficient reduction of the objective function along the di rection of search $d_k = -H_k g_k$ ("approximate line search"). The matrix H_k in some sense approximates the inverse Hessian of $F(x)$; it is specified initially as positive defi nite and then is updated according to the following matrix modification scheme:

$$H_{k+1} = H_k - \frac{p_k y_k^T H_k}{p_k^T y_k} - \frac{H_k y_k p_k^T}{p_k^T y_k} + \left(1 + \frac{y_k^T H_k y_k}{p_k^T y_k}\right) \frac{p_k p_k^T}{p_k^T y_k} \qquad (1)$$

where $y_k = g_{k+1} - g_k$, $p_k = x_{k+1} - x_k$.

This formula guarantees, at least theoretically, that successive matrix modifications are positive definite even if α_k is computed without exact line search provided that

$$y_k^T p_k > 0, \quad \forall k.$$

It has been supposed that in practice the approximate inverse Hessian matrix may loose its positive definiteness: this fact, which can be attributed to rounding error, is undesirable because then the algorithm ceases to possess the downhill property. In order to face this event first Gill and Murray [4] used an approximation J_k to the Hessian matrix of $F(x)$ (rather than the approximation H_k to the inverse Hessian) which is stored and updated in factorized form via Choleski decomposition:

$$J_k = L_k D_k L_k^T$$

where L_k = unit-lower triangular matrix and D_k = diagonal matrix.

The factorized algorithm computes the direction of search d_k solving the system of linear equations:

$$J_k d_k = -g_k$$

and then, properly searching along d_k, it comes out that $x_{k+1} = x_k + \alpha_k d_k$. The matrix updating scheme equivalent to that for updating H_k is given by the following formula:

$$J_{k+1} = J_k - \frac{J_k p_k p_k^T J_k}{p_k^T J_k p_k} + \frac{y_k y_k^T}{p_k^T y_k} \qquad (2)$$

This formula can be easily put in the form

$$J_{k+1} = J_k + \tau_1 g_k g_k^T + \tau_2 y_K y_K^T \qquad (3)$$

with τ_1 and τ_2 appropriate scalars.

The introduction of Choleski factorization of matrix J_k permits to guarantee that J_k is positive definite only ensuring that the diagonal elements of D_k are positive; furthermore, taking advantage of the particular structure (3) of the updating scheme, suitable methods for modifying the Choleski factors can be devised in such a way that the matrix J_k remains positive definite whatever the rounding error arising.

3. NUMERICAL RESULTS

The comparison of the numerical performances of the LDL^T decomposition of the BFS algorithm with the straight BFS implementation has been done on a number of test problems most of them could be expected to cause serious numerical difficulties.

The BFS factorized algorithm incorporates a factorization procedure [5] very well founded theoretically and which has proved very effective in practice; furthermore it uses an efficient line search procedure based on the Powell conditions [6].

Exactly the same algorithm (i.e. same line search procedure, same termination criterion, same starting conditions) has been used to obtain the unfactorized version.

It is worth noting that numerical experiments have been carried out in single length arithmetic to make the effects of round-off error as remarkable as possible.

In the sequel the set of test functions used is reported.

Test functions

1. Rosenbrock standard

$$F(x)= 100\,(x_1^2 - x_2)^2 + (1 - x_1)^2$$

Starting point $x_0 = (-1.2, 1.0)^T$

2. Rosenbrock repeated

$$F(x) = \sum_{i=1}^{n/2} 100\,(x_{2i-1}^2 - x_{2i})^2 + (1 - x_{2i-1})^2$$

Starting point $x_0 = (-1.2, 1.0, \ldots, -1.2, 1.0)^T$

3. Rosenbrock extended

$$F(x) = \sum_{i=1}^{n-1} 100\,(x_{i+1} - x_i^2)^2 + (1 - x_i)^2$$

Starting point $x_0 = (-1.2, 1.0, -1.2, 1.0, \ldots)^T$

4. Hilbert quadratic

$$F(x) = \frac{1}{2}(x - \bar{x})^T\, Q(x - \bar{x})$$

$$Q = [q_{ij}] \quad , \quad q_{ij} = \frac{1}{i+j-1} \quad , \quad i,j = 1 \ldots n$$

$$\bar{x} = (1,1,1,\ldots\ldots,1)^T$$

Starting point $x_0 = (0,0,0,\ldots,0)^T$

5. Least squares exponential fit

$$F(x) = \sum_{j=1}^{s} (d_j - \sum_{i=1}^{m} x_i\, e^{-x_{m+i} t_j})^2$$

$$s = 20 \quad , \quad m = 3 \quad ; \quad t_1 = 0.1 \quad , \quad t_2 = 0.2, \ldots, t_{20} = 2$$

$$\bar{d}_j = \sum_{i=1}^{3} a_i\, e^{-b_i t_j} \quad ; \quad a = (1,1,1,)^T \quad , \quad b = (\tfrac{1}{10}, 1, 10)^T \quad , \quad j = 1 \ldots 20$$

$d_j = \bar{d}_j + 0.0001 \times r\,[-1,1]$, where the random numbers $r \in [-1,1]$ are generated so that their distribution is uniform in this interval.

Starting point $x_0 = (1,1,1,\ 0.5,\ 1.5,\ 5.0)^T$

6. Powell singular

$$F(x) = (x_1 + 10x_2)^2 + 5(x_3 - x_4)^2 + (x_2 - 2x_3)^4 + 10(x_1 - x_4)^4$$

Starting point $x_0 = (3, -1, 0, 1)^T$

7. Powell penalty

$$F(x)= e^{x_1x_2x_3x_4x_5} + \frac{1}{2}\sigma \sum_{i=1}^{3} (c_i(x))^2$$

where:

$$c_1(x)= x_1^2 + x_2^2 + x_3^2 + x_4^2 + x_5^2 -10$$

$$c_2(x)= x_2x_3-5\, x_4x_5$$

$$c_3(x)= x_1^3 + x_2^3 + 1$$

Starting from x_0 compute the minimum of F(x) with $\sigma=\sigma_1$; then, starting from partial minimum $\bar{x}$ (σ_1), again minimize F(x) with $\sigma=\sigma_2$.

Repeate this process for increasing values of σ.

Starting point $x_0 = (1,1,....,1)^T$

A short comment on some of these test functions is worth while under the respect of numerical difficulties that could arise.

For the test function N.3, with n large, the changes of search direction necessary become numerous and therefore the errors made in each iteration might have a serious cumulative effect on the algorithm. Test functions N.6 and N.7 have in common the fact that the Hessian matrix G becomes very ill-conditioned when the solution $\bar{x}$ is approa ched; in fact, in the first case $G(\bar{x})$ has two zero eigenvalues whilst in the second case G has three eigenvalues that tend to infinity as σ tends to infinity (i.e. when x tends to $\bar{x}$). The addition of numerical random noise to data in function N.5 creates a severe unfavourable situation because it can lead to completely incorrect informa tion concerning the exponential's factors. Again an unfavourable feature is incorpo rated in test function N.4, a quadratic function where Q is the Hilbert matrix which is notoriously very ill-conditioned for n sufficiently high.

In all cases the minimum has been reached satisfactorily and surprising similari ties between the performances of the different algorithm's versions have been found. Numerical results are reported in Table 1, where the following notation is used:

n = number of variables of the objective function

$$\delta_{it}= N_{it} - \bar{N}_{it} \quad ; \quad \delta_{it}\% = \frac{\delta_{it}}{N_{it}} \cdot 100$$

$$\delta_{fe}= N_{fe} - \bar{N}_{fe} \quad ; \quad \delta_{fe}\% = \frac{\delta_{fe}}{N_{fe}} \cdot 100$$

where:

N_{it}= global number of iterations in the case of unfactorized implementation

$\bar{N}_{it}$= global number of iterations in the case of factorized implementation

N_{fe}= number of function evaluations in the case of unfactorized implementation

$\bar{N}_{fe}$= number of function evaluations in the case of factorized implementation

Of course negative values of δ's mean that the unfactorized version of the BFS algorithm performs better than its LDL^T decomposition.

4. CONCLUDING REMARKS

The LDL^T algorithm has proved so similar to the straightforward BFS in performance that the extra overhead required for the LDL^T algorithm appears not to be justified.

This statement is motivated from the variety of the problems used which cover many different situations having in common ill-conditioning and which in many ways could cause serious numerical difficulties.

As far as a theoretical explanation is concerned, some considerations can be done. First of all it must be stressed that numerical difficulties occasionally reported by some authors have to be imputed to early implementation of DFP algorithm. Secondly, it seems that at least one specific argument supporting the introduction of the factorized algorithm is questionable. In fact it has been reported [7] that a bonus which emerges from recurring an estimate J_k of the Hessian is the fact that matrix-vector multiplications in the updating formula are avoided. Since $J_k\, p_k = -\alpha_k\, g_k$, the formula (2) can be put in the form (3) which involves no matrix- vector multiplications in spite of formula (1) for updating H_k; no similar relation holds for the quantity $H_k\, y_k$ whose calculation can be affected by rounding errors if H_k is ill-conditioned. However it can be shown, utilizing the error analysis of matrix-vector multiplication due to Wilkinson [8], that the relative error in computing $d_k = -H_k\, g_k$ is given (dropping the index k for convenience) by:

$$\| d-\bar{d} \|_2 / \| d \|_2 \leq (1.06)2^{-t}\, n^{\frac{3}{2}}\, \| H \|_2\, \| H^{-1} \|_2 = (1.06)2^{-t}\, n^{\frac{3}{2}}\, \rho(H)$$

where $\rho(H)$ is the usual condition number of H, where the direction of search $\bar{d}$, obtained using standard floating point computation, satisfies the exact equation:

$\bar{d} = -Hg + e$ with $\| e \|_2 \leq (1.06)2^{-t}\, n \| H \|_F\, \| g \|_2$ (where floating-point arithmetic is used with a mantissa of t binary digits) and where the following relations are used:

$$\| H \|_F \leq n^{\frac{1}{2}}\, \| H \|_2 \quad , \qquad \| H^{-1}\, d \| \leq \| H^{-1} \|_2\, \| d \|_2$$

Since (4) provides only an upper bound on the relative error in d, this argument by itself is questionable in supporting a priori the numerical superiority of factorized approach. Indeed, numerical results seem to show that, in view of performances, no special advantage emerges.

It is worth noting that other experiments carried out with some different variable metric algorithms, all of these incorporating a factorization of the BFS updating formula, exhibit no superiority in performances versus the straightforward BFS algorithm. This happens, precisely, in the case of dogleg algorithm due to Powell [9] and the Davidon's projections algorithm [10] which have been extensively tested versus the standard BFS algorithm [11]. As far as the LDL^T implementation is concerned, which is the most interesting in the present context, an additional set of numerical results, documented in [11], has been reported in Table 2. These results are particularly valuable because they permit to get the same conclusions as those suggested by Table 1 on the relative merits of factorized BFS algorithm versus its unfactorized version, although in the last comparison test problems, starting conditions, line search technique and termination criterion are different from the previous one.

Combining the results of the last comparison with the special character of experiments reported in Table 1 (test problems selected in such a way to create specific numerical difficulties) and also taking into account the variety of modern complex strategies compared, it seems reasonable to infer that a straight BFS implementation is both more efficient and more robust then any other known variable metric algorithm.

TABLE 1

Test functions	δ_{it}	δ_{fe}	δ_{it}%	δ_{fe}%
Rosenbrock standard				
n = 2	3	7	9.4	15.2
Rosenbrock repeated				
n = 20	3	2	6.9	3.5
n = 40	5	11	11.9	16.9
Rosenbrock extended				
n = 21	-5	-15	-2.6	-5.1
n = 41	1	-17	0.3	-3.3
n = 101	-16	-7	-2.3	-0.7
Hilbert quadratic				
n = 10	0	2	0.0	8.0
n = 20	3	1	13.0	2.9
n = 30	-2	-3	-8.0	-9.0
n = 40	-11	-17	-50.0	-54.8
n = 100	2	2	5.7	3.9
Least squares exponential fit				
n = 6	-3	-7	-2.8	-5.6
Powell singular				
n = 4	0	0	0.0	0.0
Powell penalty				
n = 5				
σ_1= 0.01	-5	-5	-27.7	-17.2
σ_2= 0.1	0	0	0.0	0.0
σ_3= 1.0	1	1	16.6	6.6
σ_4= 10	-2	-3	-25.0	-15.0
σ_5= 100	0	0	0.0	0.0

TABLE 2

Test functions	δ_{it}	δ_{fe}	$\delta_{it}\%$	$\delta_{fe}\%$
Rosenbrock 2				
(-1.2,1)	0	0	0.0	0.0
(2,-2)	4	4	25.0	22.2
(-3.635,5.621)	-1	-4	-2.2	-7.3
(6.39,-.221)	0	0	0.0	0.0
(1.489,-2.547)	-1	-1	-4.8	-4.2
Rosenbrock 20				
(-1.2,1)	-17	-18	-13.2	-10.4
(2,-2)	-4	-4	-11.8	-9.5
(-3.635,5.621)	-7	-14	-4.5	-8.0
(6.39,-.221	-60	-65	-49.2	-47.4
(1.489,-2.547)	23	22	27.0	22.0
Wood				
(-3,-1,-3,-1)	-1	4	-1.3	4.1
(-3,1,-3,1)	-5	-1	-6.3	-0.9
(-1.2,1,-1.2,1)	9	12	13.0	13.5
(-1.2,1,1.2,1)	0	3	0.0	6.2
Weibull				
(5,.15,2.5)	-1	11	-2.7	20.3
(250,.3,5)	-6	-5	-15.4	-8.9
(100,3,12.5)	12	-1	28.5	-2.0
Power 20				
(1,1,...1)	-64	-73	-71.8	-73.0
Powell 4				
(-3,-1,0,1)	-4	-4	-10.8	-10.2
Powell 36				
(-3,-1,0,1,....)	4	2	4.6	2.2
Mancino				
n = 10	-4	0	-22.2	0.0
n = 20	0	0	0.0	0.0
n = 30	0	0	0.0	0.0

REFERENCES

[1] BROYDEN, C.G. *"The convergence of a class of double rank minimization algorithms"* J. Inst. Maths. Applics. 6, 1970, pp.76-90 and 222-231.

[2] FLETCHER, R. *"A new approach to variable metric algorithms"* The Computer Journal 13, 1970; pp.317-322.

[3] SHANNO, D.F. *"Conditioning of quasi-Newton methods for function minimization"* Mathematics of Computation 24, 1970, pp. 647-656.

[4] GILL, P.E. MURRAY,W. *"Quasi-Newton methods for unconstrained optimization"* J. Inst. Maths. Applics. 9, 1972, pp.91-108.

[5] FLETCHER, R. and POWELL, M.J.D. *"On the modification of LDL^T factorizations"* Mathematics of Computation 3, 1974.

[6] POWELL, M.J.D. *"Some global convergence properties of a variable metric algorithm for minimization without exact line search"* A.E.R.E. Technical Report C.S.S. 15, Harwell, 1975.

[7] BRODLIE, K.W. *"Unconstrained minimization"* In the State of Art in Numerical Analysis, ed. by D.A. Jacobs, Academic Press 1977.

[8] WILKINSON, J.H. *"Rounding errors in algebraic processes"* Her Majiesty's Stationery Office, pg. 83, London 1963.

[9] POWELL,M.J.D. *"A new algorithm for unconstrained optimization"* In Nonlinear Pro garmming, Ed. by J.B. Rosen, O.L. Mangasarian and K. Ritter, Academic Press 1970 pp.31-66.

[10] DAVIDON, W.C. *"Optimally conditioned optimization algorithms without line sear ches"* Mathematical Programming 9, 1975, pp. 1-30.

[11] SHANNO, D.F. and Kang-Hoh PHUA *"Numerical comparison of several variable metric algorithms"* MIS Technical Report N. 21, University of Arizona 1977.

A UNIFIED APPROACH TO NONLINEAR PROGRAMMING ALGORITHMS BASING ON SEQUENTIAL UNCONSTRAINED MINIMIZATIONS

Ch.Großmann
TU Dresden, Sektion Mathematik
DDR-8027-Dresden, Mommsenstraße 13

With the development of powerful unconstrained minimization techniques also various methods transforming a given constrained optimization problem into a sequence of unconstrained ones have been constructed. In this paper we present an unified approach to these methods involving barrier-penalty-methods and methods of centers as well as augmented Lagrangian and shifting methods. Basing on this approach the proofs of convergence can be partially unified, the rates of convergence of these methods can be estimated and general duality bounds can be derived. In this paper we concentrate our attention to duality theory.

Let us consider the following nonlinear programming problem

$$f_o(x)\text{-min! subject to } x\in G:=\{x\in X \mid f_i(x)\leqslant 0,\ i=1,..,m\} \tag{1}$$

with a continuous mapping $f \mid X \to R^{m+1}$ and a closed subset $X\subset R^n$. Now, we select a family U of generating functions $u \mid R^{m+1} \to \tilde{R}:=R^1\cup\{+\infty\}$ and define

$$F(x,u)=\inf\{u(v) \mid v \geqslant f(x)\} \quad \text{for any } x\in X,\ u\in U . \tag{2}$$

The main principle of the sequential unconstrained minimization techniques consists in substituting the given constrained problem (1) by auxiliary problems

$$F(x,u)\text{-min! s.t. } x\in X \tag{3}$$

with fixed functions $u\in U$. If $X=R^n$ the the problems (3) are really unconstrained. Let denote $x(u)$ a solution of (3). Then we are interested in finding such a function $u^*\in U$ or such a sequence $\{u^k\}\subset U$ that $x(u^*)$ or any accumulation point of $\{x(u^k)\}$ solves the constrained problem (1). Most of the various well known sequential unconstrained minimization techniques we can get from the general approach by specifying the family U and the specific principle for construction a sequence $\{u^k\}\subset U$.

Let denote $Q=\{v\in R^{m+1} \mid \exists\, x\in X \text{ such that } f(x)\leqslant v\}$ and by e^o the 0-th unit vector. To short our notation we set $w=\binom{w_o}{\underline{w}}$ with $\underline{w}=(w_1,...,w_m)^T$ for any $w\in R^{m+1}$. Now, we investigate the problem

$$u(v)\text{-min! s.t. } v\in Q . \tag{4}$$

However this problem can't be solved directly because the set Q is not explicitely available. The following proposition shows the close relation between the problems (1), (3) and (4).

<u>Theorem 1</u>: Let be $u \in U$ fixed, let exist a solution $x(u)$ of problem (3) such that $F(x(u),u) < +\infty$ and let exist a related solution $v(u)$ of the linearly constrained problem

$$u(v)\text{-min! s.t. } v \geqslant f(x(u)) \ . \tag{5}$$

Then $v(u)$ solves the problem (4) also and there holds $F(x(u),u)=u(v(u))$. If furthermore $v(u)-\varepsilon e^{o} \notin Q$ for any $\varepsilon>0$ then $x(u)$ solves the perturbed problem

$$f_o(x)\text{-min! s.t. } x \in X,\ \underline{f}(x) \leqslant \underline{v}(u) \ . \tag{6}$$

The proof of theorem 1 is given in [6]. We remark that theorem 1 generalizes the theorem of EVERETT [2] originally established for the Lagrangian related to (1). This case is contained in theorem 1 if especially $U=\{u \mid \exists y \in R^m_+ \text{ such that } u(w)=w_o+y^T\underline{w} \ \forall\, w \in R^{m+1}\}$.

To derive a duality theory we introduce the following property
(V): If $\tilde{v} \in Q$ and $\tilde{v}-\varepsilon e^{o} \in Q$ for some $\varepsilon>0$ then we have
$\inf\{u(v) \mid v \in Q\} < u(\tilde{v})$ for any $u \in U$.

Generally property (v) forms an assumption to the family U and to the given nonlinear programming problem (1). However, if U can be represented with a family Φ of functions $\varphi \mid R^m \to \tilde{R}$ by

$$U=\{u \mid \exists\ \varphi \in \Phi\ ,\ u(w)=w_o+\varphi(\underline{w})\ \forall\, w \in R^{m+1}\} \tag{7}$$

then property (V) holds without additional assumptions to the family U and to the problem (1)

Now, for any $u \in U$ we define

$$S(u)=\{t \in R^1 \mid u((t,0,..,0)^T) \leqslant \inf_{v \in Q} u(v)\} \tag{8}$$

and

$$\tau(u)=\sup\{t \mid t \in S(u)\} \ . \tag{9}$$

Thus τ forms a functional $\tau \mid U \to \bar{R}:=\tilde{R} \cup \{-\infty\}$. We get the following theorem characterizing weak duality.

<u>Theorem 2:</u> Let property (V) hold. Then for any $x \in G$ and $u \in U$ we have the inequality $f_o(x) \geqslant \tau(u)$.

On the base of theoren 2 the problem

$$\tau(u)\text{-sup! s.t. } u \in U \tag{10}$$

can be interpreted as a dual problem with respect to the primal problem (1).

To derive strong duality and saddle point results, now, we restrict us to families U given by (7) with

$$\varphi(0)=0 \quad \text{for any } \varphi\in\Phi . \tag{11}$$

Let denote $u_\varphi \in U$ the function defined by $\varphi\in\Phi$, that means

$$u_\varphi(w)=w_0+\varphi(\underline{w}) \quad \text{for any } w\in R^{m+1} .$$

Then (7)-(9), (11) lead to

$$S(u_\varphi)=\{t\in R^1 \mid t\leq \inf_{v\in Q} [v_0+\varphi(\underline{v})]\}$$

and consequently

$$\tau(u_\varphi)=\inf_{v\in Q} [v_0+\varphi(\underline{v})] \quad \text{for any } \varphi\in\Phi .$$

Let denote $g\mid\Phi\to\overline{R}$ and $T\mid X\times\Phi\to\overline{R}$ the functionals $g(\varphi)=\tau(u_\varphi)$ and $T(x,\varphi)=F(x,u_\varphi)$ for any $x\in X$, $\varphi\in\Phi$. Thus (10) is equivalent to

$$g(\varphi)\text{-sup! s.t. } \varphi\in\Phi. \tag{12}$$

Furthermore we define $\underline{T}\mid\Phi\to\overline{R}$, $\overline{T}\mid X\to\overline{R}$ by

$$\underline{T}(\varphi)=\inf\{T(x,\varphi)\mid x\in X\} \quad \text{for any } \varphi\in\Phi ,$$
$$\overline{T}(x)=\sup\{T(x,\varphi)\mid \varphi\in\Phi\} \quad \text{for any } x\in X .$$

<u>Lemma 1</u>: Let be the family U given by (7), (11). Then the functionals g and $\underline{T}$ are identical.

<u>Proof</u>: We denote

$$K(x,\varphi,v)=\begin{cases} v_0+\varphi(\underline{v}) & \text{, if } v\geq f(x)\\ +\infty & \text{, otherwise}\end{cases}$$

for any $x\in X$, $\varphi\in\Phi$, $v\in R^{m+1}$. Then we get

$$T(x,\varphi)=\inf\{K(x,\varphi,v)\mid v\in R^{m+1}\} \quad \forall\, x\in X,\ \varphi\in\Phi .$$

This results in

$$g(\varphi)=\inf_{v\in Q}[v_0+\varphi(\underline{v})]=\inf_{v\in R^{m+1}}\ \inf_{x\in X} K(x,\varphi,v)=$$
$$=\inf_{x\in X}\ \inf_{v\in R^{m+1}} K(x,\varphi,v)=\inf_{x\in X} T(x,\varphi)=\underline{T}(\varphi) \quad \text{for any } \varphi\in\Phi . \ //$$

Thus the dual problem (10) can be expressed by

$$\underline{T}(\varphi)\text{-sup! s.t. } \varphi\in\Phi . \tag{13}$$

If the function T satisfies the following condition

$$\bar{T}(x) = \begin{cases} f_o(x) & , \text{ if } x \in G \\ +\infty & , \text{ otherwise} \end{cases} \tag{14}$$

then the primal problem (1) is equivalent to

$$\bar{T}(x)\text{-min! s.t. } x \in X . \tag{15}$$

The relation (14) generalizes a well known property of ordinary Lagrangians. This property plays an essential role for the introduction of generalized Lagrangians (see [4], [12], [13] e.g.) and related saddle point results. The properties proposed by MANGASARIAN [8] guarantee (14) also.

Theorem 3: Let the family U possess the structure (7), (11). Furthermore let (14) hold. If $(\hat{x}, \hat{\varphi}) \in X \times \Phi$ forms a saddle point of T, that means

$$T(\hat{x}, \varphi) \leq T(\hat{x}, \hat{\varphi}) \leq T(x, \hat{\varphi}) \quad \text{for any } x \in X, \ \varphi \in \Phi, \tag{16}$$

then $\hat{x}$ solves the primal problem (1) and $\hat{\varphi}$ solves the dual problem (12).

Proof: From the definitions of $\bar{T}$, $\underline{T}$ and the inequality (16) we get

$$\bar{T}(x) \geq T(x, \hat{\varphi}) \geq T(\hat{x}, \hat{\varphi}) = \bar{T}(\hat{x}) \quad \text{for any } x \in X$$

and

$$\underline{T}(\varphi) \leq T(\hat{x}, \varphi) \leq T(\hat{x}, \hat{\varphi}) = \underline{T}(\hat{\varphi}) \quad \text{for any } \varphi \in \Phi .$$

Thus $\hat{x}$ solves (15) and $\hat{\varphi}$ solves (13). The property (14) and the lemma complete the proof. //

Due to [1] a function $c \mid R^m \to \bar{R}$ is called Φ-subdifferentiable at the point $\hat{z} \in R^m$ if some function $\hat{\varphi} \in \Phi$ exists such that

$$c(z) + \hat{\varphi}(z) \geq c(\hat{z}) + \hat{\varphi}(\hat{z}) \quad \text{for any } z \in R^m .$$

Let denote $\chi \mid R^m \to \bar{R}$ the primal function of the problem (1), that means

$$\chi(z) = \inf \{ f_o(x) \mid x \in X, \ \underline{f}(x) \leq z \} \quad \text{for any } z \in R^m.$$

Similar to [1], [13] we get

Theorem 4: Under the conditions of theorem 3 a function $\hat{\varphi} \in \Phi$ satisfying

$$g(\hat{\varphi}) = \chi(0) \tag{17}$$

exists if and only if χ is Φ-subdifferentiable at 0.

Proof: As shown in the proof of lemma 1 the equality

$$g(\varphi) = \inf \{ v_o + \varphi(\underline{v}) \mid v \in Q \}$$

for any $\varphi \in \Phi$ holds. Using the definition of the set Q and of the function χ we get

$$g(\varphi) = \inf_{\underline{v} \in R^m} \ \inf_{x \in X, \ \underline{f}(x) \leq \underline{v}} \{ f_o(x) + \varphi(\underline{v}) \} = \inf_{\underline{v} \in R^m} \{ \chi(\underline{v}) + \varphi(\underline{v}) \} . \tag{18}$$

If $\hat{\varphi} \in \Phi$ fulfills (17) then we have

$$\chi(0) = g(\hat{\varphi}) \leq \chi(\underline{v}) + \hat{\varphi}(\underline{v}) \quad \text{for any } \underline{v} \in R^m \quad . \tag{19}$$

With (11) this prooves χ to be Φ-subdifferentiable at 0. On the other hand the Φ -subdifferentiability of χ at 0 guarantees (19). Now, (18) leads to (17). //

In the next theorem we formulate a sufficient condition for the primal function χ to be Φ-subdifferentiable at the origin.

Theorem 5: Let be the functions f_i, $i=0,1,\dots,m$ three times differentiable and let denote $x^* \in G$ a point satisfying the second order sufficient optimality condition (see [3] e.g.). Additionally we suppose that strict complementarity holds and that the gradients of the active constraints at x^* are linearly independent. Furthermore let the conditions of theorem 3 be fulfilled and to any $\tilde{y} \in R^m_+$, $\tilde{\mu} > 0$ a function $\tilde{\varphi} \in \Phi$ is assumed to exist such that

$$\nabla \tilde{\varphi}(0) = \tilde{y}$$

and

$$w^T \nabla^2 \tilde{\varphi}(z) w \geq \tilde{\mu} \| w \|^2 \qquad \forall\, w, z \in R^m \quad . \tag{20}$$

Then there is an $\varepsilon > 0$ such that the primal function χ additionally restricted to $X = U_\varepsilon(x^*) := \{x \in R^n \mid \|x - x^*\| < \varepsilon\}$ is Φ-subdifferentiable at the point 0.

Proof: Applying the implicit function theorem to the second order optimality condition we can show (see [10] e.g.) neighbourhoods $U_\varepsilon(x^*)$, $U_\delta(0)$ to exist such that χ with $X = U_\varepsilon(x^*)$ is twice differentiable on $U_\delta(0)$. From the compactness of X and the continuity of f_o we get a finite $\nu = \inf\{f_o(x) \mid x \in X\}$. The definition of χ leads to

$$\chi(z) \geq \nu \qquad \text{for any } z \in R^m .$$

Due to the optimality criterion some $\tilde{y} \in R^m_+$ exists satisfying

$$\nabla f_o(x^*) + \sum_{i=1}^{m} \tilde{y}_i \nabla f_i(x^*) = 0, \quad \tilde{y}^T \underline{f}(x^*) = 0 \ .$$

Furthermore the equality

$$\nabla \chi(0) = -\tilde{y} \tag{21}$$

holds. Now, we select $\tilde{\mu} > 0$ such that

$$\tilde{\mu}\delta > \|\tilde{y}\| \quad \text{and} \quad \frac{\tilde{\mu}}{2}\left(\delta - \frac{\|\tilde{y}\|}{\tilde{\mu}}\right)^2 - \frac{\|\tilde{y}\|^2}{2\tilde{\mu}} \geq \chi(0) - \nu \tag{22}$$

holds and $\nabla^2 \chi(z) + \tilde{\mu} I$ is positive definite for any $z \in U_\delta(0)$. If $\tilde{\varphi} \in \Phi$ is selected according to (20) then

$$\tilde{\varphi}(z) \geq \tilde{\varphi}(0)+\tilde{y}^T z+\frac{\tilde{\mu}}{2}\|z\|^2 = \frac{\tilde{\mu}}{2}\| z+\frac{1}{\tilde{\mu}}\tilde{y}\|^2 - \frac{1}{2\tilde{\mu}}\|\tilde{y}\|^2 \geq$$

$$\geq \frac{\tilde{\mu}}{2}\left(\|z\|-\frac{1}{\tilde{\mu}}\|\tilde{y}\|\right)^2-\frac{1}{2\tilde{\mu}}\|\tilde{y}\|^2 \quad \text{for any } z\in R^m .$$

Using (22) this results in

$$\tilde{\varphi}(z) \geq \chi(0)-\nu \geq \chi(0)-\chi(z) \quad \text{for any } z\notin U_\delta(0). \tag{23}$$

From (21) and the positive definiteness of $\nabla^2\chi(z)+\tilde{\mu}I$ on $U_\delta(0)$ we get

$$\chi(z)+\tilde{\varphi}(z) \geq \chi(0)+\tilde{\varphi}(0) \quad \text{for any } z\in U_\delta(0)$$

and with (22)

$$\chi(z)+\hat{\varphi}(z) \geq \chi(0)+\tilde{\varphi}(0) \quad \text{for any } z\in R^m . \quad //$$

On the base of theorem 5 we can derive strong duality results in a local sense.

The present paper contains only some duality results. An application of the unified approach proposed here to convergence theory and to the rate of convergence of sequential minimization algorithms is given in [5] - [7].

The author thanks Prof.J.Stoer for bringing the thesis [9] to his attention during the discussion at IFIP-conference. Results of this paper are related to the investigations described here for the special case of penalty methods.

References

[1] Dolecki,S; Kurcyusz,S.: On Φ-convexity in extremal problems. Technical report, Institute of automatic control, Warszawa, 1976.

[2] Everett,H.D.: Generalized Lagrange multiplier method for solving problem of optimum allocation of resources. Operations Res. 11(1963), 399-417.

[3] Fiacco,A.V.; McCormick,G.P.: Nonlinear programming: sequential unconstrained minimization techniques. Wiley, New York, 1968.

[4] Gol'stejn,E.G.; Tret'jakov,N.V.: Modificirovannye funkcii Lagranža. Ekon. Mat. Metody 10(1974)3, 568-591.

[5] Großmann,Ch.: Rates of convergence in methods of exterior centers. Math.OF Statist., Ser. Optimization 9(1978)3, 373-388.

[6] Großmann,Ch.: Common properties of nonlinear programming algorithms basing on sequential unconstrained minimizations. (to appear in Proc. 7th summer school on nonlinear analysis, Berlin 1979).

[7] Großmann,Ch.; Kaplan,A.A.: Strafmethoden und modifizierte Lagrangefunktionen in der nichtlinearen Optimierung. Teubner-Text, Leipzig, 1979.

[8] Mangasarian,O.L.: Unconstrained Lagrangians in nonlinear programming. SIAM J. Control 13(1975)4, 772-791.

[9] Neckermann,O.K.: Verallgemeinerte konjugierte Dualität und Penaltyverfahren. Dissertation, Würzburg, 1976.

[10] Robinson,S.M.: Perturbed Kuhn-Tucker-points and rates of convergence for a class of nonlinear programming problems. Math. Programming 7(1974)1, 1-16.

[11] Rockafellar,R.T.: Augmented Lagrange multiplier functions and duality in nonconvex programming. SIAM J. Control 12(1974)2, 268-285.

[12] Roode,J.D.: Generalized Lagrange functions in mathematical programming. Dissertation, Rotterdam, 1968.

[13] Seidler,K.-H.: Zur Dualisierung in der nichtlinearen Optimierung. Dissertation, TH Ilmenau, 1972.

[14] Wierzbicki,A.P.; Kurcyusz,S.: Projection on a cone, penalty functionals and duality for problems with inequality constraints in Hilbert space. SIAM J. Control Optimization 15(1977), 25-56.

MINIMAX OPTIMIZATION USING QUASI-NEWTON METHODS

Jørgen Hald and Kaj Madsen
Technical University of Denmark
DK-2800 Lyngby, Denmark

Summary: The problem under consideration is that of minimizing the objective function

$$F(\underline{x}) = \max_{1\le j\le m} f_j(\underline{x})$$

where $\{f_j\}$ is a set of m nonlinear, differentiable functions of n variables $\underline{x} = \{x_1, x_2, \ldots, x_n\}^T$. This problem can be solved by a method that uses linear approximations to the functions f_j, and normally this method will have a quadratic final rate of convergence. However, if some regularity condition is not fulfilled at the solution then the final rate of convergence may be very slow. In this case second order information is required in order to obtain a fast final convergence. We present a method which combines the two types of algorithms. If an irregularity is detected a switch is made from the first order method to a method which is based on approximations of the second order information using only first derivatives. It has been proved that the combined method has sure convergence properties, and that normally the final rate of convergence will be either quadratic or superlinear.

1. Introduction.

In this paper we consider the problem of minimizing the minimax objective function

$$F(\underline{x}) \quad \max_{1\le j\le m} f_j(\underline{x}) \tag{1}$$

where the functions f_j are supposed to be smooth, and $\underline{x} = (x_1, x_2, \ldots, x_n)^T$.

An excellent theoretical treatment of minimax optimization can be found in the book of Dem'yanov and Malozemov, [10]. Algorithms for minimizing (1) by using only first derivative information have been published by several authors during the past ten years. Lately it has become clear that in some situations second derivative information is necessary in order to obtain fast final convergence. Examples of algorithms based upon this are those of Hettich [15], Han [14], Charalambous and Moharram [6], Hald and Madsen [12], Watson [22] and Conn [8].

The objective function (1) is in general a non-differentiable function having directional derivatives in all directions. Normally, the minimum is situated at an edge, that is a point where two or more functions are equal, cf figure 1 which shows level curves for minimax objective functions in 2 variables

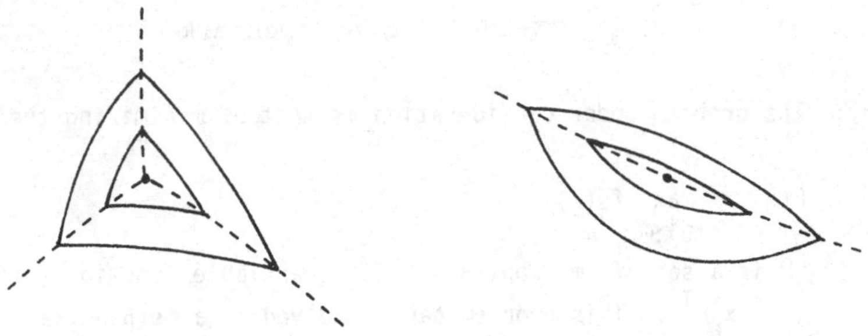

1a 1b

Figure 1

In 1a there is no smooth valey through the solution and the minimum is numerically very well determined: no second derivative information (positive definiteness) is needed, the minimum is characterized by only first derivatives of the 3 functions f_j which determine the minimum. Therefore it is possible to construct algorithms based on first derivative information, with fast final convergence in cases like 1a. It was proved in [18] that the stage 1 (see below) algorithm of this paper, which is of the type mentioned, has quadratic final rate of convergence to the solution $\underline{x}$ when any subset of the set $\{\underline{f}_i'(\underline{x}^*) \mid f_i(\underline{x}^*) = F(\underline{x}^*)\}$ has maximal rank. This condition is the co-called <u>Haar-condition</u>, which ensures that no smooth valley passes through the solution.

In 1b of figure 1 there is a smooth valley through the solution, namely along the dotted line. In this case some second order information may be needed: For directions through the valley the minimum is not characterized by first derivatives only (however this is still the case for all other directions). This suggests that in situations like 1b (where the number of functions determining the minimum is not larger than the number of unknowns) some second order information, or approximate second order information, is needed in order to obtain a fast final rate of convergence. But the fact that the level curves of F have sharp corners is still useful: In figure 1b, for example, first derivatives will, with a quadratic rate of convergence, give the information that the solution is at the dotted line, so the dimension of the problem is reduced from 2 to 1 in this case. In general such a valley is always characterized by the fact that some functions are equal. Suppose that the number of such functions is s and the functions are f_j , $j \in A(\underline{x}^*)$, i.e. $F(\underline{x}^*) = f_j(\underline{x}^*) > f_i(\underline{x}^*)$ for $j \in A(\underline{x}^*)$ and

$i \notin A(\underline{x}^*)$. Then the following must hold in the valley and at the solution,

$$f_{j_0}(x) - f_j(\underline{x}) = 0 \ , \quad j \in A(\underline{x}^*) \ , \quad \text{where} \quad j_0 \in A(\underline{x}^*) \ , \tag{2}$$

so by linearizing these, we can obtain a quadratic convergence to the valley. If the Haar-condition is satisfied at the solution then $s \geq n+1$ and the Jacobian of the system (2) has rank n, so there is no valley (figure 1a). In this case a Newton iteration applied to (2) gives quadratic convergence to the solution and it requires only first derivatives of the functions f_j. If, however, $s \leq n$ or the Jacobian of (2) is rank deficient at $\underline{x}^*$ then some information is needed in addition to (2). We use some equations that correspond to the Kuhn Tucker equations in nonlinear programming, namely the following (see for instance [15]):

$$\begin{aligned} \sum_{j \in A(\underline{x})} \lambda_j \underline{f}_j'(\underline{x}) &= \underline{0} \\ \sum_{j \in A(\underline{x})} \lambda_j &= 1 \end{aligned} \tag{3}$$

for $\underline{x} = \underline{x}^*$. Furthermore $\lambda_j \geq 0$ for $j \in A(\underline{x}^*)$. Solving (2) and (3) for the minimax problem is equivalent to solving the Kuhn Tucker equations in nonlinear programming when the number of active constraints is less than the number of unknowns, [15].

The minimax algorithm that we propose in this paper consists in two parts. Normally, the algorithm of [16] is used; this is called <u>stage 1</u>. But if a smooth valley through the solution is detected, then a switch is made to an algorithm which solves (2) and (3) through a quasi-Newton iteration. This is called <u>stage 2</u>. If it turns out that the stage 2 iteration is unsuccessful (for instance if the active set $f_1, \ldots, f_s$ has been wrongly chosen), then a switch is made back to stage 1. The algorithm may switch several times between stage 1 and stage 2. However our experiences indicate that normally only very few switches will take place, so the iteration will finish either as a quadratically convergent stage 1 iteration or as a superlinearly convergent stage 2 iteration according as the Haar condition is satisfied or not at the solution.

<u>2. Details of the algorithm.</u> In the following we give a rather complete description of the algorithm. We omit only a few computational details which are not of importance for the understanding of the algorithm. These may be found in [12].

The algorithm consists in four parts: The stage 1 iteration, the stage 2 iteration and two sets of criteria for switching.

2A. The stage 1 iteration. This is the iteration of [16]. At the k'th stage of the algorithm we have an approximation $\underline{x}_k$ of the solution and we wish to use the gradient information at $\underline{x}_k$ to find a better approximation $\underline{x}_{k+1}$. Therefore we find the increment as a solution $\underline{h}_k$ of the linear minimax problem

Minimize (as a function of $\underline{h}$)

$$\bar{F}(\underline{x}_k,\underline{h}) \equiv \max_{1\leq i\leq m} \{f_i(\underline{x}_k) + \underline{f}_i'(\underline{x}_k)^T\underline{h}\} \tag{4}$$

subject to the constraints

$$||\underline{h}||_\infty \leq \Lambda_k \ ,$$

where $\Lambda_k = c_{k-1}||\underline{h}_{k-1}||_\infty$, with $c_{k-1} = 0.5$, 1, or 2 according as the iteration number $k-1$ is unsuccessful, not unsuccessful, or successful. If the objective function F decreases then $\underline{x}_{k+1} = \underline{x}_k+\underline{h}_k$, otherwise $\underline{x}_{k+1} = \underline{x}_k$. Note that no line search is involved.

2B. The stage 2 iteration. We suppose that a set of indices A has been determined in 3C; A is called the current active set and is an approximation of the set of indices that are active at the solution $\underline{x}^*$,

$$A^* \equiv A(\underline{x}^*) \equiv \{j\,|\,f_j(\underline{x}^*) = F(\underline{x}^*)\} \tag{5}$$

Now an approximate Newton iteration is applied to the system

$$\begin{aligned} \sum_{j\in A} \lambda_j\, \underline{f}_j'(\underline{x}) &= \underline{0} \\ \sum_{j\in A} \lambda_j - 1 &= 0 \\ f_{j_0}(\underline{x}) - f_j(\underline{x}) &= 0 \quad , \quad j\in A\ ,\ j\neq j_0 \end{aligned} \tag{6}$$

where $j_0\in A$ is fixed. (The unknowns are $(\underline{x},\underline{\lambda})$). The Newton iteration is approximate because we do not use $\underline{\underline{f}}''_j(\underline{x})$, so the derivatives of the first set of equations in (6) with respect to $\underline{x}$ are approximated by finite differences on the first iteration and then updated for the subsequent iterations. Every time a switch from stage 1 to stage 2 is made a new finite difference approximation is calculated.

2C. Conditions for switching to stage 2. It is not disasterous to start a quasi-Newton iteration with an incorrect active set A, since in that case a switch back to stage 1 will be made after some iterations (see Theorem 1 in [12]). But in order to avoid unnecessary iterations we use a rather restrictive set of criteria which must all be satisfied before the quasi-Newton iteration is started: the switch is only made if (8)-(10) below are satisfied.

For each stage 1 iteration the active set $A=A_k$ is defined as

$$A_k \equiv \{j \mid F(\underline{x}_k)-f_j(\underline{x}_k)\le\varepsilon_1\} \tag{7}$$

where ε_1 is a small positive number specified by the user. Suppose that the latest 3 different iterates, $\underline{x}_k, \underline{x}_{k-j_1}, \underline{x}_{k-j_2}$, have been calculated in stage 1. Then a switch to stage 2 is made if there exist $\lambda_j \ge 0$, $j \in A_k$, with $\sum \lambda_j = 1$, such that

$$\|\underline{h}_k\|_\infty = \Lambda_k \tag{8}$$

$$A_{k-j_2} = A_{k-j_1} = A_k \tag{9}$$

$$\Big\|\sum_{j\in A_k} \lambda_j \, \underline{f}_j'(\underline{x}_k)\Big\|_2 \le \varepsilon_2 \tag{10}$$

where $\varepsilon_2 > 0$. (In practice (10) is only tested when (8) and (9) are satisfied, and $\{\lambda_j\}$ is found by a linear least squares calculation).

The condition (8) will hold near a non-Haar solution (see lemma 1 in [13]) whereas it will not hold near a solution that satisfies the Haar condition because of the quadratic convergence in this case (cf. [18]). The condition (9) tests the stability of the active set, whereas (10) holds when $\underline{x}_k$ is close to a solution $\underline{x}^*$ with $A^* = A_k$ (see equation (3)).

2D. Causes for switching back to stage 1. The rules of this section are tested for in each quasi-Newton step. These rules guarantee that the convergence properties of the method used in stage 1 is made if one of the conditions (11), (12), or (13) below fails to hold.

We denote by $\underline{r}(\underline{x},\underline{\lambda})$ the vector of left-hand sides of the nonlinear system (6), i.e. the residual vector of (6) at the point $(\underline{x},\underline{\lambda})$. In order to continue the stage 2 iteration we require that the residuals decrease:

$$\|\underline{r}(\underline{x}_{k+1},\underline{\lambda}_{k+1})\| \le \eta \|\underline{r}(\underline{x}_k,\underline{\lambda}_k)\| \tag{11}$$

where $0<\eta<1$. We have used $\eta=0.999$.

It is also tested that no function with an index from outside the active set becomes dominating, i.e.

$$F(\underline{x}_{k+1}) = \max_{j\in A} f_j(\underline{x}_{k+1}) \tag{12}$$

Finally we test that all multipliers corresponding to the active set are non-negative,

$$\lambda_i^{(k+1)} \geq 0 \quad , \quad i \in A \tag{13}$$

Notice, that it is not required that the minimax objective function F decreases in each stage 2 iteration. Often F will increase in the first stage 2 iteration.

3. Convergence properties. We use the same smoothness assumption as in [12], namely the following,

$$\begin{aligned} &f_j(\underline{x}+\underline{h}) = f_j(\underline{x})+\underline{f}_j'(\underline{x})^T\underline{h}+o(\|\underline{h}\|) \ , \\ &\underline{f}_j' \text{ is continuous} \ , \qquad j=1,\dots,m \end{aligned} \tag{14}$$

A stationary point of F is a point for which the generalized derivative, [7],

$$\partial^*F(\underline{x}) \equiv \text{conv}\{\underline{f}_i'(\underline{x}) \mid f_i(\underline{x}) = F(\underline{x})\} \ , \tag{15}$$

conv being the convex hull, contains $\underline{0}$. It follows that a stationary point is a point where the equalities (3) are satisfied. It is shown, for instance in [18], that any local minimum of F is a stationary point, and on the other hand that if the Haar-condition is satisfied at a stationary point $\underline{x}$, then $\underline{x}$ is a strict local minimum.

In [12] we examined the global convergence properties of the algorithm presented here and proved the following result:

Theorem 1 If the sequence $\underline{x}_k$ generated by the algorithm is convergent then the limit point is a stationary point.

In the local convergence theorems we assume that $\underline{x}_k \to \underline{x}^*$ and that $\underline{x}^*$ is non-degenerate, which means that every active function at $\underline{x}^*$ is necessary for $\underline{x}^*$ being a stationary point:

$$f_i(\underline{x}^*) = F(\underline{x}^*) \Rightarrow \underline{0} \notin \text{conv}\{\underline{f}_j'(\underline{x}^*) \mid f_j(\underline{x}^*)=F(\underline{x}^*),\ j\neq i\}$$

The following two theorems are proved in [13]:

Theorem 2 Suppose that $\underline{x}_k \to \underline{x}^*$ and that $\underline{x}^*$ is non-degenerate. Then $A_k=A^*\equiv A(\underline{x}^*)$ for all large values of k provided ε_1 in (7) is chosen small enough. Furthermore, if there is only a finite number of stage 2 iterations then $s^*=(n+1)$ and the rate of concergence is quadratic.

Theorem 3 Suppose that $\underline{x}_k$ converges to the non-degenerate stationary point $\underline{x}^*$ and that the quasi-Newton iteration used is locally and linearly convergent at $(\underline{x}^*,\underline{\lambda}^*)$. If there is an infinite number of stage 2 iterations then every iteration from a certain point is a quasi-Newton iteration with the active set $A = A^*$.

Theorems 2 and 3 show that when $\underline{x}_k$ converges to a non-degenerate point then the rate of convergence is at least as good as the rate of convergence of the method used in stage 2 provided we use a locally and linearly convergent quasi-Newton iteration. Broyden's method, [2], has the required properties, and if $\underline{r}'(\underline{x}^*,\underline{\lambda}^*)$ is regular and $\underline{r}$ satisfies at Lipschitz condition at $(\underline{x}^*,\underline{\lambda}^*)$ then the rate of convergence is superlinear as shown in [3]. Since equations (6) can be set up such that the Jacobian is symmetric it is also possible to use Powell's symmetric Broyden update [19] which has superlinear convergence under the conditions mentioned above (see [3]). Under a few additional assumptions we will also have superlinear convergence when using the DFP or the BFGS update, [3].

Numerical examples. In [12] and [13] a number of numerical experiments are reported. Here the space only allows us to give one example.

Example 1. We formulate the Rosen-Suzuki problem [20] as a minimax problem following [5]. This gives a problem with 4 unknowns and 4 functions. Two starting points were used, $\underline{x}_0^I = (0,0,0,0)^T$ and $\underline{x}_0^{II} = (2,2,5,0)^T$. The solution $\underline{x}^* = (0,1,2,-1)^T$ was found to 14 decimals accuracy using 19 calculations of each function when $\underline{x}_0 = \underline{x}_0^I$, and 21 function calculations when $\underline{x}_0 = \underline{x}_0^{II}$. In both cases 1 switch was made to stage 2 where the calculation was finished.

References

1. J.W. Bandler and C. Charalambous, "New algorithms for network optimization", IEEE Trans. Microwave Theory Tech., vol. MTT-21, pp. 815-818, 1973.
2. C.G. Broyden, "A class of methods for solving nonlinear simultaneous equations", Math. Comp., 19, pp. 577-593, 1965.
3. C.G. Broyden, J.E. Dennis and J.J. Moré, "On the local and superlinear convergence of quasi-Newton methods", Ibid., 12, pp. 223-246, 1973.
4. C. Charalambous and A.R. Conn, "Optimization of microwave networks", IEEE Trans. Microwave Theory Tech., vol. MTT-23, pp. 834-838, 1975.
5. C. Charalambous and A.R. Conn, "An efficient method to solve the minimax problem directly", SIAM J. Num. Anal., Vol. 15, No. 1, pp. 162-187, 1978.

6. C. Charalambous and O. Moharram, "A new approach to minimax optimization", Department of Systems Design, University of Waterloo, Ontario, Canada, pp. 1-4, 1978.

7. F.H. Clarke, "Generalized gradients and applications", Transactions of the American Mathematical Society 205, pp. 247-262, 1975.

8. A.R. Conn, "An Efficient Second Order Method to solve the (Constrained) Minimax Problem", Department of Combinatorics and Optimization, University of Waterloo, Ontario, Canada. Report CORR 79-5, pp. 1-48, 1979.

9. W.C. Davidon, "Optimally conditioned optimization algorithms without line searches", Math. Progr. 9, pp. 1-30, 1975.

10. V.F. Dem'yanov and V.N. Malozemov, "Introduction to minimax", Wiley, New York, 1974). Translated from: Vvedenie v minimaks (Izdatel'stvo "Nauka", Moscow, 1972).

11. J.E. Dennis and J.J. Moré, "Quasi-Newton Methods, Motivation and Theory". SIAM Review, vol. 19, pp. 46-89, 1977.

12. J. Hald and K. Madsen, "An 2-stage algorithm for minimax Optimization". Proceedings of the "Colloque International sur l'Analyse et l'Optimization des Systemes, IRIA, France, 1978.

13. J. Hald and K. Madsen, "Combined LP and quasi-Newton methods for minimax optimization". Report No. NI-79-05, Institute for Numerical Analysis, Technical University of Denmark, DK-2800 Lyngby, Denmark.

14. S.P. Han, "Superlinear convergence of a minimax method", Cornel Tech. Report. TR 78-336, 1978.

15. R. Hettich, "A Newton-method for nonlinear Chebyshev approximation". In approximation Theory, Lect. Notes in Math. 556, R. Schaback, K. Scherer, eds., Springer, Berlin-Heidelberg-New York, pp. 222-236, 1976.

16. K. Madsen, "An algorithm for minimax solution of overdetermined systems of non-linear equations", Journal of the Institute of Mathematics and its Applications 16, pp. 321-328, 1975.

17. K. Madsen and H. Schjær-Jacobsen, "Singularities in minimax optimization of networks", IEEE Trans. on Circuits and Systems, Vol. CAS-23, No. 7, 1976, pp. 456-460.

18. K. Madsen and H. Schjær-Jacobsen, "Linearly constrained minimax optimization", Math. Progr. 14, 1978, pp. 208-223.

19. M.J.D. Powell, "A new algorithm for unconstrained optimization", Nonlinear Programming, J.B. Rosen, O.L. Mangasarian and K. Ritter, eds., Academic Press, New York, 1970.

20. J.B. Rosen and S. Suzuki, "Construction of non-linear programming test problems", Comm. A.C.M., 8, pp. 113, 1965.

21. H.H. Rosenbrock, "An automatic method for finding the greatest or least value of a function". Comput. J., Vol. 3, pp. 175-184, 1960.

22. G.A. Watson, "The minimax Solution of an Overdetermined System of Non-linear Equations", J. Inst. Maths. Applics. 23, pp. 167-180, 1979.

ALGORITHMS FOR THE SOLUTION OF A DISCRETE MINIMAX PROBLEM: SUBGRADIENT METHODS AND A NEW FAST NEWTON - METHOD

Roland Hornung

Institut für Angewandte Mathematik und Statistik der Universität Würzburg, Am Hubland, D - 8700 Würzburg

Federal Republic of Germany

1. Introduction

Let be given the following " discrete minimax problem "

$$(P) \qquad \min_{x \in R^n} \max_{1 \leq j \leq m} f_j(x)$$

where $f_j : R^n \to R$ are convex and sufficiently smooth functions

The convexity condition is only a technical assumption to avoid troubles with the convergence, the smoothness assumption is the more important one.

Denote

$$f(x) := \max_{1 \leq j \leq m} f_j(x), \quad x \in R^n$$

This " maximum - function " has the following wellknown properties (see Dem'yanov, Malozemov [5]):

(1a) f is convex, continuous, non - smooth, but

(1b) f is differentiable at any $x \in R^n$ in any direction $s \in R^n$, i.e. there exists the directional derivative

$$f'(x;s) := \lim_{t \downarrow 0} \frac{f(x+ts) - f(x)}{t} = \sup_{g \in \partial f(x)} (g, s)$$

where the subdifferential $\partial f(x)$ ([17]) here can be represented by

$$\partial f(x) = \text{conv} \, (\nabla f_i(x) \mid i \in I(x))$$

(" conv " = convex hull)

and $$I(x) := \{ j \mid f_j(x) = f(x) \}$$

denotes the set of the " active " indices at x.

(1c) <u>optimality - condition</u> of problem (P):

$\bar{x}$ is solution of (P) $\Longleftrightarrow$ $0 \in \partial f(\bar{x})$ $\Longleftrightarrow$

$$0 \in \left\{ \sum_{i \in I(\bar{x})} l_i \nabla f_i(\bar{x}) \;\middle|\; l_i \geq 0,\ i \in I(\bar{x}),\ \sum_{i \in I(\bar{x})} l_i = 1 \right\}$$

Let be $\epsilon > 0$: x^+ is ϵ- optimal $\Longleftrightarrow 0 \in \partial_\epsilon f(x^+) \Longleftrightarrow f(x^+) \leq f(\bar{x}) + \epsilon$

\- - -

2. Application to nonlinear constrained optimization problems (Bandler - Charalambous - Method ([1]))

Given

$$(\,P^+\,) \quad \min \{ F(x) \mid x:\ F_j(x) \leq 0,\ j = 1,2,\ldots,l,\ x \in R^n \}$$

Set (in (P)): $f_1(x) := F(x)$

$$f_j(x) := F(x) + c_{j-1} F_{j-1}(x), \quad j = 2,3,\ldots,m := l+1$$

Then, for suitable values $c_j > 0$, $j = 1,2,\ldots,l$, (P^+) is equivalent to a problem like (P).

\- - -

3. Algorithms for (P)

There are many algorithms for the solution of (P) (and, partly, for other nondifferentiable problems, too). We give a brief survey for two reasons: The first reason we want to compare the cited algorithms with a method developed by me and described later, the second one we can use some of these methods to improve the behavior of global convergence of the algorithm of mine.

(3a) Descent algorithms

These subgradient - algorithms are generalizations of the <u>steepest</u> descent method (Lemaréchal, Mifflin ([11] , [12]), Dem'yanov ([5]), Wolfe ([18])). A general formulation is given in ([9]).

<u>Advantage</u> of these methods: <u>global</u> convergence (to an ϵ - optimal solution.

<u>Disadvantage</u>: slow (sublinear) asymptotic convergence rate

(3b) Gradient - projection method (Charalambous, Conn ([3]))
<u>global</u> convergence, unknown convergence rate

(3c) Linear approximation method (Madsen, Schjaer-Jacobsen ([13]))
<u>global</u> convergence; under (relatively) strong conditions (Haar - condition for the " active " gradients at the optimal solution) (local) <u>quadratic</u> convergence rate.

- - -

Now I will represent a method developed by me, which has a (local) quadratic convergence rate under milder conditions.

4. Newton - Algorithm

The basic idea of this method is to solve the optimality - condtion (1c) of problem (P) by a Newton - method. Several Newton - methods to solve (P) were developed by me ([9]); one of these I will represent here. Hettich ([7]) gives a survey of a number of numerical methods, among other things a Newton-method, for the nonlinear Chebyshev approximation, a problem related to that of (P). In the special case of (P) Madsen and Hald ([6]) developed a method similar to that of mine.
The Newton - method can be formulated in two stages:
First we present the proper Newton - method, which is a local algorithm and has an asymptotically quadratic convergence rate. Then we mention, how the domain of convergence of that method can be extended.

We consider again the optimality - conditions (1c) and we assume that we know the index set $I := I(\bar{x})$ ($= \{1,2,\dots,p\}$ w.l.o.g.), the indices of the " active " functions at the optimal solution $\bar{x} \in R^n$; let be $\bar{x}$ the unique solution of (P).
Because of $\mathrm{I}_1 := 1 - \sum_{i \in I \setminus \{1\}} \mathrm{I}_i$, we receive with $I_0 := I \setminus \{1\}$:

(4a) $g(\bar{x},\mathrm{I}) := (1 - \sum_{i \in I_0} \mathrm{I}_i) \nabla f_1(\bar{x}) + \sum_{i \in I_0} \mathrm{I}_i \nabla f_i(\bar{x}) = 0$

(4b) $\sum_{i \in I_0} \mathrm{I}_i \leq 1, \ \mathrm{I}_i \geq 0, \quad i \in I_0$

and, additionally a trivial condition (at $\bar{x}$!)

(4c) $\Delta f_i(\bar{x}) := f_i(\bar{x}) - f_1(\bar{x}) = 0, \quad i \in I_0$

Note that the index " 1 " is not distinguished, we can choose any other one of I . $\bar{l} := (\bar{l}_2, \bar{l}_3, \ldots, \bar{l}_p) \in R^{p-1}$ is an optimal multiplier. If we neglect the condition (4b) we can interprete $\bar{z} := (\bar{x}, \bar{l}) \in R^{n+p-1}$ as a solution of a nonlinear equation system (we will see later that condition (4b) is satisfied automatically close to the solution, if a condition of non - degeneracy holds).
Combinig (4a) and (4c) we get the nonlinear equation system

(4d)
$$F(z) = F(x,l) = \begin{pmatrix} g(x,l) \\ \Delta f(x) \end{pmatrix} = 0$$

(where $l := (l_2, l_3, \ldots, l_p) \in R^{p-1}$, $z = (x, l) \in R^{n+p-1}$ and $\Delta f(x) := (\Delta f_2(x), \ldots, \Delta f_p(x))^T$)

with $n+p-1$ equations and the same number of unknown variables.
We want to solve the system (4d) by the wellknown Newtonmethod ([14])

(4e)
$$0 \overset{!}{=} F(z') \overset{\circ}{=} F(z) + DF(z)(z'-z) \text{ , i.e.}$$

$$z' - z = \begin{pmatrix} x' - x \\ l' - l \end{pmatrix} = - DF(x,l)^{-1} F(x,l)$$

with
$$D^2 f(x,l) := \Big(1 - \sum_{i \in I_0} l_i \Big) D^2 f_1(x) + \sum_{i \in I_0} l_i \, D^2 f_i(x)$$

$$G(x) := (\nabla f_2(x) - \nabla f_1(x), \ldots\ldots, \nabla f_p(x) - \nabla f_1(x))$$

and
$$DF(z) = DF(x,l) := \begin{pmatrix} D^2 f(x,l) & G(x) \\ G(x)^T & 0 \end{pmatrix}$$

(4e) generates a sequence of iterationpoints $z_k = (x_k, l^k) \in R^{n+p-1}$, starting with a " suitable " point $z_0 = (x_0, l^0)$.
To guarantee convergence we have to assume the following mild conditions:

(4f) <u>Regularity condition</u>

(i) $\exists_{U(\bar{x})} \; \forall_{i \in I := I(\bar{x})} \; \forall_{x \in U(\bar{x})}$ $D^2 f_i(x)$ exist and are " lipschitz - continuous "

(ii) $\forall_{i \in I}$ $\nabla f_i(\bar{x})$ are affinely independent, i.e. from

$$\sum_{i \in I} \lambda_i \nabla f_i(\bar{x}) = \sum_{i \in I} \mu_i \nabla f_i(\bar{x}) = 0, \quad \sum_{i \in I} \lambda_i = \sum_{i \in I} \mu_i = 1$$

follows $\lambda = \mu$ (and $\bar{\lambda}$ is <u>unique</u> at the optimum),

(iii) The matrix $\sum_{i \in I} \bar{l}_i \, D^2 f_i(\bar{x})$ is nonsingular

(4g) <u>Condition for non - degeneracy</u>

At $\bar{x}$ there exists $\bar{l} > 0$, $\bar{l} \in R^p$, with $\sum_{i \in I} \bar{l}_i \nabla f_i(\bar{x}) = 0$, $\sum_{i \in I} \bar{l}_i = 1$.

Then we can state the following

(4h) <u>Theorem:</u>

(i) The optimality condition (4b) , i.e.

$$\sum_{i \in I_0} \bar{l}_i \leq 1, \; \bar{l}_i \geq 0, \; i \in I_0$$

is (automatically) satisfied for sufficiently large k :

$$l_i^k > 0, \quad i \in I_0$$

$$l_1^k := 1 - \sum_{i \in I_0} l_i^k > 0$$

(ii) <u>Newton - Kantorovich</u> :

For all starting points $z_0 \in U(\bar{z})$ (a suitable neigbour - hood of $\bar{z} = (\bar{x}, \bar{l})$) the sequence $\{z_k\}$ converges quadratically to $\bar{z}$.

\- - -

(4i) <u>Remarks</u> :

(i) Remember the preliminary assumption above that we know the index set $I = I(\bar{x})$, the indices of the active functions at the solution $\bar{x}$. If we use a wellknown result of Dem'yanov, we can forget that assumption, because it holds

$$\exists_{\bar{\epsilon} > 0} \; \forall_{0 < \epsilon \leq \bar{\epsilon}} \; \exists_{V(\bar{x})} \; \forall_{x \in V(\bar{x})} \quad I_\epsilon(x) = I(\bar{x})$$

where

$$I_\epsilon(x) = \{ j \mid f(x) - f_j(x) \leq \epsilon \}$$

i.e. the index set $I := I_\epsilon(x)$ is a suitable estimate for $I(\bar{x})$ close to the solution.

(ii) $s_k := \Delta x_k = x_{k+1} - x_k$ (in (4e) above) is a <u>descent direction</u> of f at x_k, if (besides the regularity - and non-degeneracy - conditions) holds :

$$I(x_k) = I \ (= I(\bar{x}) \ ); \quad D^2 f(x_k, l^k) \text{ is positive definite };$$

(iii) <u>Quasi - Newton method</u> : We can replace the exact Hessian matrices $D^2 f(x_k, l^k)$ in (4e) by Quasi - Newton - update - matrices B_k , i.e. we have

$$D_k = \begin{pmatrix} B_k & G(x_k) \\ G(x_k)^T & 0 \end{pmatrix} \quad (\cong DF(x_k, l^k) \ )$$

I had chosen the Powell - symmetric - Broyden formula ([2]). Then the asymptotic convergence rate (in the direct prediction case, i.e. no line-search, the step-length parameter $t_k = 1$) is <u>Q - superlinear</u> ([9]).

The Newton-method (and Quasi-Newton-method) described above has the following properties of the convergence : local convergence, quadratic (or Q - superlinear) convergence rate. To extend the area of convergence we consider the heuristic idea: Combination of the (Quasi-) Newton - method and a <u>subgradient method</u> of the Dem'yanov - type or Lemaréchal - type (globally convergent!) and application of <u>armijo - type line-searches</u> ([14]) in the Newton steps (" damped " Newton).

Extension of the domain of convergence - the 2-stage algorithm :

N e w t o n - s u b g r a d i e n t method

→ outer algorithm: subgradient - method

<u>Testcriterion:</u> If $I_\epsilon(x_{k+1}) = I_\epsilon(x_k)$, then

inner algorithm: (Quasi -) Newton - method

If some testcriterions are <u>not</u> satisfied, <u>Restart</u>.

We call the 2-stage algorithm above again " Newton-method." We have <u>not</u> a global convergence theorem, but we can hope that the Newton-steps start (sufficiently) close to the solution.
Some numerical examples illustrate the theoretical results.

Numerical examples

We make a numerical comparison between the subgradient methods (represented by the methods of Dem'yanov ([5]) and Lemaréchal ([11])), our Newton algorithm and some other wellknown methods by mean of several examples (academic examples as well as wellknown problems of nonlinear constrained optimization transformed to the minimax problem (P) by exact penalty minimax technique). We use own ALGOL 60 codes of the Lemaréchal - , Dem'yanov - , Newton - and Quasi - Newton - methods. The results of all the other methods are cited of the literature.

Some technical remarks:

<u>line - search</u>: gradient steps: a golden section type procedure of Kiekebusch - Müller ([10])
Newton steps: armijo type method ([14])

<u>stop criterion</u>: STOP, if $\|g_k\| \leq$ eps, where g_k is a (ϵ -) subgradient in the iterationpoint x_k ,and eps is a given, small positive number. Instead of that we can stop, if the (extended) optimality condition (4a - 4c) is nearly satisfied, i.e. $\|F(z_k)\| \leq$ eps.

The results were computed on the TR440 computer of the computing center of the Würzburg University.

<u>Problem 1</u> (Charalambous, Conn ([3])):

Dimension $n = 2$, number of functions $m = 3$.

$f_1(x) = x_1^4 + x_2^2$, $f_2(x) = (2-x_1)^2 + (2-x_2)^2$, $f_3(x) = 2\cdot\exp(-x_1+x_2)$

Solution: $\bar{x}_1 = \bar{x}_2 = 1$, $f(\bar{x}) = 2$ ($= f_1(\bar{x}) = f_2(\bar{x}) = f_3(\bar{x})$)

Starting from $x_1 = 1$, $x_2 = -0.1$ only we print out here the norm - differences $\|x_k - \bar{x}\|$ of the last steps to give a first impression of the final rate of convergence.

Charalamb.,Conn	Dem'yanov	Lemaréchal	Newton
:	:	:	:
$5\ 10^{-2}$	$3\ 10^{-2}$	$5.5\ 10^{-1}$	$5\ 10^{-1}$
$9\ 10^{-3}$	$2\ 10^{-2}$	$9\ 10^{-2}$	$1\ 10^{-1}$
$8\ 10^{-3}$	$7\ 10^{-3}$	$1.4\ 10^{-3}$	$2\ 10^{-3}$
$1\ 10^{-4}$	$2\ 10^{-3}$	$1\ 10^{-4}$	$3\ 10^{-6}$

\- - -

<u>Problem 2</u> (Rosen - Suzuki - Problem ([8])):

Dimension $n = 4$, number of functions $m = 4$; this nonlinear constraint optimization problem is transformed to (P) by exact penalty minmax-technique ([1]).

Method	Iterations	Number of evaluations $f_i(x)$	$\nabla f_i(x)$	$D^2 f_i(x)$	CPU-time (sec)
Dem'yanov	15	239	72	0	1.58
Lemaréchal	17	270	89	0	1.67
Charal., Conn	18	111	?	0	?
Newton	13	73	45	17	0.88
Quasi-Newton	12	82	68	0	0.81
Conn ([4], [16])	?	247	247	0	2.83
VF02 (Powell)	?	48	48	0	1.277

"?" means " not available " .

- - - - -

<u>Problem 3</u> (Madsen, Schjaer-Jacobson ([13])):

This is a constraint minimax problem: Dimension $n = 2$, number of functions $m = 3$:

$f_1(x) = x_1^2 + x_2^2 + x_1 x_2 - 1$, $f_2(x) = \sin x_1$, $f_3(x) = -\cos x_2$,

with an additional constraint $h(x) = 3\,x_1 + x_2 + 2.5 \leq 0$.

Note that the Haar - condition is <u>not</u> satisfied, and therefore the method of Madsen, Schjaer-Jacobsen ([13]) has only a sublinear convergence rate.

$\lVert x_k - \bar{x} \rVert$:	Madsen, Schjaer-Jacobsen	Newton	Quasi - Newton
	⋮	⋮	⋮
	$0.786_{10}{}^{-1}$	$4\ 10^{-1}$	$4\ 10^{-1}$
	$0.786_{10}{}^{-1}$	$4\ 10^{-1}$	$4\ 10^{-1}$
	$0.286_{10}{}^{-1}$	$4\ 10^{-1}$	$6\ 10^{-7}$
	$0.214_{10}{}^{-1}$	$9\ 10^{-11}$	$6\ 10^{-9}$

- - - - -

Method	Iterations	Number of evaluations $f_i(x)$	$\nabla f_i(x)$	$D^2 f_i(x)$	CPU-time (sec)
Dem'yanov	6	33	39	0	0.53
Newton	5	25	13	8	0.32
Quasi-Newton	6	27	16	0	0.31
Madsen, Schj.	≥ 10	≥ 30	≥ 30	0	?

- - - - -

In the case of <u>satisfied</u> Haar - condition the results of the Madsen -

Schjaer-Jacobsen method are equivalent to those of our Newton - method.

References

1. Bandler, Charalambous: " Nonlinear programming using minimax techniques ", Journal Optim. Theory Appl. 13 (1974), pp. 607 - 619
2. Broyden, Dennis, Moré: " On the local and superlinear convergence of Quasi - Newton - methods ", J.Inst. Math. Appl. 12 (1973), pp. 223 - 245
3. Charalambous, Conn: " An efficient method to solve the minimax problem directly ", SIAM Journal on Numerical Analysis 15 (1) (1978) pp. 162 - 187
4. Conn: " Constrained optimization using a nondifferentiable penalty function ", SIAM J.Num.Anal. 10 (1973), pp. 764 - 784
5. Dem'yanov, Malozemov: " Introduction to minimax ", John Wiley and sons 1974
6. Hald, Madsen: " A 2 - stage algorithm for minimax optimization ", Report No NI-78-11, Sept. 1978, Danmarks Tekniske Højskole, Lyngby
7. Hettich: " Numerical Methods for Nonlinear Chebyshev Approximation ", in: Meinardus, ed., "Approximation in Theorie und Praxis", Bibliographisches Inst., Mannheim, 1979 (to appear)
8. Himmelblau: " Applied Nonlinear Programming ", McGraw - Hill, 1972
9. Hornung: " Algorithmen zur Lösung eines diskreten Minimax Problemes ", Thesis, University of Würzburg, 1979
10. Kiekebusch - Müller: " Eine Klasse von Verfahren zur Bestimmung von stationären Punkten, insbesondere Sattelpunkten ", Thesis, University of Würzburg, 1976
11. Lemaréchal: " An extension of Davidon methods to nondifferentiable problems ", Math. Progr. Study 3 (Nondifferentiable Optimization) (1975), pp. 95 - 109
12. Lemaréchal, Mifflin: " Nonsmooth optimization ", Proceedings of the IIASA Workshop, March 28 - April 8, 1977
13. Madsen, Schjaer-Jacobsen: " Constrained minimax optimization ", Report No 77-03, 1977, Danmarks Tekniske Højskole, Lyngby
14. Ortege, Rheinboldt: " Iterative solution of nonlinear equations in several variables ", New York, London: Academic Press 1970
15. Pankrath: Diploma thesis on algorithms of Dem'yanov, University of Würzburg, 1979 (to appear)
16. Reinhardt: " Untersuchung eines Verfahrens zur Minimierung einer nichtdifferenzierbaren exakten Penaltyfunktion " Diploma - thesis University of Würzburg, 1977
17. Rockafellar: " convex analysis ", Princeton University Press, Princeton, N.Y. 1970
18. Wolfe: " A method of conjugate subgradients ", Math. Progr. Study 3 (1975), pp. 190 - 205

Acknowledgment

I wish to express my thanks to Professor J. Stoer, University of Würzburg, for many helpful discussions and suggestions.

ALGORITHM OF SEARCH FOR GLOBAL EXTREMUM OF FUNCTION FROM VARIABLES MEASURED IN DIFFERENT SCALES

G.S.Lbov
Institute of Mathematics, Novosibirsk, USSR

INTRODUCTION

An algorithm of the search for global extremum of a function is considered. The algorithm realizes an adaptive strategy of planning of experiments or function calculations.

In the course of the experiment analysis a class of logical functions is used on every step of adaptation to realize the mentioned strategy for the case of variables measured in different scales.

§ 1. STATEMENT OF A PROBLEM

The real function $y=f(x)$ is considered. A point $x=(x_1,\dots,x_j,\dots,x_n)$ of n-dimensional space is an element of the set $\mathcal{D}=\mathcal{D}_1\times\dots\times\mathcal{D}_j\times\dots\times\mathcal{D}_n$ where $\mathcal{D}_j$ is a range of values of the variable X_j measured in this or that scal. The main definitions, description and research of scales can be found in [1]. If the variable X_j is Boolean one, then $\mathcal{D}_j=\{0,1\}$. If X_j is measured in the scale of denominations, the range $\mathcal{D}_j$ is a set of some names. If X_j is measured in the scale of order, the range $\mathcal{D}_j$ is an ordered discrete set. In this scale the exhibiting degree of some property is measured.

In the case of a quantative variable (temperature is an example of such variable) or a scale of quotients (e.g. length), the range $\mathcal{D}_j$ is the interval $[a_j, b_j]$ on the real variable X_j. Divide this interval into ℓ_j equal subintervals and consider a set of values X_j which are the midpoints of the subintervals chosen as the range $\mathcal{D}_j$.

We shall associate to each set $\mathcal{D}_j$ $(j=1,\dots,n)$ a set of natural numbers $\{1,\dots,\beta_j,\dots,\ell_j\}$ retaining the natural order of values for the variables measured in the scale of order, intervals and quotients. The capacity of the set $\mathcal{D}$ is $N_0=\prod_{j=1}^{n}\ell_j$. The metric suggested in [2] for the case of variables measured in different scales is assigned on this set.

Let $x^*=(x_1^*,\dots,x_j^*,\dots,x_n^*)$ be a point in which the function $f(x)$ reaches its maximum

$$f(x^*)=\max_{x\in\mathcal{D}} f(x).$$

We introduce a notion of $\mathcal{E}$-vicinity of the point of global maximum of the function. The value $f(x)$ is called $\mathcal{E}$-adjacent to $f(x^*)$, ($\mathcal{E} \geq 0$, integer; $x, x^* \in \mathcal{D}$) if the interval $[f(x), f(x^*))$ contains $\mathcal{E}$ different function values. If $f(x)=f(x^*)$ the value $f(x)$ will be 0-adjacent one. A set of points $B(\mathcal{E},x^*) \subseteq \mathcal{D}$ in which function values do not exceed $\mathcal{E}$-adjacent to $f(x^*)$ is called $\mathcal{E}$-vicinity of a point x^*. At any arbitrary point $x \in \mathcal{D}$ the function value can be determined either experimentally or by calculation. In the course of the search it is permitted to carry out a fixed number T of experiments (it is considered that to define the function value is concerned with great resource expenditure, that is why the number T is usually small). We shall call the principle of arrangement T of points $x^1,\dots,x^t,\dots,x^T$ in the range $\mathcal{D}$ - the strategy of the search. It is necessary to choose such planning strategy T of various experiments as to obtain the function value $\mathcal{E}$-adjacent to $f(x^*)$. Here, the value $\mathcal{E}$ should be minimal. The value $\mathcal{E}$ achieved during the search depends on the chosen planning strategy of the experiment and on the degree of function complexity. The problem of introducing such measure of ordering, due to complexity, multiextremal functions hasn't been solved yet. At present, as such measure Lipshits constant characterizing the function variability degree is used, as a rule. However, we may give examples where the function is "simple" for the global extremum search, but is characterized by a large Lipshits constant.

The algorithms of search for function greatest values given below use the following adaptive strategy of experiment planning [3]. The set of T experiments is divided into R groups

$$T = \tau^{(1)} + \dots + \tau^{(\varphi)} + \dots + \tau^{(R)}.$$

Basing on the results of all the previous experiments, the probability distribution function $P_\varphi(x)$ is introduced after every group of experiments. The value $P_\varphi(x)$ characterizes the probability of the function value in the point x, if the experiment was carried out at this point, to be the greatest of all function values obtained in all the previous experiments.

Introducing the function $P_\varphi(x)$ we use the following empirical hypothesis: it is considered that the probability $P_\varphi(x)$ in the vicinity of points $x^1,\dots,x^t$ $(t \leq T)$ where function values $f(x^1),\dots,f(x^t)$ have already been obtained, is the more the greater the function value in the corresponding point x^i $(i=1,\dots,t)$ is. Here, the more experiments were carried out the greater the deviation of distribution from the uniform one will be. It is done in the following way. Let's calculate the entropy

$$H^{\varphi}=-\sum_{x\in\mathcal{D}} P_{\varphi}(x)\log P_{\varphi}(x),\quad \varphi=0,1,\ldots,R.$$

At the beginning of the search when we have no information about the preference of some element of the set $\mathcal{D}$ to other ones, we introduce a uniform distribution of probabilities $P_0(x)=\frac{1}{N_0}$ where $N_0=|\mathcal{D}|$. In this case the entropy is maximal and equals $H^0=\log N_0$. Introducing the value H^{φ} as some monotonically decreasing function $H(t,K)$ from a number of experiments carried out $t=\sum_{\nu=1}^{\varphi}\tau^{(\nu)}$ gives the restriction of the search area as far as the experiments are carried out. Decreasing rate of this function is given by some parameter K called an adaptation coefficient. The greater the parameter value K (the degree of adaptation) the more restrictions for a class of functions $\{f(x)\}$ considered.

The arrangement of the $(\varphi+1)$-th group of experiments is done according to the function $P_{\varphi}(x)$. We assume that the arrangement of $\tau^{(\varphi+1)}$ experiments on the set $\mathcal{D}$ is done according to the function $P_{\varphi}(x)$, if for an arbitrary subset $\mathcal{D}'\subseteq\mathcal{D}$ the number of experiments is proportional to the value $\sum_{x\in\mathcal{D}'}P_{\varphi}(x)$. Apparently, the fulfilment of this condition gives the best arrangement of experiments. The problem of the best arrangement of experiments needs theoretical treatment. Planning of experiments with this condition being fulfilled is connected with mathematical and technical difficulties.

In the present paper the function $P_{\varphi}(x)$ is assigned in the following way.

Let the range $\mathcal{D}_j$ of every variable be some set of names $\{x_{j1},\ldots,x_{j\beta},\ldots,x_{jl_j}\}$. The capacity of the set $\mathcal{D}$ is $N_0=\prod_{j=1}^{n} l_j$. We choose the function

$$H^{\varphi}=H(t,K)=\log\left(1-K\frac{t}{T}-\frac{t}{N_0}\right)+H^0$$

as a monotonically decreasing function $H(t,K)$ where

$$H^0=\log N_0,\quad t=\sum_{\nu=1}^{\varphi}\tau^{\nu},\quad 0\le K\le 1-\frac{T}{N_0}-a^{-H^0},$$

is the logarithm basis.

If $K=0$, the entropy is $H^{\varphi}=\log(N_0-t)$, i.e. the search area does not practically narrow (if $t\ll N_0$) as far as the experiments are carried out. If the complete sorting out ($T=N_0$) is possible, then $t\to T$ $H^{\varphi}\to 0$. For $K=1-\frac{T}{N_0}-a^{-H^0}$ the rate of "restriction" of search area will be the largest: after carrying out T experiments $H^R=0$. If $T\ll N_0$ and N_0 is large, then $\frac{t}{N_0}\simeq 0$, $\frac{T}{N_0}\simeq 0$, $a^{-H^0}\simeq 0$ and the function

$$H^{\varphi}=\log\left(1-K\frac{t}{T}\right)+\log N_0,\quad (0\le K\le 1). \tag{1}$$

If, for example, on each step of search we introduce a uniform distribution on the set $\mathcal{D}^{\varphi} \subseteq \mathcal{D}$ $P_{\varphi}(x) = \frac{1}{N_{\varphi}}$ $(N_{\varphi} = |\mathcal{D}^{\varphi}|)$, then accumulating the results of the experiments, we obtain linear "contraction" of the search area $(N_0 > N_1 > \ldots > N_R)$. The adaptation coefficients is $K = 1 - \frac{N_R}{N_0}$. We can determine the adaptation coefficient assigning the capacity of the finite set $\mathcal{D}^R$.

It should be noted that area $\mathcal{D}^{\varphi}$ can be sufficiently complex.

In the algorithm considered below for the sake of simplicity the function is assigned as follows

$$P_{\varphi}(x) = P_{\varphi}(x_1, \ldots, x_j, \ldots, x_n) = \prod_{j=1}^{n} P_{\varphi}(x_j).$$

Then

$$H^{\varphi} = \sum_{j=1}^{n} H_j^{\varphi} .$$

where

$$H_j^{\varphi} = -\sum_{x \in \mathcal{D}_j} P_{\varphi}(x_j) \log P_{\varphi}(x_j).$$

After carrying out φ groups of experiments the entropy is decreased as compared to the initial one by $\frac{H^0}{H^{\varphi}}$ times. Let's consider that the entropy along each of the variables also decreases by the same degree, then

$$H_j^{\varphi} = \frac{H_j^0}{H^0} \cdot H^{\varphi} = \frac{\log \ell_j}{\log N_0} \cdot H^{\varphi}.$$

The value H^{φ} is determined by formula (1). One of the ways to assign the distribution $P(x_j)$ is described in [3]. For that it is necessary to know, besides the value of the magnitude H_j^{φ}, the order of preference determined on a set of variable values X_j (see § 2).

§ 2. OPTIMIZATION ALGORITHM

The search algorithms of the approximate value of the function extremum consists of three blocks: the block of planning of experiment groups, the block of detection of logical regularities in tables of the experimental data and the block of adaptation. Let's consider the mentioned blocks in turn.

Planning of experiments. The algorithms of planning a group of experiments will preliminarily be described for the case where the number of each variable is ℓ, and the probability distribution function $P(x) = \frac{1}{\ell^n}$. In the sequel the point of n-dimensional space is called a realization. The given block selects from a set of possible realizations a subset A consisting of the minimal number of realizations. In addition to that, complete choosing of all the possible combinations is fulfilled for any selected subsystem consisting of m variables. We consider here the case when $m=2$. Apparently such a choice of a subset from τ realizations satisfies fully enough the introduced uniform distribution. However, the complexity of the algorithm of such choice and the number of realizations increase with the number m growth. Below we consider the selection of a subset A for $m=2$. We are interested in a subset A of the minimal capacity. Consider the complete set $\{A_n\}$ of such subsets of the set A. Let $\tilde{A}_n$ be a subset of realizations having the least of $\{A_n\}$ capacity. In every particular subset A_n one may indicate a subset of minimal capacity. We designate it as $\hat{A}_n$. Here, $|\tilde{A}_n| \leq |\hat{A}_n| \leq |A_n|$. Formulate an algorithm constructing the set A_n and $\hat{A}_n$ and obtain some upper estimate for

Statement: $|\tilde{A}_2| = |\tilde{A}_3| = |\hat{A}_2| = |\hat{A}_3| = \ell^2$.

Proof. It is obvious that $|\tilde{A}_2| = |\hat{A}_2| = \ell^2$ and $|\tilde{A}_2| \leq |\tilde{A}_3| \leq |\hat{A}_3| \leq$ $\leq |A_3|$. To prove it it is sufficient to construct such A_3 that $|A_3| = \ell^2$.

We write a particular realization of $\tilde{A}_2$ as (a_1, a_2) and develop it to the realization of A_3 in the following way: the corresponding realization in A_3 has the form (a_1, a_2, a_3) where $a_3 = \Psi(a_1, a_2) =$ $= (\lambda + a_1 + a_2)$ (mod ℓ) and λ is an arbitrary number. Now we need to show that all such realizations really form a totality which realizes all the value pairs. We restrict our proof to a case where $\lambda = 0$; the proof is the same as $\lambda \neq 0$ as well.

Let's check up whether the pair (a_1, a_3) belongs to some realization from A_3:

a) at $a_3 > a_1$ we have

$$(a_1, a_3 - a_1) \in \tilde{A}_2 \quad \text{and} \quad (a_1, (a_3 - a_1), \Psi(a_1, a_3 - a_1)) = (a_1, a_3 - a_1, a_3) \in A_3$$

b) at $a_3 \leq a_1$ $(a_1, (\ell + a_3 - a_1)) \in \tilde{A}_2$ and $(a_1, (\ell + a_3 - a_1),$

$$\Psi(a_1, (\ell + a_3 - a_1))) = (a_1, a_3 + \ell - a_1, a_3) \in A_3$$

Now we check up whether the pair (belongs to some realization from A_3:

a) at $a_3 > a_2$, it is obvious that $(a_3 - a_2, a_2, a_3) =$

$$= ((a_3 - a_2), a_2, \Psi(a_3 - a_2, a_2)) = (a_3 - a_2, a_2, a_3) \in A_3$$

b) at $a_3 \leq a_2$ $(a_3 - a_2 + \ell, a_2, a_3) =$

$$= ((a_3 - a_2 + \ell), a_2, \Psi(a_3 - a_2 + \ell, a_2)) \in A_3 .$$

The statement is proved.

Further we construct some A_n and show that $|A_n| < ([\frac{n}{2}]-1)\cdot \ell^2$. We construct $([\frac{n}{2}]-1)$ -series of realizations, each of them containing ℓ^2 realizations. First we consider the construction of the 1st group. Let's take all the realizations $\{(a_1,a_2)\}$ along the variables X_1, X_2, and develop them to realizations along all the variables.

$$a_3 = \varphi_{a_3}(a_1,a_2);\ a_4 = \varphi_{a_4}(a_1,a_2), \ldots, a_n = \varphi_{a_n}(a_1,a_2),$$

where $\varphi_{a_i}(a_1,a_2) = (a_1+a_2+i) \pmod{\ell}$. Then we take a pair of variables X_3, X_4 and deal them similarly. For this it makes no difference how the realization continues on the first two variables. Continue doing so until we reach the last pair of variables in the case of the even number of variables, and one before last if the number is odd. The number of series obtained is $[\frac{n}{2}]-1$, and each group contains ℓ^2 realizations. Obviously, these realizations exhaust all pairs of variable values. The system of realizations obtained can be reduced if we exclude the ones where all the pairs of variable values occur among the rest of realizations. Looking through all the realizations we obtain a set of realizations $\tilde{A}_n \subseteq A_n$, i.e. the minimal subset of ones. We act similarly in the case where the number of variable values is not constant. In this case we artificially supplement the variable values, repeating some of them up to the maximal number of values in the system of variables. If we have a probability distribution $P(x_j)$ on the values of a variable, then we may get a totality of the former values - by $[P(x_{j1})\cdot 10]$ number, second values - by $[P(x_{j2})\cdot 10]$ number, etc. Values with the probability weight $P(x_{j\cdot}) < 0.1$ will not get into that totality. Then we may bring the amount of values in the totality to ℓ, successively excluding occurence of a value which is the most representative in a totality at that moment.

<u>Detection of logical regularities.</u> Let us by means of algorithm of planning the experiments have a table $\{x_j^i\}$ ($i=1,\ldots,r$; $j=1,\ldots,n$; r - number of realizations in a group; n - number of variables), and also we obtain a set of function values $\{y_i\}$ where $y_i = f(x^i)$.

At first, on the basis of a set of values $\{y_i\}$ we divide the dependent variable y into L intervals. For this purpose we may use one of the known algorithms of the automatical classification of the value set $\{y_i\}$. Division is carried out with the realization of the condition

$$\bar{P}_\omega = \frac{r_\omega}{r} \geqslant \delta,$$

where r_ω is a number of values from $\{y_i\}$ contained in the interval ω $(\omega=1,\ldots,L)$, δ is some "threshold".

Further, on the basis of the table $\{x_j^i\}$ we choose logical statements which characterize the chosen intervals ω of the variable Y.

Let's introduce some necessary definitions. We say the expression of the following form to be an elementary statement on the variable :

1) $X_j(a)=x$ for Boolean variables, and variables measured in the scale of denominations (a is the name of an arbitrary realization);

2) $\alpha_j < X_j(a) \leq \beta_j$ for variables measured in scales of order, relations, intervals (α_j and β_j - some values of the variable X_j).

In the first case a set of elementary statements coincides with a set of values of the variable considered, in the second - such set consists of various combinations of neighbouring values of the corresponding variable.

We denote the elementary statements as $\mathcal{J}$, and the statement which supplements it - as $\bar{\mathcal{J}}$.

We call the conjunction

$$S=\mathcal{J}_1 \wedge \mathcal{J}_2 \wedge \ldots \wedge \mathcal{J}_j \wedge \ldots \wedge \mathcal{J}_m \quad (m \leq n)$$

a logical statement S based on some subset of variables. We say that a statement S is held on some realization if every elementary statement contained in S is true on this realization.

The set of elementary statements in S, their number and parameters α_j, β_j are the functions of the previous experiments.

As an informative statement characterizing the interval ω we regard a statement for which

$$\frac{z_{S\omega}}{z_\omega} \geq \delta \quad \text{and} \quad \frac{\sum_{i \in \mathcal{F}} \nu_i}{z - z_\omega} \leq \gamma ,$$

where $\mathcal{F}$ is a set of realizations on which the proposition S is fulfilled and simultaneously the variable Y values corresponding to these realizations do not belong to the interval ω; $z_{S\bar{\omega}} = |\mathcal{F}|$, ν_i is a "weight" coefficient of the i-th realization; δ and γ are parameters evaluating the degree of statement informability.

The coefficient ν_i is chosen for the following reasons: the longer the distance between the value y_i and the nearest interval ω bound is, the more is the "weight" of the corresponding realization and the less informative is the statement S. A particular example of selecting the value ν_i as a distance function is given in [4].

There exists the algorithm of selection of all informative conjunctions.

<u>Block of adapatation</u>. The variation of probability distribution function while the search is carried out is done in the following way. To assign probabilities $P_\varphi(x_j)$ to various values of the variable X_j,

it is necessary to determine the order of their preference at the present moment of the search. In our case, this order is determined basing on a collection of informative statements $\underset{\omega}{\cup}\{S_i\}$ $(\omega=1,\dots,L)$ obtained after processing the experiments of the φ-th group. The interval of the greatest function values corresponds to the value $\omega=1$.

To order the values of the variables under consideration, we associate each value of the variable with a L-digit binary code. The code is constructed according to the rule: if in a group of statements $\{S_i^{\omega}\}$ the given value of the variable is not contained in the elementary statement, zero is put into the code position ω, otherwise π it is 1. Let's denote the number of the first zero digit of the code as b_1^{π}, if it is looked through from the left, and the number of the last 1 in the same examination ($b_1^{\pi} \leq b_p^{\pi}$) - as b_p^{π}. The code ordering is done on the basis of the following euristic considerations. The smaller the number b_p^{π}, the better the code provided all other conditions are equal: the corresponding variable value enters the regularity which characterizes the interval of greater function values. The greater the difference $b_p^{\pi}-b_1^{\pi}$, the worse the code, provided all other conditions are equal: the corresponding variable value enters the regularities which characterize both greater and smaller function values. Taking this into account, it is proposed to consider the code π_2 to be better than the one π_1, if

$$b_1^{\pi_2}-b_1^{\pi_1} \leq b_p^{\pi_1}-b_p^{\pi_2} .$$

The codes incomparable according to the given rule are ordered arbitrarily. The code whose all digits are zero, is identified with the code where the $[\frac{L}{2}]$-th digit is 1, and all the other digits are zero.

For example, let's consider two codes: $\pi_1=(1,0,\dots,0)$ and $\pi_2=(0,0,\dots,0,1)$. The code π_1 is the best since its corresponding variable value enters the statement which characterizes the interval of the best function values. The code π_2 is the worst since its corresponding variable value enters statements characterizing the interval of the function worst values ($\omega=L$).

According to the introduced code order we obtain some linear ordering on values or intervals of each of the variables X_j.

The next area for planning of experiments is chosen by reducing the range of each of the variables for the values which are the worst in the given ordering.

CONCLUSION

The program which realizes the aforedescribed algorithm was used

to solve an applied problem. It was necessary to obtain the values of ten variables assigning the technological process of sedimentation of a special alloy during the reconstruction of large-size parts of agricultural machines [5]. The use of classical methods of optimization proved to be difficult, since, besides the quantitative variables (e.g. the electrolyte temperature), there were variables measured in the scale of denominations (e.g. the type of the anode plug material). The conditions were obtained that provided high physical-mechanical properties of falling out alloy. Simultaneously, non-informative variables were excluded.

REFERENCES

1. J.Pfanzagl. Theory of measurement. Physica-Verlag. Wiirzburg-Wien, 1971.
2. Voronin J.A., Introduction of a measure of similarity for solutions of geological problems, (in Russian) Doklady AN USSR vol.199, No 5, (1971) 1011-1014.
3. Lbov G.S. Training for extremum determination of function of variables measured in names scale. Sec.Int.Conf. on Artificial Intelligence, London, (1972), p.418-423.
4. Lbov G.S., On a nonparametric approach to problems of empiricial prediction in Addaptive systems and their Applications, in Russian , Nauka, Novosibirsk (1978) 78-81 .
5. Revjakin V.P., Lbov G.S., Shishkin G.M., Analysis of Random Searching Process, in Electrochemisty and its Application in Industry, Tiumen (1971), 31-38 .

A METHOD FOR SOLVING EQUALITY CONSTRAINED OPTIMIZATION PROBLEMS BY UNCONSTRAINED MINIMIZATION

G.Di Pillo*, L.Grippo**, F.Lampariello**

* Istituto di Automatica, Università di Roma, Via Eudossiana, 18 - 00184 Roma, Italy.

** Centro di Studio dei Sistemi di Controllo e Calcolo Automatici del CNR, Via Eudossiana,18 00184 Roma, Italy.

ABSTRACT

In this paper we consider a new augmented Lagrangian function which allows to solve equality constrained optimization problems by a single unconstrained minimization. The main computational problems arising in the minimization of the augmented Lagrangian are discussed and a procedure for the automatic selection of the penalty coefficient is described. Numerical results obtained for a set of standard test problems are reported.

1. INTRODUCTION

In this paper we consider the following minimization problem:

$$\min\{f(x):g(x)=0\},\ x\in R^n, f:R^n\to R^1, g:R^n\to R^m, m\leq n \ ; \tag{1}$$

we assume, unless otherwise stated, that f and g are two times continuously differentiable in R^n.

An important class of algorithms for this problem is that based on the consideration of augmented Lagrangian functions, originally introduced by Hestenes and Powell. We refer, e.g. to [1-5] for an exposition of the method and of related refinements and extensions.

The main drawback of the augmented Lagrangian method is that, in principle, it requires a sequence of unconstrained minimization problems to be solved. To overcome this difficulty, further developments were proposed in [2,6,7] by introducing in the augmented Lagrangian a multiplier vector continuously dependent on x. These methods require a single unconstrained minimization; however a matrix inversion is needed at each function evaluation and this may limit somewhat their applicability.

In [8] we proposed a different approach based on the consideration of a new class of augmented Lagrangians obtained by adding to the augmented Lagrangian of Hestenes a penalty term on the first order necessary condition $\nabla f + \frac{\partial g'}{\partial x}\lambda = 0$. It was shown that, under suitable hypotheses, a local solution of the constrained problems and the corresponding value of the Lagrange multiplier can be found by performing a single local unconstrained minimization with respect to both x and λ, for finite values of the penalty coefficient and without requiring matrix inversions.

In this paper we introduce the augmented Lagrangian function:

$$S(x,\lambda;c)=L(x,\lambda)+\tau[\lambda' g(x)]^2+c\|g(x)\|^2+\|M(x)\nabla_x L(x,\lambda)\|^2 \ , \tag{2}$$

where

$$L(x,\lambda) = f(x) + \lambda' g(x),$$

$$\nabla_x L(x,\lambda) = \nabla f(x) + \frac{\partial g(x)'}{\partial x}\lambda,$$

$c > 0$ is a penalty coefficient, $\tau \geq 0$, and $M(x)$ is an appropriate $(p\times n)$ weighting matrix with continuously differentiable elements and $m\leq p\leq n$.

In Section 2 we show how the results obtained in [8] for $\tau = 0$ can be extended to the case $\tau \geq 0$ considered in (2). For $\tau > 0$, the function $S(x,\lambda;c)$ is bounded from below in $R^n \times R^m$, provided that $\inf f(x) > -\infty$, and this can be computationally advantageous with respect to the case $\tau = 0$.

In Section 3 we consider the main computational problems arising in the unconstrained minimization of $S(x,\lambda;c)$, that is the presence of second order derivatives of f and g in the gradient formulas of $S(x,\lambda;c)$ and the selection of the penalty coefficient c.

Finally, in Section 4, we report the numerical results obtained for a set of standard test problems.

2. PROPERTIES OF THE AUGMENTED LAGRANGIAN $S(x,\lambda;c)$

This section contains the main results concerning the relationships between optimal solutions of problem (1) and unconstrained minima of $S(x,\lambda;c)$.

It will be proved that the function (2) introduced here enjoys the same properties established in [8] for the case $\tau = 0$. We remark, however, that some of the assumptions employed in [8] are slightly weakened and, as a consequence, some proofs are given in a different form.

Preliminarly, we give the gradient formulas of $S(x,\lambda;c)$ in the product space $R^n \times R^m$:

$$\begin{aligned}\nabla_x S(x,\lambda;c) &= \nabla_x L(x,\lambda) + 2\tau[\lambda' g(x)]\frac{\partial g(x)'}{\partial x}\lambda \\ &+ 2c\,\frac{\partial g(x)'}{\partial x}g(x) + 2\,\nabla_x^2 L(x,\lambda)M'(x)M(x)\nabla_x L(x,\lambda) \\ &+ 2\left[\sum_{j=1}^{p}\left(\frac{\partial m_j'(x)}{\partial x}\right)'\nabla_x L(x,\lambda)e_j'\right]M(x)\nabla_x L(x,\lambda)\end{aligned} \tag{3}$$

$$\nabla_\lambda S(x,\lambda;c) = g(x) + 2\tau[\lambda' g(x)]g(x) + 2\frac{\partial g(x)}{\partial x}M'(x)M(x)\nabla_x L(x,\lambda) \tag{4}$$

where e_j is the j-th column of the $(p \times p)$ identity matrix and $m_j(x)$ is the j-th row of $M(x)$. The above expressions can be easily obtained by employing the dyadic expansion

$$M(x) = \sum_{j=1}^{p} e_j m_j(x)$$

We consider first the relationship between stationary points of the Lagrangian function $L(x,\lambda)$ and stationary points of $S(x,\lambda;c)$.

THEOREM 1. *Let* $(\bar{x},\bar{\lambda})$ *be a stationary point for* $L(x,\lambda)$; *then*

(a) $(\bar{x},\bar{\lambda})$ *is a stationary point for* $S(x,\lambda;c)$
(b) $S(\bar{x},\bar{\lambda};c) = f(\bar{x})$

PROOF. By assumption, $\nabla_x L(\bar{x},\bar{\lambda}) = 0$, $\nabla_\lambda L(\bar{x},\bar{\lambda}) = g(\bar{x}) = 0$, which imply, from (3) (4), $\nabla_x S(\bar{x},\bar{\lambda};c) = 0$, $\nabla_\lambda S(\bar{x},\bar{\lambda};c) = 0$, that is (a), and from (2), (b).□

THEOREM 2. *Let* $X \times \Lambda$ *be a compact subset of* $R^n \times R^m$ *and assume that* $\frac{\partial g(x)}{\partial x}M'(x)$ *is an* $(m \times m)$ *nonsingular matrix for any* $x \in X$. *Then, for every* $\tau \geq 0$, *there exists a* $\bar{c} > 0$ *such that for all* $c \geq \bar{c}$, *if* $(\bar{x},\bar{\lambda}) \in X \times \Lambda$ *is a stationary point of* $S(x,\lambda;c)$, $(\bar{x},\bar{\lambda})$ *is also a stationary point of* $L(x,\lambda)$.

PROOF. Let $(\bar{x},\bar{\lambda}) \in X \times \Lambda$ be a stationary point of $S(x,\lambda;c)$. Then, by (4), $\nabla_\lambda S(\bar{x},\bar{\lambda};c) = 0$ implies

$$M(\bar{x})\nabla_x L(\bar{x},\bar{\lambda}) = -\frac{1}{2}\left[\frac{\partial g(\bar{x})}{\partial x}M'(\bar{x})\right]^{-1}(1+2\tau[\bar{\lambda}' g(\bar{x})])g(\bar{x})$$

Therefore, since $\nabla_x S(\bar{x},\bar{\lambda};c) = 0$, we have:

$$0 = M(\bar{x}) \nabla_x S(\bar{x},\bar{\lambda};c) = \left\{ -\frac{1}{2}\left[\frac{\partial g(\bar{x})}{\partial x} M'(\bar{x})\right]^{-1} (1+2\tau[\bar{\lambda}'g(\bar{x})]) + 2M(\bar{x})\frac{\partial g(\bar{x})}{\partial x}'(cI+\tau\bar{\lambda}\bar{\lambda}')\right.$$

$$\left. -M(\bar{x})\left[\nabla_x^2 L(\bar{x},\bar{\lambda})M'(\bar{x}) + \sum_{j=1}^{p}\left(\frac{\partial m_j'(\bar{x})}{\partial x}\right)'\nabla_x L(\bar{x},\bar{\lambda})e_j'\right]\left[\frac{\partial g(\bar{x})}{\partial x} M'(\bar{x})\right]^{-1}(1+2\tau[\bar{\lambda}'g(\bar{x})])\right\} g(\bar{x})$$

Hence, by the continuity assumptions and the compactness of $X \times \Lambda$, for any $\tau \geq 0$ there exists a $\bar{c} > 0$ such that for all $c \geq \bar{c}$ and any $(\bar{x},\bar{\lambda}) \in X \times \Lambda$ the matrix multiplying $g(\bar{x})$ is nonsingular, so that, for $c \geq \bar{c}$, $g(\bar{x}) = 0$. On the other hand, $\nabla_\lambda S(\bar{x},\bar{\lambda};c) = 0$ and $g(\bar{x}) = 0$ imply $M(\bar{x})\nabla_x L(\bar{x},\bar{\lambda}) = 0$, so that from $\nabla_x S(\bar{x},\bar{\lambda};c) = 0$ we get $\nabla_x L(\bar{x},\bar{\lambda}) = 0$. □

We state now the following global optimality result.

THEOREM 3. *Let* $(\bar{x},\bar{\lambda})$ *be a stationary point for* $L(x,\lambda)$ *and assume that:*

(i) $\bar{x}$ *is the unique global minimum point of problem* (1) *on a compact set* $X \subset R^n$, *and* $\bar{x} \in \text{int}(X)$

(ii) $\frac{\partial g(\bar{x})}{\partial x} M'(\bar{x})$ *is an* $(m \times m)$ *nonsingular matrix.*

Then, for every compact set $\Lambda \subset R^m$ *such that* $\bar{\lambda} \in \text{int}(\Lambda)$ *and every* $\tau \geq 0$, *there exists a* $c^* > 0$ *such that, for all* $c \geq c^*$, $(\bar{x},\bar{\lambda})$ *is the unique global minimum point of* $S(x,\lambda;c)$ *on* $X \times \Lambda$.

PROOF. Let $\Lambda \subset R^m$ be a compact set such that $\bar{\lambda} \in \text{int}(\Lambda)$, and assume that the theorem is false. Then, for any integer k, there exists a $c_k \geq k$ and a point $(x_k,\lambda_k) \in X \times \Lambda$, $(x_k,\lambda_k) \neq (\bar{x},\bar{\lambda})$, which affords a global minimum to $S(x,\lambda;c_k)$ on $X \times \Lambda$. Therefore

$$S(x_k,\lambda_k;c_k) \leq S(\bar{x},\bar{\lambda};c_k) = f(\bar{x}) \tag{5}$$

where the last equality follows from theorem 1.

We prove first that there exist values $\rho > 0$ and k_ρ such that, for all $k \geq k_\rho$, either

$$\|x_k - \bar{x}\| \geq \rho \tag{6}$$

or

$$\|x_k - \bar{x}\| < \rho \quad \text{and} \quad \|\lambda_k - \bar{\lambda}\| \geq \rho \tag{7}$$

In fact, being $\bar{x} \in \text{int}(X)$, by (ii) and the continuity assumptions, there exists a neighbourhood $\Omega \subset X$ of $\bar{x}$ such that $\frac{\partial g(x)}{\partial x} M'(x)$ is nonsingular for any $x \in \Omega$. Let now $\rho > 0$ be such that

$$\bar{B} \triangleq \{(x,\lambda): \|x-\bar{x}\| \leq \rho,\ \|\lambda-\bar{\lambda}\| \leq \rho\} \subset \Omega \times \text{int}(\Lambda)$$

Then, $(x_k,\lambda_k) \in \bar{B}$ would imply that (x_k,λ_k) is a stationary point for $S(x,\lambda;c_k)$, so that, by theorem 2, there should exist a value $c_\rho > 0$ such that for all $c_k \geq c_\rho$, (x_k,λ_k) is a stationary point of $L(x,\lambda)$. In this case $g(x_k) = 0$ and, by theorem 1, $S(x_k,\lambda_k;c_k) = f(x_k)$, so that (5) would contradict assumption (i). Therefore, for all $c_k \geq c_\rho$, either (6) or (7) must hold.

Consider now the sequence $\{(x_k,\lambda_k)\}$; since $X \times \Lambda$ is compact there exists a convergent subsequence (relabel it $\{(x_k,\lambda_k)\}$ such that $\lim_{k\to\infty}(x_k,\lambda_k) = (\hat{x},\hat{\lambda}) \in X \times \Lambda$. By (5), we have $\limsup_{k\to\infty} S(x_k,\lambda_k;c_k) \leq f(\bar{x})$, which, since $c_k \to \infty$, implies:

$$g(\hat{x}) = 0 \tag{8}$$

and

$$f(\hat{x}) + \|M(\hat{x})\nabla_x L(\hat{x},\hat{\lambda})\|^2 \leq f(\bar{x}) \tag{9}$$

By assumption (i), (8) (9) imply $\hat{x} = \bar{x}$ and

$$M(\bar{x})\nabla f(\bar{x}) + M(\bar{x})\frac{\partial g(\bar{x})'}{\partial x}\hat{\lambda} = 0$$

from which, by (ii), $\hat{\lambda} = \bar{\lambda}$. Therefore we get a contradiction either with (6) or with (7).□

A direct consequence of the preceding theorem is the following local result.

THEOREM 4. *Let* $(\bar{x},\bar{\lambda})$ *be a stationary point for* $L(x,\lambda)$ *and assume that:*

(i) $\bar{x}$ *is an isolated local minimum point of problem* (1)

(ii) $\frac{\partial g(\bar{x})}{\partial x}M'(\bar{x})$ *is an* $(m \times m)$ *non singular matrix.*

Then, for every $\tau \geq 0$, *there exists a* $\bar{c} > 0$ *such that, for all* $c \geq \bar{c}$, $(\bar{x},\bar{\lambda})$ *is an isolated local minimum point of* $S(x,\lambda;c)$.

A global result which, in a certain sense, represents the converse of theorem 3 can easily be stated making use of theorem 2.

THEOREM 5. *Let* $X \times \Lambda$ *be a compact subset of* $R^n \times R^m$ *and assume that:*

(i) $\frac{\partial g(x)}{\partial x}M'(x)$ *is an* $(m \times m)$ *nonsingular matrix for any* $x \in X$

(ii) *for any global minimum point* x^* *of problem* (1) *on* X, *there exists a multiplier* $\lambda^* \in \Lambda$ *such that* $\nabla_x L(x^*,\lambda^*) = 0$.

Then, for every $\tau \geq 0$, *there exists a* $c^* > 0$ *such that, for all* $c \geq c^*$, *if* $(\bar{x},\bar{\lambda}) \in \text{int}(X \times \Lambda)$ *is a global minimum point of* $S(x,\lambda;c)$ *on* $X \times \Lambda$, $\bar{x}$ *is a global minimum for problem* (1) *on* X.

PROOF. By theorem 2, there exists a $c^* > 0$ such that, for all $c \geq c^*$, if $(\bar{x},\bar{\lambda}) \in \text{int}(X \times \Lambda)$ minimizes $S(x,\lambda;c)$ on $X \times \Lambda$, $(\bar{x},\bar{\lambda})$ is a stationary point of $L(x,\lambda)$. Therefore, by (b) of theorem 1, $S(\bar{x},\bar{\lambda};c) = f(\bar{x}) \leq S(x,\lambda;c)$, $\forall (x,\lambda) \in X \times \Lambda$. Therefore it results, in particular:

$$f(\bar{x}) \leq f(x), \quad \forall (x,\lambda) \in X \times \Lambda: \; g(x) = 0, \; \nabla_x L(x,\lambda) = 0$$

and this implies, by (ii) that $\bar{x}$ is a global minimum point for problem (1) on X. □

REMARK. We note that assumption (ii) of theorem 5 is satisfied whenever $\frac{\partial g(x)}{\partial x}$ has full rank on X and any global minimum point of problem (1) on X is an interior point of X.□

The next theorem establishes a relationship between local unconstrained minimum points of $S(x,\lambda;c)$ and local solutions of problem (1).

THEOREM 6. *Let* $X \times \Lambda$ *be a compact subset of* $R^n \times R^m$ *and assume that* $\frac{\partial g(x)}{\partial x}M'(x)$ *is an* $(m \times m)$ *nonsingular matrix for any* $x \in X$. *Then, for every* $\tau \geq 0$, *there exists a* $\bar{c} > 0$ *such that, for all* $c \geq \bar{c}$, *if* $(\bar{x},\bar{\lambda}) \in X \times \Lambda$ *is a local unconstrained minimum point of* $S(x,\lambda;c)$, $\bar{x}$ *is a local minimum point for problem* (1).

PROOF. By theorem 2, there exists a $\bar{c} > 0$ such that, for all $c \geq \bar{c}$, if $(\bar{x},\bar{\lambda}) \in X \times \Lambda$ is a local unconstrained minimum point of $S(x,\lambda;c)$, then $g(\bar{x}) = 0$ and $\nabla_x L(\bar{x},\bar{\lambda}) = 0$. This implies $S(\bar{x},\bar{\lambda};c) = f(\bar{x})$. Moreover, there exist neighbourhoods Ω, Γ of $\bar{x},\bar{\lambda}$ such that $f(\bar{x}) \leq S(x,\lambda;c)$, $\forall x \in \Omega$, $\lambda \in \Gamma$, and this yields:

$$f(\bar{x}) \leq f(x) + \left\| M(x)\nabla f(x) + M(x)\frac{\partial g(x)'}{\partial x}\lambda \right\|^2, \quad \forall x \in \Omega \cap \{x: g(x)=0\}, \forall \lambda \in \Gamma \qquad (10)$$

On the other hand, by the continuity assumptions, there exists a neigh bourhood Ω' of $\bar{x}$, $\Omega' \subset \Omega$, such that

$$\lambda = -\left[M(x)\frac{\partial g(x)'}{\partial x}\right]^{-1} M(x)\nabla f(x) \in \Gamma, \quad \forall x \in \Omega' \tag{11}$$

By combining (10) (11), we conclude $f(\bar{x}) \leq f(x)$, $\forall x \in \Omega' \cap \{x : g(x)=0\}$. □

3. UNCONSTRAINED MINIMIZATION OF $S(x,\lambda;c)$

We consider here the main computational aspects of the unconstrained minimization of $S(x,\lambda;c)$. In particular we first discuss the problems related to the presence of first order derivatives of $f(x)$ and $g(x)$ in the augmented Lagrangian, and then we describe a procedure for the automatic selection of the penalty coefficient c.

The minimization of $S(x,\lambda;c)$ by methods which make use of the gradient $\nabla S(x,\lambda;c)$ requires the evaluation of second order derivatives of $f(x)$ and $g(x)$. If these are not easily available, it is possible to approximate the term which contains $\nabla_x^2 L(x,\lambda)$ in the gradient component $\nabla_x S(x,\lambda;c)$ by a finite difference formula; in particular, letting $u(x,\lambda) \triangleq M'(x)M(x)\nabla_x L(x,\lambda)$ we can take

$$\nabla_x^2 L(x,\lambda)u(x,\lambda) \simeq \frac{1}{t}\left[\nabla_x L(x+tu(x,\lambda),\lambda) - \nabla_x L(x,\lambda)\right] \tag{12}$$

where t is a sufficiently small number.

A similar expedient can be adopted for the last term of $\nabla_x S(x,\lambda;c)$ if the weighting matrix $M(x)$ is taken as $M(x) = \mu\partial g(x)/\partial x$ which can be a convenient choice, as indicated in [8].

The minimization of $S(x,\lambda;c)$ by methods which make use of the Hessian matrix requires the evaluation of third order derivatives of $f(x)$ and $g(x)$. This can be avoided by employing a Newton-type algorithm based on a suitable approximation of the Hessian matrix.

Assume that f and g are three times continuously differentiable functions on R^n and $M(x)$ is a two times continuously differentiable matrix.

Consider the following symmetric matrix:

$$\hat{H}(x,\lambda;c) = \begin{pmatrix} \hat{H}_{11}(x,\lambda;c) & \hat{H}_{12}(x,\lambda;c) \\ \hat{H}_{12}'(x,\lambda;c) & \hat{H}_{22}(x,\lambda;c) \end{pmatrix}$$

$$\hat{H}_{11}(x,\lambda;c) = \nabla_x^2 L(x,\lambda;c) + 2\tau\frac{\partial g(x)'}{\partial x}\lambda\lambda'\frac{\partial g(x)}{\partial x} + 2c\frac{\partial g(x)'}{\partial x}\frac{\partial g(x)}{\partial x} + 2\nabla_x^2 L(x,\lambda)M'(x)M(x)\nabla_x^2 L(x,\lambda)$$

$$\hat{H}_{12}(x,\lambda;c) = \frac{\partial g(x)'}{\partial x} + 2\nabla_x^2 L(x,\lambda)M'(x)M(x)\frac{\partial g(x)'}{\partial x}$$

$$\hat{H}_{22}(x,\lambda;c) = 2\frac{\partial g(x)}{\partial x}M'(x)M(x)\frac{\partial g(x)'}{\partial x}$$

Denoting by $H(x,\lambda;c)$ the Hessian matrix of $S(x,\lambda;c)$, it can be easily verified that, if $(\bar{x},\bar{\lambda})$ is a stationary point of $L(x,\lambda)$, then $\hat{H}(\bar{x},\bar{\lambda};c) = H(\bar{x},\bar{\lambda};c)$. Furthermore, the following result holds, whose proof is similar to that of theorem 1 in [8].

THEOREM 7. *Let* $(\bar{x},\bar{\lambda})$ *be a stationary point for* $L(x,\lambda)$ *and assume that:*

(i) $\bar{x}$ *is a local minimum point of problem* (1) *satisfying the second order sufficiency condition:*

$$x'\nabla_x^2 L(\bar{x},\bar{\lambda})x > 0, \quad \forall x: x \neq 0, \ \frac{\partial g(\bar{x})}{\partial x}x = 0$$

(ii) $\frac{\partial g(\bar{x})}{\partial x} M'(\bar{x})$ *has full rank.*

Then, for every $\tau \geq 0$, *there exists a* $\bar{c} > 0$ *such that, for all* $c \geq \bar{c}$, $H(\bar{x},\bar{\lambda};c)$ *is positive definite and* $(\bar{x},\bar{\lambda})$ *is an isolated local minimum point for* $S(x,\lambda;c)$.

On the basis of the preceding results it is possible to define, for the minimization of $S(x,\lambda;c)$, Newton-type algorithms which employ the search direction

$$d = -\hat{H}^{-1}(x,\lambda;c)\nabla S(x,\lambda;c) \tag{13}$$

provided that $\hat{H}(x,\lambda;c)$ is nonsingular. This enables second order convergence to be ensured. Of course, as for the Newton methods, suitable precautions must be taken for the case in which $\hat{H}(x,\lambda;c)$ cannot be guaranteed to be definite positive.

As regards the automatic selection of the penalty coefficient c, we describe a procedure which extends to the augmented Lagrangian introduced here, a result stated in [9].

We assume that it is available an unconstrained minimization algorithm, defined by an iteration map $A: R^n \times R^m \to \Pi(R^n \times R^m)$, which for given values of the penalty coefficient, converges (in the usual sense) to a stationary point of $S(x,\lambda;c)$.

The algorithm below makes use of a preselected increasing sequence $\{c_j\}$, with $c_{j+1} \geq c_j > 0$ and $c_j \to \infty$.

ALGORITHM MODEL

Initial guess: $z_o = (x_o,\lambda_o)$

Step 0: set $j = 0$

Step 1: set $i = 0$ and set $(x_o,\lambda_o) = z_j$

Step 2: If $\nabla_x S(x_i,\lambda_i;c_j) = 0$ and $\nabla_\lambda S(x_i,\lambda_i;c_j) = 0$ go to step 3; else go to step 4.

Step 3: If $g(x_i) = 0$ stop; else go to step 6.

Step 4: If $\nabla_\lambda S'(x_i,\lambda_i;c_j)\frac{\partial g(x_i)}{\partial x}M'(x_i)M(x_i)\nabla_x L(x_i,\lambda_i)$
$+\nabla_x S'(x_i,\lambda_i;c_j)M'(x_i)M(x_i)\left[\frac{\partial g(x_i)}{\partial x}\right]' g(x_i)$
$\geq \|g(x_i)\|^2$ go to step 5; else go to step 6.

Step 5: compute $(x_{i+1},\lambda_{i+1}) \in A\left[(x_i,\lambda_i)\right]$, set $i=i+1$ and go to step 2.

Step 6: set $z_{j+1} = (x_i,\lambda_i)$, $j = j+1$ and go to step 1.

The convergence properties of this algorithm are given in the following theorem.

THEOREM 8. *Assume that:*

(i) $\frac{\partial g(x)}{\partial x}M'(x)$ *is an* $(m \times m)$ *non singular matrix,* $\forall x \in R^n$

(ii) *for every* $c > 0$ *and* $(x_o,\lambda_o) \in R^n \times R^m$, *any accumulation point of the sequence* $\{(x_i,\lambda_i)\}$, *generated by the iteration map A, is a stationary point of* $S(x,\lambda;c)$.

Then:

(a) *if the algorithm constructs a finite sequence* $\{(x_i,\lambda_i)\}_{i=o}^{\nu}$ *and stops, then* (x_ν,λ_ν) *is a stationary point of* $L(x,\lambda)$;

(b) *if the algorithm constructs an infinite sequence $\{(x_i,\lambda_i)\}$, then any accumulation point $(\bar{x},\bar{\lambda})$ is a stationary point of $L(x,\lambda)$;*

(c) *if the algorithm constructs an infinite sequence $\{z_j\}$, then this sequence has no accumulation point.*

PROOF. If the algorithm terminates at (x_ν,λ_ν), then, by construction, (x_ν,λ_ν) is a stationary point of S and $g(x_\nu)=0$. Therefore, by (i), (a) follows from the proof of theorem 2.

As regards (b), suppose that $\{z_j\}$ is finite, with last element z_μ, and let $(\bar{x},\bar{\lambda})$ be an accumulation point of $\{(x_i,\lambda_i)\}$. By (ii) we have $\nabla_x S(\bar{x},\bar{\lambda};c_\mu)=0$, $\nabla_\lambda S(\bar{x},\bar{\lambda};c_\mu)=0$ and, because of the test in step 4 and the continuity assumptions, it results $g(\bar{x})=0$. Then, again by (i) and the proof of theorem 2, (b) is proved. Finally, suppose that the algorithm constructs an infinite sequence $\{z_j\}$, which has an accumulation point $\bar{z}$. We show that this leads to a contradiction. In fact, let $z_j \to \bar{z}$, $j \in K$ and consider the compact set $C = \{z_j,\ j \in K\}$. Define the function:

$$\theta(x,\lambda;c) = \nabla_\lambda S'(x,\lambda;c)\frac{\partial g(x)}{\partial x}M'(x)M(x)\nabla_x L(x,\lambda) + \nabla_x S'(x,\lambda;c)M'(x)M(x)\left[\frac{\partial g(x)}{\partial x}\right]' g(x) - \|g(x)\|^2$$

Recalling (3) (4) and assumption (i), it can be easily verified that there exists a value $\hat{c}>0$ such that for any $c \geq \hat{c}$ and any $(x,\lambda)\in C$, it results:

$$\theta(x,\lambda;c) \geq 0$$

Let $j^* \in K$ be such that $c_{j^*} \geq \hat{c}$, then the algorithm could not have constructed any point $z_j = (x_j,\lambda_j)$ with $j \in K$, $j \geq j^*$, on account of a failure to satisfy the test in step 4. Therefore the points z_j, $j \in K$, $j \geq j^*$, should have been produced because of the transfer in step 3 and this implies $\nabla_x S(x_j,\lambda_j;c_{j-1})=0$, $\nabla_\lambda S(x_j,\lambda_j;c_{j-1})=0$, $\forall j \in K$, $j \geq j^*$. But, according to theorem 2, there must exist a value $\bar{c}>0$ such that $c \geq \bar{c}$ implies that any stationary point of $S(x,\lambda;c)$ in the compact set C, is also a stationary point for $L(x,\lambda)$. It follows that for $c_j \geq \bar{c}$, the algorithm should have terminated at step 3. Thus we get a contradiction with the assumption that $\{z_j\}$ is infinite. □

We remark that the algorithm above is a special case of Algorithm Model 4 considered in [9]; the proof of theorem 8 parallelizes that given in [7] for the exact penalty function proposed there.

4. NUMERICAL RESULTS

In this section we investigate the performance of Newton-type and Quasi-Newton algorithms for the unconstrained minimization of $S(x,\lambda;c)$.

Four test problems are considered. For all problems we assumed in (2): $M(x)=\eta I$, $\eta>0$, a fixed value for the penalty coefficient c and two values for the coefficient τ : $\tau=0$ and $\tau=1$.

The unconstrained minimization algorithms considered were a Quasi-Newton method employing the BFS formula (Algorithm A) and a Newton-type method employing the search direction (13) (Algorithm B). Moreover for TP1 and TP2 the Quasi-Newton method was also tested with the finite difference approximation formula (12) for $t=0.001$ (Algorithm A1).

For every test problem and every algorithm employed we report the values of the parameters c, η, τ used, the number LS of line searches needed to obtain the specified accuracy (measured in terms of $\|g\|$ and

$\|\nabla_x L\|$), and the corresponding number NS of function evaluations. Each function evaluation requires the computation of f, g and $\nabla_x L$. It is to be noted that each gradient evaluation requires only the additional computation of $\nabla_x^2 L$; obviously this is not needed for algorithm A1.

TP1 (Powell, [10]):

$$f(x) = \exp(x_1 x_2 x_3 x_4 x_5)$$

$$g_1(x) = x_1^2 + x_2^2 + x_3^2 + x_4^2 + x_5^2 - 10$$

$$g_2(x) = x_2 x_3 - 5 x_4 x_5$$

$$g_3(x) = x_1^3 + x_2^3 + 1$$

The solution is $\bar{x} = (-1.71714, 1.59571, 1.82725, -0.76364, -0.76364)$ and the corresponding Lagrange multiplier is $\bar{\lambda} = (0.04016, -0.03796, 0.00522)$. The starting point was $x_0 = (-2,2,2,-1,-1)$, $\lambda_0 = (0,0,0)$.

The numerical results are shown in Table 1. It appears that a positive value for τ, which improves the robustness of the method, does not affect the convergence rate. Furthermore, we observe that Algorithm A1, which avoids the evaluation of $\nabla_x^2 L$, yields almost the same results as Algorithm A. As expected, Algorithm B is much faster in terms of line searches and functions evaluations.

TP2 (Miele et al., [11]):

$$f(x) = (x_1-1)^2 + (x_1-x_2)^2 + (x_3-1)^2 + (x_4-1)^4 + (x_5-1)^6$$

$$g_1(x) = x_1^2 x_4 + \sin(x_4-x_5) - 2\sqrt{2}$$

$$g_2(x) = x_2 + x_3^4 x_4^2 - 8 - \sqrt{2}$$

The solution is $\bar{x} = (1.1661, 1.1821, 1.3802, 1.5060, 0.6109)$ and the corresponding Lagrange multiplier is $\bar{\lambda} = (-0.08553, -0.03187)$. The starting point was $x_{oi} = 2$, $i = 1,\dots,5$, $\lambda_{oj} = 0$, $j = 1,2$.

The numerical results are reported in Table 2. It can be seen that the same remarks made for TP1 can be repeated for the present case.

TP3 (Bartholomew-Biggs, [12]):

$$f(x) = x_1^4 + \sum_{i=2}^{20} x_i^2$$

$$g_i(x) = x_i + x_{21-i} - i, \quad i = 1,2$$

$$g_j(x) = x_j^2 + x_{21-j} - j - 1, \quad j = 3,5,7,9$$

$$g_k(x) = x_k + x_{21-k}^2 - k - 1, \quad k = 4,6,8,10$$

The solution is $\bar{x} = (0.58976, 1, 1.87083, 0.5, 2.34521, 0.5, 2.73861, 0.5, 3.08221, 0.5, 3.24037, 0.5, 2.91548, 0.5, 2.54951, 0.5, 2.12132, 0.5, 1, 0.41024)$ and the corresponding Lagrange multiplier is $\bar{\lambda} = (-0.82047, -2, -1, -1, -1, -1, -1, -1, -1, -1)$. The starting point was $x_{oi} = 1$, $i = 1,\dots,20$, $\lambda_{oj} = 0$, $j = 1,\dots,10$.

The numerical results are reported in Table 3. Since the solution of TP3 takes a significant amount of computer time, we report also the times taken to solve the problem on the UNIVAC 1110 computer of the Rome University. It appears that algorithm B is much faster than A also in terms of computing times.

TP4 (Rockafellar, [3]):

Table 1.- TP1

ALGORITHM	c	η	τ	$\|g\|$	$\|\nabla_x L\|$	LS	NS
A	1	1	1	2.3 E-5	1.2 E-5	20	26
	1	1	0	3.1 E-5	7.0 E-6	17	22
	10	1	1	2.1 E-5	2.3 E-5	31	38
	10	1	0	2.4 E-5	6.1 E-5	31	39
	1	0.1	1	6.6 E-5	7.8 E-4	21	25
	1	0.1	0	7.0 E-5	3.8 E-4	21	25
A1 (t=0.001)	1	1	1	1.7 E-5	9.1 E-6	20	26
	1	1	0	3.1 E-5	1.6 E-5	17	22
	10	1	1	2.6 E-5	7.4 E-5	31	38
	10	1	0	3.1 E-5	1.1 E-4	32	39
	1	0.1	1	5.6 E-5	7.2 E-4	21	25
	1	0.1	0	6.7 E-5	3.6 E-4	21	25
B	1	1	1	2.6 E-6	6.4 E-7	7	9
	1	1	0	2.0 E-6	4.6 E-7	7	9
	10	1	1	6.1 E-8	8.8 E-9	8	10
	10	1	0	7.3 E-8	6.6 E-9	8	10
	1	0.1	1	9.2 E-5	9.5 E-4	7	10
	1	0.1	0	1.8 E-6	4.5 E-7	7	9

Table 2.- TP2

ALGORITHM	c	η	τ	$\|g\|$	$\|\nabla_x L\|$	LS	NS
A	1	1	1	2.5 E-5	5.9 E-5	45	51
	1	1	0	9.6 E-6	1.3 E-5	43	55
	10	1	1	3.7 E-6	1.1 E-5	52	59
	10	1	0	5.1 E-6	7.7 E-5	49	56
	10	0.1	1	5.9 E-6	6.8 E-5	61	68
	10	0.1	0	7.9 E-6	4.3 E-4	56	64
A1 (t=0.001)	1	1	1	7.2 E-6	2.7 E-5	45	52
	1	1	0	2.1 E-5	2.3 E-5	40	51
	10	1	1	5.1 E-6	2.0 E-5	52	64
	10	1	0	5.1 E-6	3.0 E-5	51	62
	10	0.1	1	6.2 E-6	4.6 E-4	59	65
	10	0.1	0	1.7 E-6	1.3 E-4	54	62
B	1	1	1	3.0 E-5	3.2 E-5	7	10
	1	1	0	3.8 E-6	4.5 E-5	8	11
	10	1	1	5.7 E-7	4.9 E-6	8	11
	10	1	0	9.4 E-7	1.1 E-5	8	12
	10	0.1	1	4.7 E-6	5.8 E-5	10	12
	10	0.1	0	2.0 E-6	2.4 E-5	9	11

Table 3.- TP3

ALGORITHM	c	η	τ	$\|g\|$	$\|\nabla_x L\|$	LS	NS	time (sec)
A	10	1	1	6.2 E-5	2.5 E-4	36	72	15.1
	10	1	0				>100	
	100	1	1	6.0 E-6	6.2 E-5	37	73	15.6
	100	1	0	2.3 E-6	3.1 E-5	37	73	15.6
	100	0.1	1	1.5 E-5	4.2 E-3	37	64	14.4
	100	0.1	0				>100	
B	10	1	1	7.1 E-7	1.9 E-5	5	11	6.6
	10	1	0	3.0 E-7	3.8 E-4	4	9	5.6
	100	1	1	3.3 E-7	3.8 E-6	5	10	6.7
	100	1	0	6.4 E-7	7.1 E-5	4	8	5.5
	100	0.1	1	5.3 E-7	1.1 E-4	7	18	8.9
	100	0.1	0	5.9 E-7	6.7 E-5	4	8	5.4

Table 4.- TP4

ALGORITHM	c	η	τ	$\|\nabla_x L\|$	LS	NS	x_1	$\|x_2\|=\|g\|$
A	1	1	1	5.6 E-10	33	35	1.7 E-4	5.8 E-10
	1	1	0	7.9 E-10	27	30	3.7 E-4	5.9 E-10
	10	1	1	2.7 E-10	33	35	1.3 E-4	2.8 E-10
	10	1	0	2.7 E-10	33	35	1.2 E-4	2.6 E-10
	100	0.1	1	2.5 E-11	34	37	1.5 E-4	3.8 E-11
	100	0.1	0	1.5 E-8	34	37	1.5 E-4	4.8 E-12
B	1	1	1	1.3 E-11	20	21	1.5 E-4	3.4 E-13
	1	1	0	3.1 E-12	21	22	9.7 E-5	5.4 E-13
	10	1	1	4.1 E-12	21	22	1.0 E-4	1.2 E-13
	10	1	0	1.3 E-11	20	21	1.5 E-4	8.7 E-14
	100	0.1	1	1.3 E-11	20	21	1.5 E-4	5.0 E-15
	100	0.1	0	1.3 E-11	20	21	1.5 E-4	5.2 E-15

$$f(x) = x_1^4 - x_2 + x_1 x_2$$
$$g(x) = x_2$$

The solution is $\bar{x} = (0,0)$, $\bar{\lambda} = 1$, and the starting point was $x_o = (0.5, 0.5)$, $\lambda_o = 0$.

This problem is one that does not satisfy the second order sufficiency conditions. As illustrated in [3], the approximate solution of this problem by the ordinary augmented Lagrangian method requires quite large values of the penalty coefficient and a considerable amount of function evaluations. In comparison with the results given in [3], Table 4 shows that the augmented Lagrangian considered here allows to obtain a close approximation of the solution $(\bar{x}_1, \bar{x}_2)$, for comparatively very small values of the penalty coefficient and with much less computational effort.

In conclusion, although a numerical comparison with existing techniques was not performed, the method proposed here seems to be competitive with current alternatives. Moreover, from the experience gained, it seems that the use of algorithm B is advisable, whenever second order derivatives are available.

REFERENCES

[1] R.T.ROCKAFELLAR: *Penalty Methods and Augmented Lagrangians in Nonlinear Programming*. 5th IFIP Conference on Optimization Techniques, Part I, R. Conti, A. Ruberti eds., Springer-Verlag (1973), pp. 418-425.

[2] R.FLETCHER: *Methods Related to Lagrangian Functions*. Numerical Methods for Constrained Optimization, P.E.Gill, W.Murray eds., Academic Press (1974), pp. 219-239.

[3] D.A.PIERRE, M.J.LOWE: *Mathematical Programming via Augmented Lagrangians: an Introduction with Computer Programs*. Addison-Wesley (1975).

[4] D.P.BERTSEKAS: *Multiplier Methods: a Survey*. Automatica (1976), v. 12, pp. 133-145.

[5] M.R.HESTENES: *Optimization Theory. The Finite Dimensional Case*. John Wiley & Sons (1975).

[6] R.FLETCHER: *An exact penalty function for nonlinear programming with inequalities*. Math. Progr. (1973) v.5, pp. 129-150.

[7] H.MUKAI, E.POLAK: *A Quadratically Convergent Primal-Dual Algorithm with Global Convergence Properties for solving Optimization Problems with equality constraints*. Math. Programming, vol.9, n. 3, dec. 1975, pp. 336-349.

[8] G.DI PILLO, L.GRIPPO: *A new class of Augmented Lagrangians in Nonlinear Programming*. SIAM J. on Control and Optimization, (1979), vol.17, n.5, pp. 618-628.

[9] E.POLAK: *On the Stabilization of Locally Convergent Algorithms for Optimization and Root Finding*. Automatica (1976), Vol. 12, pp. 337-342.

[10] M.J.D.POWELL: *A method for nonlinear constraints in minimization problems*. Optimization, R.Fletcher, ed., Academic Press (1969), pp. 283-298.

[11] A.MIELE, P.E.MOSELEY, A.V.LEVY, G.M.COGGINS: *On the Method of Multipliers for Mathematical Programming Problems*. J. Optimization Theory and Appl. (1972), v. 10, pp. 1-33.

[12] M.C.BARTHOLOMEW-BIGGS: *A Matrix Updating Technique for Estimating Lagrange Multipliers when Solving Equality Constrained Minimization Problems by the Recursive Quadratic Programming Method*. The Hatfield Polytechnic, N.O.C., Technical Rep. No. 96, July 1978.

RANDOMLY GENERATED NONLINEAR PROGRAMMING TEST PROBLEMS

K. Schittkowski
Institut für Angewandte Mathematik und
Statistik
Universität Würzburg
87 Würzburg, W. Germany

1. Introduction

Any development or comparison of nonlinear programming software for solving the problem

$$
x \in \mathbb{R}^n: \quad
\begin{array}{ll}
\min & f(x) \\
g_j(x) = 0 \ , & j=1,\dots,m_e \\
g_j(x) \geq 0 \ , & j=m_e+1,\dots,m \\
x_l \leq x \leq x_u &
\end{array}
\tag{1}
$$

with continuously differentiable functions $f, g_j: \mathbb{R}^n \to \mathbb{R}$, $j=1,\dots,m$, has to be based on extensive numerical tests. This requires to search for test problems of the form (1), to know as much as possible about their mathematical structure, and to implement them in an appropriate way. Most test problems which are used in the past to test and compare optimization programs consist of so called 'real life' problems which are believed to reflect typical structures of practical nonlinear programming problems, for example the Colville problems, confer Himmelblau [2] or Hock and Schittkowski [3,4]. But this class of test examples has some disadvantages especially since the precise solution is not known a priory preventing to relate the efficiency of a code to the achieved accuracy. In this paper, a completely different approach is presented: the construction of randomly generated test problems with predetermined solutions.

2. Fundamentals of the test problem generator

A test problem generator has to be presented satisfying the following conditions:

a) It is possible to produce several classes of test problems like small and dense problems, big and sparse problems, problems with equality or inequality constraints only, and so on.

b) Each class of test problems is completely described by very few parameters, for example dimension, number of constraints, upper and lower bounds.

c) A repeated execution of the generator yields arbitrarily many different problems of the class randomly.

d) An optimal solution, i.e. a point satisfying the Kuhn-Tucker and a second order condition, is known a priori. The corresponding precise objective function value is zero.

e) It is possible to construct test problems with special properties like convex, linearly constrained, ill-conditioned, degenerate, or indefinite problems.

f) Each problem can be provided with different randomly generated starting points.

Indeed, a test problem generator following these guidelines allows to produce a wide range of different problems for general purpose tests on the one side and problems with special features on the other side. First we have to define the Lagrangian function of problem (1):

$$L(x,u) := f(x) - \sum_{j=1}^{m} u_j \, g_j(x) \,, \tag{2}$$

$x \in \mathbb{R}^n$, $u = (u_1,\dots,u_m)^T \in \mathbb{R}^m$. To construct test problems with a predetermined optimal solution, we have to formulate a second order sufficient condition, confer McCormick [5]:

<u>Theorem:</u> Let $f, g_1, \dots, g_m$ be twice differentiable functions. A point $x^* \in \mathbb{R}^n$ with $x_l < x^* < x_u$ is an isolated local minimizer of (1), if there exists a vector $u^* = (u_1^*,\dots,u_m^*)^T$, such that the following conditions are valid:

a) (Kuhn-Tucker condition)

$$\begin{aligned} & g_j(x^*) = 0 \,, \quad j=1,\dots,m_e. \\ & g_j(x^*) \geq 0 \,, \quad j=m_e+1,\dots,m. \\ & u_j^* \geq 0 \,, \quad j=m_e+1,\dots,m. \\ & u_j^* \, g_j(x^*) = 0 \,, \quad j=m_e+1,\dots,m. \end{aligned} \tag{3}$$

$D_x L(x^*,u^*) = 0.$

b) (Second order condition) For every nonzero vector y where $y^T D_x g_j(x^*) = 0$, $j=1,\dots,m_e$, and $y^T D_x g_j(x^*) = 0$ for all j with $u_j^* > 0$, $j=m_e+1,\dots,m$, it follows that

$$y^T D_x{}^2 L(x^*,u^*) y > 0 \ . \tag{4}$$

The symbols $D_x f$ and $D_x{}^2 f$ represent the first and second derivatives of a function f with respect to the variable x.

The construction of a test problem is based on a series of m+1 arbitrary twice continuously differentiable functions s_j, $j=0,1,\dots,m$, defined on any subset of $\mathbb{R}^n$ containing the interval $[x_l,x_u]$. Furthermore we need a randomly chosen $x^* \in \mathbb{R}^n$ with $x_l < x^* < x_u$ which will define a (at least local) minimizer of the optimization problem.

First we have to establish that x^* is feasible with exactly m_a active constraints, where m_a is a predetermined integer. In addition, it should be allowed for constructing special types of test problems to predetermine the gradients $D_x g_j(x^*)$, $j=1,\dots,m_e+m_a$. Define therefore the restrictions by

$$\begin{aligned} g_j(x) &:= s_j(x) - s_j(x^*) + d_j{}^T(x^* - x) \ , \ j=1,\dots,m_e+m_a \\ g_j(x) &:= s_j(x) - s_j(x^*) + \mu_j \ , \ j=m_e+m_a+1,\dots,m \ , \end{aligned} \tag{5}$$

where $d_j \in \mathbb{R}^n$, $j=1,\dots,m_e+m_a$, and the real numbers μ_j, $j=m_e+m_a+1,\dots,m$, are randomly chosen within the interval $(0,m)$.

The objective function is defined by

$$f(x) := s_o(x) + \frac{1}{2} x^T H x + c^T x + \alpha \tag{6}$$

with an n by n matrix H, $c \in \mathbb{R}^n$, and $\alpha \in \mathbb{R}$. The quadratic term of f has to be determined so that x^* satisfies the Kuhn-Tucker condition (3), the second order condition (4), and the condition $f(x^*) = 0$. Therefore we determine optimal Lagrange multipliers $u^* = (u_1^*,\dots,u_m^*)^T$ with

$$\begin{aligned} u_j^* &\geq 0 \ , \ j=m_e+1,\dots,m_e+m_a \ , \\ u_j^* &= 0 \ , \ j=m_e+m_a+1,\dots,m \ , \end{aligned} \tag{7}$$

furthermore an n by n matrix P with

$$y^T P \ y > 0 \tag{8}$$

for all nonzero vectors y with $y^T D_x g_j(x^*) = 0$, $j=1,\dots,m_e$, and

$y^T D_x g_j(x^*) = 0$ for all j with $u_j^* > 0$, $j=m_e+1,\ldots,m$. It is easy to see that the definition of the matrix

$$H := - D_x^2 s_o(x^*) + \sum_{j=1}^{m} u_j^* D_x^2 g_j(x^*) + P \qquad (9)$$

leads to $D_x^2 L(x^*,u^*) = P$ implying that the second order condition (4) is always satisfied. The Kuhn-Tucker condition $D_x L(x^*,u^*) = 0$ requires to define c by

$$c := - D_x s_o(x^*) - Hx^* + \sum_{j=1}^{m} u_j^* D_x g_j(x^*) \ . \qquad (10)$$

Finally, the constant term α is given by

$$\alpha := - s_o(x^*) - \frac{1}{2} x^{*T} H x^* - c^T x^* \ , \qquad (11)$$

and guarantees that $f(x^*) = 0$. This completes the construction of a test problem provided that one knows how to choose the following data:

a) The series of twice continuously differentiable functions $s_o,\ldots,s_m$.

b) The linear terms of the restrictions, i.e. the vectors $d_1,\ldots,d_{m_e+m_a}$.

c) The optimal Lagrange multipliers, i.e. any $u^* = (u_1^*,\ldots,u_m^*)^T$ satisfying (7).

d) The Hessian of the Lagrangian with respect to x^* and u^*, i.e. a matrix P satisfying (8).

These data are specialized in the following sections to allow the construction of optimization problems in accordance to the individual purpose of the test designer.

3. General test problems

We consider now the construction of test problems for general purpose tests, i.e. for tests determining the overall efficiency, global convergence, and reliability of an optimization program. In this case, one could define the functions $s_o, s_1, \ldots, s_m$ by signomials, generalized polynomial functions of the kind

$$s(x) = \sum_{j=1}^{k} c_j \prod_{i=1}^{n} x_i^{a_{ij}} \ , \ x > 0, \qquad (12)$$

where the coefficients c_j and the exponents a_{ij} are real numbers. Functions of this kind are considered because of their simple structure and the observation that many 'real life' problems are defined by signomials, for example geometric programming problems, confer Duffin, Peterson, Zener [1]. Since each signomial is completely described by the data c_j and a_{ij}, $j=1,\dots,k$, $i=1,\dots,n$, it is possible to produce these data randomly using predetermined bounds. In accordance with 'real life' geometric programming problems, it should be allowed to have the exponents of the signomials integer. Furthermore one should implement the possibility to vary the density of the coefficient matrix (a_{ij}).

Since it is not required to predetermine the gradients of the active constraints, we let $d_j = 0$ for $j=1,\dots,m_e+m_a$. The optimal Lagrange multipliers u_j^* are given by the instructions

$$\begin{aligned} &u_j^* \in (b_1,b_2) \text{ randomly chosen, } j=1,\dots,m_e, \\ &u_j^* \in (0,b_3) \text{ randomly chosen, } j=m_e+1,\dots,m_e+m_a, \\ &u_j^* = 0\ , \quad j=m_e+m_a+1,\dots,m. \end{aligned} \tag{13}$$

To satisfy the sufficient second order optimality condition, consider an upper triangular matrix U whose elements are randomly chosen within the interval (b_4,b_5) and compute the positive definite matrix

$$P := U^T U\ . \tag{14}$$

These definitions satisfy the requirements of the last section for the construction of a test problem. The bounds $b_1,\dots,b_5$ for determining the Lagrange multipliers or the elements of U are predetermined by the user. The reader should be aware that the signomials are not convex functions in general implying that the given solution x^* is only a local one. In other words, it is possible that an optimization code approximates a solution with a function value less than zero. This situation is not considered as a disadvantage since these test runs can be used to determine the global convergence of an optimization program. In [6] we present the data for constructing 80 test problems as described in this section and, in addition, detailed numerical results for comparing 13 qualified optimization programs.

4. Linearly constrained test problems

Test problems with linear equality and inequality constraints are easily obtained by defining

$$s_j(x) := a_j^T x\ , \quad j=1,\dots,m,$$

with randomly chosen vectors $a_j \in \mathbb{R}^n$, $j=1,\dots,m$. If it is not required to predetermine the gradients of the active constraints, let $d_j = 0$ for $j=1,\dots,m_e+m_a$. For the construction of the objective function f, one could use any signomial s_o of the kind (12), furthermore the instructions (13) and (14) for determining the Lagrange multipliers u^* and the Hessian $D_x^2L(x^*,u^*)$.

In addition, it is possible to generate convex linearly constrained test problems. In this case, one could replace s_o by a convex exponential sum of the form

$$s(x) = \sum_{j=1}^{k} c_j \exp(\sum_{i=1}^{n} a_{ij} x_i) \qquad (15)$$

with randomly generated $c_j > 0$ and $a_{ij} \in \mathbb{R}$. The optimal Lagrange multipliers are given by the instructions (13). To guarantee the convexity of f on $\mathbb{R}^n$, one should set H = 0 or, equivalently, $P = D_x^2 s_o(x^*)$. This matrix is at least positive semi-definite and positive definite, if any positive definite matrix is added to P. In this convex case, the local minimizer x^* is a global one.

5. Degenerate test problems

An optimization problem of the kind (1) is called a degenerate one, if at least one of the Lagrange multipliers u_j^*, $j=1,\dots,m_e+m_a$, vanishes, i.e. degeneracy occurs if at least one of the active constraints is redundant at the optimal solution x^*. If in the worst case all Lagrange multipliers are zero, the constrained local minimizer x^* is identical with an unconstrained local minimizer of f. We denote a test problem nearly degenerate, if the Lagrange multipliers differ widely in their order of magnitude. Both situations arise in practical applications and by numerical experiments we try to get an answer to the main question: How does an optimization code behave under different degrees of degeneracy. Especially, we are interested in the following questions:

a) Are there any numerical difficulties when solving degenerate problems?

b) Does an optimization program take any advantage of redundant constraints or nearly degenerate problems?

c) Does degeneracy influence the final accuracy of an optimization code?

Proceeding from a set of signomials $s_o,s_1,\dots,s_m$ and $d_j = 0$, $j=1,\dots,m_e+m_a$, the matrix P could be determined by (14) guaranteeing the second order condition. Test problems with varying degree of degeneracy are obtained for example by the following conditions:

$$
\begin{aligned}
&\text{a)} \quad u_j^* = 1 \;, \; j=1,\ldots,m_e+m_a . \\
&\text{b)} \quad u_j^* = 10^{(-2(j-1))} \;, \; j=1,\ldots,m_e+m_a . \\
&\text{c)} \quad u_j^* = 1 \;, \; j=1,\ldots,[\tfrac{1}{2}(m_e+m_a)] \\
&\phantom{\text{c)}} \quad u_j^* = 0 \;, \; j=[\tfrac{1}{2}(m_e+m_a)]+1,\ldots,m_e+m_a . \\
&\text{d)} \quad u_j^* = 0 \;, \; j=1,\ldots,m_e+m_a .
\end{aligned}
\tag{16}
$$

To allow intermediate comparisons, the test problems should be distinguished only by these Lagrange multipliers. All other data like dimension, number of constraints, signomials s_j, $j=0,\ldots,m$, should be identical leading to a series of test problems with an increasing degree of degeneracy. The data for determining 24 test problems and numerical results obtained by 16 optimization programs are contained in [7].

6. Ill-conditioned test problems

It is well-known from optimization theory that (at least for convex problems) the optimal solution of (1) defines a saddle point of the Lagrangian (2) and vice versa. In this case, the solution x^* is a minimizer of the function $L(x,u^*)$, where u^* denotes the optimal Lagrange multipliers. Numerical experience in unconstrained optimization shows that the local convergence of a standard unconstrained nonlinear programming code depends heavily on the condition number of the Hessian matrix at the optimal solution, in our case on cond $D_x{}^2L(x^*,u^*)$ = cond P. Since many programs designed for the solution of the constrained problem are based on minimizing an augmented Lagrangian, we intend to construct test problems with different condition numbers of the Hessian matrix $D_x{}^2L(x^*,u^*)$ and we are concerned with the question how ill-conditioning influences the final accuracy and the efficiency (CPU-time, number of function and gradient evaluations) of an optimization code.

For generating ill-conditioned test problems, we use a set of signomials $s_o,\ldots,s_m$, furthermore $d_j = 0$ for $j=1,\ldots,m_e+m_a$, and randomly chosen multipliers u^*, see (13). The matrix P is defined by

$$
P := \begin{pmatrix} H_\nu & : & 0 \\ 0 & : & I_{n-\nu} \end{pmatrix} , \tag{17}
$$

where $I_{n-\nu}$ denotes the $(n-\nu)$ by $(n-\nu)$ unit matrix and H_ν the ν by ν Hilbert matrix

$$H_\nu := \left(\frac{1}{i+j-1}\right)_{\nu,\nu} .$$

It is obvious that the sufficient optimality criteria are satisfied. By varying ν it is possible to produce test problems with an increasing condition number of the Hessian of the Lagrangian. This condition number is approximately given by $\exp(3.5\nu)$, confer Zielke [8], and to give some examples, consider $\nu = 3,5,8$:

$$\text{cond } H_3 \approx 3.6_{10}4$$

$$\text{cond } H_5 \approx 4.0_{10}7$$

$$\text{cond } H_8 \approx 1.4_{10}12 .$$

Numerical results for comparing 16 optimization programs executed for solving ill-conditioned test problems are presented in [7]

7. Indefinite test problems

Until now we proceeded from the fact that the matrix $D_x^2L(x^*,u^*)$ is positive definite. This is a stronger assumption than required by the second order condition (4) and not always satisfied in practice. Therefore, we intend to construct indefinite test problems to check if an indefinite Hessian matrix of the Lagrangian leads to numerical difficulties, to another final accuracy, or to an increased efficiency.

First we define again a set of signomials $s_o,\ldots,s_m$, and the linear terms of the restrictions are given by

$$d_j := D_x s_j(x^*) - e_j \ , \ j=1,\ldots,m_e+m_a \ , \qquad (18)$$

where e_j denotes the j-th axis vector. As a consequence, the gradients of the first m_e+m_a restrictions at x^* are axis vectors, i.e.

$$D_x g_j(x^*) = e_j \ , \ j=1,\ldots,m_e+m_a \ ,$$

confer (5). The Lagrange multipliers u_j^* are randomly chosen as described by (13) with the additional assumption that $u_j^* \neq 0$, $j=1,\ldots,m_e+m_a$. The matrix P is given by

$$P := \begin{pmatrix} \sigma U_1^T U_1 & : & 0 \\ 0 & : & U_2^T U_2 \end{pmatrix} \begin{matrix} \} \ m_e+m_a \\ \} \ n-m_e-m_a \end{matrix}$$

with upper triangular matrices U_1 and U_2 whose elements are randomly chosen between predetermined bounds. The matrix is indefinite if and only if $\sigma \le 0$. For a

$$y := \begin{pmatrix} z_1 \\ z_2 \end{pmatrix}, \quad z_1 \in \mathbb{R}^{m_e+m_a}, \quad z_2 \in \mathbb{R}^{n-m_e-m_a}, \quad y \neq 0,$$

the condition $y^T D_x g_j(x^*) = 0$ for $j=1,\ldots,m_e+m_a$, is equivalent with $z_1 = 0$ leading to

$$y^T D_x{}^2 L(x^*,u^*)y = z_2{}^T U_2{}^T U_2 z_2 > 0.$$

This implies the validity of the second order sufficient condition for all values of σ. The data for the construction of 24 test problems with $\sigma < 0$, $\sigma = 0$, or $\sigma > 0$ and numerical results are presented in [7].

8. Convex test problems

An optimization problem (1) is called a convex one if the objective function f(x) is strictly convex and if the set of all feasible points is convex. The last condition is satisfied if there are no equality constraints ($m_e = 0$) and if all restriction functions $g_j(x)$, $j=1,\ldots,m$ are concave. The main attribute of convex problems is the fact that every local minimizer is a global one preventing difficulties with alternate local solutions. This allows to provide a test problem with a starting point far away from the solution x*. Solving the same problem with a starting point close to the solution, gives the possibility to test the sensitivity of an optimization program with respect to the position of the starting point. Furthermore, it is possible to test if a code is able to take advantage of the convex structure of an optimization problem.

First we have to look for a method to generate convex functions. One possible way is to define exponential sums of the form (15) with randomly chosen $c_j \ge 0$ and $a_{ij} \in \mathbb{R}$, $j=1,\ldots,k$, $i=1,\ldots,n$. It is easy to see that these functions are derived from signomials by simple exponential transformations $y_i = \exp(x_i)$, $i=1,\ldots,n$. Consider now m+1 convex exponential sums $t_o,t_1,\ldots,t_m$ and let

$$s_o(x) := t_o(x)$$

$$s_j(x) := -\, t_j(x), \quad j=1,\ldots,m.$$

Using the functions $s_o,\ldots,s_m$ and the instructions of section 2, we get concave restrictions g_j, $j=1,\ldots,m$. In this case, we may set $d_j = 0$, $j=1,\ldots,m_a$. The optimal Lagrange multipliers are randomly chosen positive numbers and to achieve H = 0, define

$$P := D_x^2 s_o(x^*) - \sum_{j=1}^{m} u_j^* D_x^2 s_j(x^*) .$$

This matrix is positive semi-definite, since s_o is convex, since $u_j^* > 0$, and since the functions s_j are concave. A strictly convex objective function and a positive definite matrix P are obtained by adding a positive definite matrix to P. In progress of our comparative study of optimization programs, we constructed 25 convex test problems and tested 16 codes numerically.

References

[1] R.J. Duffin, E.L. Peterson, C. Zener, Geometric Programming - Theory and applications, John Wiley & Sons, New York, London, Sydney, 1967.

[2] D.M. Himmelblau, Applied Nonlinear Programming, McGraw-Hill, 1972.

[3] W. Hock, K. Schittkowski, Test examples for the solution of nonlinear programming problems. Part 1, Preprint No.44, Institut für Angewandte Mathematik und Statistik, Universität Würzburg, 1979.

[4] W. Hock, K. Schittkowski, Test examples for the solution of nonlinear programming problems. Part 2, Preprint No.45, Institut für Angewandte Mathematik und Statistik, Universität Würzburg, 1979.

[5] G.P. McCormick, Second order conditions for constrained minima, SIAM Journal on Applied Mathematics, Vol.15, No.3 (1967), 641-652.

[6] K. Schittkowski, A numerical comparison of 13 nonlinear programming codes with randomly generated test problems, to appear: Numerical Optimisation of Dynamic Systems, L.C.W. Dixon, G.P. Szegö eds., North-Holland Publishing Company.

[7] K. Schittkowski, The construction of degenerate, ill-conditioned and indefinite nonlinear programming problems and their usage to test optimization programs, submitted for publication.

[8] G. Zielke, Test matrices with maximal condition number, Computing, Vol.13 (1974), 33-54.

METHOD OF REGULARIZED APPROXIMATIONS AND ITS APPLICATION TO CONVEX PROGRAMMING

J.S. Sosnowski

Polish Academy of Sciences
Systems Research Institute
Newelska 6, 01-447 Warszawa, POLAND

1. INTRODUCTION

In the paper, we will consider algorithms for solution of the following convex problem

minimize $f_o(x)$ /P/

subject to $g(x) \leqslant 0$

where $f_o: R^n \longrightarrow R$ and $g: R^n \longrightarrow R^m$ are differentiable convex functions.

The paper demonstrates that for convex programming problem /P/ , the penalty and multiplier methods and their modifications can be considered as applications of algorithms of regularized aproximations to the dual, and the Lagrange functions respectively.

In the paper convergence properties are obtained for modifications of penalty and multiplier method which are called regularized penalty function method and regularized multiplier method.

We also suggest an application of the quasi-Newton update formula in place of the ordinary multiplier update.

2. ALGORITHMS OF REGULARIZED APPROXIMATIONS

By a regularization of a convex, semicontinuous and proper function $f: R^n \longrightarrow (-\infty,+\infty]$ we mean the addition to f of a strongly convex function $r: R^n \longrightarrow R$.

In the paper the function r will assume the following form

$$r(x) = \frac{1}{2\eta} < x - \tilde{x}, H(x - x) > = \frac{1}{2\eta} \| x - \tilde{x} \|_H^2 \qquad /1/$$

where H is symmetric and positive defined $n \times n$ matrix, $\tilde{x}$ is a fixed point, and $\eta \in R$, $\eta > 0$.

Let us consider the following two types of the algorithms of regularized approximations to minimize convex function f.

A. Proximal Point Algorithm

Assuming given the parameters: $0<\rho_k \nearrow \rho_\infty \leq \infty$, and initial point $x^o \in R^n$, the sequence of points is defined by

$$x^{k+1} = \underset{x\in R^n}{\operatorname{argmin}} \left\{ f(x)+ \frac{1}{2\rho_k} \| x - x^k \|^2_{H_k} \right\} \qquad /2/$$

B. Reference Point Algorithm

Assuming given the parameters: $0<\rho_k \nearrow \infty$ and reference point $\bar{x} \in R^n$, the sequence of points is defined by

$$x^{k+1} = \underset{x\in R^n}{\operatorname{argmin}} \left\{ f(x) + \frac{1}{2\rho_k} \| x - \bar{x} \|^2_{H_k} \right\} \qquad /3/$$

In both algorithms, in each iteration, a strongly convex function is minimized. The terms which ensure strong convexity /regularize a convex, not necessarily strictly convex function f/ assume various forms. In the reference point algorithm the minimized function, in each iteration k, is independent of minimizer, determined in the previous iteration. The reference point algorithm provides convergence to the minimizing point of f only if $\rho_k \to \infty$. If f is a convex polyhedral function one can choise ρ_k finite which ensures finite convergence of the algorithm [6]. In the reference point algorithm $\{x^k\}$ is bounded and converges to an optimal solution with minimal distance from $\bar{x}$, i.e.

$$x^k \to x^\infty = \underset{\hat{x}\in \hat{X}}{\operatorname{argmin}} \| \hat{x} - \bar{x} \|^2_{H_\infty} \qquad /4/$$

where: $\hat{X}$ is the set of optimal solution and $H_k \to H_\infty$.
The proximal point algorithm provides convergence to the minimizing point even for a given constant $\rho_k = \rho > 0$.
The proximal point algorithm was investigated by Rockafellar [4] with $H_k = I$, where I is a unit matrix.
The reference point algorithm is due to Tikhonov [7].
For the problem finding a saddle point of a convex-concave function $l: R^n \times R^m \to [-\infty, +\infty]$ we can consider a combination of the mentioned algorithms.
Let us consider the following types of algorithms of regularized approximations:

- one side regularization

$$(x^{k+1}, y^{k+1}) = \underset{x \quad y}{\operatorname{argminmax}} \left\{ l(x,y) - \frac{1}{2\rho_k} \| y - y^k \|^2_H \right\} \qquad /5/$$

$$(x^{k+1},y^{k+1}) = \underset{x\quad y}{\text{argminmax}} \left\{ l(x,y) - \frac{1}{2\rho_k} \|y-\bar{y}\|_H^2 \right\} \qquad /6/$$

- two sides regularization

a/ symmetric

$$(x^{k+1},y^{k+1}) = \underset{x\quad y}{\text{argminmax}} \left\{ l(x,y) - \frac{1}{2\rho_k} \|y-\bar{y}\|_H^2 + \frac{1}{2\eta_k} \|x-\bar{x}\|_Q^2 \right\} \qquad /7/$$

b/ asymmetric

$$(x^{k+1},y^{k+1}) = \underset{x\quad y}{\text{argminmax}} \left\{ l(x,y) - \frac{1}{2\rho_k} \|y-y^k\|_H^2 + \frac{1}{2\eta_k} \|x-\bar{x}\|_Q^2 \right\} \qquad /8/$$

where $0 < \eta_k \in R$.

Let us assume that l is Lagrange function of (P):

$$l(x,y) = f_o(x) + \langle y, g(x)\rangle + \delta(y|y \geqslant 0) \qquad /9/$$

where

$$\delta(y|y\geqslant 0) = \begin{cases} 0 \text{ if } & y \geqslant 0 \\ -\infty \text{ if } & y \not\geqslant 0 \end{cases}$$

Let us assume that $Q = I$ and $H = I$ I - unit matrix . One can prove that a pair which is the minimax point in /5/ can be derived in the two following steps:

$$1^o \quad x^{k+1} = \text{argmin} \left\{ f_o(x) + \frac{1}{2\rho_k} \|(y^k + \rho_k g(x))_+\|^2 - \frac{1}{2\rho_k} \|y^k\|^2 \right\} \qquad /10/$$

$$2^o \quad y^{k+1} = (y^k + \rho_k g(x^{k+1}))_+ \qquad /11/$$

where $(d)_+$ is a projection on R^n_+ (vector with the i-th coordinate equal to $\max(0,d_i)$) .

The steps 1°, 2° are the standard steps of the ordinary multiplier method and represent minimization, with given Lagrange multiplier y^k and given parameter ρ_k, of the augmented Lagrange function $L: R^n \times R^m \times (0,\infty) \longrightarrow R$.

$$L(x,y,\rho) = f_o(x) + \frac{1}{2\rho} \|(y + \rho g(x))_+\|^2 - \frac{1}{2\rho} \|y\|^2 \qquad /12/$$

One can prove, that the sequence of multipliers defined in the multiplier method is obtained as a results of application of the proximal point algorithm for maximizing a dual function φ ($\varphi(y) = \inf_x l(x,y)$) , i.e.

$$y^{k+1} = \underset{y}{\text{argmax}} \left\{ \varphi(y) - \frac{1}{2\rho_k} \|y-y^k\|^2 \right\} \qquad /13/$$

Similarly we can obtain a pair which is the minmax point in /6/

$$1^{o}\quad x^{k+1} = \operatorname{argmin}\left\{f_o(x) + \frac{1}{2\varrho_k}\|(\bar{y} + \varrho_k\, g(x))_+\|^2 - \frac{1}{2\varrho_k}\|\bar{y}\|^2\right\} \qquad /14/$$

$$2^{o}\quad y^{k+1} = (\bar{y} + \varrho_k\, g(x^{k+1}))_+ \qquad /15/$$

If $\bar{y} = 0$ then the function in brackets in /14/ is as a matter of fact an external penalty function.

In this case sequence of multipliers is obtained as a results of application of the reference point algorithm

$$y^{k+1} = \operatorname*{argmax}_{y}\left\{\varphi(y) - \frac{1}{2\varrho_k}\|y - \bar{y}\|^2\right\} \qquad /16/$$

and

$$y^k \longrightarrow y^{\infty} = \operatorname*{argmin}_{\hat{y} \in \hat{Y}} \|\hat{y} - \bar{y}\| \qquad /17/$$

where $\hat{Y}$ is the optimal solution set of the dual problem to (P).

Now, let us consider methods which follow from two sides regularization.

One can prove that a pair which is the minmax point in (7) can be derived by the following algorithm.

Regularized Penalty Function Method

Assuming given the parameters $0<\varrho_k = \eta_k \nearrow \infty$ and reference points $\bar{x} \in R^n$, $\bar{y} \in R^m$ the sequences $\{x^k\}$, $\{y^k\}$ are defined by

$$1^{o}\quad x^{k+1} = \operatorname*{argmin}_{x}\left\{L(x,\bar{y},\varrho_k) + \frac{1}{2\eta_k}\|x - \bar{x}\|^2\right\} \qquad /18/$$

$$2^{o}\quad y^{k+1} = (\bar{y} + \varrho_k\, g(x^{k+1}))_+ \qquad /19/$$

And similarily we can devire

Regularized Multiplier Method

Assuming given the parameters: $0<\varrho_k \nearrow \varrho_\infty \leq \infty$, $0<\eta_k \nearrow \infty$, $\sum_{k=0}^{\infty}(\varrho_k/\eta_k)^{1/2} < \infty$ and reference point $\bar{x} \in R^k$, and initial multiplier $y^o \in R^m$ we have

$$1^{o}\quad x^{k+1} = \operatorname*{argmin}_{x}\left\{L(x,y^k,\varrho_k) + \frac{1}{2\eta_k}\|x - \bar{x}\|^2\right\} \qquad /20/$$

$$2^{o}\quad y^{k+1} = (y^k + \varrho_k\, g(x^{k+1}))_+ \qquad /21/$$

For algorithm (18)-(19) the following properties hold true (compare with [1]):

$$y^k \longrightarrow y^\infty = \underset{\hat{y} \in \hat{Y}}{\operatorname{argmin}} \|\hat{y} - \bar{y}\| \ , \ x^k \longrightarrow x^\infty = \underset{\hat{x} \in \hat{X}}{\operatorname{argmin}} \|\hat{x} - \bar{x}\| \quad /22/$$

and for (20)-(21)

$$y^k \longrightarrow y^\infty \in \hat{Y} \text{ and } x^k \longrightarrow x^\infty = \underset{\hat{x} \in \hat{X}}{\operatorname{argmin}} \|\hat{x} - \bar{x}\| \quad /23/$$

The regularized multiplier method can be useful to solving problems with practical lack of solution uniqueness and with a set of optimal solutions containing more than one optimal point.

The problem of selection of one element which can be assumed the solution to the optimization problem remains very crucial. In general, solving an optimization problem one can not find all solutions. For instance, applying simplex method in case when a unique minimum solution does not exist one should derive all optimal basic solutions and then define their all convex combinations. The simplex algorithm is usually terminated when one of optimal solution is found. Often one does not even know properties of this solution. For problems lacking unique minimum Tikhonov [7] suggested to formulate an optimization problem in a slightly different way. He introduces an additional condition which make solution of an optimization problem unique and simultaneously enables the selection of an element with given properties from a set of optimal solutions. In many practical cases the conditions can be formulated as follows: select, from a set of optimal points a point which is closest relative to given a priori "reference point". Selection of the reference point depends on a model. For instance, in the optimal planning model [5] , [6] it could be a point which ensures balanced economic growth.

3. LINEAR APPROXIMATION AND REGULARIZATION

For convex programming the ordinary method of multipliers can be viewed as a gradient method for maximization of the dual function [2] which corresponds to the augmented Lagrangian

$$\Psi(y) = \inf_x \ L(x,y,\varrho) \quad /24/$$

In the paper we consider the possbility of application of a quasi-Newton method for maximization of Ψ . The first let us consider the problem

$$\text{minimize } f(x), \quad x \in X \subset R^n \quad /25/$$

where $f: R^n \longrightarrow R$ is continuously differentiable convex function having gradient Lipschitz-continuous with constant L

$$|\nabla f(x') - \nabla f(x'')| \leqslant L\|x' - x''\| \qquad /26/$$

Let us consider the following algorithm for minimization of f: Given an initial point x^o, the sequence of points $\{x^k\}$ is defined by

$$x^{k+1} = \underset{x \in X}{\operatorname{argmin}} \{ f(x^k) + < \nabla f(x^k), x - x^k > + \frac{1}{2\alpha_k} \|x - x^k\|^2_{H_k} \} \qquad /27/$$

In fact we use the linear approximation of the function f and add to f a strongly convex function, where H_k is symmetric positive defined matrix. When $X=R^n$ the formula (27) is equivalent to the foldowing step

$$x^{k+1} = x^k - \alpha_k H_k^{-1} \nabla f(x^k) \qquad /28/$$

For α_k satisfying

$$0 < \underline{\alpha} \leqslant \alpha_k < \bar{\alpha} < \frac{2\|\nabla f(x^k)\|^2}{L\| \nabla f(x^k)\|^2_{H_k^{-1}}} \qquad /29/$$

the sequence $\{x^k\}$ converges to $\hat{x} \in \hat{X}$. This is a generalization of the Golstein and Tretyakov [2] results. In [2] was assumed $H_k = I$. The procedure (28) - (29) is a quasi-Newton method when H_k is in some sense an approximation to $\nabla^2 f(\hat{x})$.

The dual function which corresponds to augmented Lagrangian is concave, continuous, differentiable and $\nabla\Psi$ has Lipschitz constant equal to ϱ [2].

We substitute the gradient multiplier uptade/11/ for the following quasi-Newton update:

$$y^{k+1} = y^k - \alpha_k H_k^{-1} \nabla\Psi(y^k) \qquad /30/$$

where H_k will be an approximation of $\nabla^2\Psi(y)$. Since

$$\nabla\Psi(y^k) = \frac{1}{\varrho}(y^k + \varrho g(x^{k+1}))_+ - \frac{1}{\varrho} y^k \qquad /31/$$

then

$$y^{k+1} = y^k - \frac{\alpha_k}{\varrho} H_k^{-1} \left[(y^k + \varrho g(x^{k+1}))_+ - y^k \right] \qquad /32/$$

where α_k satisfies

$$0 < \underline{\alpha} \leqslant \alpha_k \leqslant \bar{\alpha} < \frac{2\|\nabla\Psi(y^k)\|^2}{\varrho\| \nabla\Psi(y^k)\|^2_{H_k^{-1}}} \qquad /33/$$

Under suitable conditions on the problem (P) the function Ψ have two continuous derivatives and Hessian $\nabla^2\Psi(y)$ is negative definite in some neighboorhood of $\hat{y}$. For updating of H_k^{-1} we can use for example the BFGS formula. The H_{k+1}^{-1} depends on H_k^{-1} and on the difference in gradients

$$\gamma = \nabla\Psi(y^{k+1}) - \nabla\Psi y^k \qquad /34/$$

and on the change in variables

$$\delta = y^{k+1} - y^k \qquad /35/$$

The linear approximation and regularizarion with respect to x can be also applied to Lagrange function. This is consistent with the methods proposed in [3] and [8].

REFERENCES

[1] Antipin A.C., Method of regularization in convex programming. Economics and Mathematical Methods 11 (2)(1975) 336-342 (in Russian).

[2] Gol'shtein E.G., Tretyakov N.V., The gradient method of minimization and algorithms of convex programming based on modefied Lagrangian functions. Economics and Mathematical Methods 11 (4)(1975) 730-742 (in Russian).

[3] Powell M.J.D., Algorithms for nonlinear constraints that use Langrange functions. Mathematical Programming 14 (1978) 224-248.

[4] Rockafellar R.T., Monotone operators and the proximal point algorithm. SIAM J. Control Opt. 14 (1976) 877-898.

[5] Sosnowski J.S., Linear programming via augmented Lagrangians and conjugate gradient method, presented at International Conference on Methods of Mathematical Programming, Zakopane, Poland, 1977.

[6] Sosnowski J.S., Dynamic optimization of multisector linear production model. Systems Research Institute, Warszawa, Ph.D. Thesis 1978 /in Polish/.

[7] Tikhonov A.A., Arsenin V.Y., Methods of solution in correct problems. Nauka Moscow 1974.

[8] Wierzbicki A.P., A quadratic approximation method based on augmented Lagrangian functions for nonconvex nonlinear programming problems. IIASA WP-78-61, Laxenburg, Austria, 1978.

METHODS OF HIERARCHICAL OPTIMIZATION FOR INTERCONNECTED SYSTEMS

Piotr Tatjewski
Institute of Automatic Control
Technical University of Warsaw
00-665 Warszawa, Poland

1. Introduction

Hierarchical approach to optimization and control of complex, interconnected systems is nowadays a standard, justified technique, see, e.g., (Findeisen 1974, Findeisen et al. 1980, Singh 1977, Cohen 1978, Tatjewski 1979). The main subject of this paper is to present some recently obtained hierarchical optimization algorithms for such systems. First, the optimization problem will be described. After a short review of the main (classical) approaches to this problem the input prediction method will be introduced. Coordination algorithms of this method which is a representative of the prediction type methods based on augmented Lagrangians, are the main subject of this paper. Finally, another prediction method based on augmented Lagrangian will be briefly commented.

It is assumed that the system consists of N interconnected subsystems, each of them described by subsystem output mapping F_i, constraint set CU_i and performance index Q_i. Interconnections between subsystems are described by coupling matrices H_i composed of zeros and ones. The optimization problem is therefore as follows

$$\begin{cases} \text{find such } (\hat{c}, \hat{u}) \text{ that} \\ (\hat{c}, \hat{u}) = \arg\min \; \Psi(Q_1(c_1,u_1),\dots,Q_N(c_N,u_N)) \\ \text{subject to } \; y_i = F_i(c_i,u_i),\; (c_i,u_i) \in CU_i,\; u_i = H_i y,\; i=1,\dots,N, \end{cases} \quad (1)$$

where c_i, u_i, y_i are controls, inputs and outputs of subsystem i, and $c^T \triangleq (c_1^T, c_2^T, \dots, c_N^T)$, etc. Ψ is some function coupling local performance indices into the overall one - it is assumed to be a summation in the main part of the paper.

A great number of results concerning the hierarchical approach for solving (1) is now available. The review of the most important methods, with some precise theoretical results, comparisons and references, can be found in (Tatjewski 1979). One of the most widely known hierarchical optimization methods is the prediction method introduced by Takahara (1965), see also (Singh 1977). This method, called the classical prediction method (CMP) in the sequel, is a special case of prediction methods based on augmented Lagrangians, which were proposed, for various types of problems and in various versions, by Tatjewski (1976, 1977, 1979), Hakkala and Hirvonen (1977), Watanabe and Matsubara (1978). The primal-dual method of Wierzbicki (1976) is also of that class.

Some prediction methods can be based on normal Lagrangian with quite good results for a certain class of interconnected systems. The use of the augmented Lagrangian is a more general approach, thus resulting in methods applicable to wider class of interconnected systems. The main reasons are threefold. First, methods using the normal Lagrangian base on the existence of a saddle-point of it. However, this saddle-point generally does not exist for nonconvex problems - duality gaps can arise; but it usually exists when the augmented Lagrangian is used instead of the normal one . Second, even if the saddle point of the normal Lagrangian exists, price method or classical prediction method can be also not applicable due to the fact that the lower-level solutions can be nonunique. It is easily seen on simple example problem: minimize $(c_1+c_2^2+u_2)$ subject to $y_1 = 2c_1$, $u_2 = y_1$, $0 \leq c_1 \leq 1$, $0 \leq c_2 \leq 1$, $0 \leq u_2 \leq 1$ (the analysis is left to the reader). This drawback can also be eliminated by the use of the augmented Lagrangian, as it is in the case of the above example. And, third, the use of the augmented Lagrangian enables to derive new, efficient coordination algorithms.

Due to the lack of space we concentrate on some aspects of the input prediction method only, some extensions will be briefly mentioned in the conclusions.

2. The input prediction method

The augmented Lagrangian for problem (1) can be formulated as follows

$$L_a(c,u,\lambda,\varrho)=\sum_{i=1}^{N} Q_i(c_i,u_i) + \sum_{i=1}^{N} \langle \lambda_i, u_i - H_i F(c,u) \rangle + \sum_{i=1}^{N} \tfrac{1}{2}\varrho \| u_i - H_i F(c,u) \|^2 \qquad (2)$$

where $F^T \triangleq (F_1^T, F_2^T, \ldots, F_N^T)$ is the whole system output mapping. Rearranging the terms in (7) it can be easily seen that

$$L_a(c,u,\lambda,\varrho) = \sum_{i=1}^{N} L_{ai}(c_i,u,\lambda,\varrho) \qquad (3)$$

Thus the local problem i (LP_i) is

$$\begin{cases} \text{for given } u, \lambda \text{ and } \varrho \text{ find control} \\ \hat{c}_i(u,\lambda,\varrho) = \arg\min\limits_{C_i(u)} L_{ai}(\cdot,u,\lambda,\varrho) \\ \text{where } C_i(u) \triangleq \{ c_i: \ (c_i,u_i) \in CU_i \}, \quad i=1,\ldots,N, \end{cases} \tag{4}$$

The <u>coordinator function</u> has the form

$$\hat{L}_a(u,\lambda,\varrho) = \sum_{i=1}^{N} L_{ai}(\hat{c}_i(u,\lambda,\varrho),\ u,\lambda,\varrho) \tag{5}$$

The task of the coordinator is to find the following saddle-point $(\hat{u};\hat{\lambda})$ of $\hat{L}_a(\cdot,\cdot,\varrho)$

$$\hat{L}_a(\hat{u},\lambda,\varrho) \leq \hat{L}_a(\hat{u},\hat{\lambda},\varrho) \leq \hat{L}_a(u,\hat{\lambda},\varrho), \quad u \in U, \tag{6}$$

where $\quad U \triangleq \{ u \quad : \ \forall\ i=1,\ldots,N \quad C_i(u) \neq \emptyset \ \}.$ (7)

The thorough theoretical analysis of the existence conditions of saddle-point (6) such that $(\hat{c}(\hat{u},\hat{\lambda},\varrho),\hat{u})$ is a solution to initial optimization problem (1) is not the aim of this paper, and can be made using quite well developped techniques of general augmented Lagrangians theory, as presented e.g. in brillant paper of Rockafellar (1974).

Our attention will be focused on coordination strategies, i.e. algorithms finding saddle point $(\hat{u};\hat{\lambda})$. We will look for it by finding points satisfying its necessary optimality conditions. They can be easily formulated for the case when local constraint sets CU_i are separable, i.e., $CU_i = C_i \times U_i$, $i = 1,\ldots,N$, since then the derivatives of coordinator function (6) are

$$(\hat{L}_a)'_{\lambda_i}(u,\lambda,\varrho) = (L_a)'_{\lambda_i}(\hat{c}_i(u,\lambda,\varrho),\ u,\lambda,\varrho) \tag{8}$$

$$(\hat{L}_a)'_{u_i}(u,\lambda,\varrho) = (L_a)'_{u_i}(c_i(u,\lambda,\varrho),\ u,\lambda,\varrho),\ i=1,\ldots,N. \tag{9}$$

Assuming now, for simplicity, that set $U \triangleq \times_{i=1}^{N} U_i$ is not active at the solution we have the following necessary conditions for the saddle-point

$$(L_a)'^{T}_{\lambda_i}(\hat{c}_i(u,\lambda,\varrho),\ u,\lambda,\varrho) = 0 \tag{10}$$

$$(L_a)'^{T}_{u_i}(\hat{c}_i(u,\lambda,\varrho),\ u,\lambda,\varrho) = 0 \quad i=1,\ldots,N, \tag{11}$$

where superscript T denotes transposition. Assuming additionally, for simplicity of presentation, that local constraints sets CU_i are not present at all we can characterize points $\hat{c}_i(u,\lambda,\varrho)$ as those satisfying equations

$$(L_a)'^{T}_{c_i}(c_i,u,\lambda,\varrho) = 0, \quad i = 1,\ldots,N. \tag{12}$$

Taking into account the form of augmented Lagrangian (2) and using global (overall) notation $F^T \triangleq (F_1^T,F_2^T,\ldots,F_N^T)$, $H^T \triangleq H_1^T H_2^T \ldots H_N^T$,

$Q \triangleq \sum_{i=1}^{N} Q_i$ we get the compact, overall system form of the equations (10), (11) and (12)

$$u - HF(c,u) = 0 \tag{13}$$

$$Q_u'^T(c,u) + [I - HF_u'(c,u)]^T(\lambda + \varrho(u - HF(c,u))) = 0 \tag{14}$$

$$Q_c'^T(c,u) - [HF_c'(c,u)]^T(\lambda + \varrho(u - HF(c,u))) = 0 \tag{15}$$

The <u>coordination strategies</u> (coordinator algorithms) are as follows: after solving N local problems (4) (i.e., satisfying equation (15)) u and λ are modified using some step formulae derived on the basis of equations (13) and (14); then the local problems are solved with modified values of u and λ, etc; until the equations (13) and (14) are satisfied.

Note that equations (13), (14), (15) are, precisely, the necessary optimality conditions for initial optimization problem (1) when $\varrho = 0$ is set (with sets CU_i omitted for simplicity). The whole discussion made in this section can be viewed upon as an indication how to use these necessary optimality conditions - how to modify them (to use the augmented Lagrangian) and how to treat the obtained relations.

Let us assume that after step k (after iteration k) of the coordinator algorithm the points $c^k \triangleq \hat{c}(u^k, \lambda^k, \varrho)$, u^k and λ^k have been obtained. Then iteration k+1 consists in finding points c^{k+1}, u^{k+1}, λ^{k+1} satisfying

$$g(c^{k+1}, u^{k+1}, \lambda^k) = 0 \tag{16}$$

$$h(c^{k+1}, u^k, \lambda^{k+1}) = 0 \tag{17}$$

$$Q_c'^T(c^{k+1}, u^k) - [HF_c'(c^{k+1}, u^k)]^T(\lambda^k + \varrho(u^k - HF(c^{k+1}, u^k))) = 0 \tag{18}$$

where operators g and h define, basing on (13) and (14), the way u^k and λ^k are modified. E.g., they can be as follows:

$$u^{k+1} - HF(c^{k+1}, u^k) = 0 \tag{19}$$

$$Q_u'^T(c^{k+1}, u^k) + \lambda^{k+1} - [HF_u'(c^{k+1}, u^k)]^T \lambda^k + [I - HF_u'(c^{k+1}, u^k)]^T \varrho(u^k - HF(c^{k+1}, u^k)) = 0 \tag{20}$$

Algorithm (19), (20) will be called the generalized <u>Takahara algorithm</u> (THA), since it was proposed by Takahara (1965), for a dynamic problem without constraints (using normal Lagrangian, i.e., $\varrho = 0$). Classical prediction method, as introduced in the previous section, used only this algorithm as coordination strategy. Note that in this algorithm gradient w.r.t. λ is used to obtain equation for u^{k+1}, and gradient w.r.t. u is used to obtain equation for λ^{k+1} - moreover, Eqs. (19), (20) are equivalent to

$$u^{k+1} = u^k - (L_a)_\lambda'^T (c^{k+1}, u^k, \lambda^k) \quad (21)$$

$$\lambda^{k+1} = \lambda^k - (L_a)_u'^T (c^{k+1}, u^k, \lambda^k). \quad (22)$$

Eq. (13) does not depend on multiplier λ, therefore it was difficult to derive an efficient formula for λ^{k+1} from it, when $\varrho = 0$ was set. But when the augmented Lagrangian is used we can adjust multipliers according to the known, efficient Hestenes-Powell multiplier rule $\lambda^{k+1} = \lambda^k + \varrho(u^k - HF(c^{k+1}, u^k))$. Deriving from (14) equation for u^{k+1} the following <u>multiplier algorithm</u> (MA) can be proposed

$$Q_u'^T(c^{k+1}, u^k) + [I - HF_u'(c^{k+1}, u^k)]^T(\lambda^k - \varrho HF(c^{k+1}, u^k)) + \\ + \varrho(u^{k+1} - [HF_u'(c^{k+1}, u^k)]^T u^k) = 0 \quad (23)$$

$$\lambda^{k+1} = \lambda^k + \varrho(u^k - HF(c^{k+1}, u^k)) \quad (24)$$

It is easy to show that Eqs. (23), (24) are equivalent to

$$u^{k+1} = u^k - \frac{1}{\varrho}(L_a)_u'^T (c^{k+1}, u^k, \lambda^k) \quad (25)$$

$$\lambda^{k+1} = \lambda^k + \varrho(L_a)_\lambda'^T (c^{k+1}, u^k, \lambda^k) \quad (26)$$

The multiplier algorithm is therefore a gradient-search algorithm with specifically chosen step coefficients. Note that this algorithm is defined only for $\varrho > 0$.

The THA and MA algorithms can be treated as some iteration processes of the form

$$G(c^{k+1}, u^{k+1}, \lambda^{k+1}, c^k, u^k, \lambda^k) = 0, \quad k=0,1,2,\ldots (27)$$

where eqs. (16), (17), (18) define the mapping G. Then, under appropriate assumptions (Ortega and Rheinboldt 1970), the root-convergence factor σ of the iteration process (27) is

$$\sigma \triangleq sr\left[-G_1'(\hat{x}, \hat{x})^{-1} G_2'(\hat{x}, \hat{x})\right] < 1, \quad (28)$$

where $sr[\cdot]$ denotes the spectral radius and $x^T \triangleq (c^T, u^T, \lambda^T)$. Let us denote the operator $-G_1'(\hat{x}, \hat{x})^{-1} G_2'(\hat{x}, \hat{x})$ for THA and MA algorithms as Γ_{THA} and Γ_{MA}, respectively; and the root-convergence factor as σ_{THA} and σ_{MA}. It is possible to evaluate general formulae for Γ_{THA} and Γ_{MA}; they are, however, very complicated and practically do not allow general conclusions about σ_{THA} and σ_{MA}. Therefore, we took a simple (linear quadratic) example problem and evaluated $\sigma_{THA}(\varrho)$ and $\sigma_{MA}(\varrho)$, for various values of parameters of this problem (i.e., coefficients in the performance function and output equations), ranging ϱ from 0 to $+\infty$. The results were as follows:

- for each set of the chosen problem coefficients there was some range of values of ϱ such that $\sigma_{MA}(\varrho) < 1$,
- for almost the half of chosen sets of the problem coefficients there were no values of ϱ with $\sigma_{THA}(\varrho) < 1$.

Two typical cases are shown in Fig. 1 and Fig. 2.

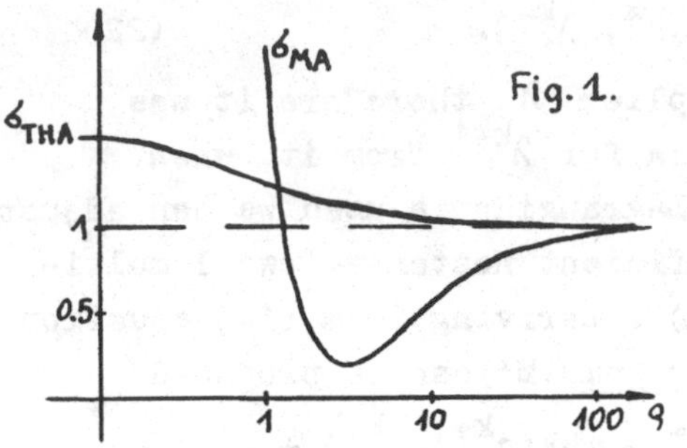

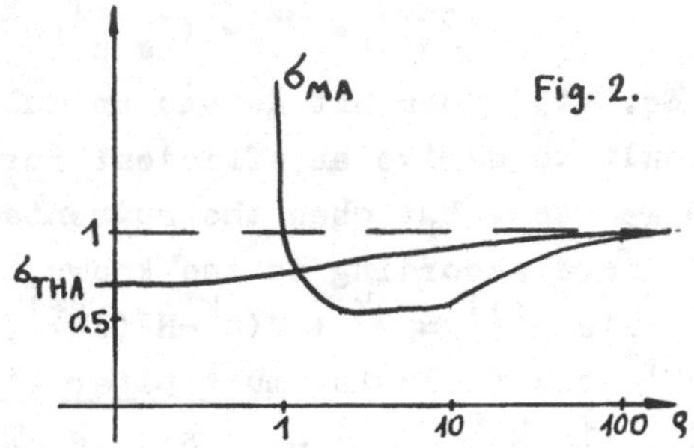

The conclusions obtained for one simple example problem (however, for various values of its coefficients) cannot be simply generalized to all cases, of course. But the obtained results seem to bring to light some fundamental features of both algorithms.

Algorithms THA and MA are the simplest, but not the only possible for derivation from Eqs. (13) and (14). E.g., we can modify THA using instead of (20) the equation

$$Q_u'^T(c^{k+1},u^k) - [I - FH_u'(c^{k+1},u^k)]^T(\lambda^{k+1} + \varrho(u^k - HF(c^{k+1},u^k))) = 0 \qquad (29)$$

or modify MA using instead of (23) the equation

$$Q_u'^T(c^{k+1},u^k) + [I - HF_u'(c^{k+1},u^k)]^T(\lambda^k + \varrho(u^{k+1} - HF(c^{k+1},u^k))) = 0. \qquad (30)$$

We have found that such modifications do not change significantly basic properties of both algorithms, only slightly different convergence factors can be generally achieved.

The way of modifying u and λ as discussed in this section, i.e., using some step formulae (16), (17), is not the only one possible. We could, e.g., modify inputs u minimizing function $\hat{L}_a(\cdot,\lambda,\varrho)$ for each value of λ, and using some step formula for adjusting of λ only. Such algorithms were used by Tatjewski (1976), Watanabe and Matsubara (1978). We found, however, that they are less effective (Tatjewski and Michalak 1980).

3. Some related methods

It is easily seen that the interconnection inputs u and interconnection outputs y are two equivalent ways of describing the system interconnections (u=Hy, and $y = H^{-1}u$, H is always invertible). Using inputs u we write the output and structure equations as $u_i - H_iF(c,u) = 0$, $i=1,\dots,N$, as it was the case in the input prediction method (IPM). They can be used, however, in the equivalent form $y_i - F_i(c_i,H_iy) = 0$, $i=1,\dots,N$. The prediction method using these equations is called the output prediction method (OPM), to make a distinction. IPM and OPM are equivalent as far as function Ψ coupling local performance indices Q_i, see (1), is a summation. For problems with nonadditive functions Ψ IPM cannot be directly applied, and OPM can be applied if only Ψ is strictly order-preserving. The detailed description of OPM for such cases can be found in (Tatjewski 1979, Findeisen et al. 1980).

When local constraints $(c_i,u_i) \in CU_i$ are binding simultaneously the controls c_i and interactions u_i (are not only on controls $c_i \in C_i$, or are not separable - $CU_i = C_i \times U_i$) then the well known price method (see, e.g., Findeisen et al. 1980) is often computationally superior to the classical prediction method, if only both are applicable to a given problem. However, when using the augmented Lagrangian we must realize that it is not separable with respect to interactions u_i, due to the existance of the cross-terms $\langle u_j, H_{ji}F_i(c_i,u_i)\rangle$. This does not matter for prediction methods where only controls c_i are local variables, but destroys the separability needed for the price method where both controls c_i and interactions u_i are local decision variables. Nevertheless, generalization of the price method for the use of the augmented Lagrangian through some approximation of its cross-terms is possible, see.,e.g.,(Findeisen et al.1980).

4. Conclusions

Brief review of the main reasons for using the augmented Lagrangians in hierarchical optimization was presented at the beginning of the paper. Then the input prediction method based on augmented Lagrangian (2) was presented. Various coordination algorithms of the method were introduced in a unified manner as some iteration processes of the type (27). This way of presentation, a rather general one (Looze and Sandell 1979), seems to be well suited to analize various coordination algorithms of prediction methods. The Takahara algorithm (Takahara 1965) was generalized to the augmented Lagrangian case, and the multiplier algorithm (based on Hestenes-Powell multiplier rule) was presented. These two basic algorithms were then, for the first time, compared to each other - showing that the multiplier algorithm is much more universally applicable. Some new coordination algorithms,being versions of the two presented above, were also introduced. Finally, the output prediction method, closely related to the input prediction method and beeing superior to it in some cases was briefly commented. Since the prediction methods happen to be some of the most efficient, further research towards deeper understanding of various coordination algorithms and relations between them seems advisable.

The author would like to express his gratitude to Prof.W.Findeisen and to the colleagues from his Hierarchical Control Group in Warsaw, for encouragement and valuable discussions.

References

Cohen, G. (1978). Optimization by decomposition and coordination: a unified approach. IEEE Trans.Autom.Control, 23, 222-232.

Findeisen, W. (1974). Multilevel Control Systems (in Polish).PWN, Warszawa. (German translation: Hierarchische Stenerungssysteme. Verlag Technik, Berlin 1977).

Findeisen. W.,F.N. Bailey,M.Brdyś, K.Malinowski, P.Tatjewski, and A.Woźniak (1980). Control and Coordination in Hierarchical Systems. J.Wiley, London, to be published.

Hakkala, L., and J. Hirvonen (1977). Gradient-based dynamical coordination strategies for interaction prediction method. Report B-38, Helsinki University of Technology.

Looze, D.P., and N.J.Sandell,Jr. (1979). A decomposition theory of hierarchical control. Manuscript.

Ortega, J.M., and W.C. Rheinboldt (1970). Iterative Solution of Nonlinear Equations in Several Variables. Academic Press, New York.

Rockafellar, R.T. (1974). Augmented Lagrange multiplier functions and dualility in nonconvex programming. SIAM J.Control, 12, 268-285.

Singh,M.G. (1977) Dynamical Hierarchical Control. North Holland, Amsterdam.

Takahara,Y. (1965). A multi-level structure for a class of dynamical optimization problems. M.S. Thesis, Case Western Reserve University, Cleveland, Ohio.

Tatjewski,P. (1976). Properties of multilevel dual optimization methods. Ph.D.Thesis, Technical University of Warsaw, Warsaw (in Polish).

Tatjewski,P. (1977). Dual methods of multilevel optimization. Bull. Acad.Pol.Sci.Ser.Sci.Tech., 25, 247-254.

Tatjewski,P. (1979). Multilevel optimization techniques.In: Second Workshop on Hierarchical Control. Institute of Automatic Control, Technical University of Warsaw, 241-266.

Tatjewski,P., and P.Michalak (1980). Algorithms of prediction methods in multilevel optimization. Systems Science, No 1, to be published.

Watanabe,N., and M.Matsubara (1978). An infeasible method of large-scale optimization by direct coordination of subsystem inputs. J.Optimiz.Theory & Appl., 24, 437-448.

Wierzbicki,A. (1976). A primal-dual large-scale optimization method based on augmented Lagrange Functions and iteraction shift prediction. Ricerche di Automatica, 7, 35-59.

STRUCTURAL ANALYSIS OF LARGE NONLINEAR PROGRAMMING PROBLEMS

Eugeniusz Toczyłowski
Institute of Automatic Control
Technical University of Warsaw

Special purpose strategies which take most of structural properties of the nonlinear programming problems are considered and structural algorithms with emphasis on the regular output set assignment are discussed.

1. Strategies.

Let us consider the large-scale equality constrained problem

minimize $f(y)$

subject to

$$y \in R^n, \quad h_i(y) = 0 \quad i = 1,\dots,m \tag{1}$$

where $f\colon R^n \longrightarrow R$, $h = (h_1,\dots,h_m)\colon R^n \longrightarrow R^m$.
Then restate the problem (1) by partitioning the variables into two sets $x \in R^m$ and $u \in R^{n-m}$:

minimize $f(x,u)$

subject to

$$h(x,u) = 0$$
$$x \in R^m, \quad u \in R^{n-m} \tag{2}$$

We will discuss how to take advantage of the problem's structure in optimization algorithms. The following essential four strategies are possible:

Strategy 1. Fast algorithms for constrained optimization calculations can be obtained by applying Newton's or variable metric methods for constrained optimization [3]. The major advantage of that algorithms is that they can be easily modified to structured problems. In [1] it was shown that the elimination of the linearized equality constraints

$$h_x(y^k)\,\delta x + h_u(y^k)\,\delta u + h(y^k) = 0 \tag{3}$$

with respect to δx at each step of the optimization algorithm reduces storage requirements and permits to take computational advantage of the sparsity and structure of the Jacobian matrix $h_y(y^k)$. The goal of structural analysis in case of that strategy is to find the best partition of the variables y into sets of dependent variables x and independent u in such a way that the square matrix $h_x(y^k)$ is nonsingular, reasonably sparse and has a structure convenient for efficient use by existing codes for solving large sparse systems of equations.

Strategy 2. The technique which has been quite succesful for solving large scale constrained and highly nonlinear programming problems is based on a computational elimination of equality constraints and a considerable number of variables. Under appropriate assumptions, all equality constraints can be satisfied at each iteration by solving $h(x,u) = 0$ with respect to x for given u. This results in the following transformation of the general optimization problem

$$\begin{array}{lcl} \min\limits_{(x,u)} f(x,u) & & \min\limits_{u} f(x(u),u) \\ h(x,u) = 0 & \Longrightarrow & g(x(u),u) \leqslant 0 \\ g(x,u) \leqslant 0 & & \end{array} \qquad (4)$$

Since the equations $h(x,u) = 0$ may be reasonably sparse and structured, the structural techniques can be used to simplify the computational requirements of iterative methods and can assure regularity assumptions. In Quasi-Newton methods, for instance, the structural analysis can minimize the computational complexity of the iterative process. Additionally, the cost of numerical evaluation of the derivative $\frac{dx}{du}$ can be reduced.

Stregegy 3. Nonlinear equality constraints may not be solved at each step of an optimization algorithm but they can be transformed by structural enalysis to the most appropriate form.
This results in the general transformation:

$$\begin{array}{lcl} \min\limits_{(x,u)} f(x,u) & & \min\limits_{(v,u)} f(x(v,u),u) \\ h(x,u) = 0 & \Longrightarrow & h(x(v,u),u) = 0 \\ g(x,u) \leqslant 0 & & g(x(v,u),u) \leqslant 0 \end{array} \qquad (5)$$

where v is the reduced set of dependent variables. Structural techniques used here can reduce the number of dependent variables v and can improve sparsity or measures of linearity of the constraints (Lipschitz constants of Jacobian matrix).

Strategy 4. Many methods of finding the solution of (1) are based on seeking a saddle point $(\hat{y},\hat{\lambda})$ of the augmented Lagrangean function

$$\Lambda(y,\lambda) = f(y) + \lambda^T h(y) + \frac{1}{2}\varrho \| h(y) \|^2 \qquad (6)$$

We will show how to replace the saddle-point seeking by unconstrained minimization.

Assume the existence of the restated problem (2) with such partitioning $y=(x,u)$, that λ can be calculated from the first order Lagrangean conditions with respect to x, i.e from the equation

$$f_x(x,u) + \lambda^T h_x(x,u) = 0 \qquad (7)$$

Let us substitute λ^T in (6) by

$$\lambda^T(y) = f_x(y)\, h_x^{-1}(y) \qquad (8)$$

As the result, the following unconstrained problem

$$\underset{y \in R^n}{\text{minimize}} \quad Q(y) = f(y) + \lambda^T(y)\, h(y) + \frac{1}{2}\varrho \| h(y) \|^2 \qquad (9)$$

can be used to solve the constrained problem (1). This proposal is justified by the following theorem.

Theorem.

If there exists a local minimizing point $\hat{y}$ of the constrained problem (1) which satisfies standard second-order sufficiency conditions for an isolated local minimum, i.e.

(i) f, h_i $i=1,\dots,m$ are twice Lipschitz continuously differentiable in a neighborhood of $\hat{y}$

(ii) $\nabla h_i(\hat{y})$ $i=1,\dots,m$ are linearly independent

(iii) there exists unique Lagrange multiplier vector $\hat{\lambda} \in R^m$ such that $\nabla L(\hat{y},\hat{\lambda}) = 0$ and $\delta y^T \nabla^2 L(\hat{y},\hat{\lambda}) \delta y > 0$ for all $\delta y \in R^n$ with $\delta y \neq 0$, $\nabla h(\hat{y}) \delta y = 0$ (where ∇L, $\nabla^2 L$ denote the partial derivative and Hessian matrix with respect to y of $L(y,\lambda) = f(y) + \lambda^T h(y)$),

then there exists a partition $y = (x,u)$ such that for sufficiently large ϱ $\hat{y}$ is a local minimizing point of the unconstrained problem (9).

Proof.

From (i) and (ii) there exists a partition $y=(x,u)$, $x \in R^m$, $u \in R^{n-m}$, such that $h_x(\hat{y})$ is nonsingular. Thus for the equation

$$\partial_x f(y) + \lambda^T \cdot h_x(y) = 0$$

the assumptions of the implicite function theorem are satisfied in the point $(\hat{y},\hat{\lambda})$. Hence is appears that

$$\lambda(\hat{y}+\delta y) = \hat{\lambda} + \lambda_y \cdot \delta y + o(\|\delta y\|),$$

where λ_y denotes the derivative $\frac{d\lambda}{dy}(\hat{y})$.

Let Q be defined as in (9), then after simple computations

$$Q(\hat{y}+\delta y) = Q(\hat{y}) + Q_y(\hat{y})\cdot\delta y + \frac{1}{2}\delta y^T Q_{yy}(\hat{y})\cdot\delta y + o\|\delta y\|^2)$$

where

$$Q(\hat{y}) = L(\hat{y},\hat{\lambda})$$

$$Q_y(\hat{y}) = \nabla L(\hat{y},\hat{\lambda})$$

$$Q_{yy}(\hat{y}) = \nabla^2 L(\hat{y},\hat{\lambda}) + 2\lambda_y^T \nabla h(\hat{y}) + \varrho \nabla h(\hat{y})^T \cdot \nabla h(\hat{y})$$

Now we will use the following lemma which may be easily proved.

Lemma 1. Let $H \in \mathcal{L}(R^n, R^m)$ be onto and $A \in \mathcal{L}(R^n)$ be symmetric. If for all $y \in \ker H$, $y \neq 0$ $y^T A y > 0$, then for every $T \in \mathcal{L}(R^m, R^n)$ there exists ϱ_0 such that for all $\varrho \geq \varrho_0$ matrix $B = A + TH + \varrho H^T H$ is positively defined. From lemma 1, after putting $H = \nabla h(\hat{y})$ and $T = 2\lambda_y^T$ it follows that for sufficiently large ϱ $Q_{yy}(\hat{y})$ is positively defined.

Since $Q_y(\hat{y}) = 0$, $\hat{y}$ is a local minimizing point of the unconstrained problem (9), Q.E.D.

Relevance of the last strategy to large-scale programming is hampered by the fact that penelty function (6) tends to destroy sparsity.

2. Structural analysis

The structural analysis of large nonlinear programming problems may be divided into the following steps:

(i) partitioning of y into x and u in such a way that the existence of the implicite function x(u) is provided.

(ii) output set assignment, i.e. assignment to every separate equation a distinct dependent variable which can be calculated from that equation

(iii) finding a block triangular structure, i.e. partitioning the set of equations into smaller subsets that may be solved independently in a proper sequence [2], [4].

(iv) splitting the irreducible subsets by independent variables adjacent with these subsets [8].

(v) reordering of the equations within each partitioned subsystem by theoring [4] or partial teoring procedure [5].

Since almost all steps of structural analysis have been described and discussed elsewhere (see for instance [4], [6], [7]), in next section we will present only an algorithm for a regular output set assignment, i.e. the algorithm that provides a regular partitioning of the variables in conjunction with an output set assignment.

3. A regular output set assignment algorithm.

Let $H = [h_{ij}] \in \mathcal{L}(R^n, R^m)$ be the Jacobian matrix of the constraint function $h: R^n \longrightarrow R^m$ in a given point y^o such, that $h(y^o) = 0$. Matrix H provides the information which can be applied in the general output set assignment algorithm [7] in order to comply with the regularity assumption. The algorithm will follow after some definitions and comments.

Let $B \in \mathcal{L}(R^n)$ be a given permatation matrix. Then the linear system

$$Hy = 0 \tag{10}$$

is equivalent to

$$Gv = 0 \tag{11}$$

where $G = HB^T$ and $v = Bx$. Thus B permutes variables.

Definition. The digraph $(\mathcal{V}, E,)$ of the system (11) consists of a set of n vertices $\mathcal{V} = \{1, \ldots, n\}$ and a set of arcs E, where $(j,i) \in E$ if and only if $g_{ij} \neq 0$, where $[g_{ij}] = G$.

Definition. Given a permutation matrix B, the labeled digraph of (10) is the digraph of the system (11), where $G = H \cdot B^T$;

A regular output set assigument is equivalent to finding a permutation matrix B, for which the labeled digraph of the system (11) has loops $(i,i) \in E$ and moreover m x m matrix

$$S = \begin{bmatrix} g_{11} & \cdots\cdots & g_{1m} \\ \\ g_{m1} & \cdots\cdots & g_{mm} \end{bmatrix}$$

is nonsingular.

The algorithm is based on a sequence of simple path shiftings of the labeled graphs of (10) defined as follows.

Assume, that in the labeled graph of (10) there exists a simple path R from vertex i_1 to i_k, i.e. a sequence of distinct edges $(i_1,i_2), (i_2,i_3), \ldots, (i_{k-1},i_k) \in R \subset E$.

The shifting of the path R is performing such variable permutations that for every $(i,j) \in R$ the variable v_i becomes the output variable of the j-th equation (additionally i_k-th variable becomes the output variable of the i_1-th equation). Thus the shifting of the path R can be defined by the following n x n permutation matrix $D = [d_{ij}]$

$$d_{ji} = 1 \qquad (i,j) \in R \qquad \text{or } i = i_k,\ j = i_1$$
$$d_{ii} = 1 \qquad i \in R$$
$$d_{ij} = 0 \qquad \text{otherwise}$$

The shifting implements the following repermutation

$$B^T \longrightarrow B^T \circ D^T$$

<u>The algorithm.</u>

Step 1° (Initialization) $k := 0$, $S_k = I$, $B_k = I$

$$\mathcal{V}_1^k = \emptyset,\ \mathcal{V}_2^k = \{1,\dots,m\},\ \mathcal{V}_o^k := \{m+1,\dots,n\}\ .\ \text{Go to Step } 2^o$$

Step 2° (finding a simple path)

If $\mathcal{V}_2^k = \emptyset$, then Stop (a regular output set was found)

Otherwise, for given B_k find in the labeled graph of (10) a simple path R from $j \in \mathcal{V}_o^k \cup \mathcal{V}_2^k$ to $r \in \mathcal{V}_2^k$ with intermediate vertices belonging to $\mathcal{V}_1^k$ and additionally with $\eta_{jr} \neq 0$, where $\eta_j = S_k^{-1} g_j$ and g_j is the j-th column of $G_k = H \cdot B_k^T$.

If such j and r do not exist , a regular output set does not exist (Stop). Otherwise go to Step 3°

Step 3^o(simple path shifting). Perform the shifting of the path R i.e. recompute $B_{k+1} = B_k \circ D^T$, and recompute S_{k+1}^{-1} from

$$S_{k+1}^{-1} = T_{k+1}^{-1} S_k^{-1} \quad \text{where}$$
$$T_{k+1} = I + [0,\dots,\ \eta_j - e_r,\dots,\ 0]$$

(inverse may be kept in PFI form)

$$\mathcal{V}_1^{k+1} := \mathcal{V}_1^k \cup \{r\} \quad \mathcal{V}_2^{k+1} := \mathcal{V}_2^k \setminus \{r\}.\quad k := k+1,\ \text{Go to Step } 2^o$$

Remarks. Condition $\eta_{jr} \neq 0$ in Step 2 preserves the regularity of the output set . The efficiency of this algorithm seems to be comparable with the efficiency of the simplex algorithm with a sparse matrix technique (PFI).

References

[1] Berna T.J, Locke M.H., Westerberg A.W "A new approach to optimization of Chemical Processes" Techn. Rep, Carnegie-Mellon University (1979)

[2] Harrary P. Num. Math. (1962) p. 128-135.

[3] Powell M.J.D - these Proceedings.

[4] Steward D.W. SIAM J. Num. Anal. p. 345 (1965)

[5] Toczyłowski E. in IFAC Workshop on System Analysis, Bielsko B., published by Pergamon Press (1978)

[6] Toczyłowski E "SCS Conference" July 24-28, 1978 Newport Beach, California

[7] Toczyłowski E. - Proceedings of IMACS Congress on Simulation of Systems, North Holland (1979)

[8] Toczyłowski E. Techn. Rep.(1977) in Polish.

ON THE USE OF STATISTICAL MODELS OF MULTIMODAL FUNCTIONS FOR THE CONSTRUCTION OF THE OPTIMIZATION ALGORITHMS

A. Žilinskas
Institute of Mathematics and Cybernetics
Academy of Sciences of the Lithuanian SSR
Vilnius, USSR

Introduction. Optimization algorithms are usually constructed on the base of a model of an objective function. Quadratic models are useful in the local optimization theory [1] . However, in order to construct multimodal algorithms the model must be adequate to more uncertain behaviour of the function. Rather general assumptions on the complicated objective function f(x), $x \in A \subset R^n$, may be formalized as the axioms describing comparative probabilities of the possible values of f(x). The family of random variables Y_x, $x \in A$, with densities $p_x(\cdot)$ represents these axioms as shown in [2] . Further characterization of the statistical model is considered as a problem of extrapolation under uncertainty.

The choice of x for a point of current evaluation of the objective function by the optimization algorithm may be interpreted as a choice between $p_x(\cdot)$ on the base of the accepted statistical model. The global strategy of the search for the minimum is characterized by the axioms on rationality of the choice; it is shown that such a strategy may be reduced to an optimal procedure in respect of the statistical model. The problems of termination of the minimization based on global strategy and transition to precise evaluation of the main local minima are considered. The results of the minimization of some test functions are presented.

The Bayesian approach for the optimization theory is suggested in [3, 4] . The present paper developes it basing on the ideas close to that known as neoBayesian [5] .

Construction of the statistical model. If the evaluation (calculation) of the value of the objective function is very cheep (consumes only a little of computer time) and n is not large, the global minimum may often be found by a simple combination of the deterministic or stochastic uniform grid with a local algorithm. Such a simple method, however requires a great number of the objective function evaluations [6] and therefore it is out of use for the minimization of practical functions whose evaluation usually requires much computer time. It is intuitively obvious that the global strategy based on the non-uniform grids which are more compact in the neighbouhoods of the best points found may be more efficient than that based on uniform grids. To construct rationally such non-uniform grids the model of the objective function is necessary.

Let an objective information on $f(\cdot)$ be only $x_i, f(x_i)$, where $x_i \in A$, $i=\overline{1,k}$. Besides, we have a subjective information (for example, the experience of the solution of similar problems in the past) about the complexity and multimodality of $f(\cdot)$. Let us consider some natural assumptions on the uncertainty of the value $f(x)$, $x \in A$, $x \neq x_i$, $i=\overline{1,k}$. The minimal a priory information on $f(\cdot)$ in consideration seems sufficient to compare any two intervals of possible values of $f(x)$ according to their probabilities. We suppose that a priory information on $f(\cdot)$ generates CP - a binary relation of comparative probability for the intervals of the possible values of $f(x)$; i.e. for any two intervals I_1, I_2 we may state either $f(x) \in I_1$ is no less probable than $f(x) \in I_2$ or $f(x) \in I_2$ is no less probable than $f(x) \in I_1$. Rather natural assumptions on CP are formulated axiomatically and it is shown in [2] that there exists a unique probability density $p_x(\cdot)$ compatible with the axioms. The results of a psychological experiment show that most important of these axioms are intuitively acceptable. Therefore the unknown value $f(x)$ may be interpreted as a random variable Y_x with density $p_x(\cdot)$ and the family of random variables Y_x, $x \in A$ may be taken for a statistical model of $f(x)$. Further characterization of CP with the aim of obtaining a statistical model corresponding to a stochastic function is given in [7]; the axioms of [7], however, are more complicated and not so natural as those of [2]. The study of the structure of CP leads to the proof of the existence and uniqueness of $p_x(\cdot)$ but to construct an optimization algorithm it is necessary to have the constructive form of $p_x(\cdot)$. The discussion in [2] shows that $p_x(\cdot)$ may be considered Gaussian. Therefore it is necessary to define only the mean value and variance of Y_x.

The expected (most likely, representative) value the function at the point x, $m_k(x,(x_i,y_i),i=\overline{1,k})$, corresponds to the mean of Y_x and the conditional (in respect of the known values $y_i=f(x_i)$) uncertainty of $f(x)$, $s_k(x,(x_i,y_i),i=\overline{1,k})$, may be taken for a variance of Y_x. Then it is the problem of extrapolation under uncertainty. The most common way of considering such a problem appears to be the axiomatic characterization of an extrapolator $m_k(x,(x_i,y_i),i=\overline{1,k})$ and the expected deviation $f(x)$ from $m_k(x,(x_i,y_i),i=\overline{1,k})$ (i.e. conditional uncertainty of extrapolation) $s_k(x,(x_i,y_i),i=\overline{1,k})$. Let us postulate [8]: 1) the independence of the extrapolation from the scale factor and the zero point of the scale of y_i, $i=\overline{1,k}$; 2) its independence of the numeration of (x_i,y_i), $i=\overline{1,k}$; 3) the equality $m_k(x_i,(x_j,y_j),j=\overline{1,k})=y_i$, $i=\overline{1,k}$; 4) the invariance of the extrapolator in respect of an aggregation of the data (x_i,y_i), $i=\overline{1,k}$. The unique extrapolator satisfying these axioms is

$$m_k(x,(x_i,yi),i=\overline{1,k})=\sum_{i=1}^{k} y_i w_i(x,x_j,j=\overline{1,k}), \qquad (1)$$

where the weights have some natural properties [8]. The conditional uncertainty of extrapolation which may be characterized by similar axioms is $s_k(x,(x_i,y_i),i=\overline{1,k})= \|x-x_i\| w_i(x,x_j,j=\overline{1,h})$ [9] where $\|\cdot\|$ denotes the Euclidean distance in R^n. In the case $n=1, x_i \leqslant x < x_{i+1}$, the weights $w_i(x,x_j,j=\overline{1,k})=(x_{i+1}-x)/(x_{i+1}-x_i)$, $w_{i+1}=1-w_i(x,x_j,j=\overline{1,k})$ are very natural. The experimental investigation shows that for $n \geqslant 2$ the appropriate expression of the weights is $w_i(x,x_j,j=1,k)= d(x,x_i)/\sum_{j=1}^{k} d(x,x_j)$, $d(x,x_i)=\exp(-c\|x-x_i\|^2)/\|x-x_i\|$, $c>0$ [10].

The discussion of the suggested axioms and comparison with those of [11, 12] is given in [10] . Some additional axioms characterize a special case of (1) corresponding to the conditional mean of a Gaussian random function [8]. It is well known that conditional mean is an optimal mean-root-square error extrapolator with respect to the Gaussian random function chosen as a statistical model of the function considered. The last result establishes the relation between the axiomatic approach to the extrapolation problem and that based on the use of random functions as statistical models. In fact it shows that the extrapolation is rather a part of the construction of a statistical model than the usual primary choice of the statistical model and then its use for the construction of an extrapolator.

Summarising these results we may regard the family of Gausian random variables Y_x with means $m_k(x,(x_i,y_i),i=\overline{1,k})$ and variances $s_k(x,(x_i,y_i), i=\overline{1,k})$ as a statistical model of an objective function. Such a model is not only well grounded theoretically but also it is simpler from computational point of view than Gaussian random functions [10] usually used as statistical models of complicated functions under uncertainty.

<u>Construction of the algorithm.</u> Using the statistical model the results obtained by a minimization procedure may be interpreted and the result of the current evaluation of an objective function may be forecast. But even in such a situation the definition of a rational optimization algorithm is not trivial.

If a random function is chosen as a statistical model of an objective function then the Bayesian algorithms [3, 4] (i.e. the algorithms with a minimal mean error) seem most well-grounded. But these algorithms are defined by a system of multidimensional Bellman equations [3, 4] and therefore their realization is difficult. To avoid the difficulties of consideration of the consequences of the current optimization step to the final decision, only optimal one-step algorithms are considered in [13, 14, 15, 16]. But the rationality of such algorithms needs some grounding.

The choice of x for the point of the current evaluation of $f(\cdot)$ by the minimization algorithm may be interpreted as a choice between $p_x(\cdot)$ on the base of the accepted statistical model. From rather general assumptions on the rationality of the choice follows the existence of value function $u(\cdot)$ such that the choice of $p(\cdot)$ is preferable to the choice of $g(\cdot)$ if and only if $\int_{-\infty}^{\infty} u(t)p(t)dt \geq \int_{-\infty}^{\infty} u(t)g(t)dt$ [17]. In order to construct the value function corresponding to the conception of rational search for the global minimum let us characterise preferences ($\succ$) between $p_x(\cdot)$ i.e. between the vectors of their parameters $(m_k(x,(x_i,y_i),i=\overline{1,k}),\ s_k(x,(x_i,y_i),i=\overline{1,k})$.

A1) For arbitrary $m^1 < m^2$, $s^1 > 0$ there exists s such that $(m^1,s^1) \succ (m^2,s^2)$ if $s^2 \leq s$. This axiom states that to choose the point for the current observation at which the expected value of the function is comparatively large may be rational only in the case of great uncertainty.

A2) For arbitrary $m^1, s^1, m^2 \geq y_{ok} = \min_{1 \leq i \leq k} y_i$, the relation $(m^1,s^1) \succ (m^2,0)$ holds, i.e. to choose the point at which the function value is certainly larger than the minimal value found is not rational.

A3) $(m^1,s^1) \succ (m^2,s^2)$ if and only if $(m^1,ks^1) \succ (m^2,ks^2)$, $k>0$, i.e. the preference relation is invariant with respect to the scale of the uncertainty.

A4) The value function is non-negative and piecewise constant. This assumption seems rather strong but it is not restrictive because every value function may be approximated by such a function with desirable accuracy.

It may be shown that the unique (up to the factor) function satisfying A1-A4 is $u(t)=I(z_{ok}-t)$, where $z_{ok} < y_{ok}$ and $I(\cdot)$ is the unit-step function. Therefore the utility of the choice of x for the k+1-th evaluation is proportional to the probability $v_{k+1}(x)=P(Y_x < z_{ok})= =G((z_{ok}-m_k(x,(x_i,y_i),i=\overline{1,k}))/(s_k(x,(x_i,y_i),i=\overline{1,k})^{1/2})$ where $G(\cdot)$ is the Gaussian distribution function.

Algorithm corresponding to these conceptions evaluates the objective function at the k+1-th minimization step at the point of $\max_{x\in A} v_{k+1}(x)$. Since the maximum point depends on rather arbitrary choice of the weights $w_i(\cdot)$ and the level z_{ok} it seems reasonable to maximize $v_{k+1}(x)$ only with rough accuracy orienting ourselves towards excluding the evaluations of $f(\cdot)$ at the points of small utility. Such a strategy distributes the evaluation points over the whole set A but it does that more compactly in the neighbourhoods of the best points found. Therefore $m_k(x,(x_i,y_i),i=\overline{1,k})$ becomes rather a good approximation of $f(\cdot)$ at these subsets and the local minima of $m_k(\cdot)$ may be used for the approximation of minima of $f(\cdot)$. In a one-dimensional case the local minima of $m_k(\cdot)$ may be evaluated very simply [18, 19] but in case $n \geq 2$ the iterative minimization procedure is necessary. Since a more exact definition of a one-dimensional local minimum requires only several evaluations of $f(\cdot)$ it is reasonable to compute all the minima found by means of the local algorithm [18, 19]. In a multidimensional case the interactive decision which minima ought to be defined more exactly seems to be very useful. The alternative automatic choice includes two best minima found: $f(x_{o1}) \leq f(x_{o2})$, and those which differ from $f(x_{o2})$ no more than E%. The termination condition of the one-dimensional algorithm is described in [18, 19]. The multidimensional algorithm terminates the global search if the number of evaluated local minima exceeds the given value L. The global search terminates also if the maximal allowable number of function evaluations N is exhausted or the criterion of the evaluation points compactness in the neighbourhood of the best point found exceeds the compactness anywhere no less than M times.

<u>Results of experimental testing.</u> The results of testing of the one-dimensional algorithm [18] and its earlier version [19] show that it is more efficient than other algorithms of similar destination in the sense of required number of the objective function evaluations [18, 20]. Analogous conclusions are given in [21].

Many multidimensional algorithms are tested using the functions given in [21]. These functions were minimized by the suggested algorithm as well. The test functions are [21]:

1. Shekel's family (three functions with m=5,7,10):

$$f_1(x)=-\sum_{i=1}^{m} 1/((x-a_i)(x-a_i)^t+c_i),$$

$$x=(x_1,\dots,x_n),\ a_i=(a_{i1},\dots,a_{in}),$$

$$0\leqslant x_j\leqslant 10,\ j=\overline{1,n},\ n=4.$$

i	1	2	3	4	5	6	7	8	9	10
a_i^t	4.	1.	8.	6.	3.	2.	5.	8.	6.	7.
	4.	1.	8.	6.	7.	9.	5.	1.	2.	3.6
	4.	1.	8.	6.	3.	2.	3.	8.	6.	7.
	4.	1.	8.	6.	7.	9.	3.	1.	2.	3.6
c_i	.1	.2	.2	.4	.4	.6	.3	.7	.5	.5

2. Hartman's family (two functions with n=3 and n=6):

$$f_2(x)=-\sum_{i=1}^{m} c_i \exp(-\sum_{j=1}^{n} a_{ij}(x_j-p_{ij})^2),\ m=4,\ x=(x_1,\dots,x_n),$$

$$p_i=(p_{i1},\dots,p_{in}),\ a_i=(a_{i1},\dots,a_{in}),\ 0\leqslant x_j\leqslant 1,\ j=\overline{1,n}.$$

n	i	a_i						p_i						c_i
3	1	3.	10.	30.				.3689	.1170	.2673				1.
	2	.1	10.	35.				.4699	.4387	.7474				1.2
	3	3.	10.	30.				.1091	.8732	.5547				3.
	4	.1	10.	35.				.0381	.5743	.8828				3.2
6	1	10.	3.	17.	3.5	1.7	8.	.1312	.1696	.5569	.0124	.8283	.5886	1.
	2	.05	10.	17.	.1	8.	14.	.2329	.4135	.8307	.3736	.1004	.9991	1.2
	3	3.	3.5	1.7	10.	17.	8.	.2348	.1451	.3522	.2883	.3047	.6650	3.
	4	17.	8.	.05	10.	.1	14.	.4047	.8828	.8732	.5743	.1091	.0381	3.2

3. Branin:

$$f_3(x_1,x_2)=a(x_2-bx_1^2+cx_1-d)^2+e(1-f)\cos x_1+e,$$

$a=1$, $b=5.1/(4\pi^2)$, $c=5/\pi$, $d=6$, $e=10$, $f=1/(8\pi)$, $-5\leq x_1\leq 10$, $0\leq x_2\leq 15$.

4. Goldstein and Price:

$$f_4(x_1,x_2)=(1+(x_1+x_2+1)^2(19-14x_1+3x_1^2-14x_2+6x_1x_2+3x_2^2))(30+ (2x_1-3x_2)^2(18-32x_1+12x_1^2+48x_2-36x_1x_2+27x_2^2)),\ -2\leq x_1\leq 2,\ -2\leq x_2\leq 2.$$

The algorithm was coded in FORTRAN and the computer BESM-6 was used. The results of minimization are given in the table. The termination of global search is defined by $N=500$, $M=3$ and L given in the table. The number of an objective function evaluations by global strategy is denoted by N_g. The minima for more exact evaluations were chosen automatically ($\bar{E}=\max(e_1,e_2)$, $e_i=|f(x_{oi})-f(\bar{x}_{oi})|\,/\,|f(\bar{x}_{oi})|\cdot 100(\%)$, $\bar{x}_{oi}$ is the point found by the local algorithm from the initial point x_{oi}, $i=1,2$), and local algorithm [22] which is a modification of [23] was used. The local search is terminated if the norm of gradient is less than 10^{-4} or the function decrease in two iterations is less than 10^{-4} or the number of iterations reaches $\max(4,1.5n)$. The number of function evaluations is denoted by N and the computer processor time does by T. For comparison of T with the results obtained by means of other computers let us note that the time of a thousand evaluations of $f_1(x)$, $n=4$, $m=5$, $x_j=4.$, $j=\overline{1,4}$, is equal to 2.0 sec. In the table the rough evaluations of three local minima (if $L>3$ then three minima found at first) from the results of global search are presented as well as their exact values (if they were chosen for exact evaluation with the help of the local algorithm).

Conclusions. The information on a multimodal function is formalized as a binary relation of comparative probability (CP) between intervals of possible values of the function. Some naturals assumptions on CP imply that the family of random variables represents such an information. The problem of further characterization of the statistical model is considered as a problem of extrapolation under uncertainty. The algorithm of minimization is characterized by axioms of rational choice. The comparison of the results of minimization of some test functions with those of [21] show that this algorithm is rather efficient in the sense of the required number of the objective function evaluations.

Acknowledgements. The author greatly appreciates Mrs. J. Valevičienė's help while programming and testing this algorithm and Mr. V. Tiešis' assistance with the local optimization algorithm.

Table.

		i	Rough evaluations of local minima 1	2	3	Precise evaluations of local minima 1	2	3	Global minimum	L N_g N T
f_1	m=5	f	-0.797	-0.434	-0.999	-2.615	-0.609	-5.100	-10.15	5
		x	3.030	6.639	8.999	2.973	7.610	7.998	4.000	246
			6.864	7.955	8.522	7.015	6.760	7.999	4.000	950
			2.352	6.781	7.913	2.993	7.511	8.001	4.000	245.0
			6.182	6.711	8.102	7.035	7.627	8.001	4.000	
	m=7	f	-0.922	-1.197	-0.893	-2.685	-3.588	-5.522	-9.910	5
		x	3.030	5.464	4.989	2.954	4.925	3.923	3.923	276
			6.864	5.028	5.431	7.064	5.015	4.004	4.038	1017
			2.352	2.680	4.039	2.969	3.071	4.296	4.017	319.5
			6.182	3.754	4.164	7.079	2.949	3.978	4.050	
	m=10	f	-1.148	-0.964	-0.485	-5.322	-2.797	-0.892	-10.53	10
		x	5.444	3.030	1.918	3.743	2.968	3.067	3.999	336
			4.117	6.864	4.828	4.138	6.993	4.274	3.998	2224
			3.819	2.352	4.369	4.125	2.980	4.395	4.000	340.5
			3.908	6.182	6.279	3.916	7.005	4.890	3.996	
f_2	n=3	f	-3.702	-3.763	-3.742	-3.862	-3.863	-3.853	-3.863	5
		x	0.6433	0.1669	0.5220	0.1354	0.1134	0.2319	0.1134	212
			0.5505	0.5583	0.5418	0.5586	0.5557	0.5574	0.5557	363
			0.8424	0.8841	0.8372	0.8528	0.8525	0.8485	0.8525	197.3
	n=6	f	-2.615	-1.307	-2.528	-3.322		-3.322	-3.322	5
		x	0.1368	0.6500	0.3323	0.2017		0.2017	0.2017	262
			0.3140	1.0000	0.2680	0.1500		0.1500	0.1500	627
			0.4908	0.4346	0.3639	0.4768		0.4768	0.4768	321.7
			0.2465	0.4684	0.2741	0.2753		0.2753	0.2753	
			0.2814	0.2072	0.3665	0.3117		0.3116	0.3117	
			0.7809	0.1226	0.7681	0.6573		0.6573	0.6573	
f_3	f		0.9342	7.178	7.296	0.3979	0.3979	0.3979	0.3979	3
	x		9.608	3.400	-3.392	9.425	3.142	-3.142	3.142	86
			3.247	2.093	13.06	2.475	2.275	12.275	2.275	164
										54.7
f_4	f		4.653	32.45	89.50	3.000	3.295		3.000	3
	x		-0.035	-0.604	0.874	0.000	0.029		0,000	84
			-0.948	-0.352	-0.420	-1.000	-0.977		-1.000	165
										50.9

References

1. Numerical Methods for Constrained Optimization, eds. P.E. Gill and W.Murrey, Academic Press, 1974.

2. Žilinskas A. On Statistical Models for Multimodal Optimization. Math.Operat.Stat. ser. Statistics, Vol.9, No 2 (255-266), 1978.

3. Mockus J.B. On Bayesion Methods of Seeking the Extremum, (in Russian). Avtomatika i Vyczislitelnaja Technika, No 3, (53-62), 1972.

4. Mockus J. On Bayesian Methods of Seeking the Extremum and their Applications. In Information processing 77, ed. B.Gilchrist. North-Holland, (195-200), 1977.

5. Savage L. The Foundations of Statistics Reconsidered. In Risk and Uncertainty, eds. K.Borch and J.Mossin, Mc Millan, NY, (174-188), 1968.

6. Mc Cormick G.P. Attempts to calculate Global Solutions of Problems that may have Local Minima. In Numerical Methods for Non-linear Optimization, ed. F.A.Lootsma, Academic Press, (209-222), 1972.

7. Žilinskas A., Katkauskajte A., Construction of Models of Complex Functions with Uncertainty, (in Russian) Proceedings of the 7-th All-union Conference on Coding and Information Communication. Part I. Moskow-Vilnus, (70-74), 1978.

8. Žilinskas A., Axiomatic Approach to Extrapolation in Uncertainty, (in Russian), Avtomatika i Telemechanika (in press).

9. Žilinskas A., On Axiomatic Characterisation of Statistical Models of Multimodal Functions, (in Russian), in Proceedings of Seminar on Applications of Random Searching (in press).

10. Žilinskas A., An Analysis of Multidimensional Extrapolation in Uncertainty, (in Russian), Teoria Optimalnych Reshenij, No 4, Vilnus, (27-53), 1978.

11. Fine T. Extrapolation when Very Little is known about the Source. Inform. Contr. Vol.16, (331-359), 1970.

12. Goldman J., An Approach to Estimation and Extrapolation with Possible Applications in an Incomplitely Specified Environment. Inform.Contr., Vol.30, (203-223), 1976.

13. Kushner H., A New Method of Locating the Maximum Point of an Arbitrary Multipeak Curve in the Presence of Noise. Trans. ASME, ser.D., J.Basic Eng., Vol.86, No 1, (97-105), 1964.

14. Shatjanis V., On one Method of Multimodel Optimization, in Russian , Avtomatika i Vyczislitelnaja Technika, No 3, (53-62), 1971.

15. Žilinskas A., The One Step Bayesian Method for Searching an Extremum of a Function of One Variable, (in Russian), Kibernetika, No 1, (139-144), 1975.

16. Strongin R., Numerical Methods for Multimodal Problems, (in Russian), Nauka, 1978.

17. Fishburn P.C., Theory for Decision Making, J.Wiley, New York,1970.

18. Žilinskas A., Two Algorithms for One-dimensional Multimodal minimization, (to appear).

19. Žilinskas A., Optimization of One-dimensional Multimodal Functions, Algorithms AS 133. Applied Statistics, Vol.27 , (367-375), 1978.

20. Žilinskas A., On One-dimensional Multimodal Minimization. In Trans of Eighth Prague Conf. on Inform. Theory, Stat.Dec.Func.,Rand. Proc., Vol.B, (392-402), 1978.

21. Dixon L.C.W., Szego G., The Global Optimization Problem: An Introduction. In Towards Global Optimization 2, eds.L.C.W.Dixon, G.P. Szego, North-Holland, (1-15), 1978.

22. Teshic V., Variable Metrics Method for Local Minimization in Presence of Bounds. Vyczislitelnaja Technika, Kaunas, (11-114), 1975.

23. Biggs M.C. Minimization Algorithm Making Use of Nonquadratic Properties of the Objective Function, J.Inst.Math. and Appl., Vol.8, No 3, (315-327), 1971.

STABILITY ANALYSIS IN PURE AND MIXED-INTEGER LINEAR PROGRAMMING

B. Bank

Sektion Mathematik

Humboldt-Universität zu BERLIN

Unter den Linden 6

DDR - 1086 Berlin

Abstract: We examine the stability behaviour of pure and mixed-integer linear programs for the case that the coefficients of the objective function and the right-hand sides of the constraints change their value. The stability of the program will be defined by using continuity properties of the extremal value, the constraints-set-mapping and the optimal set. The considerations are independent of compactness requirements imposed on the feasible set. It is, however, necessary to assume rationality of the constraints matrix.

1. Introduction

Similar as it is made in continuous nonlinear programming we may use the concept of point-to-set maps in order to examine the stability of integer programs. MEYER did so in order to prove general properties of integer programs. RADKE gave a first stability analysis for mixed-integer programs under compactness requierements. The aim of our contribution is to present stability results for pure and mixed integer linear programms for the case that the coefficients of the objective function and the right-hand sides of the constraints change their value only. The results are independent of compactness requierements imposed on the constraint set. The statements are given without proofs, they are available in BANK and BANK et al.

We are concerned with the program

$P_s(\mu,\lambda)$: $\max\{ f(\mu,z) \mid z \in G_s(\lambda)\}$

where

(1) $f(\mu,z) = \mu^T z$

(2) $G_s(\lambda) = \{z \in R^n \,/\, Az = \lambda,\ z \geq 0,\ z_1,\dots,z_s \in Z\}$,

(3) $\mu \in R^n$, $\lambda \in R^m$ arbitrary parameter vectors,

(4) Z the set of all integers,

(5) s ($0 \leq s \leq n$) number of integer variables,

(6) A-(m x n) matrix of rank m.

In order that our results hold we have the general assumption:

(G) A is comprised by rational data. Dropping out the assumption will violate most of the results. By $\mathfrak{A}_s$ we denote the set of all pairs (μ,λ) such that $P_s(\mu,\lambda)$ has a solution, which is called the <u>set of solvability</u>. The <u>feasible parameter set</u> is denoted by

$$\mathcal{L}_s = \{ \lambda \in R^m \,/\, G_s(\lambda) \neq \emptyset \}.$$

The maps related to the stability statements are

(i) the <u>constraints-set-map</u> $G_s : \mathcal{L}_s \longrightarrow 2^{R^n}$

with $G_s(\lambda)$ defined by (2) ,

(ii) the <u>extremal value</u> $\varphi_s : \mathfrak{A}_s \longrightarrow R$

with $\varphi_s(\mu,\lambda) = \max\limits_{z \in G_s(\lambda)} f(\mu, z)$,

(iii) the <u>optimal set</u> $\psi_s : \mathfrak{A}_s \longrightarrow 2^{R^n}$

with $\psi_s(\mu,\lambda) = \{ z \in G_s(\lambda) \,/\, \varphi_s(\mu,\lambda) = f(\mu, z) \}$.

The problem $P_s(\mu,\lambda)$ is said to be <u>stable</u> at $(\mu^o,\lambda^o) \in \mathfrak{A}$ if

1° G_s is continuous at λ^o

2° φ_s is continuous at (μ^o, λ^o),

3° ψ_s is upper semicontinuous at (μ^o, λ^o).

We call a point-to-set map $\Omega : \Lambda \longrightarrow 2^{R^n}$ continuous at a point λ^o if Ω is both lower semicontinuous and upper semicontinuous at λ^o, where we use lower semicontinuity in the sense of BERGE, and we take upper semicontinuity in the sense of HAUSDORFF (defined by DOLECKI).

A subset $W \subset \mathfrak{A}_s$ is called a <u>stability-set</u> of $P_s(\mu,\lambda)$ if the restrictions of G_s, φ_s and ψ_s to W fulfil the requirements 1°, 2° and 3° for any point $(\mu,\lambda) \in W$.

2. Stability-Results

Theorem 1

The set of solvability $\mathfrak{A}_s$ is non empty, closed and

$$\mathcal{O}\!\mathcal{L}_s = \mathcal{U}^* \times \mathcal{L}_s$$

holds for any s $(o \leq s \leq n)$, where $\mathcal{U}^*$ is the dual of the recession cone $\mathcal{U}$ of the polyhedron $G_o(\lambda)$.

Theorem 2

The set of solvability $\mathcal{O}\!\mathcal{L}_s$ is connected if both $o \leq s < n$ and $\dim \{ d \in R^m / d = \mathcal{B} y,\ y \geq o \} = m$ hold, where y is the vector of continuous variables and $\mathcal{B}$ is the corresponding submatrix of A.

Theorem 3

(i) Let $s = n$. The set of solvability $\mathcal{O}\!\mathcal{L}_n$ may be divided into a countable number of maximal stability sets

$$V(\bar{\lambda}) = \mathcal{U}^* \times \{\bar{\lambda}\} \quad , \quad \bar{\lambda} \in \mathcal{L}_n .$$

Each stability-set $V(\bar{\lambda})$ is a non-empty polyhedral cone with a vertex at $(O, \bar{\lambda}) \in R^n \times R^m$.

(ii) The restriction of the extremal value φ_n to $V(\bar{\lambda})$ is convex and piecewise linear over $V(\bar{\lambda})$. Each stability set $V(\bar{\lambda})$ has a partition into a finite number of polyhedral cones P_v such that

(1) $\dim P_v = \dim V(\bar{\lambda})$

(2) φ_n is linear on P_v

hold for any cone P_v.

Theorem 4

(i) Let $o < s < n$. There exists a partition of $\mathcal{O}\!\mathcal{L}_s$ into a countable number of starshaped stability-sets

$$W(\bar{\lambda}) = \mathcal{U}^* \times Q(\bar{\lambda}), \quad \bar{\lambda} \in \mathcal{L}_s .$$

The sets $Q(\bar{\lambda})$ are in general the largest subsets of $\mathcal{L}_s$ where the continuity of G_s holds. $W(\bar{\lambda})$ is in general neither open nor closed.

(ii) Moreover the extremal value φ_s is a upper semicontinuous function on $\mathcal{O}_s$.

(iii) The restriction of φ_s to $V(\bar{\lambda}) = \mathcal{U}^* \times \{\bar{\lambda}\}$ is convex and piecewise linear for each fixed $\bar{\lambda} \in \mathcal{L}_s$. For the linear behaviour of φ_s over subsets of $V(\bar{\lambda})$ the same as in theorem 3 holds.

The statements (ii) of Theorem 3 and (iii) of Theorem 4 are classical and were proven by NOLTEMEIER. Part (ii) of Theorem 3 was proven by MEYER also under an integer boundness condition.

3. References

B.BANK, Qualitative Stabilitätsuntersuchungen rein- und gemischt-ganzzahliger linearer parametrischer Optimierungsprobleme Seminarbericht Nr. 6, Humboldt-Universität zu Berlin, 1978.

B.BANK, J.GUDDAT, D.KLATTE, B.KUMMER, K.TAMMER, Nichtlineare parametrische Optimierung, Seminarbericht Nr. , Humboldt-Universität zu Berlin, 1980 (to appear).

C. BERGE, Espaces topologiques.Fonctions multivoques Dunod Paris,1959.

S. DOLECKI, Constraints stability and moduli of semicontinuiuy. 2nd IFAC Symp. on Distributed Parameter Systems, Warwick 1977 (Preprint)

R.R. MEYER, Integer and mixed integer programming models: General properties, IOTA, Vol.16, (1975), No. 3/4

H. NOLTEMEIER, Sensitivitätsanalyse bei diskreten linearen Optimierungsproblemen, Lecture Notes in Operations Research and Mathematical Systems No. 30, Springer Verlag, Berlin-Heidelberg-New York (1970)

M. RADKE, Sensitivity Analysis in Discrete Optimization. Western Management Science Institute. University of California, Los Angeles, Working Paper 240, 1975

ALTERNATIVE GROUP RELAXATION OF INTEGER PROGRAMMING PROBLEMS

P. Bertolazzi - C. Leporelli - M. Lucertini
Istituto of Automatica and CSSCCA of C.N.R.
Via Eudossiana 18 - 00184 - ROMA (ITALY)

ABSTRACT

The classical group approach to integer linear programming problems (IP) can be generalized in order to obtain group minimization problems with different computational load and different relaxation.

The aim of this work is to analyze some group problems, associated to the same (IP), both from the point of view of the relaxation of the (IP) and of the complexity of the group solution algorithm; evaluation criteria for these group problems are pointed out.

1. INTRODUCTION

The group theoretical approach for solving integer programming problems (IP), from a computational point of view often appears unsuitable, because of the large size of the abelian group.

In fact, the complexity of algorithms that solve the group problem (GP) generally increases with the size of the group (shortest path algorithms, knapsack algorithms [1,2,3,4,5,8,9]).

It should be noticed that the solution of GP may not only be useful in finding the optimal solution of IP (GP is generally a relaxation of IP), but can also be useful as a lower bound in a branch and bound procedure.

It is therefore important to find methods for a more efficient solution of GP.

This paper presents some procedures(exact and approximated) to reduce the computational load in solving GP , by reducing the size of the group. A set of problems obtained from GP is defined. They are in general a relaxation of GP easier to solve then GP.,giving a bound better than the LP one.

As it's well known, in the classical approach [10], the size of the group depends on the value of the determinant of the chosen dual feasible LP basis matrix (usually the optimal one).

In order to reduce the size of the group many procedures have been proposed [10], all of them based essentially on the reduction of the determinant of the basis matrix. For example [6] presents number theoretic procedures based on dividing for a suitable constant all the coefficients of a constraint; this procedure reduces the value of the determinant and thus generally the size of the group. On the other hand, Jeroslow [7] has shown some generalizations of the usual group problem construction, that suggest a different way of reducing the size of the group.Given an IP problem, let $\{A\}$ indicate the group generated by the columns of the matrix A of equality constraints of IP.The corresponding GPH problem is the minimum cost problem on the finite quotient group $(\{A\}:\{H\})$, where H is an $(m\times t)$ matrix of full rank m whose columns are elements of $\{A\}$ (and $\{H\}$is the subgroup of $\{A\}$generated by H).Of course the size of the group depends on how the matrix H spans A. The classical group approach gives H=B,where B is the optimal LP basis.

The GPH problem can be written as a set of congruence constraints. Generally these congruence constraints are expressed utilizing the Smith Normal Form of H. As stated earlier the GPH problem is in general a relaxation of the IP. For example if H is a submatrix of A the relaxation corresponds to dropping the nonnegativity constraints on the variables associated to the columns of H.

A simple way of defining a group {H} that strictly contains {B}, is to take H as [B ¦ h], where h is a nonbasic column of A such that $h \notin \{B\}$. In this paper the properties of the problem deriving from this definition of H, dropping the nonnegativity constraint on x_h, are firstly analyzed (section 2). In section 3 an approach based on reintroducing the dropped constraint by a Lagrangean method is presented.

2. A REDUCED GROUP PROBLEM

The problem IP can be written in the following form

$$\text{(IP)} \quad \begin{aligned} &\min\ (z-z^{LP}) = c_h x_h + c_N x_N \\ &\Delta' x_B + h x_h + N x_N = b \\ &C' x_B, x_h, x_N \geq 0, \text{ integers} \end{aligned}$$

where Δ' is the (diagonal) optimal LP basic matrix expressed in Smith normal form ($\Delta' = R'BC'$), [h ⋮N] is the corresponding non basic matrix, $c_h \geq 0$, $c_N \geq 0$ are the optimal LP costs, z^{LP} is the optimal value of the LP associated to IP, R' and C' are integer unimodular (m×m), matrices and $C'x_B$ are the original variables of the problem.

The classical group problem GP, formulated with respect to B, expressed in Smith Normal Form [10] is:

$$\text{(GP)} \quad \begin{aligned} &\min\ z_G = c_h x_h + c_N x_N \\ &[h \mid N] \begin{bmatrix} x_h \\ x_N \end{bmatrix} \equiv b \pmod{\delta'} \\ &x_h, x_N \geq 0, \text{ integers} \end{aligned}$$

where $\equiv$ indicates a congruence relation, $\delta'_i = \Delta'_{ii}$ and $\prod_i \delta'_i = |\det B| = \det \Delta'$. Let (z^*_G, x^*_h, x^*_N) be the optimal solution of GP. The reduced group problem GPH, formulated with respect to $H = [\Delta' \mid h]$, expressed in Smith Normal Form is:

$$\text{(GPH)} \quad \begin{aligned} &\min\ z_H = \tilde{d} + \tilde{c}_h x'_h + \tilde{c}_N x_N \\ &R N x_N \equiv R b \pmod{\delta} \\ &x_N \geq 0, \text{ integer} \\ &x'_h \text{ integer} \end{aligned}$$

where R and C are integer unimodular matrices and (see [11]):

$$\begin{bmatrix} \delta_1 & & 0 & \vdots & 0 \\ & \ddots & & \vdots & \vdots \\ 0 & & \delta_m & & 0 \end{bmatrix} = \begin{bmatrix} \Delta & \vdots & 0 \end{bmatrix} = R \begin{bmatrix} \Delta' & \vdots & h \end{bmatrix} C$$

(2.1) $\det \Delta = \prod_i \delta_i =$ G.C.D. {determinants of order m minors of H} = G.C.D.$\{(\det \Delta'), (\det \Delta')h_1/\delta'_1, \ldots, (\det \Delta')h_m/\delta'_m\}$.

$$\begin{bmatrix} x_B \\ x_h \end{bmatrix} = \begin{bmatrix} C_{11} & C_{12} \\ C_{21} & C_{22} \end{bmatrix} \begin{bmatrix} x'_B \\ x'_h \end{bmatrix}$$

$$\tilde{d} = c_h C_{21} \Delta^{-1} Rb$$

$$\tilde{c}_h = c_h C_{22} ,$$

$$\tilde{c}_N = c_N - c_h C_{21} \Delta^{-1} RN$$

LEMMA 1. If $\det C = 1$ then $C_{22} = \frac{\det \Delta'}{\det \Delta}$ and $1 \leq C_{22} \leq \delta'_m$.

PROOF. This result follows directly from (2.1) and from the expression of the determinant of a partitioned matrix.◀

The optimal solution of GPH is in general unbounded. In fact if $\tilde{c}_h \neq 0$ then, as x'_h is only constrained to be integer, $z^*_H = -\infty$; on the other hand if some entries of $\tilde{c}_N$ are negative also $z^*_H = -\infty$.
However, supposing that $\tilde{c}_N \geq 0$, it is possible to give an approximate algorithm to solve GP as follows:

ALGORITHM 1.

1. Solve GPH with $x'_h = 0$; let $\bar{x}_N$ be the solution
2. Given $\bar{x}_N$, find $\bar{x}'_h$ such that the constraint

(2.2) $$x_h = C_{21}x'_B + C_{22}x'_h = C_{21}\Delta^{-1}R(b - N\bar{x}_N) + C_{22}x'_h \geq 0$$

is satisfied; the value of x'_h , that minimize z_H is:

(2.3) $$\bar{x}'_h = \lceil \Gamma(\bar{x}_N) \rceil, \quad \Gamma(\bar{x}_N) = \frac{C_{21}\Delta^{-1}R(N\bar{x}_N - b)}{C_{22}}$$

and then

$$\bar{x}_h = C_{22}(\lceil \Gamma(\bar{x}_N) \rceil - \Gamma(\bar{x}_N))$$ ◀

This algorithm gives a feasible solution of GP $(\bar{x}_h, \bar{x}_N)$ with objective function $\bar{z}_G = c_h\bar{x}_h + c_N\bar{x}_N = \tilde{d} + \tilde{c}_h\bar{x}'_h + \tilde{c}_N\bar{x}_N$.

Remark that the following non linear group problem is equivalent to GP:

$$\min z_N = \tilde{d} + \tilde{c}_h \lceil \Gamma(x_N) \rceil + \tilde{c}_N x_N$$

$$RNx_N \equiv Rb \pmod{\delta}$$

$$x_N \geq 0, \text{ integer}$$

3. A LAGRANGEAN APPROACH TO SOLVE GP

3.1. The constraint $x_h \geq 0$ (see 2.2) can be reintroduced in the objective function of GPH using Lagrangean technique ; in this way a tighter relaxation of GP is obtained.

The new problem (GPL) can be written:

$$\text{(GPL)} \quad L(\lambda) = \min z_L = z_H + \lambda\left[C_{22}x_h' - C_{21}\Delta^{-1}R(Nx_N - b)\right]$$

$$RNx_N \equiv Rb \pmod{\delta}$$

$$x_N \geq 0, \text{ integer}$$

where $\lambda \geq 0$ is a scalar, and x_h' doesn't appear in the constraints of GPL, but is only constrained to be integer.

Obviously the optimal value of the Lagrangean is given by:

$$L^* = \max_{\lambda \geq 0} L(\lambda)$$

Remark that

$$z^*_{PL} \leq L^* \leq z^*_G.$$

LEMMA 2. A necessary optimality condition for GPL is $\lambda = c_h$.

PROOF. In fact $(c_h C_{22} - \lambda C_{22})$ is the cost coefficient of x_h' in GPL. As x_h' doesn't appear in the constraints if the cost coefficient is different from zero the objective function is unbounded. From $C_{22} \geq 1$ the result follows. ◀

By substituting the optimal value of λ in z_L, we obtain:

$$L(\lambda) = \min z_L = c_N x_N$$

where x_N belongs to the feasible region of GPH, and x_h' is not constrained and doesn't affect the objective function of GPL.

Let $(\overset{*}{L}, x_N^L)$ be the optimal solution of GPL.

LEMMA 3. (x_h', x_N^L) leads to a feasible solution (x_h, x_N^L) of GP if (see 2.2, 2.3)

$$x_h' \geq \Gamma(x_N^L), \text{ integer} ◀$$

Obviously, as $c_h C_{22} \geq 0$, among these feasible solutions, the best one is obtained from:

$$\hat{x}_h' = \left\lceil \Gamma(x_N^L) \right\rceil \qquad \left(\hat{x}_h = C_{22}\left(\left\lceil \Gamma(x_N^L) \right\rceil - \Gamma(x_N^L)\right)\right)$$

It is possible now to give an approximate algorithm as follows:

ALGORITHM 2.

1. Solve GPL, let x_N^L be the solution

2. Given x_N^L, calculate $\hat{x}_h'$ (or $\hat{x}_h$); let $\hat{z}_G$ be the corresponding value of the objective function $\hat{z}_G = \tilde{d} + \tilde{c}_h \hat{x}_h' + \tilde{c}_N x_N^L = c_h \hat{x}_h + c_N x_N^L$

3.2. A sufficient optimality condition can now be obtained with the following consideration:

1. $L^* \leq z_G^*$;
2. from x_N^L it is possible to construct a feasible solution of $GP((\hat{x}_h', x_N^L)$, see Lemma 3.);
3. the value of z_G associated with this feasible solution $\hat{z}_G$ is always not better than z_G^* $(z_G^* \leq \bar{z}_G)$;
4. if $L^* = \hat{z}_G$ then the feasible solution is optimal for $GP(x_N^L = x_N^*$, $\hat{x}_h = x_h^*)$.

THEOREM 1. The maximum value of the difference $(\hat{z}_G - z_G^*)$ is given by:

$$c_h \hat{x}_h = c_h C_{22}(\lceil \Gamma(x_N^L) \rceil - \Gamma(x_N^L)) \leq c_h(C_{22} - 1) \leq c_h(\delta_m' - 1).$$

PROOF. As $(\hat{z}_G - z_G^*) \leq (\hat{z}_G - L^*)$, from the expression of $\hat{z}_G$ and L^* follows the result; remark that the value of $(\lceil \cdot \rceil - \cdot)$ belongs to the interval $[0,1)$. ◀

COROLLARY 1. If $\hat{x}_h = 0$, i.e. $\Gamma(x_N^L)$ is integer or $c_h = 0$, then $\hat{z}_G = z_G^*$, and $(\hat{x}_h, x_N^L)$ is the optimal solution of GP. ◀

3.3. It is now possible to give a procedure of choosing the column h in order to satisfy (if possible) a given level of approximation. The procedure is based on the following result.

Let α be the maximum accepted value of $(\hat{z}_G - z_G^*)$ and $C_{22}(h)$ the value of C_{22} corresponding to a given column h (remember that $C_{22} = \det \Delta' / \det \Delta$):

THEOREM 2. If $C_{22}(h) \leq (\alpha + c_h)/c_h$ then $(\hat{z}_G - z_G^*) \leq \alpha$

PROOF. $(\hat{z}_G - z_G^*) \leq (\hat{z}_G - L^*) \leq c_h(C_{22}(h) - 1)$, then if $c_h(C_{22}(h)-1) \leq \alpha$ then $(\hat{z}_G - z_G^*) \leq \alpha$. ◀

ALGORITHM 3.

1. Given α, find h such that the equation

$$C_{22}(h) \leq (\alpha + c_h)/c_h$$

is satisfied; let $\hat{h}$ be the result (if no $\hat{h}$ exists it means that it is not possible to guarantee "a priori" the given approximation).

2. Calculate $(x_N^L, \hat{x}_{\hat{h}})$ utilizing Algorithm 2; $(c_{\hat{h}} \hat{x}_{\hat{h}})$ is the maximum value of the error (in general less than $[c_{\hat{h}}(C_{22}(\hat{h}) - 1)]$). ◀

Note that the same procedure can be applied if we fix the maximum percentage error $\alpha' = ((\hat{z}_G - z^*_G)/z^*_G)$ utilizing instead of z^*_G the optimal value of the LP associated to the IP (hypothized positive).

4. CONCLUSIONS

In this paper some algorithms are proposed for the exact or approximated solution of an IP problem over a cone. In particular some sufficient conditions such that Algorithm 2 leads to an optimal solution, are given. It should be noticed that if some of nonbasic variables of the LP optimal solution have zero cost coefficient, it is always possible to find the optimal solution utilizing one of these columns as column h. A method of finding approximated solutions with a given level of accuracy, is also briefly presented; the crucial point of the method is a fast computation of $C_{22}(h)$.

REFERENCES

[1] DER-SAN CHEN, ZIONTS: Comparison of some algorithms for solving the Group Theoretic Integer Programming Problem.Operations Research, vol. 24, n.6. Nov -Dec. 1976.

[2] J.F.SHAPIRO: Group theoretic algorithms for the integer programming problem II: extension to a general algorithm - Operations Research, vol. 16, pp.928-947, 1968.

[3] G.A.GORRY, J.F.SHAPIRO: An adaptive group theoretic algorithm for integer programming problems. Management Science vol.17 n. 5 Gen. 1971.

[4] J.F.SHAPIRO: Dynamic programming algorithms for the integer programming problem I: the integer programming problem as a knapsack type problem. Operations Research vol. 16, pp. 103-121, 1968.

[5] D.E.BELL: Constructive group relaxation for integer programs. Siam Journal on Applied Mathematics, vol. 30, n. 4, June 1976.

[6] G.A.GORRY, J.F.SHAPIRO, L.A.WOLSEY: Relaxation methods for pure and mixed integer programming problems. Management Science vol. 18, n. 5, Jan. 1972.

[7] R.JEROSLOW: Cutting plane theory: algebraic methods. Discrete Mathematics, vol. 23, n. 2, Aug. 1978.

[8] E.V.DENARDO, B.L.FOX:"Shortest Route Methods: 1. Reaching, Pruning and Buckets" Op. Res., Jan. 1979.

[9] E.V.DENARDO, B.L.FOX: "Shortest Route Methods: 2. Group knapsacks, Expanded Networks and branch and bound" Univ. of Montreal, Pub. 263-2, March 1978.

[10] H.M.SALKIN:"Integer programming" Addition Wesley 1975.

[11] F.R.GANTMACHER: "The theory of matrices" Chelsea Pub. Comp. 1959.

EFFICIENT METHOD APPLYING INCOMPLETE ORDERING FOR SOLVING THE BINARY KNAPSACK PROBLEM

Miklós Biró
Computer and Automation Institute Hungarian Academy of Sciences, 1111 Budapest, Kende u. 13. HUNGARY

0. Introduction

The binary knapsack problem is one of the simplest models in integer programming. It is formulated as follows:

$$\max \ \underline{c}' \underline{x}$$
$$\underline{a}' \underline{x} \leqslant b$$
$$\underline{x} \in \{0,1\}^n$$

We can assume without loss of generality, that $\underline{c} \geqslant \underline{0}$ and $\underline{a} > \underline{0}$. Most of the authors still assume that the variables have been rearranged in the order of decreasing $c_j \ / \ a_j$ ratios.

Beeing quite simple, the knapsack problem serves as relaxation or subproblem for a lot of more complex ones. In these cases a great number of knapsack problems are to be solved. This fact underlines the importance of elaborating very fast knapsack algorithms.

At the effectiveness reached by the use of reduction methods, the initial sorting of the variables becomes the most expensive step of any algorithm. [3]

In this paper we propose a method, which eliminates the initial sorting and can be combined with any knapsack algorithm. Because this method is of more general interest, we attempt to formulate it more generally.

1. Role of the sorting

Let $q_j = c_j / a_j$, $j = 1,\dots,n$ and define the following index sets:

$$N = \{1,\dots,n\}$$
$$E_k = \{j \in N : q_j = q_k\}, \quad k = 1,\dots,n$$
$$G_k = \{j \in N : q_j > q_k\} + \{j \in E_k : j \leqslant k\}, \quad k = 1,\dots,n$$
$$\bar{G}_k = N - G_k$$

Evidently the set of G_k sets is completely ordered according to the inclusion relation. The following assertions are equivalent:

(i) $G_i \subset G_j$

(ii) $i \in G_j$

(iii) $q_i \geqslant q_j$

Considering a 0-1 knapsack problem, the corresponding continuous problem ($x \in \{0,1\}^n$ replaced by $0 \leqslant x \leqslant 1$) can be solved easily by filling the knapsack after the variables have been sorted, i.e. assume that $q_1 \geqslant q_2 \geqslant \dots \geqslant q_n$, then:

$$(1.1) \quad x_j^c = \begin{cases} 1 & j \in G_p - \{p\} \\ \frac{1}{a_p}\left(b - \sum_{j \in G_p - \{p\}} a_j\right) & \text{where } p = \min\left\{k : \sum_{j \in G_k} a_j > b\right\} \\ 0 & j \in \bar{G}_p \end{cases}$$

where x_j^c denotes the continuous optimum value of the j-th variable. But the ordering is also expedient when looking for a good feasible solution required by every reduction method.

When using Lagrangean relaxation to reduce the problem size, the choice of the multiplier strongly influences the effectiveness of the reduction, which is based on the following theorem:

Theorem 1.1 [2,4]

Let $M(q) = \max\{c'x + q(b - a'x) : x \in \{0,1\}^n\}$, $q \geqslant 0$ and z_L : a feasible objective value.

If $M(q) - |c_j - q a_j| < z_L$

then $c_j - q a_j < 0$ implies $x_j^D = 0$

and $c_j - q a_j > 0$ implies $x_j^D = 1$

where x_j^D denotes the discrete optimum value of the j - th variable.

In practice the best multiplier is the Geoffrion's one, i.e. the number $q_p \left(M(q_p) = \min_{q>0} M(q) \right)$ where p is the so called pivot index of the continuous solution, which can be easily found after the variables have been sorted.

First we will eliminate the task of finding a good feasible solution.

2. Method of the estimated optimum

Let us consider a problem, for which some solving method is already given.

Let us suppose that the solving effort could be considerably reduced if we knew some quantitative information, which unfortunately cannot be produced before having solved the problem.

In this case it may be suitable to make a reasonable estimation and to solve the problem applying this one. If the information what we get after solving is independent of our estimation and supports its fitness then the problem is solved. If the fitness of the estimation is not supported then the procedure must be repeated with a modified estimation, and of course applying all the usefull information we have got from the previous iterations.

In the case of the 0-1 knapsack problem the unknown information is the integer optimum objective value, and the reduction method is the Lagrangean relaxation.

Algorithm:

1. $i := 1$, Estimate the integer optimum value /denote by $\hat{z}_1$/
2. Reduce the problem size using this estimated value.
3. If the remaining problem has no feasible solution then GO TO 6.
4. Solve the remaining small-size problem. /denote by z_{R_i} the best objective value/
5. If $z_{R_i} \geqslant \hat{z}_i$ then the problem is solved, and z_{R_i} is the discrete optimum value. STOP.
6. Modify the estimation by an appropriate value

   i.e. $\hat{z}_{i+1} := \hat{z}_i - \varepsilon_i$ where $\lim_{i \to \infty} \varepsilon_i > 0$

   $i := i + 1$, GO TO 2.

The following inequalities hold during the algorithm

$$z_{R_1} \leqslant z_{R_2} \leqslant \dots \leqslant z_{R_i}$$

$$\hat{z}_1 > \hat{z}_2 > \dots > \hat{z}_i$$

It can be easily seen that $Z_{R_i} \geq \hat{Z}_i$ will hold in a finite number of steps which proves the finitness of the method. The efficiency of the algorithm depends strongly on the initial estimation and the choice of ε_i in each iteration.

3. How to find the best multiplier?

As we saw before, the simplest method for finding the pivot variable, from which the best multiplier can be determined seems to fill the knapsack after sorting.

D.Fayard and G. Plateau proposed an algorithm (NKR) which finds the pivot by a quicksort-type method, where the sorting can be stopped as soon as the pivot is localized [3]. We have to mention that E. Balas and E. Zemel developed a procedure based on the same idea [1].

We present another approach, which does not need complete ordering. This method initializes the estimated optimum algorithm, that is its 3. step can be omitted.

Denote by $\hat{q}$ an estimation of the multiplier q_p. We can produce the following sets without ordering, applying theorem 1.1, utilizing $\hat{q}$ and an arbitrary estimation $\hat{z}$:

$$R_1 = \{j \in N : \text{ variable } x_j \text{ have been fixed at } 1\}$$

$$R_o = \{j \in N : \text{ variable } x_j \text{ have been fixed at } 0\}$$

$$F = \{j \in N : \text{ variable } x_j \text{ is free}\}$$

These sets form a partition of N, i.e. $N = R_1 + R_0 + F$.

Denote by r the pivot index of the remaining problem /we have to suppose, that it is feasible, i.e. $\sum_{j\in R_1} a_j \leq b$ /. If we assume that the variables of the set N have been sorted i.e. $q_j \geq q_{j+1}$, $j \in N$, and we define

$$(3.1) \qquad r' = \min\left\{k \in N : \sum_{j\in R_1+(F\cap G_k)} a_j > b\right\}$$

then we can easily prove that $r' \in F$, thus $r = r'$ and this means that for finding r we have to search among the elements of F only. If the relation in (3.1) does not hold for any $k \in N$ then all the succeeding assertions are obvious taking $r = n+1$ and $q_{n+1} = 0$, thus we disregard this case.

Let $\quad h = \max\{k \in N : q_k > \hat{q}\}$

/the sorting is still supposed in this definition but henceforward it is not/

If the above relation does not hold for any $k \in N$ then let $h=0$ and $G_h = \emptyset$.

Lemma 3.1.

a./ $G_h = R_1 + (F \cap G_h)$

b./ If $G_h \subset G_k$ then $R_1 + (F \cap G_k) = G_h + (F \cap \bar{G}_h \cap G_k)$

Proof: According to the definitions and to theorem 1.1

$$R_1 \subset G_h \quad \text{and} \quad R_0 \subset \bar{G}_h \tag{3.2}$$

which implies a/.

$$R_1 + (F \cap G_k) = R_1 + (F \cap G_k \cap G_h) + (F \cap G_k \cap \bar{G}_h) = G_h + (F \cap \bar{G}_h \cap G_k)$$

since $G_h \subset G_k$ and since, a/ holds.

Lemma 3.2 : The Geoffrion's multiplier (q_p) lies between an arbitrary multiplier $(\hat{q})$ which helps to reduce the size of the problem, and the Geoffrion's multiplier of the remaining problem (q_r) i.e. if $q' = \max\{\hat{q}, q_r\}$ $q'' = \min\{\hat{q}, q_r\}$ then $q' \geqslant q_p \geqslant q''$

Proof: We decompose the proof into three cases.

1. case: $\hat{q} > q_p$

then $G_h \subset G_p - \{p\}$

and there exists an index k for which

$$G_p - \{p\} = G_k , \quad \text{thus} \quad G_h \subset G_k$$

according to the lemma 3.1/b

$$R_1 + (F \cap G_k) = G_h + (F \cap \bar{G}_h \cap G_k) \subset G_h + (\bar{G}_h \cap G_k) = G_k$$

then from (1.1) we have:

$$\sum_{j \in R_1 + (F \cap G_k)} a_j \leqslant \sum_{j \in G_k} a_j \leqslant b$$

this means according to (3.1) that $G_k = G_p - \{p\} \subsetneqq G_r$ thus $q_p \geqslant q_r$.

2. case: $q_p > \hat{q}$

then $G_p \subset G_h$ and according to (3.2)

$$R_1 + (F \cap G_p) = (R_1 \cap G_h) + (F \cap G_p) \supset (R_1 \cap G_p) + (F \cap G_p) = G_p$$

then
$$\sum_{j \in R_1 + (F \cap G_p)} a_j \geqslant \sum_{j \in G_p} a_j > b$$

This means according to (3.1) that $G_r \subset G_p$ thus $q_r \geqslant q_p$.

<u>3. case:</u> $q_p = \hat{q}$ then see theorem 3.1 .

The following lemma can be easily seen by slightly modifying the above proof.

<u>Lemma 3.3:</u>

If $q' = \max\{\hat{q}, q_p\}$, $q'' = \min\{\hat{q}, q_p\}$

and $\{j \in N : q' \geqslant q_j \geqslant q''\} \subset F$ then $q_r = q_p$.

<u>Theorem 3.1:</u> A mutiplier $(\hat{q})$ which helps to reduce the problem-size equals to the Geoffrion's multiplier of the remaining problem (q_r) if and only if it equals to the Geoffrion's multiplier of the whole problem (q_p).

i.e. $\hat{q} = q_r$ iff $\hat{q} = q_p$.

<u>Proof:</u> The "only if" part of the theorem is a direct consequence of lemma 3.2. We have to prove the other part.

Let us suppose that $\hat{q} = q_p$

then $p \notin G_h$ which means $\sum_{j \in G_h} a_j \leqslant b$

consequently according to (3.1) and lemma 3.1/a

$$r \in \bar{G}_h$$

According to the definition of h and to the supposition

$$\bar{G}_h \cap G_p \subset E_p$$

For every $j \in E_p$, $c_j - \hat{q}a_j = 0$ thus according to theorem 1.1 we have $E_p \subset F$.

Applying lemma 3.1/b and the above relations

$$G_p \subset (G_p \cap \bar{G}_h) + G_h = G_h + (F \cap \bar{G}_h \cap G_p) = R_1 + (F \cap G_p)$$

thus

$$\sum_{j \in R_1 + (F \cap G_p)} a_j \geqslant \sum_{j \in G_p} a_j > b$$

which means $r \in G_p$ consequently $r \in E_p$ thus $q_r = q_p$.

4. <u>The pivot locating algorithm</u>

The skeleton of the algorithm derives easily from the above theorems:

<u>Algorithm</u>

1. Estimate the value of the best multiplier $(\hat{q})$

   $q_1 = +\infty$, $q_2 := 0$

2. Apply the method of the estimated optimum while the remaining problem does not possess a feasible solution (while $\sum_{j \in R_1} a_j > b$).

3. Find the best multiplier of the remaining problem (q_r).
/the problem to be examined is of a quite small size, because the estimated objective value can be "better" than the optimum when reducing/

4. If the best multiplier of the remaining problem equals to our estimation, then it equals to the best multiplier of the whole problem i.e. if $\hat{q} = q_r$ then $\hat{q} = q_p$ STOP.

5. Set the new estimation in the bisecting point of the shortest interval which on the basis of the previous iterations contains the best multiplier.
i.e.
if $\hat{q} > q_r$ then let $q_1 = \min \{\hat{q}, q_1\}$; $q_2 = \max \{q_r, q_2\}$
else let $q_1 = \min \{q_r, q_1\}$; $q_2 = \max \{\hat{q}, q_2\}$

and $\hat{q} := \frac{1}{2}(q_1 + q_2)$

<u>GOTO2.</u>

The theorem 3.1 guarantees that the algorithm will stop in step 4. as soon as the estimation equals to the best multiplier.

We can see applying lemma 3.2 that the sequence of the estimations $\hat{q}$ converges exponentially to the value of the best multiplier.

The algorithm can be made finite by inserting the following tests:

<u>Test 1:</u> If after an iteration we find that one of the two extremities of the shortest interval containing the best multiplier have not changed, and this extremity originates from a remaining problem, then this extremity as a new estimation is suitable to choose instead of the bisecting point.

<u>Test 2.:</u> If the interval to be divided is sufficiently short, then either we can stop without any action accepting a small error, or test whether the interval contains some variables with different q_j.

The first test is justified by the lemma 3.3 and the following assertion:

- If the best multiplier of the remaining problem equals to the best of the whole problem, then the corresponding extremity of the interval does not change any more.

The first test alone garantees the finitness by a condition, which is often satisfied in practice but cannot be verified before the algorithm ends.

<u>Proposition:</u> If the estimated objective value $(\hat{z})$ which gives a feasible remaining problem is less than the continuous optimum ob-

jective value (z_c) of the whole problem, then the best multiplier (q_p) has a neighbourhood $(V(q_p))$ where choosing an arbitrary estimation $(\hat{q})$ the best multiplier of the remaining problem (q_r) equals to the preceding one.

i.e. If $\hat{z} < z_c$ then there exists a $V(q_p)$ such that for every $\hat{q} \in V(q_p)$ we obtain $q_r = q_p$.

Proof: It is obvious that q_p has a neighbourhood $V'(q_p)$ such that $\{j \in N: q_j \in V'(q_p)\} = E_p$ We will prove that q_p has neighbourhood $V''(q_p)$ where for every $\hat{q} \in V''(q_p)$ we obtain $E_p \subset F$.
Then let $V(q_p) = V'(q_p) \cap V''(q_p)$ and the proposition will be proved applying lemma 3.3.

Define $f_j(q) = M(q) - |c_j - q a_j|$, $j=1,\dots,n$. f_j is a continuous function of q.

thus $\forall \varepsilon > 0, \ \exists \delta_j > 0 \ : \ |q - q_p| < \delta_j \Rightarrow |f_j(q_p)| - |f_j(q)| \leqslant \dots < \varepsilon$

It is easy to see that $z_c = f_j(q_p)$ for every $j \in E_p$. Then for $\varepsilon = z_c - \hat{z}$ we obtain

$\hat{z} < f_j(q)$ for every $j \in E_p$ and q such that $|q - q_p| < \delta_j$,

thus $E_p \subset F$.

Then let $V''(q_p) = \bigcap_{j \in E_p} \{ q : |q - q_p| < \delta_j \}$

5. Computational experiences

The tested algorithms were coded in SIMULA-67 language on the CDC-3300 computer of the HAS.

The first method /METHOD 1./ consists of the sorting, followed by a simple Ingargiola and Korsh type reduction [6] and a Greenberg and Hagerich type algorithm [5].

The other methods were based on the estimated optimum algorithm /ESTOPT/. The initial estimation $\hat{z}_1$ was $z_c = M(q_p)$ given by the procedure determining the pivot. ε_j was choosen $\min\{|c_j - \hat{q} a_j| : j \notin F\}$. The remaining problems were solved by the first method for make the comparison possible.

30 test problems were generated for each size from uniform distribution between 1 and 999, and each problem was solved with three right hand sides: 25 %, 50 %, 75% of $\sum_{j \in N} a_j$.

The estimated optimum method preceded whether by the pivot locating or the NKR algorithm, proved to be linear in the number of variables.

/On the CDC-3300 the solving-time was approximately 4xn milliseconds, where n is the size of the problem./The last estimation $\hat{z}$ fitted in with the discrete optimum in almost every case.However the pivot locating algorithm proved not to be better than the NKR algorithm,when the initial pivot estimation was choosen 0.75 without any calculation. We observed that the algorithms were faster for larger right hand sides.

The following table contains some characteristic average values of the 3x30 problems solved from each size.

SIZE	CPU-time ratios		NKR+ESTOPT	
	[SORT/METHOD 1.] x 100	[(NKR+ESTOPT)/METHOD 1.] x 100	MAXIMUM SIZE OF REMAINING SORTED AND SOLVED PROBLEMS	NUMBER OF ITERATIONS
50	46%	108%	7.1	4.2
100	57%	55%	8.7	4.7
150	60%	50%	10.6	4.7
200	63%	45%	11.5	4.7
250	69%	44%	12.5	4.8
500	74%	23%	15.6	4.7
750	83%	13%	16.5	4.4
1000	88%	10%	17.0	4.2

Acknowledgments

The author would like to thank Professors L.B. Kovács and B. Vizvári for their helpful remarks,and Professors S. Walukiewicz and R. Sosinski for putting at his disposal their FORTRAN coded program which could not yet be combined with the above ones because of software difficulties.

References

[1] Balas, E. and Zemel, E. "Solving large zero-one knapsack problems", Management Sciences Research Report No. 408 R Carnegie-Mellon University /1977/

[2] Fayard, D. and Plateau, G. "Techniques de resolution du Probleme du Knapsack en variables bivalents". Bulletin de la Direction des Etudes et Recherches E.D.F. Serie C, No.1. /1976/

[3] Fayard, D. and Plateau, G. "Techniques de resoltuion du probleme du knapsack en variables bivalentes: partie 3" Publications n° 91 du Laboratoire de Calcul de l'Université des Sciences et Techniques de Lille /1977/

[4] Geoffrion, A.M. "Lagrangean Relaxation for Integer Programming", Mathematical Programming Study 2 /1974/ 82-114.

[5] Greenberg, H. and Hegerich, R.L. "A branch search algorithm for the knapsack problem", Management Science 16/5/ /1970/ 327-332.

[6] Ingargiola, G.P. and Korsh, J.F. "Reduction algorithm for zero-one single knapsack problems", Management Science 20/4/ /1973/ 460-463.

[7] Lauriere, M. "An algorithm for the 0-1 knapsack Problem", Mathematical Programming 14 /1978/ 1-10.

[8] Nauss, R.M. "An efficient algorithm for the 0-1 knapsack problem" Management Science 23 /1/ /1976/ 27-31

[9] Suhl, U., "An algorithm and efficient data structures for the binary knapsack problem", European Journal of Operational Research 2/1978/ 420-428.

[10] Zoltners, A.A, "A direct descent binary knapsack algorithm" Journal of the Association for Computing Machinery, 25/2/ /1978/ 304-311.

WEIGHTED SATISFIABILITY PROBLEMS AND SOME IMPLICATIONS

P.M. Camerini and F. Maffioli
Centro di Studio per le Telecomunicazioni Spaziali of CNR
and Istituto di Elettrotecnica ed Elettronica, Politecnico
20133 MILANO (Italy)

1. Summary

Three problems are considered in this paper, namely (literal) weighted satisfiability (WSAT), quadratic assignment (QA) and parity matroid (PM) problems. These problems are linked via a set of reductions, which imply (i) a border line between "easy" and "hard" WSAT problems and (ii) the possibility of utilizing subgradient techniques for bounding optimal solutions to many "hard" problems.

2. Definitions

Let us define the following recognition problems. For the notation and terminology used the reader is referred to [1,2]. The weight of a set is the sum of the weights of its elements.

Weighted Satisfiability (WSAT)

INPUT: Clauses $C_1, C_2, \ldots, C_r$ of literals from the set $Y = \{y_1, y_2, \ldots, y_s; \bar{y}_1, \bar{y}_2, \ldots, \bar{y}_s\}$, integer h, weighting function $w : Y \to \mathbb{Z}$.

PROPERTY: There is a maximal subset $S \subseteq Y$ of weight $w(S) \leq h$ which satisfies the conjunction of the given clauses, i.e.

(a) S does not contain a complementary pair of literals

(b) $S \cap C_j \neq \emptyset$, $j = 1,2,\ldots,r$.

Quadratic Assignment (QA)

INPUT: Integer ℓ, 4-dimensional integer weight matrix $(c_{ij,pq})$, $i,j,p,q = 1,2,\ldots,t$.

PROPERTY: There is a boolean matrix (x_{ij}) s.t. its weight

$$c = \sum_{ij} \sum_{pq} c_{ij,pq}\, x_{ij}\, x_{pq} \leq \ell, \quad \sum_i x_{ij} = 1 \text{ and } \sum_j x_{ij} = 1.$$

Matroid with Parity-conditions (MP)

INPUT: Matroid $M = (E, \mathfrak{J})$ integer v, weighting function $u : E \to \mathbb{Z}$, proper partition $\mathfrak{B} = \{B_1, B_2, \ldots, B_b\}$ of E.

PROPERTY: There is a subset $I \subseteq E$ of weight $u(I) \geq v$, which is an independent parity set of M, i.e. $I \in \mathfrak{J}$ and either $B_j \cap I \neq \emptyset$ or $B_j \cap I = B_j$, $j = 1,2,\ldots,b$.

In the sequel it will be found useful to speak of:

(a) WSAT-m/n, where m is the maximum number of literals in a clause and n is the maximum number of clauses containing a literal or its complement;

(b) MP-g, where g is the maximum cardinality of the parity blocks $B_1, B_2, \ldots, B_b$;

(c) PMP-g as a particular case of MP-g, where M is a partition matroid.

3. Reductions

Reduction 1: QA $\propto$ WSAT-2/n

QA $\propto$ LQA (see [3]) where $z_{ij,pq} = x_{ij}\, x_{pq}$ and

Linearized Quadratic Assignment (LQA)

INPUT: Integer ℓ, 4-dimensional integer matrix $(c_{ij,pq})$, $i,j,p,q = 1,2,\ldots,t$

PROPERTY: There is a boolean matrix $(z_{ij,pq})$ s.t.

$$\sum_{ij} \sum_{pq} c_{ij,pq}\, z_{ij,pq} \leq \ell \qquad \text{and}$$

$$\sum_{ip} z_{ij,pq} = 1 \; ; \; \sum_{jp} z_{ij,pq} = 1 \; ; \; \sum_{pq} z_{ij,pq} = 0 \text{ or } t \; ;$$

$$z_{ij,pq} = z_{pq,ij}.$$

LQA $\propto$ WSAT-2/n:

$$Y = \{z_{ij,pq}\} \cup \{\bar{z}_{ij,pq}\} = \{y_1,\ldots,y_s\} \cup \{\bar{y}_1,\ldots,\bar{y}_s\} \quad .$$

For each $i,j,p,q = 1,2,\ldots,t$ there are clauses $(\bar{z}_{ij,pq} \vee z_{ij,ij})$, $(\bar{z}_{ij,pq} \vee z_{pq,pq})$.

For each $i,j,p = 1,2,\ldots,t$ there are clauses $(\bar{z}_{ij,ij} \vee \bar{z}_{ip,ip})$, $(\bar{z}_{ij,ij} \vee \bar{z}_{pj,pj})$.

Let K be a large positive integer (f.i. $K > \sum_{ij} \sum_{pq} c_{ij,pq}$) and $h = (t^4-t^2)\, K + \ell$.

For each $i,j,p,q = 1,2,\dots,t$ let

$w(z_{ij,pq}) = c_{ij,pq}$ and

$w(\bar{z}_{ij,pq}) = K.$

Reduction 2: WSAT-m/n $\propto$ PMP-n

Modify WSAT by adding a dummy clause of the kind $(y_h \vee \bar{y}_h)$ whenever for some h, $1 \leq h \leq s$ both y_h and $\bar{y}_h$ appear in the given instance of WSAT. Let r' be the new total number of clauses ($r' \leq r+s$). Let there be an element of E for each occurrence of a literal. Let $\pi = \{P_1, P_2, \dots, P_r\}$ be the partition of E induced by the clauses. Let $M = (E, \mathfrak{I})$ be a partition matroid, where

$$\mathfrak{I} = \{I \subseteq E : |I \cap P_p| \leq |P_p|-1, \quad p = 1,2,\dots,r'\} .$$

For each y_h $(\bar{y}_h)$, $1 \leq h \leq s$, there is a parity block B_h (B_{h+s}) of elements of E corresponding to all the occurrences of y_h $(\bar{y}_h)$. (Possibly $B_h = \emptyset$ for some h.)

Let H be a large positive integer (f.i. $H > w(Y)$. For each i, $1 \leq i \leq 2s = b$, and for each $e \in B_i$, let

$$u(e) = \begin{cases} \dfrac{w(y_i)+H\delta_i}{|B_i|} & \text{if} \quad i \leq s \\[2ex] \dfrac{w(\bar{y}_{i-s})+H\delta_{i-s}}{|B_i|} & \text{if} \quad i < s \end{cases}$$

where

$$\delta_i = \begin{cases} 1 & \text{if there is a dummy clause } (y_i \vee \bar{y}_i) \\[2ex] 0 & \text{otherwise.} \end{cases}$$

Let $v = w(Y) - h + (r'-r)\,H$.

Reduction 3: WSAT-m/n $\propto$ WSAT-m/3

For each i, $1 \leq i \leq s$ do the following.
If the i-th variable occurs more than three times, i.e. either y_i or $\bar{y}_i$ occurs in clauses $C^{(1)}, C^{(2)}, \dots, C^{(n_i)}$, where $n_i > 3$,

(a) replace y_i $(\bar{y}_i)$ with new variables $y_i^{(1)}, y_i^{(2)}, \dots, y_i^{(n_i)}$ $(\bar{y}_i^{(1)}, \bar{y}_i^{(2)}, \dots, \bar{y}_i^{(n_i)})$ i.e. in each $C^{(j)}$, $1 \leq j \leq n_i$, replace y_i

with $y_i^{(j)}$, or $\bar{y}_i$ with $\bar{y}_i^{(j)}$, depending on whether the i-th variable occurs in $C^{(j)}$ as y_i, or $\bar{y}_i$ respectively;

(b) add n_i new clauses of the form

$(y_i^{(1)} \vee \bar{y}_i^{(2)})\ (y_i^{(2)} \vee \bar{y}_i^{(3)}) \ldots (y_i^{(n_i-1)} \vee \bar{y}_i^{(n_i)})$
$(y_i^{(n_i)} \vee \bar{y}_i^{(1)})$.

(Note that above clauses insure that $y_i^{(1)} = y_i^{(2)} = \ldots = y_i^{(n_i)}$.)

4. Implications

The reductions exhibited in the previous section imply the following.

Implication 1. WSAT-m/2 $\in \mathcal{P}$. In fact, by reduction 2, WSAT-m/2 $\propto$ PMP-2. Consider the following problem:

Degree Constrained Subgraph (DECS)

INPUT: Graph $G = (N,A)$, weighting function $f : A \to \mathbb{Z}$, positive integer vector (d_i), $i = 1,2,\ldots, |N|$, integer β.

PROPERTY: There is a subgraph $G' = (N,A')$ such that for each i, $1 \le i \le |N|$ the degree of the i-th node in G' is less that d_i and $f(A') \ge \beta$.

Then PMP-2 $\propto$ DECS where

$$N = \{P_1,P_2,\ldots,P_{r'}\},$$

$$A = \{B_1,B_2,\ldots,B_b\},$$

$$d_i = |P_i| \quad , \quad 1 \le i \le r' \quad (=|N|),$$

$$f(B_j) = u(B_j) \quad 1 \le j \le b \quad (=|A|),$$

$$\beta = v \ ,$$

with B_j connecting the two (not necessarily distinct) nodes P_p, P_q s.t. $|P_p \cap B_j| = |P_q \cap B_j| = 1$.

Since DECS $\in \mathcal{P}$ (see [2]), the result follows.

Implication 2. Since QA is well known to be NP-complete, from reductions 1 and 3, it follows that WSAT-2/3 is NP-complete.

Comment: implication 1 and 2 identify the border-line between "easy" and "hard" WSAT problems in terms of parameters m,n. Note that a similar result holds for the usual unweighted versions of satisfiability problems [1].

Implication 3. Consider now the optimization formulations of the problems defined in Section 2.

Let c_o, w_o and u_o be the weights of the optimum solutions to three instances of QA, WSAT and PMP respectively. Assume these instances are linked via reductions 1 and 2. Then $w_o = w(Y) + (r'-r)\ H-u_o$ and $c_o = w_o-(t^4-t^2)\ K$.

Similarly, if u_b is an upper bound to u_o, $w_b = w(Y)+(r'-r)H-u_b$ and $c_b = w_b - (t^4-t^2)\ K$ are lower bounds to w_o and c_o, respectively.

A promising method for finding a possibly tight upper bound u_b is thoroughly described in [5]. This method is based upon the observation that the weights of all parity sets are invariant if the weighting u is replaced by a new weighing u', provided

$$u'(B_j) = u(B_j) \quad , \text{ for } j = 1,2,\ldots,b. \tag{3.1}$$

A-fortiori, such a change in the weighting function does not affect the optimality of a solution. Let $\omega(u')$ be the weight of an independent set of M obtained by the greedy algorithm [4] under weighting u' (ignoring parity conditions). Obviously $\omega(u)$ is an upper bound to u_o and a tigher bound is given by

$$u_b = \min_{u'} \omega(u') \tag{3.2}$$

where minimization is taken over all weighting u' satisfying (3.1).

The minimization problem (3.2) can be effectively attached by any of the existing subgradient optimization techniques [6], [7] since $\omega(u')$ is easily proved to be a convex function, so that a local optimum is also a global optimum.

Implication 4. Moreover note that if for some u' the subgradient

$$\nabla\omega(u') = 0 \ , \tag{3.3}$$

not only the minimum of (3.2) is obtained, but $u_b = \omega(u') = u_o$, i.e. u_b is an exact estimate of the optimum. In most cases (3.3) does not hold, thus implying that some parity conditions are not satisfied by the independent set I_g obtained at the end of the greedy algorithm. However it is possible that I_g satisfies most of the parity conditions. This suggests the following iterative heuristic approach to solve MP. Compute the maximal subset S_g of I_g respecting parity conditions and add to S_g the solution to a new MP problem where M is the original matroid contracted in S_g.

Comment. Implications 3 and 4 may be considered for obtaining upper and lower bounds to other NP-complete problems than QA. In fact many NP-complete problems(such as for instance Steiner Tree, Plant Location and Constrained Transportation Problems) can be easily reduced to WSAT.

References

[1] Karp R.M. (1972), "Reducibility among combinatorial problems" in R.E. Miller and J.W. Thatcher eds. Symp. on Complexity of Computations, Plenum Press, N.Y.

[2] Lawler E.L. (1976), Combinatorial Optimization: Networks and Matroids, Holt, Rinehard & Winston, N.Y.

[3] Lawler E.L. (1963), "The quadratic assignment problem", Mgt. Sci. 9, 586-589.

[4] Edmonds J. (1971), "Matroids and the greedy algorithm",Math. Programm.1, 127-136.

[5] Camerini P.M., and Maffioli F. (1978), "Heuristically guided algorithm for K parity matroid problems", Discrete Math. 21, 103-116.

[6] Camerini P.M., Fratta L., and Maffioli F. (1975), "On improving relaxation methods by modified gradient techniques", Math. Programm. Study 3, 26-34.

[7] Wolfe P., Held M. and Crowder H. (1974), "Validation of sub-gradient optimization", Math. Programm. 6, 62-88.

ON TWO METHODS FOR SOLVING THE BOTTLENECK MATCHING PROBLEM

Ulrich Derigs, Universität zu Köln

Abstract

We present two methods for solving the bottleneck matching problem

- the "Hungarian" method and
- the shortest augmenting path method.

1. Introduction

Let $G = (V,E,\psi)$ be a loopless graph with vertex set V and edge set E. For every $e \in E$ $\psi(e) \subseteq V$ is the pair of vertices which meets e. Whenever we can do so without loss of clarity we will omit the incidence function ψ and write $G = (V,E)$ resp. $e = \{i,j\}$ with $i,j \in V$. Now a matching M is a subset of E with the property that the edges in M are pairwise disjoint. M is called perfect if $\bigcup_{e \in M} e = V$ holds. Let $\mathcal{M}$ denote the set of all perfect matchings in G. With every edge $e_{ij} \in E$ we associate a real/integer cost value c_{ij}. Then we obtain the well known *perfect matching problem with sum objective* (SMP):

$$(1.1) \qquad \min_{M \in \mathcal{M}} \sum_{e_{ij} \in M} c_{ij} \, .$$

A related problem is the socalled *perfect bottleneck matching problem* (BMP)

$$(1.2) \qquad \min_{M \in \mathcal{M}} \max_{e_{ij} \in M} c_{ij} \, .$$

Here the cost of a perfect matching is determined by the most expensive edge in the matching.
Both problems are special cases of a socalled *algebraic matching problem* (AMP) which covers even some more relevant objective functions. AMP is treated in [2], [3] and [4] where efficient algorithms are developed.

In this paper we will specialize two algorithms for solving AMP to the BMP. Section 2 summarizes well known combinatorial properties of matchings in graphs. In Section 3 we present the "Hungarian" method for solving BMP and in Section 4 a shortest augmenting path method

is introduced. Section 5 reports computational experience.
It is obvious that not every graph admits perfect matchings. But in this paper we are only concerned with complete graphs with an even number $|V|$, which allow perfect matchings. This can be assumed w.l.o.g. introducing an artificial node resp. artificial edges with sufficiently high cost values if necessary.

2. Combinatorial Properties

In this section we consider some combinatorial properties of matchings in graphs which are used in the next sections.

For any $V_1 \subseteq V$ we define the coboundary of V_1

$$\delta(V_1) := \{e \in E \quad |\psi(e) \cap V_1| = 1\}$$

and the set of edges having both ends in V_1

$$\gamma(V_1) := \{e \in E \quad |\psi(e) \subseteq V_1\} .$$

The graph $G[V_1] := (V_1, \gamma(V_1))$ is the subgraph induced by V_1.
A subgraph H of G having the same nodeset as G is said to span G.

An alternating path with respect to a matching M in G is a path the edges of which are alternately elements of M and not.
Alternating trees and circles are defined analogously.
An augmenting path is an alternating path between two unmatched nodes.
With an augmenting path P we define

$$M \oplus P := (M \setminus P) \cup (P \setminus M) .$$

$M \oplus P$ is again a matching in G.
For alternating circles K we define $M \oplus K$ analogously.

The following theorem is fundamental for every algorithmic treatment of matchings in graphs.

(2.1) Theorem (BERGE [1])
A matching M contains a maximum number of edges iff it admits no augmenting path.

Such matchings are called maximum cardinality matching (m.c.-matching).

For $B \subseteq V$ we define the graph $G \times B = (V_B, E_B, \psi_B)$ obtained from G by shrinking B by

$$V_B := (V \setminus B) \dot{\cup} \{v_B\}$$

$$E_B := E \setminus \gamma(B)$$

$$\psi_B(e) := \begin{cases} \psi(e) & \text{if } e \in E_B \setminus \delta(B) \\ (\psi(e) \setminus B) \cup \{v_B\} & \text{if } e \in \delta(B) \end{cases}$$

The node v_B is called pseudonode of B. For $M \subset E$ we define $M_B := M \cap E_B$.
Let $\mathcal{A}$ be a set of subsets of V. We say $\mathcal{A}$ is nested if $|A| \geq 3$ for all $A \in \mathcal{A}$ and $A, B \in \mathcal{A}$ with $A \cap B \neq \emptyset \Rightarrow A \subset B \vee B \subset A$.
For any $A \in \mathcal{A}$ we define $\mathcal{A}[A] := \{B \in \mathcal{A} \mid B \subsetneq A\}$.
If $\{A_1, \ldots, A_n\}$ is the set of maximal elements of $\mathcal{A}$ we define

$$G \times \mathcal{A} := (\ldots(G \times A_1) \times A_2) \ldots \times A_n) .$$

The order of the sets $A_1, A_2, \ldots, A_n$ has no effect on $G \times \mathcal{A}$. With E the edge set of $G \times \mathcal{A}$ we denote $M_{\mathcal{A}} := M \cap E_{\mathcal{A}}$ for $M \subseteq E$.

We are interested in nested families $\mathcal{A}$ having the additional property

(2.2) $\quad G[A] \times \mathcal{A}[A]$ is spanned by an odd circle for each $A \in \mathcal{A}$.

Those nested families are called shrinking families. A (maximal) set A from a shrinking family with $|M \cap \gamma(A)| = \frac{1}{2}(|A| - 1)$ is called (outermost) blossom with respect to M. The role of blossoms and shrinking families in connection with matching problems is studien in [13].

The following property is fundamental for an algorithmic treatment of matchings in nonbipartite graphs.

(2.3) Theorem (EDMONDS [7])

Let M be a matching in G and $\mathcal{A}$ a shrinking family such that every $A \in \mathcal{A}$ is a blossom with respect to M. Then every augmenting path P in $G \times \mathcal{A}$ with respect to $M_{\mathcal{A}}$ induces an augmenting path P with respect to M.

(2.4) Corollary

Let $\mathcal{A}$ be a shrinking family of G and $M_{\mathcal{A}}$ a m.c.-matching in $G \times \mathcal{A}$. Then $M_{\mathcal{A}}$ induces a (m.c.) matching M in G such that every $A \in \mathcal{A}$ is a blossom with respect to M.

To determine an augmenting path with startnode s we construct an alternating tree with root s. This tree leads to a bicoloring of its nodes. Therefore we can attach an "S"-label to nodes which bear the same color as the root and a "T"-label to the other nodes of the tree.

Now a blossom is detected whenever there is an edge $e \in M$ between two "S"-labeled nodes. This blossom is then shrunken to a pseudonode. The pseudonode receives an "S"-label and we try to enlarge the tree by adding appropriate nodes and edges. If an "S"-labeled node is joined with an unmatched node an augmenting path has been detected and the matching can be augmented by changing the role of matching and non-matching edges on this path. For this purpose pseudonodes have to be expanded and the path through these blossoms has to be restored. LAWLER [12] and GABOW [9] describe appropriate labeling techniques to provide backtracing through nested blossoms.

If all "S"-labeled nodes are only connected with "T"-labeled nodes, the tree is said to be hungarian and no augmenting path starting from node s exists.

To determine a m.c.matching alternating trees are built from each unmatched node s as root using the above mentioned labeling and shrinking steps. An augmenting path is detected if two "S"-labeled nodes out of two different trees are connected by an edge.

If all trees become hungarian the actual matching is a m.c.matching in the shrunken graph. After expanding all pseudonodes this matching can be extended to an m.c.matching in the original graph due to Corollary (2.4).

At the end of the algorithm the nodeset V is partitioned into three classes $V = \mathcal{S} \cup \mathcal{T} \cup \mathcal{U}$ where $\mathcal{S}$ is the set of all "S"-labeled nodes or nodes contained in "S"-labeled pseudonodes. Further every unmatched node belongs to the class $\mathcal{S}$. $\mathcal{T}$ is the set of all "T"-labeled nodes. The class $\mathcal{U}$ consists of 2k $(k \geq 0)$ unlabeled nodes which are joined by k matching edges.

3. The Hungarian method

The following procedure can be interpreted either as a generalization of a method for solving the bottleneck assignment problem [10] or as a transposition of EDMONDS' blossom-algorithm [8] to the bottleneck objective. A theoretical foundation of this method is presented in [3] where the general algebraic matching problem is treated.

The algorithm starts with the determination of a "good" lower bound $\underline{z}$ for the optimal objective value z. Then we define the admissibility graph $G(\underline{z}) = (V, E_{\underline{z}})$ with

(3.1) $$E_{\underline{z}} := \{e_{ij} \mid c_{ij} \leq \underline{z}\} .$$

Now a m.c. matching M in $G(\underline{z})$ is determined using the labeling procedure described in Section 2. If M is perfect it is optimal for BMP. Otherwise the labeling method yields a partition $V = \mathcal{S} \cup \mathcal{T} \cup \mathcal{U}$. Now we define

(3.2) $$E(S,S) := \{e_{ij} \mid i,j \in \mathcal{S}, \text{ yet not in the same blossom}\}$$

(3.3) $$E(S,U) := \{e_{ij} \mid i \in \mathcal{S}, j \in \mathcal{U}\}$$

According to Corollary (2.4) every perfect matching has to contain an edge

$$e_{ij} \in E(S,S) \cup E(S,U) .$$

Therefore we compute

(3.4) $$\underline{z}^* := \min\{c_{ij} \mid e_{ij} \in E(S,S) \cup E(S,U)\} > \underline{z} .$$

$\underline{z}^*$ is a bound for the optimal objective value, define $\underline{z} := \underline{z}^*$ and repeat the process.

This way an algorithm of $O(|V|^3)$ order can be obtained modifying LAWLER's labeling technique to the bottleneck case.

The method stated above can be improved in the following way. Instead of determining a m.c. matching in the admissibility graph $G(\underline{z})$ we construct only one alternating tree rooted at an unmatched node using the S-T labeling technique as before. If this tree becomes hungarian we define

(3.5) $$\mathcal{U} := \{i \in V \mid i \text{ not in the tree}\} .$$

With E(S,S) and E(S,U) as defined in (3.2) resp. (3.3) we compute

$$\underline{z}^* := \min\{c_{ij} \mid e_{ij} \in E(S,S) \cup E(S,U)\} > \underline{z} .$$

Define $\underline{z}$:= z* and construct a new alternating tree either with the same unmatched node as a root or with a new one. Yet computational experience have shown that whenever a tree has become hungarian it is favourable to change the root in the next iteration.

For growing one single alternating tree the labeling technique proposed by GABOW [9] and improved by HESKE [6] can be used more efficiently. On the other hand we can use the new bound in the next step and so the probability for detecting an augmenting path grows.

4. The shortest augmenting path method

Let M be a matching in G. An alternating circle K is called *negative* with respect to M if

(4.1) $c(M \oplus K) < c(M)$.

The following theorem yields an optimality criterion for BMP

(4.2) Theorem (DERIGS [2])

$M \in \mathcal{M}$ is optimal iff it allows no negative alternating circle.

Let $s \in V$ be an unmatched node with respect to M, then we define $\mathcal{P}_s(M)$ the set of all augmenting paths with startnodes s. The length of a path $P \in \mathcal{P}_s(M)$ is defined by

(4.3) $c(P) := \max_{e_{ij} \in M \oplus P} c_{ij}$.

A path $P_o \in \mathcal{P}_s(M)$ is called *shortest augmenting path* if

$$c(P_o) \leq c(P) \qquad \forall\, P \in \mathcal{P}_s(M).$$

The following theorem is basic for the algorithm presented in this section

(4.4) Theorem (DERIGS [2])

Let M_k be a matching in G which does not allow a negative alternating circle. Let $s \in V$ be an unmatched node and $P \in \mathcal{P}_s(M)$ a shortest augmenting path then $M_{k+1} = M_k \oplus P$ does not allow a negative alternating circle.

This theorem motivates the following algorithm

(4.5) Shortest augmenting path algorithm (SAP)

START: $i := 0,\ M_o := \emptyset$

① If M_i perfect → STOP!
Otherwise determine an unmatched node $s \in V$ → ②

② Determine a shortest augmenting path $P_i \in \mathcal{P}_s(M)$.

③ $M_{i+1} := M_i \oplus P_i$
$i := i + 1$ → ①

For solving the shortest path problem 2 we can use the following labeling technique.

START: $\mathcal{A} := \emptyset,\ M := M_i$

① $d_j^- := c_{s,j}$
$p(v_j) := s$ for $v_j \in V,\ v_j \neq s$
$d_j^+ := \infty$ for $v_j \in V$
s receives a S-label → ②

② Determine
$\delta_1 = \min\{d_i^- \mid v_i \text{ unlabeled}\}$
$\delta_2 = \min\{\max\{d_i^+, d_i^-\} \mid v_i \text{ S-labeled}\}$
$\delta = \min\{\delta_1, \delta_2\}$
$\delta = \delta_1$ → ③
$\delta = \delta_2$ → ④

③ If v_i is unmatched with resp. to $M_{\mathcal{A}}$ → ⑤ .
Otherwise exists $\{v_i, v_j\} \in M_{\mathcal{A}}$.
v_j receives an S-label.
Define $d_j^+ := \delta$ and scan node v_j , i.e. determine
$d_i^- := \min\{d_i^-, \max\{d_j^+, c_{ji}\}\}$ for $v_i \neq v_j$
and $p(v_i) := v_j$ if $d_i^- = \max\{d_j^+, c_{ji}\}$
→ ② .

④ Introducing edge $\{v_i, p(v_i)\}$ to the labeled tree a blossom B is detected. Define $\mathcal{A} := \mathcal{A} \cup \{B\}$ and shrink B to v_B.
v_B receives an S-label, $d_B^+ := \delta$.
Scan v_B as in step ③ node $v_j \rightarrow$ ②

⑤ An augmenting path $P_{\mathcal{A}}$ with respect to $M_{\mathcal{A}}$ has been detected. Expand all pseudo nodes to obtain the associated shortest augmenting path $P \in \mathcal{P}_s(M)$.

Using the labeling procedure and appropriate data structures the shortest path problem can be solved in $O(|V|^2)$ time.
Since $\frac{1}{2}|V|$ paths have to be determined SAP is of order $O(|V|^3)$.

5. Computational Experience

FORTRAN IV - implementations of both algorithms were tested on randomly generated problems on a CDC CYBER 76 of the Computer Center of the University of Cologne.
The improved hungarian method using Gabow's labeling technique showed to be slightly superior.
The following table shows the mean CPU-running time of randomly generated problems with 100 nodes and different ranges of cost-coefficients c_{ij}. 25 examples of each combination were generated to calculate the mean running time.

$\|V\|$ \ c_{ij}	1-100	1-1000	1-10000	$1-2^{31}-1$
100	.043	.038	.035	.042

Table 1: CPU-running time for the modified hungarian method (in sec.)

A more detailed discussion of the computational experience is presented in [2] where a FORTRAN IV Code of the modified hungarian method is listed.

References

[1] Berge, C.: Two Theorems in Graph Theory. Proc.Natl.Acad.Sci.U,S., 43, 842 - 844, (1957).

[2] Derigs, U.: Algebraische Matching Probleme. Doctoral Thesis, Mathematisches Institut der Universität zu Köln, (1978).

[3] Derigs, U.: A Generalized Hungarian Method for Solving Minimum Weight Perfect Matching Problems with Algebraic Objective. To appear in Descrete Applied Mathematics.

[4] Derigs, U.: Die Lösung minimaler perfekter Matching Probleme mittels kürzester erweiternder Pfade. Working Paper, presented at the IV. Symposium on Operations Research, Saarbrücken, (1979).

[5] Derigs, U. and G.Kazakidis: On Two Methods for Solving Minimal Perfect Matching Problems, Arbeitsbericht des Industrie-seminars der Universität zu Köln, Köln (1979).

[6] Derigs, U. and A.Heske: A Computational Study on Some Methods for Solving the Cardinality Matching Problem. Report 79-2, Mathematisches Institut der Universität zu Köln, (1979).

[7] Edmonds, J.: Paths, Trees and Flowers. Can. J. Math. 17, 449 - 467, (1965).

[8] Edmonds, J.: Maximum Matching and a Polyhedron with 0,1 Vertices. J. Res. NBS, 69B, 125 - 130, (1965).

[9] Gabow, H.: An Efficient Implementation of Edmonds' Algorithm for Maximum Matching on Graphs. JACM, 23, 221 - 234, (1975).

[10] Garfinkel, R.S.: An Improved Algorithm for the Bottleneck Assignment Problem. Op. Res., 19, 1747 - 1751, (1971).

[11] Glover, F.: Minimum Complete Matchings. ORC-Report 67-15, University of California, Berkeley, (1967).

[12] Lawler, E.L.: Combinatorial Optimization: Networks and Matroids. Holt, Rinehart and Winston Inc., New York, (1967).

[13] Edmonds, J. and W.Pulleyblank: Facets of 1-Matching Polyhedra. In: Hypergraph Seminar, Lecture Notes in Mathematics No.411, 214 - 242, Berlin, (1974).

Dr. Ulrich Derigs
Industrieseminar
Abt. Operations Research
Universität zu Köln
D-5000 Köln 41
Federal Republic of Germany

FAST APPROXIMATION ALGORITHMS FOR KNAPSACK TYPE PROBLEMS

G.V. Gens
E.V. Levner
Central Economic and Mathematical Institute
USSR Academy of Sciences, Moscow 117333

1. INTRODUCTION.

The following variations of the knapsack problem are considered: PARTITION, ARBORESCENT KNAPSACK, FIXED-CHARGE KNAPSACK, MIN-MULTIPLE-CHOICE KNAPSACK. The problems find many applications to capital budgeting, R&D project selection, decision making in multi- level economic systems. They can also be used as relaxations for solving other integer programming problems. The problems belong to the class of NP-hard problems, their computational intractability stimulating research of efficient approximation algorithms.

Purpose of this paper is to study fast, or fully polynomial [3], ϵ-approximation algorithms (i.e., ones operating in time bounded by a polynomial in the problem size and in $1/\epsilon$, ϵ being the allowable fractional error), whose first appearance is due to Babat [1], Kim and Ibarra [6], and Sahni [10].

In this paper we elaborate on the Ibarra-Kim approach, introducing some improvements which yield better time and space bounds for the partition problem. Fast algorithms for the problems mentioned are derived. We describe two methods, decomposition and binary search, for constructing bounds $\hat{f}$, satisfying $\hat{f} \leq f^* \leq c\hat{f}$, where f^* is the optimum and c is a constant, usually, $2 \leq c \leq 8$, the bounds being used for constructing fast algorithms.

In conclusion, we show that to obtain a fully polynomial ϵ-approximation algorithm for the m-dimensional 0-1 knapsack, $m > 1$, is impossible, unless $P \neq NP$.

2. THE $O(n+1/\epsilon^2)$ ALGORITHM FOR THE PARTITION PROBLEM.

The partition problem is as follows: Given n+1 positive integers $p_1, p_2, \ldots, p_n, b$, find $x_1, x_2, \ldots, x_n$ so as to maximize $f(x) = \sum_{i=1}^{n} p_i x_i$ subject to $\sum_{i=1}^{n} p_i x_i \leq b = 1/2 \sum_{i=1}^{n} p_i$, $x_i = 0$ or 1.

We first present an $O(n/\epsilon)$ algorithm APPROX-PP-1 which solves the partition problem with an arbitrary b value.

Let $\hat{f}$ be a bound such that $\hat{f} \leq f^* \leq 2\hat{f}$, f^* being the optimum. The $\hat{f}$ value can be found in $O(n)$ time [6,8].

ALGORITHM APPROX - PP-1.

Input: $p_1, p_2, \ldots, p_n, b$, $\epsilon > 0$, $\hat{f}$.

Output: an ϵ-approximate solution, x', i.e. $|f^* - f(x')| / f^* \leq \epsilon$.

Step 1. Initialize the set S^o: $S^o \leftarrow \{0\}$.

Step 2. Form $S^1, S^2, \ldots, S^n$ as follows:

Form a set T^k from S^{k-1} by the operation: $T^k \leftarrow S^{k-1} \cup \bar{S}^{k-1}$, where $\bar{S}^{k-1}$ is obtained by adding p_k to every element in S^{k-1}. Omit elements in T^k greater than b. If there are two identical elements in T^k, omit one of them. Order T^k according to increasing value of its elements:

$$T^k = \{ t_i \}_{i=1}^{M}; \quad t_1 < t_2 < \ldots < t_M.$$

Set a $s_1 \leftarrow t_1$. Let s_2 be the maximal element in T^k such that $t_i \leq s_1 + \epsilon \hat{f}$ (all the elements t_i in T^k between s_1 and s_2 being omitted); if $t_2 > s_1 + \epsilon \hat{f}$, set $s_2 \leftarrow t_2$. Let s_3 be the maximal element in T^k such that $t_i \leq s_2 + \epsilon \hat{f}$, the elements in T^k between s_2 and s_3 being omitted; if all $t_i > s_2 + \epsilon \hat{f}$, let s_3 be the element t_i next to s_2 in T^k. Continue this procedure untill we set the last element in T^k:

$s_{N(k)} \leftarrow t_M$.

All the obtained s_i form the set S^k; $|S^k| = N(k)$.

Step 3. The ϵ-approximate solution value is given by the maximal element in S^n. The corresponding vector x' is found by the ordinary trace back from S^n to S^1.

It is clear that $s_{i+2} - s_i > \epsilon \hat{f}$ $(1 \leq i \leq N(k)-2)$, hence, for any k $(1 \leq k \leq n)$ $N(k) < 2 \cdot f / \epsilon \hat{f} = 4/\epsilon$. Thus, time required to generate S^n and trace back x' is $O(n/\epsilon + n \log 1/\epsilon) = O(n/\epsilon)$, space is

$O(n/\epsilon)$. The complete proof may be found in [4,9].

Now we consider the $O(n+1/\epsilon^2)$ algorithm APPROX-PP-2 for solving the partition problem with $b = 1/2 \sum_{i=1}^{n} p_i$. Assume that for any j $p_j < b$, otherwise exact solution may be found trivially.

ALLGORITHM APPROX-PP-2.

Input: $p_1, p_2, \ldots, p_n$, $b = 1/2 \sum_{i=1}^{n} p_i$, $\epsilon > 0$.

Output: an ϵ-approximate solution, x'.

Step 1. Form the list SMALL $= \{p_i\}$, where $p_i \leq 1/2\, \epsilon b$;
LARGE $= \{1, \ldots, n\} \setminus$ SMALL.

Step 2. Set $L \leftarrow \sum_{i \in \text{LARGE}} p_i$.

If $L < b$, join all the items from LARGE with (arbitrary) items from SMALL in order that form the largest set $J \subseteq$ SMALL such that
$L + \sum_{i \in J} p_i \leq b$.

Form the ϵ-approximate solution, x', of the partition problem as follows; $x' = (x_i = 1, i \in \text{LARGE} \cup J;\ x_i = 0, \text{otherwise})$.

If $L \geq b$, use APPROX-PP-1 for solving the following "truncated" partition problem P':

Problem P': Maximize $f'(y) = \sum_{i \in \text{LARGE}} p_i y_i$

subject to $\sum_{i \in \text{LARGE}} p_i y_i \leq b = 1/2 \sum_{i=1}^{n} p_i$, $y_i = 0 \text{ or } 1$.

Let $y' = (y_i')$ be the ϵ-approximate solution of the problem P'.

Step 3. Join all the items from the $I = \{i : y_i' = 1\}$ with arbitrary items SMALL in order that form the largest set $J' \subset$ SMALL such that

$$\sum_{i \in I} p_i + \sum_{i \in J'} p_i \leq b.$$

Form the ϵ-approximate solution, x', of the original partition problem as follows:

$$x' = (x_i = y_i', i \in \text{LARGE};\ x_i = 1, i \in J';\ x_i = 0, \text{otherwise}).$$

It is clear that for any $i \in$ LARGE $p_i > 1/2\, \epsilon b$ and

$$1/2\, \epsilon b\, |\text{LARGE}| < \sum_{i \in \text{LARGE}} p_i \leq b,$$

hence, $|\text{LARGE}| \leq 2/\epsilon$.

Thus, time and space required for solving Problem P' is $O(1/\epsilon^2)$, the total time and space becoming $O(n+1/\epsilon^2)$. The proof is similar to that considered in [4,9] for the min- partition problem.

3. THE O(n^3/ϵ^2) ALGORITHM FOR THE ARBORESCENT KNAPSACK.

The problem is to maximize $c(x) = \sum_{i=1}^{n} c_i x_i$

subject to: (a) $\sum_{i \in J_k} a_i x_i \leq b_k$, $1 \leq k \leq l$, where $l \leq n$,

(b) for any m and r ($m \neq r$) $J_m \cap J_r = \emptyset$, or $J_m \subset J_r$, or $J_r \subset J_m$

(c) $x_i = 0$ or 1, $a_i, c_i \geq 0$ ($1 \leq i \leq n$).

The following notation is used:

We say that J_m is a successor to J_r, if $J_m \subset J_r$.

M_1 is the set of those sets which have no successors.

M_2 is the set of those sets, J_k, which have successors belonging only to M_1.

M_3 is the set of those sets, J_k, which have successors belonging only to M_2.

Similarly we define $M_4, M_5, \ldots, M_q$, where M_q has no predecessors.

We assume that $J_m \neq J_r$ if $m \neq r$, hence, $q \leq n$.

L_k is the set of immediate successors to the set J_k ($1 \leq k \leq l$).

ALGORITHM APPROX - AK.

Step 1. For all $k \in M_1$ consider the following problem P_k:

Problem P_k. Maximize $f_k(x) = \sum_{i \in J_k} c_i x_i$

subject to $g_k(x) = \sum_{i \in J_k} a_i x_i \leq b_k$, $x_i = 0$ or 1, $i \in J_k$.

Find the set of vectors $S^k = \{ \bar{x}_1^k, \bar{x}_2^k, \ldots, \bar{x}_{M(k)}^k \}$,

where $\bar{x}_i^k = (x_{k1}^i, x_{k2}^i, \ldots, x_{k,|J_k|}^i)$ and $\bigcup_{j \in \{1 \ldots |J_k|\}} k_j = J_k$,

such that for any feasible solution $\bar{y} = (y_1, y_2, \ldots, y_{|J_k|})$ of the problem P_k there exists a representative, $\bar{x}_l^k$, in S^k with the following properties:

(a) $f_k(\bar{y}) \leq f_k(\bar{x}_l^k) + \epsilon |J_k| / 2n \cdot \hat{f}_k$, where $\hat{f}_k \leq f_k^* \leq 2\hat{f}_k$, f_k^* being the optimum in the problem P_k,

(b) $g_k(\bar{y}) \geq g_k(\bar{x}_l^k)$.

$l \leftarrow 2$.

Step 2. For all $h \in M_l$ merge S^k, $k \in L_h$.

To carry out this, first choose any pair (S^m, S^r), $m,r \in L_h$ and merge S^m and S^r. Then merge the set obtained with other sets S^k from L_h.

[Merging S^m and S^r is carried out as follows:

(a) add all the vectors from S^m to every vector from S^r, thus obtaining the set $S^{mUr} = \{ x_1^m+x_1^r, x_1^m+x_2^r, \ldots, x_1^m+x_{|J_r|}^r, x_2^m+x_1^r, \ldots, x_2^m+x_{|J_r|}^r, \ldots$
$\ldots, x_{|J_m|}^m+x_{|J_r|}^r \}$;

(b) omit all the vectors, x, from S^{mUr} such that $g_h(x) > b_h$;

(c) find $\hat{f}_{m,r} = \max f_h(x \mid x \in S^{mUr})$ and partition the interval $[1, \hat{f}_{m,r}]$ into $\lceil 2n/\epsilon \rceil$ subintervals of size no greater than $\lceil \hat{f}_{m,r} \cdot \epsilon / 2n \rceil$;

(d) if S^{mUr} has more than one vector in any of subintervals, omit all such vectors except for one with the lowest $g_h(x)$ value.]

After merging sets $S^{(\cdot)}$ for all $h \in M_1$, set $l \leftarrow l+1$. If $l \leq q$, go to Step 2.

<u>Step 3</u>. In the set $S^{(\cdot)}$, obtained at Step 2 in the last place, the vector $\bar{x}$ with the maximal c(x) value is an ϵ-approximate solution for the original arborescent knapsack.

It is clear that we can solve the problem P_k at Step 1 by using the standard Sahni dynamic interval partitioning of the interval $[1, \hat{f}_k]$ into $\lceil 2n/\epsilon \rceil$ subintervals, with $O(n/\epsilon \cdot |J_k|)$ time and space , the total time of Step 1 being $O(n^2/\epsilon)$.

While merging S^m and S^r at Step 2, discarding the vectors introduces an error of at most $(|J_m|+|J_r|+1)/2n \cdot \epsilon \cdot \hat{f}_{r,m}$, the total error being of at most $\epsilon \cdot c^*(x)$, $c^*(x)$ being the optimum of the original arborescent problem.

Step 2 is executed no more than n times, each time taking $O(n^2/\epsilon^2)$ time and $O(n/\epsilon)$ space. Since there are at most $\lceil 2n/\epsilon \rceil$ vectors in each S^i, the trace back needed in Step 3 to find an ϵ-approximate solution, $\bar{x}$, requires $O(n \log n/\epsilon)$ time and $O(n^2/\epsilon)$ space. Consequently, the total time of APPROX-AK is $O(n^3/\epsilon^2)$ and space is $O(n^2/\epsilon)$.

Note that a version of this algorithm can be obtained with $O(n^4/\epsilon^2)$ time and only $O(n/\epsilon)$ space. The key to this is to use the time-space trade-off in just the same way as in the version of APPROX-PP-2 for solving the partition problem with $O(n/\epsilon + 1/\epsilon^2)$ time and $O(n + 1/\epsilon)$ space, described in [5].

4. THE $O(n^3/\varepsilon^2)$ ALGORITHM FOR THE FIXED-CHARGE KNAPSACK.

The problem is to maximize $f(x) = \sum_{i=1}^{n}(p_i x_i + q_i \text{ sign } x_i)$

$$\text{subject to } \sum_{i=1}^{n} (a_i x_i + b_i \text{ sign } x_i) \leq b,$$

$$x_i \text{ integer}, \quad x_i, p_i, q_i, a_i, b_i \geq 0 \ (1 \leq i \leq n).$$

We solve the problem by reducing it to a max-multiple-choice knapsack. To do this, we should know a bound $\hat{f}$ on the optimum value, f^*, such that $\hat{f} \leq f^* \leq 4\hat{f}$.

There are following well-known methodologies for finding the $\hat{f}$ values in the knapsack problems:

(a) greedy and "sophisticated" greedy heuristics [5,7];

(b) a linear programming relaxation obtained by relaxing the integrality constraint on variables (d'Atri, cited in [8]);

(c) a binary search ([5] ; h.l., Section 5).

We present in this Section another method, the decomposition, based on decomposing the original problem (in our case, the fixed charge knapsack) into two auxiliary problems.

Consider the following problems:

<u>Problem P_1</u> (0-1 knapsack). Maximize $f(y) = \sum_{i=1}^{n} (p_i + q_i) y_i$

subject to $\sum_{i=1}^{n} (a_i + b_i)\, y_i \leq b$, $y_i = 0$ or 1.

<u>Problem P_2</u> ("bottleneck" knapsack).

$$\text{Maximize } f_2(z) = \max_{1 \leq i \leq n} (p_i z_i + q_i \text{ sign } z_i)$$

$$\text{subject to } \sum_{i=1}^{n} (a_i z_i + b_i \text{ sign } z_i) \leq b, \quad z_i \text{ integer}, \quad z_i \geq 0.$$

Let $\hat{f}_1$ be a bound on the optimum, f_1^*, of the problem P_1, such that $\hat{f}_1 \leq f_1^* \leq 2\hat{f}_1$. The $\hat{f}_1$ value can be found in $O(n \log n)$ time [6,8].

Let $\hat{f}_2$ be $\max_{i:a_i>0} (p_i z_i + q_i)$, where $z_i = (b-b_i)/a_i$

It is easy to see that $1/2\ \hat{f}_2 \leq f_2^* \leq \hat{f}_2$.

We take the sum $\hat{f}_1 + 1/2\ \hat{f}_2$ as the $\hat{f}$ desired and can show that

$$\hat{f} \leq f^* \leq 4\hat{f} \quad [9].$$

Now we consider ϵ-approximation algorithm, APPROX-FCK, for solving the fixed-charge knapsack.

APPROX-FCK.

Step 1. Set $D \leftarrow 4n/\epsilon$, $d \leftarrow \max(1, \hat{f}\epsilon/n)$. Find the A_{i1} and B_{i1} values as follows:

(a) if $p_i=0$, then $A_{i1} = q_i$, $B_{i1} \equiv b_1+a_1$, $k_1=1$;

(b) if $p_i \neq 0$, then $A_{i1} = \lfloor (p_i x_i^1 + q_i)/d \rfloor d$, $B_{i1} = a_i x_i^1 + b_i$, where x_i^1 is the least integer such that $p_i x_i^1 \geq d$.

Then find A_{i2} and B_{i2} as follows:

$A_{i2} = \lfloor (p_i x_i^2 + q_i)/d \rfloor d$, $B_{i2} = a_i x_i^2 + b_i$, where x_i^2 is the least integer (greater then x_i^1) such that $p_i x_i^2 \geq 2d$.

Similarly, find $A_{i3}, B_{i3}, \ldots, A_{i,k_i+1}$, B_{i,k_i+1}, where $B_{i,k_i} \leq b$, but $B_{i,k_i+1} > b$.

Step 2. Set $A'_{ij} = A_{ij}/d$. Use the standard dynamic programming algorithm (see [8,9]) for exact solving the following max-multiple-choice knapsack problem K:

Problem K. Maximize $F(y) = \sum_{i=1}^{m} \sum_{j=1}^{k_i} A'_{ij}\, y_{ij}$

subject to $\sum_{i=1}^{m} \sum_{j=1}^{k_i} B_{ij} y_{ij} \leq b$, $\sum_{j=1}^{k_i} y_{ij} \leq 1$, $y_{ij}=0$ or 1.

Let $\bar{y}=(\bar{y}_{ij})$ be the optimal solution of the problem K.

Step 3. Using the vector $\bar{y}$ define a vector $\bar{x}=(x_1,x_2,\ldots,x_n)$ as follows:

$$x_i = (B_{ij}-b_i)/a_i, \text{ if } a_i > 0 \text{ and } \bar{y}_{ij}=1,$$

$$x_i = 1, \text{ if } a_i = 0 \text{ and } \bar{y}_{ij}=1,$$

$$x_i = 0, \text{if } \bar{y}_{ij}=0 \ (1 \leq j \leq k_i).$$

The $\bar{x}$ is the ϵ-approximate solution of the original problem [9].

It is clear that the time required to solve the problem K is $O(F^* \sum_i k_i)$. But $F^* \leq \hat{f}/d = n/\epsilon$ (since all the A_{ij} are divisible by d), and as far as $A_{ij}-A_{i,j-1} \geq d$ $(2 \leq j \leq k_i+1$, $1 \leq i \leq n)$ we have that $k_i \leq \hat{f}/d \leq D$ $(1 \leq i \leq n)$. Thus, Step 2 requires $O(n^3/\epsilon^2)$ time and $O(n^2/\epsilon)$ space. Steps 1 and 3 require $O(\sum_i k_i) = O(n^2/\epsilon)$ time and space; the $\hat{f}$ value is computed in $O(n \log n)$ time.

The complete proof can be found in [9].

5. <u>THE MIN-MULTIPLE-CHOICE KNAPSACK PROBLEM</u>;

The problem is defined as follows:

$$\text{Minimize } f(x)=\sum_{i=1}^{m}\sum_{j=1}^{k_i} c_{ij}x_{ij}$$

$$\text{subject to } \sum_{i=1}^{m}\sum_{j=1}^{k_i} a_{ij}x_{ij}\geqslant b,\quad \sum_{j=1}^{k_i} x_{ij}\leqslant 1,$$

$$c_{ij},a_{ij}\geqslant 0,\quad x_{ij}\geqslant 0,\ \text{integer},\ 1\leqslant j\leqslant k_i,\ 1\leqslant i\leqslant m.$$

Because of the minimization formulation of the problem, the general dynamic programming algorithm, <u>FRAME-MIN</u>, contains special details in forming the sets $S^o, S^1, \ldots, S^m$. Initially, $S^o = \emptyset$. At the end of iteration i ($1\leqslant i\leqslant m$) each pair $(c,a)\in S^i$ is identified with a solution x (may be unfeasible with respect to the constraint $a(x)\geqslant b$) containing items chosen from equivalence classes 1 through i.

Let $\hat{F}$ be an upper bound satisfying $\hat{F}\geqslant f^*$, f^* being the optimum. To perform iteration i, form k_i candidate items for each pair (c,a) existing in the set S^{i-1} at the end of iteration i-1. Then candidates pairs are placed in k_i separate candidate sets. The k_i+1 sets are then merged eliminating: (i) dominated entries, (ii) those entries with $c=c(x)>\hat{F}$, and (iii) all the entries with $a=a(x)$ value greater than b except for the one with the lowest $a = a(x)$ value among them.

The optimal solution corresponds to the optimal in S^m.

It is clear that the time required by FRAME-MIN is $O(\hat{F}\sum_i k_i)$, space is $O(m\hat{F})$.

We first present an algorithm <u>MCP-BOUND-1</u> for finding a bound f^o such that $f^o/_m\leqslant f^*\leqslant f^o$

<u>ALGORITHM MCP-BOUND-1</u>.

<u>Step 1</u>. Sort items in equivalence classes in nondecreasing c_{ij} value order. Omit the dominated items.

<u>Step 2</u>. Choose the item with the lowest c_{ij} value in each class: $c_{ij}=\min_j c_{ij}$, and find $C=\max_i c_{ij}$.

<u>Step 3</u>. In each equivalence class choose the item with the largest f(x) value no greater than C. If the sum of a_{ij} values of chosen items $\geqslant b$, then the sum of their c_{ij} values is the f^o desired.

<u>Step 4</u>. Choose the item (among all the classes) with the minimal a_{ij} greater than C. If the sum of a_{ij} values of chosen items (where the

last chosen element replaces in its class the previously chosen item) is greater or equals b, then the sum of their c_{ij} values is the f° desired.

Otherwise, let C be the c_{ij} of the last chosen item. Repeat Step 4.

In fact, we find $C = \min_x \max_{i,j} c_{ij}x_{ij}$ subject to the min-multiple-choice knapsack constraints. Clearly, $C \leq f^* \leq mC$ and we could take as f° the mC value.

Now consider the "truncated" min-multiple-choice knapsack problem with $c'_{ij} = \lceil c'_{ij}/K \rceil$. Let f_1^* be the optimum value in the latter problem.

We have: (a) $f^* \leq Kf_1^*$ and

(b) $f^* \geq K(f_1^* - m)$, where f^* is the optimum in the original min- multiple-choice knapsack.

We are now in position to present the $O(nm \log m)$, where $n = \sum_{i=1}^{n} k_i$, algorithm BINARY-SEARCH for finding the $\hat{f}$ value, such that $1/_4 \hat{f} \leq f^* \leq \hat{f}$.

ALGORITHM <u>BINARY-SEARCH</u>.

<u>Step 1</u>. Set $p \leftarrow 2/_m f^\circ$, $F \leftarrow 2(m+1)$.

<u>Step 2</u>. Set $K \leftarrow p/_{(m+1)}$, $c'_{ij} = \lceil c_{ij}/K \rceil$, $1 \leq j \leq k_i$, $1 \leq i \leq m$. Use the algorithm FRAME-MIN for solving the "truncated" min-multiple-choice knapsack problem with the c'_{ij} values.

If we derive a feasible solution, then $f^* \leq Kf_1^* \leq KF = 2p$, and the p can be taken as the $\hat{f}$ desired.

If we do not derive a feasible solution, this implies that

$$f_1^* > F \quad \text{and} \quad Kf_1^* > KF.$$

Since $f^* \geq Kf_1^* - Km$, we have $f^* \geq KF - Km = p(m+2)/_{(m+1)} > p$. Then we set $p \leftarrow 2p$ and go to Step 2.

As far as $f^* \leq f^\circ$ and each time after executing Step 2 a lower bound doubles, Step 2 can be executed at most $\log_2 m$ times. Hence, the total time required by BINARY-SEARCH is $O(mn \log m)$ and space is $O(m^2)$.

So an ϵ-approximate algorithm for the problem considered could take the following form:

ALGORITHM APPROX-MULT-CHOICE-KNAPSACK.

Step 1. Use BINARY-SEARCH to find $\hat{f}$ such that $1/_4\hat{f} \leq f^* \leq \hat{f}$.

Step 2. Set $p \leftarrow \epsilon \hat{f}/_4$, $K = p/_{(n+1)}$, $c'_{ij} = \lceil c_{ij}/_K \rceil$, $F = 2(m+1)$.

Use FRAME-MIN to solve (exactly) the "truncated" min-multiple-choice knapsack with the c'_{ij} values, f_1^* being its optimum value.

Then $f^* - Kf_1^* \leq \epsilon \hat{f}/_4 \leq \epsilon f^*$, and Kf_1^* can be taken as an ϵ-approximate solution of the original problem. It is clear that the algorithm requires $O(n \log n + mn \log m + nm/_\epsilon)$ time and $O(n + m^2/_\epsilon)$ space.

In conclusion, let us consider a special case of the two-dimensional knapsack problem, with the objective function $f(x) = \sum_{i=1}^{n} x_i$. If we could find its ϵ-approximate solution, $\bar{f}$, we would take $\epsilon = 1/_{(n+1)}$ and then, since $f(x)$ is integer, and $f^* - F \leq \epsilon f^* \leq n/_{(n+1)} < 1$, the $\bar{f}$ obtained should be equal to the optimum f^*. But to solve the problem exactly is NP-hard problem, the fact due to Dinic and Karzanov [2]. So, if P≠NP, it is impossible to obtain in polynomial (in the problem size) ϵ-approximation algorithm with $\epsilon = 1/_{(n+1)}$ and, therefore it is impossible to have an ϵ-approximate algorithm, polynomial both in problem size and $1/_\epsilon$.

REFERENCES.

1. Babat, L.G. Linear functions on the N-dimensional unit cube. Dokl. Akad. Nauk SSSR 222,761-762 (1975) (Russian).
2. Dinic, E.A., and Karzanov, A.V. A Boolean optimization problem. Preprint, Moscow, VNIISI, 1978 (Russian).
3. Garey, M.R., and Johnson, D.S. Approximation algorithms for combinatorial problems. In Algorithms and Complexity, Acad.Press,NY.,1976
4. Gens, G.V., and Levner, E.V. Approximation algorithms for scheduling problems. Izv. Akad. Nauk SSSR, Tehn. Kibernet.6, 38-43 (1978)
5. Gens, G.V., and Levner, E.V. Complexity of approximation algorithms. Lecture Notes in Computer Science 74, Springer Verlag (1979)
6. Ibarra, O.H., and Kim, C.E. Fast approximation algorithms for knapsack and sum of subset problems. J.ACM 22, 463-468 (1975).
7. Kannan,R.,and Korte B. Approximative combinatorial algorithms. Report WR 78107-OR, Institute of Operat. Research, Bonn, 1978.
8. Lawler, E.L. Fast approximation algorithms for knapsack problems. Memo.No UCB/ERL M77/45, Univ. of California, Berkeley, 1977.
9. Levner, E.V., and Gens, G.V. Discrete Optimization Problems and Approximation Algorithms. Moscow, CEMI (1978) (Russian)
10. Sahni, S. Algorithms for scheduling independent tasks. J.ACM 23, 114-127 (1976)

Computational Relations between Various Definitions of Matroids and Independence Systems *

D. Hausmann, B. Korte

Institut für Ökonometrie und Operations Research, Universität Bonn, W. Germany.

Abstract: The paper analyses the computational relations between well known concepts in the theory of matroids and independence systems. It is shown that these concepts although known to be theoretically equivalent are not computationally equivalent. In particular the girth concept in matroid theory is "stronger" than concepts like independence, rank or basis.

This paper is a short summary of our results in Hausmann and Korte [1,2].

Let E be a finite ground set. It is well known that a matroid on E can be defined by axioms for any one of the following nine concepts: independent set, rank, basis, circuit, spanning set, girth, closure, flat, hyperplane. Having introduced one of these concepts axiomatically the other concepts can be defined subsequently in terms of the first basic concept (cf. von Randow [5], Welsh [6]). In any case we obtain the same combinatorial structure, so they are considered to be theoretically equivalent. The purpose of this paper is to show that they are not at all equivalent if they are considered as basic steps in matroid algorithms.

For each of the nine matroid concepts we introduce a certain oracle, i.e. a mapping defined on 2^E. In a natural way INDEPENDENT, BASIC, CIRCUIT, SPANNING, FLAT, HYPERPLANE map 2^E into {YES, NO} (e.g. INDEPENDENT(S) = YES iff S is independent), RANK and GIRTH map 2^E into $\{0, \ldots, |E|, \infty\}$, and CLOSURE maps 2^E into 2^E. Now let R be one of the nine oracles and let A be an algorithm which can call R as a subroutine. If the number of calls of A on R is bounded by a polynomial of $|E|$, A is called a polynomial algorithm using R. Now we say that an oracle R_1 is polynomially reducible to an oracle R_2 ($R_1 \to R_2$) iff there exists a polynomial algorithm using R_2 which, given a subset $S \subseteq E$, produces as output $R_1(S)$. Our

* Supported by Sonderforschungsbereich 21 (DFG).

main results for matroids are now summarized in form of a graph.

Theorem 1: In the following graph there is a path from an oracle R_1 to an oracle R_2 iff R_1 is polynomially reducable to R_2.

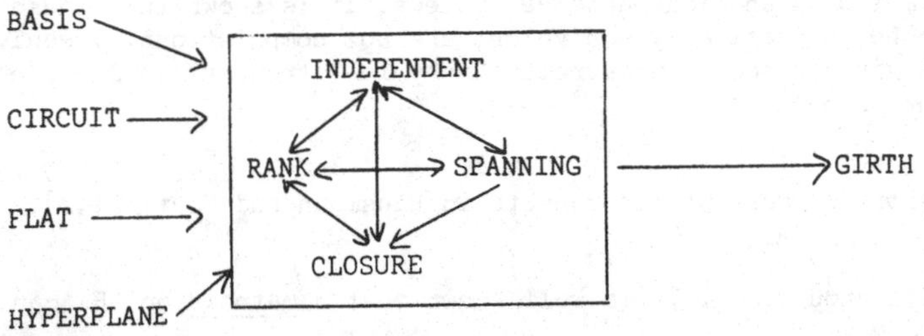

The proof of this (and also of the next) theorem can be found in Hausmann and Korte [1,2]. Similar results have also been obtained by Robinson and Welsh [4].

In applications of combinatorial algorithms we rarely encounter matroids, but much more often general independence systems. An independence system is a pair $(E,\mathcal{I})$ where E is a finite set and $\mathcal{I}$ is a collection of subsets of E satisfying $\emptyset \in \mathcal{I}$ and

$$F \subseteq G \in \mathcal{I} \Rightarrow F \in \mathcal{I}.$$

Clearly a matroid is an independence system $(E,\mathcal{I})$ satisfying the third axiom

$$F,G \in \mathcal{I},\ |F| < |G| \Rightarrow \exists g \in G \setminus F : F \cup \{g\} \in \mathcal{I}.$$

Starting with the independent sets we can define the same concepts as in the matroid case. Some of these concepts, however, split up into two different concepts. Thus we define the r-closure of a set $S \subseteq E$ to be the set $\{e \in E: \text{rank}(S \cup \{e\}) = \text{rank}(S)\}$ and the c-closure of S to be $S \cup \{e \in E: \exists \text{ circuit } C: e \in C \subseteq S \cup \{e\}\}$ (cf. Matthews [3]). A set $F \subseteq E$ is an r-flat if it equals its r-closure and an c-flat if it equals its c-closure. A set $S \subseteq E$ is b-spanning if it contains a basis (i.e. a maximal independent set) and r-spanning if $\text{rank}(S) = \text{rank}(E)$.

We will now analyse the computational relations between these concepts which are again represented in a natural way by oracles. There is a major difference between the matroid case and the independence system case. In the matroid case each oracle R_1 which we considered is at least exponentially reducible to any other oracle R_2, that means, there exists an algorithm using R_2 which, given $S \subseteq E$, after at most exponentially many calls on R_2 produces as output $R_1(S)$. But this is not true for independence systems. Therefore in this case our results are representated by two directed graphs.

Theorem 2: In the directed graph G_1(resp. G_2) there is a path from an oracle R_1 to an oracle R_2 iff R_1 is exponentially (resp. polynomially) reducible to R_2.

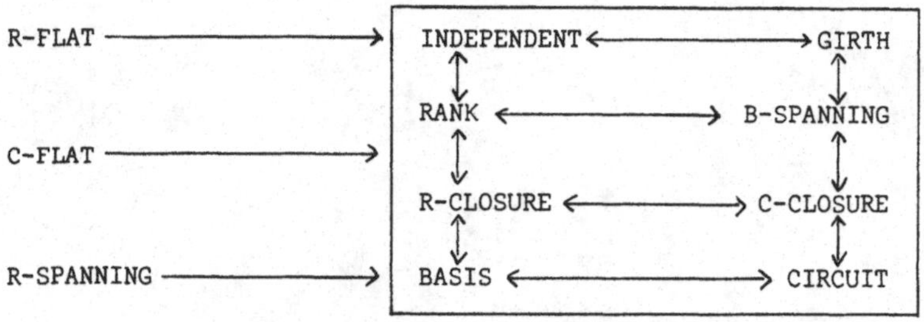

G_1: graph of exponential reducibility

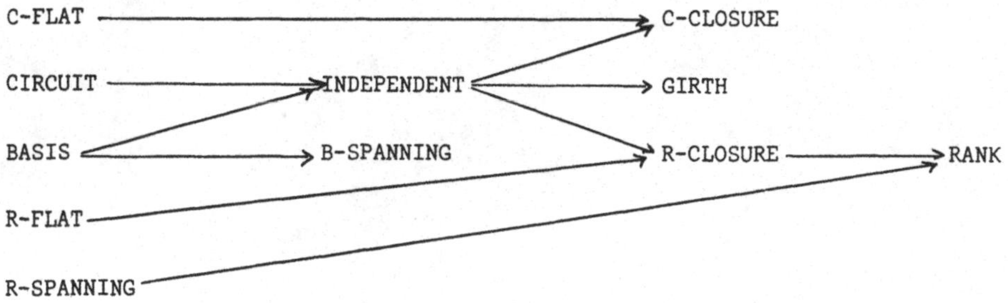

G_2: graph of polynomial reducibility

References

[1] D. Hausmann, B. Korte: Algorithmic versus axiomatic definitions of matroids. Report 79141-OR, Institut für Ökonometrie und Operations Research, University of Bonn, Bonn, W. Germany (1979).

[2] D. Hausmann, B. Korte: The relative strength of oracles for independence systems. Report 79143-OR, Institut für Ökonometrie und Operations Research, University of Bonn, Bonn, W. Germany (1979).

[3] L. Matthews: Closure in independence systems. Report 7894-OR, Institut für Ökonometrie und Operations Research, University of Bonn, Bonn, W. Germany (1978).

[4] G.C. Robinson, D.J.A Welsh: The computational complexity of matroid properties. Working paper (1979).

[5] R. von Randow: Introduction to the Theory of Matroids. Springer Verlag. Berlin, Heidelberg, New York (1975).

[6] D.J.A. Welsh: Matroid Theory. Academic Press. London, New York, San Francisco (1978).

RELATIONS AMONG INTEGER PROGRAMS

Jakob Krarup
Institute of Datalogy
University of Copenhagen

Stanislaw Walukiewicz
Systems Research Institute
Polish Academy of Sciences

The computational tractability of an integer programming problem is heavily dependent on its formulation. There is thus a permanent need for investigations of alternative formulations of a given integer programming problem; some of the results hitherto obtained have certainly proven their usefulness in practice but much more research lies ahead.

In this paper we study the relations between some linear programming relaxations of a given integer program on one side and one of its alternative formulations on the other. An estimate is provided for the maximal possible difference between the optimal values for the integer program and the corresponding linear LP-relaxations. In particular, we consider transformations of different linear nad nonlinear location problem. It is also shown how the dynamic programming procedure of the rotation of an integer constraint can reduce the duality gap. In conclusion, we demonstrate how the results of the above analyses can be implemented in algorithms for solving integer programs.

1. INTRODUCTION

It is a standard technique in mathematical programming to transform (reformlate or reduce) one problem into another equivalent in a certain sense. This technique is very important in integer programming, since any integer program has infinitely many equivalent formulations defined below. As a rule the solution time for different equivalent formulations varies substantially assuming the algorithm, implementation and the computer type fixed (see e.g., [13]) and it often seems to be true that the commonly used formulation is hardest to solve.

We will consider an integer programming problem

$$(P_1) \qquad v(P_1) = \max \sum_{j=1}^{n_1} c_j x_j$$

$$\text{s.t.} \qquad \sum_{j=1}^{n_1} a_{ij} x_j \leqslant b_i \qquad i=1,\dots,m_1$$

$$x \geqslant 0\,, \qquad x \in Z^{n_1}$$

where all data in (P_1) are integer and Z^{n_1} is a set of all n_1-dimensional integer vectors. By $F(P_1)$ we denote the set of all feasible solutions and by $v(P_1)$ the value of the objective function at an optimal point denoted by x^*.

An integer programming problem

$$(P_2) \qquad v(P_2) = \max \sum_{j=1}^{n_2} \hat{c}_j x_j$$

$$\text{s.t.} \qquad \sum_{j=1}^{n_2} \hat{a}_{ij} x_j \leqslant \hat{b}_i$$

$$x \geqslant 0\,, \qquad x \in Z^{n_2}$$

is <u>equivalent</u> to (P_1) if and only if

i) there exists an one-to-one correspondence $h : F(P_1) \longrightarrow F(P_2)$.

ii) h is order-preserving, i.e., for any $x_1, x_2 \in F(P_1)$ $cx_1 \geqslant cx_2$ if and only if $\hat{c}h(x_1) \geqslant \hat{c}h(x_2)$.

In most commonly used transformation we require $F(P_1) = F(P_2)$ and $c_j = \hat{c}_j$. In this paper we will study such equivalent transformations. We note that the transformation of an integer problem into the equivalent binary problem and the transformation of nonlinear problem into the equivalent linear problem are examples of such equivalence.

In section 2 we introduce so called tighter equivalent formulation for a given integer problem. Next we review briefly methods for constructing such formulations. In section 3 we introduce a measure of the tightness of a given formulation and show how it can be used in an estimation of the maximal number of cuts needed to solve a given problem by the method of integer forms. The other type of relations among integer programs are relaxations, which are studied in section 4.

2. TIGHTER EQUIVALENT FORMULATIONS

Computational results reported in [1,13,14] indicate that a given integer programming problem is much easier to solve or at least a good near-optimal solution is obtained relatively quicky if the feasible region to the corresponding linear programming problem is smaller, i.e., if the continuous formulation of a given problem is tighter. By $(\bar{P}_1)$ we will denote the linear programming relaxation of (P_1). Problem (P_2) is <u>tighter equivalent formulation</u> of (P_1) if $F(P_2)= F(P_1)$ and $F(\bar{P}_2) \subseteq F(\bar{P}_1)$.

A convex hull of $F(P_1)$ gives the tightest possible formulation of (P_1) and for such a formulation an optimal solution to the linear programming relaxation is at the same time an optimal solution to (P_1). Unfortunately we do not know how to find efficiently a convex hull for a given integer problem. For instance, the lifting facets procedure described by Zemel in [15] requires solving of an exponential number of integer programs in the worst-case example. Moreover, the linear programming problem obtained in this way is, in general, very difficult to solve because of its large degeneracy. Therefore, the search for the tightest possible formulation although theoreticaly interesting is impractical at present. We also note that an aggregation of $m>1$ equality constraints into one equivalent equality does not give a tighter equivalent formulation.

One of the most often used methods for obtaining a tighter equivalent formulation for a given problem is a logical analysis of $F(P_1)$. Often the knowledge of a practical problem helps very much in such nonalgorithmic approach [1,13]. As an example we will consider two formulations of a simple plant location problem SPLP .

Given a set $I=\{1,\ldots,m\}$ of possible locations for establishing new plants, SPLP deals with the supply of a single commodity from a subset of these to set $J=\{1,\ldots,n\}$ of clients with prescribed demands for the commodity. Let c_{ij} be the total cost for supplying all of client j's demand from the i-th plant, and let x_{ij} denote fraction of the client j's demand supplied from that plant. By f_i we denote the fixed cost for establishing plant i. Given the above cost structure, we seek a minimum cost production transportation plan satisfying all demands. If we assume that $y_i = 1$ if plant i is open and $y_i = 0$ otherwise, then SPLP can be formulated in two different ways:

$$\min\left(\sum_{i=1}^{m}\sum_{j=1}^{n} c_{ij}x_{ij} + \sum_{i=1}^{m} f_i y_i\right)$$

s.t. $$\sum_{i=1}^{m} x_{ij} = 1 \quad , \quad i=1,\dots,n$$

either $$y_i - x_{ij} \geqslant 0 \quad i=1,\dots,m \ , \quad j=1,\dots,n \tag{1}$$

or $$ny_i - \sum_{j=1}^{n} x_{ij} \geqslant 0 \quad i=1,\dots,m \tag{2}$$

$$x_{ij} \geqslant 0 \quad i=1,\dots,m \ , \quad j=1,\dots,n$$

$$y_i = 0 \text{ or } 1 \quad i=1,\dots,m$$

We will denote by (Q_1) the formulation of SPLP if mn "disaggregated" constraints (1) are used. In the formulation (Q_2) only m "aggregated" constraints are used. All these constraints are devices to ensure that the total fixed cost for a facility is incurred whenever positive shipments are made from it. It is easy to see that $F(\bar{Q}_1) \subseteq F(\bar{Q}_2)$ so (Q_1) is a tighter equivalent formulation of SPLP. Paper [10] gives a delailed description of methods for solving SPLP and the main conclusion from such an analysis is that (Q_1) is much easier to solve than (Q_2).

Now we describe briefly three algorithmic approaches to the construction of tighter equivalent formulation of a given integer program. In all these approaches constraints are considered separately.

2.1. Parallel shiftings

The procedure is based on the following

Theorem 1 [11].

A hyperplane with integer coefficients

$$a_1x_1 + a_2x_2 + \dots + a_nx_n = b$$

passes through integer points if and only if the greatest common divisor (gc d) of $a_1,\dots,a_n$ divides b.

If (g cd) of $a_1,\dots,a_n$ does not divide b then we may decrease b by one without changing $F(P_1)$ and again apply theorem 1 [11].

2.2. Minimal Covers

Consider a binary inequality

$$a_1x_1 + a_2x_2 + \ldots + a_nx_n \leq b \qquad (3)$$

$$x_j = 0 \text{ or } 1 \qquad j \in N = \{1,\ldots,n\}$$

Without loss of generality we may assume $0 < a_j \leq b$ and that all a_j are integer. Let F be a set of all feasible solutions to (3). By x^S we will denote a binary vector $(x_1,\ldots,x_n)$ such that $x_j = 1$ for $j \in S$ and $x_j = 0$ for $j \notin S$ where $S \subseteq N$.

A set $S \subseteq N$ is called a <u>minimal cover</u> for (3) if $x^S \notin F$ but $x^R \in F$ for any proper subset R of S. By $\bar{S}$ we denote a set of all minimal covers for (3). The following theorem is a slight modification of theorem 2 in [2].

<u>Theorem 2</u>.

Two inequalities are equivalent if and only if they have the same set of minimal covers.

As a consequence an equivalent to (3) inequality

$$\hat{a}_1x_1 + \hat{a}_2x_2 + \ldots + \hat{a}_nx_n \leq \hat{b}$$

may be constructed after solving a system of linear inequalities

$$\sum_{j \in S} \hat{a}_j \leq \hat{b} + 1 \qquad \text{for any } S \in \bar{S}$$

$$\hat{b}, \hat{a}_j \geq 0, \qquad j \in N.$$

This may be formulated as a linear programming problem after introducing an objective function, e.g., such that

$$\sum_{i=1}^{m} \hat{b}_i \longrightarrow \min$$

Various objective functions and computational aspects of such a procedure are discussed in [2].

<u>2.3. Constraint Rotation</u>

We say that a constraint $\hat{a}x \leq \hat{b}$ is <u>stronger</u> than $ax \leq b$ if $\hat{a}_j \geq a_j$, $j \in N$, $\hat{b} \leq b$ and at least for one j a strong inequality holds. Consider again inequality (3) and if for any $r \in N$ there exists $x \in F$ such that

$$\sum_{j \in (N-r)} a_jx_j + a_r = b \qquad (4)$$

then without violating F it is impossible to find a constraint stronger than (3). Therefore all constraints satisfying (4) will be called <u>strongest constraints</u>. In (4) we write $j \in (N-r)$ instead $j \in (N-\{r\})$ to simlyfy notations.

The rotation procedure is a method for finding a strongest constraint equivalent to one that is given. For any $r \in N$ we have

$$a_r x_r \leqslant b - \sum_{j \in (N-r)} a_j x_j = b - b_r$$

where $b_r \leqslant b - a_r$ for any $x \in F$ such that $x_r = 1$.

Let

$$b_r^* = \max \sum_{j \in (N-r)} a_j x_j$$

subject to $x \in F, \quad x_r = 1.$

We may substitute $\hat{a}_r = b - b_r^*$ as a new value of a_r because from the definition of b_r^* we have that the constraint

$$\sum_{j \in (N-r)} a_j x_j + \hat{a}_r x_r \leqslant b$$

is equivalent to (3).

The value b_r^* is computed as a maximal element of the set

$$B_r = \left\{ b_r \mid b_r = \sum_{j \in J} a_j, \ J \subset (N-r), \ b_r \leqslant b - a_r \right\}$$

and the sets B_r, $r=1,2,\dots,n$ are computed by a dynamic programming procedure [8,5]. At the end of this procedure we obtain a strongest constraint equivalent to a given one. In other words, a constraint which cannot be rotated is a strongest one. An integer programming in which each constraint is a strongest one is called an almost linear integer programming problem.

Theorem 3 [12].

Any integer programming problem (P_1) can be transformed into equivalent almost linear problem (P_2) and

i) $F(\bar{P}_2) \subseteq F(\bar{P}_1)$

ii) $n_2 = n_1$

iii $m_2 \leqslant m_1 + 1$

The rotation procedure can be applied to a cut constructed by any cutting plane algorithm. It can be shown by simple numerical example that the cuts constructed in the primal integer algorithm, the dual integer algorithm and in the method of integer forms are not, in general, the strongest cuts.

At the end of analysis of methods for constructing tighter equivalent formulations we compare their computational complexity. Obviously, the rotation procedure and the minimal covers approach give a stronger constraint than the parallel shifting. It was shown in [8] that the rotation procedure has the computational complexity

$0(n^2b)$ of additions and comparisons and therefore this procedure is efficient for all practical problems when range of data is small. On the other had, it is easy to construct a practical example in which the number of minimal covers grows exponentially with n.

3. MAXIMAL NUMBER OF CUTS

As we mentioned in previous section there exists a convincing empirical evidence showing that a tighter equivalent formulation is easier to solve. This means that there exists a relation between the maximal number of cuts needed to solve a given problem and its formulation. Such a relation has been established by Kolokolow in [9]. Here we give a direct proof of his result. We will consider only the method of integer forms but the result can be generalized for the other cutting plane algorithms.

Consider the following problem

$$(P) \qquad v(P) = \text{lex max} \sum_{j=1}^{n} c_j x_j$$

$$\text{s.t.} \qquad \sum_{j=1}^{n} a_{ij} x_j \leqslant b_i \;, \quad i=1,\dots,m$$

$$x \geqslant 0 \;, \qquad x \in Z^n$$

where all data are integer. We consider a lexicographic maximum to have a unique optimal solution. To avoid many pathological cases we assume that (P) is feasible and $F(\bar{P})$ is bounded. Under such assuptions the method of integer forms gives an optimal solution after constructing a finite, say r, number of cuts. We show that an estimation of r is a function of a set $S = F(\bar{P}) - F(P)$ and the duality gap $d = v(\bar{P}) - v(P)$.

We will use a simplex tableau with n+1 rows and n columns where the first row denoted as x_o corresponds to the objective function. We assume that a source row i^* for a cut generation is chosen according to the following rule

$$i^* = \min \left\{ i \;\middle|\; x_i \notin Z, \quad i=1,\dots,m \right\}.$$

Let $T = \{\bar{x}^1, \bar{x}^2, \dots, \bar{x}^r\}$ be set of all optimal linear programming solutions generated by the method of integer forms, where $\bar{x}^k = (x_o^k, x^k) \in R^{n+1}$ and $x^k \in S$. So $\bar{x}^1$ is an optimal solution to $(\bar{P}_1) = (\bar{P})$ and $\bar{x}^r$ is an optimal solution to $(\bar{P}_r)$ such that $x^r \in Z^n$ and $x^* = x^r$. We partition set T into two subsets $T_1 = \{\bar{x}^k \mid \bar{x}^k \in T, \; x_o^k \notin Z\}$

and $T_2 = \{\bar{x}^k | \bar{x}^k \in T, x_o^k \in Z\}$.

Consider a vector $\bar{x}^k \in T_1$. As $x_o^k \notin Z$ after adding the $(k+1)$-st cut an optimal solution to the corresponding linear programming problem is $\bar{x}^{k+1}$ with $x_o^{k+1} \leq \lfloor x_o^k \rfloor$, where $\lfloor x \rfloor$ is the greatest integer not greater than x and $\lceil x \rceil = \lfloor x \rfloor + 1$. Therefore $|T_1| \leq \lceil v(\bar{P}) \rceil - v(P) \leq d + 1$.

Consider now a vector $\bar{x}^k \in T_2$. Using the same arguments it is easy to show that if $x_j^k \notin Z$ and $x_i^k \in Z$ for all $i < j$, then $x_j^{k+1} \leq \lfloor x_j^k \rfloor$. Define $S^\alpha = \{x | x \in S, cx = \alpha\}$ for any $\alpha \in Z$ and $v(P) \leq \alpha \leq \lceil v(\bar{P}) \rceil$. Let $I(S^\alpha)$ be a set of all round-offs of $x \in S^\alpha$, where each element x_j is rounded down to $\lfloor x_j \rfloor$ and rounded up to $\lceil x_j \rceil$. As between $\lfloor x_j^k \rfloor$ and $\lceil x_j^k \rceil$ there is at most one cut constructed by the method of integer forms, the number of such cuts is at most equal $|I(S^\alpha)| - 1$. Therefore

$$r = |T_1| + |T_2| \leq v(\bar{P}) - v(P) + \sum_{\alpha = v(P)}^{\lceil v(\bar{P}) \rceil} \left(|I(S^\alpha)| - 1 \right)$$

If $S = \{x \in R^n | \ a \leq x \leq b \ , \quad a,b \in Z^n\}$ then

$$|I(S)| = \prod_{i=1}^{n} \left(b_i - a_i + 1 \right) - 1$$

and

$$r \leq \left(\lceil v(\bar{P}) \rceil - v(P) + 1 \right) \prod_{i=1}^{n} \left(b_i - a_i + 1 \right) - 1 \leq (d + 2) \prod_{i=1}^{n} \left(b_i - a_i + 1 \right) - 1$$

From this estimation we can see that the smaller the value $v(\bar{P})$ and the tighter the formulation of a given problem the smaller the number of necessary cuts for the worst-case example.

Results of computational experiments described in [5,6] fully support the above estimation. For all problems taken from literature the number of cuts needed is smaller for a tighter formulation. The total time of constraints and cuts rotation is rather small and never exceeds 30 per cent of the total computational time. And the main conclusion of this experiment is that for problems which are not by nature almost linear it always pays off to find a tighter equivalent formulation and next to solve it. The detailed description of the computational experiments and their results are given in [5,6].

4. RELAXATIONS

In this section we briefly review two relaxations of a given problem which have proved to be useful in the branch-and-bound procedures [3] and in the integer programming duality [3,7]. We will consider an integer programming problem in the following form

$$(P) \quad v(P) = \max \quad cx$$

s.t. $\quad Ax \leq b$

$$x \in D = \{x \mid Gx \leq h,\ x \geq 0,\ x \in Z^n\}.$$

where A is m x n matrix, G is k x n matrix and x,b,c,h are vectors of the appropriate dimension. To avoid many pathalogical cases we assume D bounded and (P) feasible. In particular we assume that $Gx \leq h$ include bounds on all variables.

The Lagrangean relaxation of (P) is defined for any $u \in R^m$, $u \geq 0$ as the integer programming problem

$$(P_L) \quad v(P_L) = \max \left(cx + u(b - Ax)\right)$$

s.t. $\quad x \in D$.

As $F(P) \subseteq F(P_L)$ and $v(P) \leq v(P_L)$ for any $x \in F(P)$ and for any $u \geq 0$, then it is reasonable to find such multipliers $u_1,\ldots,u_m$ that they solve the dual problem

$$(D_L) \quad v(D_L) = \min_{u \geq 0} v(P_L)$$

Geoffrion in [3] proved that $v(P) \leq v(D_L) \leq v(\bar{P})$, where $(\bar{P})$ is the linear programming relaxation of (P). Problem (P_L) has the integrality property if $v(P_L) = v(\bar{P}_L) = v(D_L)$ for any $u \geq 0$. Therefore, if (P_L) has the integrality property then the bounds obtained from the Lagrangean relaxation cannot be better than the bounds from (P_L). Examples given in [3,7] show that for many practical problems their Lagrangean relaxations have the integrality property.

The surrogate relaxation of (P) is defined for any $u \in R^m$, $u \geq 0$ as the integer programming problem

$$(P_S) \quad v(P_S) = \max \quad cx$$

s.t. $\quad uAx \leq ub$

$$x \in D$$

If $Gx \leq h$ include only upper bounds on all variables, then (P_S) is the knapsack problem. In this section we will consider only such a case. Similarly we define

$$(D_S) \quad v(D_S) = \min \ v(P_S)$$

s.t. $\quad u \geq 0$

The surrogate relaxation (P_S) has the integrality property if $v(P_S) = v(\bar{P}_S) = v(D_S)$, $u \geq 0$. As (P_S) is the knapsack problem it would happen for very special structured problems, when $uAX \leq ub$ gives the tightest possible formulation of (P_S). Greenberg and Pierskalla in [4] proved that the surrogate duality gap i.e., $v(D_S) - v(P)$ is at least as small and often smaller than the Lagrangean duality gap i.e., $v(D_L) - v(P)$. Therefore the surrogate relaxation is more powerful. Moreover, all results known for the knapsack problem can be used in solving (P_S). For simplicity of presentation of these results we assume that (P_S) is a binary knapsack problem

$$(P_S) \qquad v(P_S) = \max \sum_{j=1}^{n} c_j x_j$$

$$\text{s.t.} \qquad \sum_{j=1}^{n} a_j x_j \leqslant b$$

$$x_j = 0 \text{ or } 1, \quad j=1,\ldots,n\ .$$

Without loss of generality we may assume that $0 < a_j \leqslant b$,

$$\sum_{j=1}^{n} a_j > b$$

$$\frac{c_1}{a_1} \geqslant \frac{c_2}{a_2} \geqslant \cdots \geqslant \frac{c_n}{a_n}$$

Then an optimal solution to the linear programming relaxation $(\bar{P}_S)$ is given by

$$\bar{x} = \left(1,1,\ldots,1,\ \frac{1}{a_{r+1}}\left(b - \sum_{j=1}^{r} a_j\right),\ 0,0,\ldots,0\right)$$

So $0(n)$ additions and comparisons are needed to compute $\bar{x}$ in the worst -case example.

By $\hat{x}$ we denote a near-optimal solution, greedy solution to the knapsack problem computed at $0(n)$ additions and comparisons and

$$c\hat{x} \geqslant \sum_{j=1}^{r} c_j$$

In [12] the following theorem is proved

<u>Theorem 4</u>

$$\frac{v(P_S) - c\hat{x}}{v(P_S)} \leqslant \frac{\lfloor v(\bar{P}_S) \rfloor - c\hat{x}}{\lfloor v(\bar{P}_S) \rfloor} \leqslant 1 - \frac{1}{b} \sum_{j=1}^{r} a_j \leqslant 1 - \frac{r}{b}\ .$$

So the relative error of $\hat{x}$ is always less than 100 per cent and does not depend on n. Computational examples in [12] show that for practical problems this error is much smaller, about 1 per cent. So for the surrogate relaxation we have methods for computing relatively good near-optimal solutions. Finaly we note that in general the surrogate constraint can be rotated even if all original constraints $Ax \leqslant b$ in (P) are the strongest cuts. And the rotation of the surrogate constraint will decrease the duality gap and the relative error of a greedy solution to the surrogate relaxation.

REFERENCES

1. Beale E.M.L., Tomlin J.A.: An Integer Programming Approach to Class of Combinatorial Problem. Mathematical Programming, 3, (1972) 339-344.
2. Bradley G.H., Hammer P.L., Wolsey L.: Coefficient Reduction for Inequalities in 0-1 Variables. Mathematical Programming, 7, (1974) 263-282.
3. Geoffrion A.M. : Lagrangean Relaxation for Integer Programming. Math. Programming Study, 2, (1974) 82-114.
4. Greenberg H.J., Pierskalla W.P. : Surrogate Mathematical Programming. Operations Research, 18, (1970) 924-936.
5. Kaliszewski I., Walukiewicz S. : Tighter Equivalent Formulations of Integer Programming Problems. Proceedings of the IX International Symposium on Mathematical Programming, Budapest, August 1976.
6. Kaliszewski I., Walukiewicz S. : A Computationally Efficient Transformation of Integer Programming Problems. Mimeo, Systems Research Institute, Warsaw, December 1978 .
7. Karwan M.H., Rardin R.L. : Some Relationships between Lagrangian and Surrogate Duality in Integer Programming. Mathematical Programming, 17, (1979) 320-334.
8. Kianfar F. : Stronger Inequalities for 0.1 Integer Programming. Operations Research, 19, (1971) 1373-1392.
9. Kolokolow A.A. : An Upper Bound for Cutting Planes Number in the Method of Integer Forms, (in Russian), Metody Modelirowanija i Obrabotki Informacji, Novosibirsk, Nauka (1976) 106-116.
10. Krarup J., Pruzan P.M. : Selected Families of Discrete Location Problem. Part III: The Plant Location Family, Working Paper No WP-12-77, University of Calgary, August 1977 .

11. Salkin H.M., Breining P. : Integer Points on the Gomory Fractional Cut. Dept. of Operations Research, Case Western Reserve University, 1971.

12. Walukiewicz S. : Almost linear integer problems. (in Polish), Prace IOK, Zeszyt nr 23, 1975.

13. Williams H.P. : Experiments in the Formulation of Integer Programming Problems. Math. Prog. Studies 2, (1974) 180-197.

14. Williams H.P. : The Reformulation of Two Mixed Integer Programming Problems. Research Report 75-11. University of Sussex, October 1975.

15. Zemel E. : Lifting the facets of 0-1 Polytopes. Management Sciences Research Report No 354, Carnegie-Mellon University, December 1974 .

LINEAR OPTIMIZATION FOR LINEAR AND BOTTLENECK OBJECTIVES WITH ONE NONLINEAR PARAMETER

U. Zimmermann
Mathematisches Institut
Universität zu Köln
D-5000 Köln 41

Abstract

We consider for continuous coefficient functions the minimization of linear resp. bottleneck objectives of the form $c(t)^T x$ resp. $\max \{c_j(t) \mid x_j \geq 0\}$ over a given polyhedron $\{x \mid Ax = b, x \geq 0\}$ for all $t \in [\alpha,\beta] \subseteq \mathbb{R}$. For both objectives a finite sequence $(B_k \mid k = 1,2,\ldots,r)$ of feasible bases B_k optimal in $[t_k,t_{k+1}]$ for $k = 1,2,\ldots,r$ with $\alpha = t_1 < t_2 < \ldots < t_r = \beta$ is determined using the zeroes of a set of nonlinear continuous functions. Further on we discuss the relationship of t-norm to bottleneck and time cost problems for compact sets of feasible points.

Introduction

Linear optimization problems with a nonlinear parameter have recently been considered by Weickenmeyer [7], Väliaho [6] and Wüstefeld and Zimmermann [9]. A discussion of the literature can be found in these papers. Weickenmeyer and Väliaho describe methods for the case of polynomial dependence on the parameter for all coefficients of the problem. In [9] the considered polyhedron does not depend on the parameter but the coefficient functions in the objective function are arbitrary continuous functions in the parameter. This problem is of particular interest for combinatorial optimization problems for which the bases of the polyhedron are just the feasible points. It has previously been discussed by Carpentier [1], Sarkisjan [4] and Weinert [8].

We will assume that the *set of feasible solutions* is given in the form

$$(1.1) \qquad P := \{x \in \mathbb{R}^n \mid Ax = b, x \geq 0\}$$

with real m×n matrix A of rank m and real positive m-vector b. Further on we assume that P is *nonempty*, *bounded* and *nondegenerate*. Let $I := [\alpha,\beta] \subseteq \mathbb{R}$ and let $c: I \to \mathbb{R}^n$ be a continuous function. Then the linear optimization problem with one nonlinear parameter is

$$(1.2) \qquad \min_{x \in P} \; c(t)^T x \qquad , \; t \in I \; .$$

From linear programming we know that there exist optimal basic solutions for every $t \in I$. Let $\hat{P}$ denote the set of the bases of P. Then w.l.o.g. we consider

$$(1.3) \qquad \min_{B \in \hat{P}} \; c_B(t)^T x_B \qquad , \; t \in I \; .$$

For the corresponding linear bottleneck problem

$$(1.4) \qquad \min_{x \in P} \max \{c_j(t) \mid x_j > 0\} \qquad , \; t \in I \; ,$$

the existence of optimal basic solutions is shown in [2] and [11]; therefore again w.l.o.g. we consider

$$(1.5) \qquad \min_{B \in \hat{P}} \max \{c_j(t) \mid j \in B\} \qquad , \; t \in I \; .$$

We are only interested in finite solutions of (1.3) and (1.5). A finite sequence $(B_k \mid k = 1,2,\ldots,r)$ of feasible bases B_k optimal in $[t_k,t_{k+1}]$ for $k = 1,2,\ldots,r$ with $\alpha = t_1 < t_2 < \ldots < t_r = \beta$ is called a *finite optimal solution* of (1.3) resp. (1.5).

In section 2 we give criteria for the existence of finite optimal solutions and develop local optimality criteria for certain closed subintervals of I. In section 3 we formulate an algorithm for the determination of a finite solution. In both sections we proceed under consideration of (1.3) as well as (1.5). As (1.3) is discussed in [9] we give only proofs with regard to (1.5). In section 4 a special parametric problem, the so called t-norm problem is related to nonparametric versions of bottleneck and time-cost problems. In particular a result of Steinberg [5] is generalized for compact subsets of $\mathbb{R}^n_+$. Section 5 contains some concluding remarks on possible extensions and difficulties.

2. Finite solutions and local optimality criteria

During this section it will become clear that the theoretical structures of problems (1.3) and (1.5) are very similar. As far as possible we will use the same denotations for both problems. Differences in assumptions, results and interpretations will be discussed. Proofs will only be given with regard to (1.5). For a detailed discussion of (1.3) we refer to [9].

As mentioned in the introduction it suffices to consider the finite set $\hat{P}$ of feasible bases. A basis $B \in \hat{P}$ is the indexvector of *basic* variables. Then N denotes the index vector of *nonbasic* variables. Partitions of vectors and matrices will be indexed by these vectors in the usual manner.

The *set of all optimal bases* with respect to $t \in I$ is denoted by $V(t)$. The set of all bases optimal in an open interval $(\tau,\tilde{\tau})$ resp. a closed interval $[\tau,\tilde{\tau}]$ is denoted by $V(\tau,\tilde{\tau})$ resp. $V[\tau,\tilde{\tau}]$. Then $I(B):= \{t\in I \mid B\in V(t)\}$ is the set of all parameters for which B is an optimal basis. This set is easily characterized by the *objective value function* $z_B: I \to \mathbb{R}$, i.e. $z_B(t)=c_B(t)^T x_B$ resp. $z_B(t) = \max\{c_j(t) \mid j\in B\}$ for (1.3) resp. (1.5). Obviously $B \in V(t)$ iff $z_B(t) \leq z_{B'}(t)$ for all $B' \in \hat{P}$. As $\hat{P}$ is finite and z_B is a continuous function for all $B \in \hat{P}$ the set I(B) is a closed subset of I. In general it may be disconnected. Thus I(B) is the union of mutually exclusive closed subintervals of I which we call the *optimality intervals* of B.

In [9] we find necessary and sufficient conditions for the existence of a finite solution (cf. section 1). Further on a finite construction of such a solution exists. The proposed method is only of theoretical value; in order to develop a practicable solution method stronger assumptions are necessary.

In [9] it is assumed for (1.3) that the *reduced cost coefficient func-*

tions

(2.1) $$\overline{c}_j(t) := c_j(t) - c_B(t)^T A_B^{-1} A_j$$

for all $j = 1,2,\ldots,n$ and for all $B \in \hat{P}$ have only finitely many zeroes or vanish identically in I. For (1.5) we assume the same for the *difference functions*

(2.2) $$d_{\mu\nu}(t) := c_\mu(t) - c_\nu(t)$$

for all $\mu,\nu = 1,2,\ldots,n$. For both problems it can easily be seen that endpoints of optimality intervals in (α,β) are characterized as certain zeroes of the respective functions.

The following is an outline of the solution method for both problems. At first we find $B \in V(\alpha)$. Then if $B \in V(\tau)$ for $\tau \in [\alpha\ \beta)$ we determine a certain zero $\tilde{\tau} > \tau$ of the above defined functions (or $\tilde{\tau}=\beta$) such that $B \in V(t)$ for a single parameter $t \in (\tau,\tilde{\tau})$ implies $B \in V[\tau,\tilde{\tau}]$. If $B \notin V(t)$ then we perform a pivot step introducing an appropiate nonbasic variable in order to find a "better" basis. A sequence of pivot steps yields $B \in V[\tau,\tilde{\tau}]$. Then we proceed to $\tau := \tilde{\tau}$ and repeat the procedure until we reach the end of the considered interval I.

In the performance of this method we have to specify optimality criteria ($B \in V(t)$), the set of considered zeroes defining $\tilde{\tau}$ and the pivoting rule for the respective problem.

For (1.3) we know from linear programming that $B \in V(t)$ iff $\overline{c}_N(t) \geq 0$. In order to find such a simple criterion for (1.5) we extend the problem to a lexicographic optimization problem with two components. With $E(t) := \{j \mid z_B(t) = c_j(t)\}$ we consider

(2.3) $$\operatorname{lex}\min_{B \in P} \Big(z_B(t), \sum_{j \in E(t)} x_j\Big) \qquad , t \in I \ .$$

An optimal basis B for (2.3) is optimal for (1.5), too. Furthermore the sum of its basic variables corresponding to bottleneck values is minimum among all $B' \in V(t)$. Let $L(t) := \{j \mid z_B(t) > c_j(t)\}$ and $U(t) := \{j \mid z_B(t) < c_j(t)\}$. Then we define

$$(2.4)\qquad \tilde{c}_j(t) := \begin{cases} 1 & \text{for } j \in E(t) \cup U(t) \\ 0 & \text{for } j \in L(t) \end{cases}$$
$$\bar{c}_j(t) := \tilde{c}_j(t) - \tilde{c}_B(t) A_B^{-1} A_j$$

for j=1,2,...,n. With these reduced cost coefficient functions $\bar{c}_j$ we find the following result from [11].

<u>(2.5) Theorem</u>

Let $t \in I$ and $B \in \hat{P}$. Then $B \in V(t)$ with respect to (2.3) iff $\bar{c}_j(t) \geq 0$ for all $j \in N \setminus U(t)$.

Thus we have established optimality criteria of the same structure for (1.3) and (1.5). This is not surprising in view of the results in [11] which show that the corresponding nonparametric problems can be treated as special examples of the same algebraic optimization problem.

Now we will consider the respective definitions of $\tilde{\tau}$. Its general form is

$$(2.6)\qquad \tilde{\tau}(F,\gamma) := \min(\{t \in (\tau,\gamma] \mid f \in F,\ f(t) = 0\} \cup \{\gamma\})$$

with $\gamma \in (\tau,\beta]$ and with a family of functions F. This means that $\tilde{\tau}$ is the least zero greater then τ of the functions in the family F if it lies in the interval $(\tau,\gamma]$. If $(\tau,\gamma]$ does not contain a zero of any of these functions then $\tilde{\tau} = \gamma$.

With respect to (1.3) in [9] the zero $\tilde{\tau} = \tilde{\tau}(F_1,\beta)$ with

$$(2.7)\qquad F_1 := \{\bar{c}_j \mid \bar{c}_j \neq 0\}$$

is chosen; F_1 is the set of all nonvanishing nonbasic reduced cost coefficient functions with respect to the current basis $B \in V(\tau)$. As $(\tau,\tilde{\tau})$ does not contain a zero of any of these functions optimality of B at a single point $t \in (\tau,\tilde{\tau})$ implies $B \in V[\tau,\tilde{\tau}]$.

The same result can be seen with respect to (1.5) if we choose the set of all nonvanishing difference functions (cf. 2.2) and $\gamma=\beta$. In view of the computational efficiency the considered set of functions should be as small as possible. Therefore we proceed in a little bit more complicated manner. Let $\sigma := \tilde{\tau}(F_2,\beta)$ with

(2.8) $F_2 := \{d_{\mu\nu} \mid d_{\mu\nu} \neq 0,\ \mu \in B \cap E(\tau),\ \nu \in B\}$.

We choose $k \in E(t)$ for an arbitrary $t \in (\tau,\sigma)$. Then $z_B(t') = c_k(t')$ for all $t' \in [\tau,\sigma]$. Now we define $\tilde{\tau} := \tilde{\tau}(F_3,\sigma)$ with

(2.9) $F_3 := \{d_{k\nu} \mid \nu \in N \setminus E(t)\}$.

Then $(\tau,\tilde{\tau})$ does not contain a zero of any of the nonvanishing functions $d_{k\nu}$ for $\nu=1,2,\ldots,n$. Therefore $U(t)$, $E(t)$ and $L(t)$ are constant on $(\tau,\tilde{\tau})$. This shows that optimality in a single point $t \in (\tau,\tilde{\tau})$ implies optimality in $(\tau,\tilde{\tau})$; this implication holds for (1.5) as well as for (2.3). For (1.5) optimality in $(\tau,\tilde{\tau})$ implies even optimality in $[\tau,\tilde{\tau}]$.

On the other hand if the current basis $B \in V(\tau)$ is not optimal for some $t \in (\tau,\tilde{\tau})$ then we have to find a 'better' basic solution. Due to the structure of the optimality criteria in this case there exists a nonbasic variable x_s with $\bar{c}_s(t)<0$. We introduce such a nonbasic variable into the basis.

In [9] it is shown for (1.3) that the new basis B' is again optimal at τ and that a finite sequence of such pivot steps in $V(\tau)$ leads to an optimal solution in $[\tau,\tilde{\tau}]$.

The same result for (1.5) can be proved in the following way. As $U(t)$, $E(t)$ and $V(t)$ are constant in $(\tau,\tilde{\tau})$ we find that $\bar{c}_j(t)$ is constant in $(\tau,\tilde{\tau})$. In [11] the nonparametric case of (2.6) is contained as a special case. In particular it is shown that a pivot step as proposed above leads to a new basis B' with strictly lower objective value, i.e.

$$(z_B(t), \sum_{j \in E(t)} x_j) \underset{\neq}{\succ} (z_{B'}(t), \sum_{j \in E'(t)} x_j)$$

Let $\tilde{\tilde{\tau}} := \tilde{\tau}(F,\beta)$ with respect to the set F of all nonvanishing difference functions. Then in particular the above inequality holds for all $t \in (\tau,\tilde{\tilde{\tau}})$. Thus although $\tilde{\tau}$ is possibly changed after each pivot step the above inequality guarantees that after finitely many steps we find an optimal solution in $(\tau,\tilde{\tilde{\tau}})$. Due to the definition of the final $\tilde{\tau}$ this is an optimal solution in $[\tau,\tilde{\tau}]$, too. Furthermore we know in every step that $L(t) \cup E(t) \subseteq L(\tau) \cup E(\tau)$ for all $t \in (\tau,\tilde{\tau})$. Therefore a pivot step introducing a nonbasic variable x_s with $s \in N \setminus U(t) \subseteq L(t) \cup E(t)$ leads to a new basis B' with $B' \in V(\tau)$. This shows again that a finite sequence of pivot steps in $V(\tau)$ leads to an optimal solution in $[\tau, \tilde{\tau}]$.

Finally we consider the determination of a solution of the extended problem (2.3). This is a special case of a so-called time-cost problem and the solution of its nonparametric version has been discussed in nearly all papers on the nonparametric bottleneck problem, too. Solutions of (2.3) are often preferred to those of (1.5) in applications. At the end of the algorithm which is stated in the next section summarizing the results of this section the determined solution of (1.5) is optimal with respect to (2.3) in all points of I with exception of the calculated zeroes $\tilde{\tau}$. At these finitely many points it is possible that the second component of the optimal objective value function of (2.3) is discontinuous. Therefore at these points optimality with respect to (2.3) should be checked using (2.5). If the criterion is not fulfilled a separate optimal basis in these points has to be determined by appropriate pivot steps.

3. The solution method

We summarize the results of section 2 and describe an algorithm for the solution of (1.3) as well as (1.5). At some points we refer to the differing definitions of formal parameters with respect to the considered problems.

SOLUTION METHOD FOR (1.3) [(1.5)]

① (INITIAL SIMPLEX STEP)
Determine an initial basis $B \in V(\alpha)$; $t_1 := \tau := \alpha$; $k := 1$.

② (LEAST ZERO CALCULATION)
Determine $\tilde{\tau}$ according to (2.7) [resp. (2.8) and (2.9)]; choose $t \in (\tau, \tilde{\tau})$; if $\bar{c}_j(t) \geq 0$ for all $c_j \in F_1$ [resp. $j \in N \setminus U(t)$] then goto ④.

③ (PIVOTING IN $V(\tau)$)
Choose a nonbasic variable x_s with $c_s \in F_1$ [resp. $s \in N \setminus U(t)$] and $\bar{c}_s(t) < 0$; perform a pivot step introducing x_s into the basis, redefine x, B, N; goto ②.

④ (ITERATION)

$B^k := B$; $k := k+1$; if $\tilde{\tau} \geq \beta$ then $t_k := \beta$; $r := k$; stop.
Otherwise $t_k := \tau := \tilde{\tau}$, goto ②.

This algorithm has been proposed for (1.3) in [9]. Computational results for transportation problems with t-norm objective (cf. section 4) are discussed which show that one of the highly time consuming parts has to be seen in the determination of the least zeroes $\tilde{\tau}$. This is not only a problem in view of computational efficiency but also a critical theoretical point. The calculation of the least zero $\tilde{\tau}$ with respect to a family of functions F is a rather difficult numerical problem. For the computations in [9] we used a rather safe bisection method. For the general case we assumed only continuity of the considered functions. Then we know of no method which guarantees to find this very zero $\tilde{\tau}$. For special functions suitable numerical methods should be chosen with great care in view of the necessary accuracy and the resulting computation time.

4. Analysis of t-norm objectives

In this section we discuss the relationship of a parametric problem with special objective and the nonparametric versions of (1.5) and (2.3). In the following the set of feasible points P is assumed to be a compact subset of $\mathbb{R}^n_+$. Thus integer and nonlinear cases are covered as well.

For $c_j \in \mathbb{R}_+$, $j=1,2,\ldots,n$ we define $z(x,t) := \Sigma (c_j)^T x_j$ for all $x \in P$. Then $V(t)$ denotes the set of optimal solutions at t for the *t-norm problem*

(4.1) $$\min\{[z(x,t)]^{1/t} \mid x \in P\}, \; t \in [1,\infty).$$

The equivalent problem $\min\{z(x,t) \mid x \in P\}$ is a special case of (1.2) if P is a polyhedron. Let $d(x) := \max\{c_j \mid x_j > 0\}$ for $x \in P$ and in particular $d(0) := 0$. Then $\overline{V}$ denotes the set of optimal solutions of the nonparametric version $d := \min\{d(x) \mid x \in P\}$ of problem (1.5). Furtheron $\overline{\overline{V}}$ denotes the set of optimal solutions of

$$(d,\delta) := \operatorname*{lex\,min}_{x \in P} \Big(d(x), \sum_{c_j = d(x)} x_j\Big),$$

the nonparametric version of (2.3).

For quadratic assignment problems Steinberg [5] showed the existence of a parameter $\bar{\beta} \geq 1$ such that

(4.2) $$t > \bar{\beta} \Rightarrow V(t) \subseteq \bar{V}.$$

This result has been extended to general 0-1 problems by Zimmermann [10] and Krarup and Pruzan [3]. In [9] the existence of parameters $\bar{\beta}, \bar{\bar{\beta}} \geq 1$ such that (4.2) and

(4.3) $$t > \bar{\bar{\beta}} \Rightarrow V(t) \subseteq \bar{\bar{V}}$$

is shown for transportation problems. The main result of this section describes the existence of such parameters even for compact sets $P \subseteq \mathbb{R}^n_+$.

If $0 \in P$ or $\bar{V} = P$ then (4.2) is fulfilled with $\bar{\beta} = 1$. Thus for a discussion of (4.2) we assume in the following $0 \notin P$ and $\bar{V} \neq P$. Then $C_+ := \{c_j \mid c_j > d\} \neq \emptyset$. Let $d_+ := \min C_+$ and

$$\varepsilon := \min \{ \sum_{j \in C_+} x_j \mid x \notin \bar{V} \}.$$

In the case $\varepsilon = 0$ we can give examples such that there exists no parameter $\bar{\beta}$ fulfilling (4.2).

As $\bar{\bar{V}} \subseteq \bar{V}$ the existence of $\bar{\beta} \geq 1$ fulfilling (4.2) is a necessary condition for the existence of $\bar{\bar{\beta}} \geq 1$ fulfilling (4.3). Therefore it will be assumed throughout the discussion of (4.3). If $\bar{\bar{V}} = \bar{V}$ or $C_- := \{c_j \mid c_j < d\} = \emptyset$ then $\bar{\bar{\beta}} = \bar{\beta}$ fulfills (4.3). Therefore we assume furtheron $\bar{\bar{V}} \neq \bar{V}$ and $C_- \neq \emptyset$. Let $d_- = \max C_-$ and

$$\delta_+ := \min \{ \sum_{c_j = d} x_j \mid x \in \bar{V} \setminus \bar{\bar{V}} \}.$$

In the case $\delta = \delta_+$ we can give examples such that there exists no parameter $\bar{\bar{\beta}}$ fulfilling (4.3).

With the denotations introduced in the above discussion we find the following theorem.

<u>(4.4) Theorem</u>

Let $\emptyset \neq P \subseteq \mathbb{R}^n_+$ compact, let $c_j \in \mathbb{R}_+$ for $j = 1, 2, \ldots, n$ and let $\alpha := \max \{\sum x_j \mid x \in P\}$. Then

(4.4.1) If $0 \in P$ or $\bar{V} = P$ then $\bar{\beta} = 1$ fulfills (4.2). If (4.2) holds for $\bar{\beta} \geq 1$ and $\bar{V} = \bar{\bar{V}}$ or $C_- = \emptyset$ then $\bar{\bar{\beta}} = \bar{\beta}$ fulfills (4.3).

(4.4.2) If $0 \notin P$, $\bar{V} \neq P$ and $\varepsilon > 0$ then (4.2) holds with

$\bar{\beta} := (\ln \alpha - \ln \varepsilon)/(\ln d_+ - \ln d)$.

(4.4.3) If (4.2) is fulfilled for $\bar{\beta} \geq 1$, $\bar{V} \neq \bar{\bar{V}}$, $C_- \neq \emptyset$ and $\delta < \delta_+$ then (4.3) is fulfilled for $\bar{\bar{\beta}} = \max(\bar{\beta}, \gamma)$ with

$\gamma := [\ln(\alpha - \delta) - \ln(\delta_+ - \delta)]/(\ln d - \ln d_-)$.

Proof. The trivial cases (4.4.1) are easily verified. For (4.4.2) we consider $\bar{x} \in \bar{V}$ and $x \in P \setminus \bar{V}$. If $t > \bar{\beta}$ then

$$z(\bar{x},t) \leq \alpha \cdot d^t < \varepsilon \cdot d_+^t \leq z(x,t)$$

implies $V(t) \subseteq \bar{V}$. For (4.4.3) we consider $\bar{\bar{x}} \in \bar{\bar{V}}$ and $\bar{x} \in \bar{V} \setminus \bar{\bar{V}}$. If $t > \bar{\bar{\beta}}$ then

$$z(\bar{\bar{x}},t) \leq \delta \cdot d^t + (\alpha - \delta) \cdot d_-^t < \delta_+ \cdot d^t \leq z(\bar{x},t) \ .$$

Together with $V(t) \subseteq \bar{V}$ this inequality implies $V(t) \subseteq \bar{\bar{V}}$. ∎

(4.4) implies in particular the results in [3], [5], [9] and [10]. In the case $P \subseteq \mathbb{Z}_+^n$ in the nontrivial cases we find $1 \leq \varepsilon$ and $\delta + 1 \leq \delta_+$. Therefore the assumptions $\varepsilon > 0$ resp. $\delta < \delta_+$ are superfluous and the existence of $\bar{\beta}, \bar{\bar{\beta}} \geq 1$ fulfilling (4.2) and (4.3) is always valid. Similar arguments yield the existence in the case of finite sets P. In particular this holds for the consideration of the set $\hat{P}$ of feasible bases for a given polyhedron P.

At first (4.4) shows that nonparametric bottleneck problems can be solved by solving problem (4.1) for a single parameter $t > \bar{\beta}$, i.e. a linear problem with coefficients $(c_j)^t$. This is mainly a theoretical statement as computational experience shows that special procedures for bottleneck objectives are usually faster.

Secondly (4.4) shows that it is sufficient to solve (4.1) on $[1, \bar{\beta}]$ resp. $[1, \bar{\bar{\beta}}]$ only. Solutions with respect to larger values of the parameter belong to $\bar{V}$ resp. $\bar{\bar{V}}$. Then a solution of (4.1) describes the "continuous" change from a solution of the linear problem (t=1) via the solutions of the bottleneck problem ($t \geq \bar{\beta}$) to the solutions of the time-cost problem ($t \geq \bar{\bar{\beta}}$).

5. Concluding remarks

The algorithm proposed in this paper for a solution of (1.3) and (1.5) again shows the similarity of these problems. A generalization to problems of the form min $\{f(B,t) \mid B \in \hat{P}\}$, $t \in I$ with continuous functions $f(B,\cdot): I \to \mathbb{R}$ can be obtained without difficulties if the difference functions $f(B,\cdot)-f(B',\cdot)$ for $B, B' \in \hat{P}$ have only finitely many zeroes or vanish identically. Further on it would be very useful to know a local optimality criterion in order to avoid enumeration of all bases.

The critical point in the method is the necessary determination of the least zero of a set of continuous functions. This leads in the general case to theoretical as well as computational difficulties. Computational results [9] show that the solution of the parametric t-norm transportation problem is highly time consuming. Reasonable results can only be obtained for special problems for which it is possible to develop efficient codes for the pivot step as well as for the determination of least zeroes.

REFERENCES

[1] CARPENTIER, J.: A method for solving linear programming problems in which the cost depend nonlinearly on a parameter, Electricité de France, Diréction des Études et Recherches, May 1959.

[2] DERIGS, U., ZIMMERMANN, U.: Duality principles in algebraic linear programming, Report 1978-6, Mathematisches Institut, Universität zu Köln.

[3] KRARUP, J.; PRUZAN, P.M.: On the equivalence of minimax and minisum 0-1 programming problems, Proceedings of the Polish-Danish Mathematical Programming Seminar, ed. by J. Krarup and S. Walukiewiez, Part one (1978) 48-65.

[4] SARKISJAN, S.D.: On a parametric linear optimization problem with nonlinearly parameter dependent objective, Trudy vyčislit. Centra Akad. Nauk Armjan SSR. Erevan gosudarst. Univ. 2 (1964) 10-16 (Russian).

[5] STEINBERG, L.: The backboard wiring problem: a placement algorithm, SIAM Review 3 (1961) 37-50.

[6] VÄLIAHO, H.: A procedure for one-parametric linear programming, BIT 19 (1979) 256-279.

[7] WEICKENMEYER, E.: Zur Lösung parametrischer linearer Programme mit polynomischen Parameterfunktionen, ZOR 22 (1978) 131-149.

[8] WEINERT, H.: Probleme der linearen Optimierung mit nichtlinear-einparametrischen Koeffizienten in der Zielfunktion, Mathematische Operationsforschung und Statistik 1 (1970) 21-43.

[9] WÜSTEFELD, A., ZIMMERMANN, U.: Nonlinear one-parametric linear programming and t-norm transportation problems, to appear in NRLQ.

[10] ZIMMERMANN, U.: Boole'sche Optimierungsprobleme mit separabler Zielfunktion und matroidalen Restriktionen, doctoral thesis, University of Cologne, 1976.

[11] ZIMMERMANN, U.: Duality for algebraic linear programming, to appear in Linear Algebra and its Applications.

Acknowledgement

These investigations were performed in the Department of Operations Research, Stanford University, USA under sponsorship of the Deutsche Forschungsgemeinschaft, Federal Republic of Germany.

SELECTED ASPECTS OF A GENERAL ALGEBRAIC MODELING LANGUAGE

Johannes Bisschop and Alexander Meeraus
Development Research Center
World Bank
1818 H Street, N.W.
Washington, D.C. 20433

Abstract: The paper touches on the role of models in a policy/planning environment, and establishes the need for a general algebraic modeling system. Its main purpose, however, is to develop a notation which can be understood by both man and machine. The language is part of a general algebraic modeling system currently under development. A more extensive version of this paper can be obtained from the authors.

1. INTRODUCTION

In the early days of mathematical modeling, large applications were mostly of a military or industrial nature. Models were used to describe and solve well-defined problems in the areas of production and distribution, and they were employed on a routine basis. In many instances it was considered cost-effective to establish a small group of technical people whose sole responsibility was to maintain and to improve the existing package of models. In recent years the scope of mathematical modeling applications has widened, and modeling environments different from those described above have emerged. The U.S. Government, for instance, has supported the development of a large number of models, and many planning agencies around the world use mathematical models as their major tool for analysis.

In the policy/planning environment the role of models is often extended beyond their traditional use as a way to get numerical solutions to well-defined problems. Models are used to express perceptions and abstractions of reality, and they continuously change as their developers learn more about the uncertain real-world problem. Models provide the model builder/decision maker with a formal framework for data collection and analysis. They are seldomly used to get definite answers, but are employed as guides in planning and decision making, or as moderators between groups of people with conflicting knowledge and/or interests. Usually a system of many loosely connected models of different types is developed, and very few models, if any, are used on a routine basis.

The cost of building and maintaining these models is high, while the benefits are not always clearly defined. A study by the National Science Foundation on the development and use of mathematical models within the U.S. Government provides some interesting figures. The total development cost of the 650 models surveyed was US$100 million ($154,000 per model), and it took on the average 17 months to make a model operational. It was observed that 75% of all models can be operated only by the original development team, despite strong efforts in model and program documentation. Actual policy use of these models by groups other than the model designers has been minimal. Given the median size of 25 equations (only 6 models had more than 1,000 equations), the above figures look rather depressing as it takes 3 weeks and $6,000 to develop one equation on the average. Based on our own experience, we find that eighty to ninety percent of total resources currently spent on large modeling exercises is for the generation, manipulation, and reporting of these models. It is evident that this percentage must be reduced significantly if models are to become effective tools in planning and decision making in a large variety of disciplines and institutions.

This paper is the by-product of an ongoing effort to build a general algebraic modeling system (GAMS) at the World Bank. In the description of the language we have mostly emphasized the syntactical aspects, and only touch upon some of the semantic issues. A more complete description of both the modeling system and the modeling language is forthcoming.

2. THE NEED FOR A GENERAL ALGEBRAIC MODELING SYSTEM

One way to establish the need for a modeling system is to examine some of the problems that the modeling community is currently faced with. Based on our own experience, mostly from attempts to disseminate previous and ongoing research in a planning environment, we have encountered several problem areas.

The documentation of large models and their modifications is one such problem. If a project is large, and continues for one or two years, the cost of complete documentation becomes horrendous. A decision is usually made to maintain a few versions of a model. In practice this means that some basic experiments can be repeated. In the long run, however, the value of the available software becomes essentially zero as people change jobs, and any changes to existing versions require extensive set-up time. A related problem is the communication of models to interested persons that are not part of the development team. As there are no standards in notation, it is often difficult to judge from any write-up what exactly the model is. Experimentation with the model may enhance ones understanding, but this requires the use of both the model and report generators. As these programs are nontrivial, they in turn require the use of a technical person. The extensive time and money requirements prohibit many outsiders from even attempting to satisfy their own curiosity with regard to the model. No effective dissemination knowledge can therefore take place.

With the existing technology in modeling software there is no common interface with the various solution routines modelers can use for their family of models. As each solution package usually requires different data structures, it becomes both time and money consuming to switch back and forth between solution algorithms. As a result models tend to get locked into one solution package which at times limits their development. There is also no general-purpose software for the linking of models, an activity that has become more prevalent with the increased use of models.

Although the above problem areas tend to discourage large-scale modeling exercises, they are certainly not the major obstacle to the effective use of modeling in a policy/planning environment. It is the extensive time, skill and money resources currently required for the building of models that hinders their effective use. The heart of the problem is the fact that solution algorithms need a data structure and a problem representation which is impossible to comprehend by humans. At the same time, problem representations that are meaningful to humans, are not acceptable to machines. The two translation processes required can be identified as the main source of difficulties and errors. With today's technology, each translation process is broken down into a number of interrelated steps where most of the coordination and control has to be done by humans, and is therefore subject to error. That's why extensive time, skill and money resources are required for the completion of large-scale modeling exercises. In addition, it is not surprising that the overall reliability (the probability of no mistakes) of our modeling practice is embarrassingly low.

A remedy to all of the above problems is the evolution of a new modeling technology. We will need to move away from the existing labor/skill intensive approach to model building, and replace it with a machine intensive approach. That is why we have begun the development of a general algebraic modeling system (GAMS). This system provides the model builder with a notation that can be understood by both humans and machines. As a result, only one document is needed for the representation and generation of models. Most tasks that were previously performed by humans, will now be completed by the machine. In addition to providing a unified notation, the system will interface automatically with existing solution routines. It will also have the capability of linking various models. As the main purpose of this paper is to develop the language used in GAMS, we will elaborate on the modeling system in subsequent papers. The next section will serve as a first introduction to key components in the language.

3. AN ILLUSTRATION OF THE KEY WORDS IN GAMS

Most mathematical models today are specified using index sets, some data tables, some English describing manipulations involving sets and/or data, and a system of symbolic equations. As there are no guiding standards, the notation often lacks clarity, is incomplete, and shows inconsistencies. With just a little more rigor, namely replacing English with algebraic set and data mappings, existing knowledge and skills can be employed to build models. The language in GAMS stays as close as possible to existing algebraic conventions, but has a few additions to handle complexities inherent in large models. The result is a powerful notation, which allows for a complete and unambiguous representation of models.

For illustrative purposes, consider the cannery transportation problem taken from the book Linear Programming and Extensions by G.B. Dantzig. A company desires to supply its three warehouses from two canneries with given inventories in each, and wants to minimize the total shipping cost. The GAMS representation of this problem is stated on the next page.

As can be noted from the model description, we have restricted ourselves to a small character set which is available on most computers. In addition, we have assumed that there is no carriage control available (i.e., no subscripts or superscripts), and that there are only capital letters. Within these few limitations, we have adhered as much as possible to existing mathematical conventions.

The above model statement can be viewed as an integrated data base. In addition to the data tables and assignment statements, there are the symbolic equations which represent data that can only be obtained via some solution algorithm. Both data and symbolic equations are needed for a complete model representation in GAMS.

There are several key words used in the above model description. They are (in order of occurrence) SET, PARAMETER, TABLE, VARIABLE(S), EQUATION(S), SUM, MODEL, SOLVE..USING..MINIMIZING, and DISPLAY. We will comment on each of them.

Sets are used as driving indices in many mathematical models. They usually have a short name followed by a description. Following the description is a listing of the set elements contained between two "slashes." The set elements are names with up to ten characters (no blanks inside them), all separated by a comma or an end of line. Each name can have an associated description if needed (e.g. the element KANSAS has a description KANSAS CITY).

A parameter can be defined in a similar fashion, with a number following each label as we did for parameter A. An algebraic definition using an assignment statement is also possible, and this was done for parameter R (each warehouse requirement is 300 units). A third way to define a parameter is via some tabular arrangement as we did for the parameter UTCOST. Both row and column descriptions of the parameter are required. As we shall see in the next section, this two-dimensional framework can be used to represent parameters with more than two dimensions attached to them. As the table name description following the name UTCOST is restricted to one line in GAMS, we extended it using a comment statement. Any statement with a * in the first column is a comment statement in GAMS.

Variable and equation names must be defined first before they can appear in any symbolic equations. One can recognize a symbolic equation by the two dots following the equation name. Note that the availability constraint SUPPLY is defined over the domain (set) C. It is a short-hand notation for two availability constraints, namely one for each cannery. In the next section we shall see how one can control this domain of definition in equation statements. The summation in the SUPPLY equation is indicated by SUM, and followed by the set name W to which the summation operation is to be applied. Each symbolic equation in GAMS has a type. In the above example we have =L= (a less than or equal to constraint), =G= (a greater than or equal to constraint), and =E= (an equality constraint).

```
SET  C  CANNERIES / SEATTLE, SAN-DIEGO /;
SET  W  WAREHOUSES /
        NEW YORK
        CHICAGO
        KANSAS  KANSAS CITY /;

PARAMETER  A  AVAILABLE INVENTORIES (CASES OF TINS PER DAY) /
        SEATTLE    350
        SAN-DIEGO  650 /;

PARAMETER  R  REQUIRED INVENTORIES (CASES OF TINS PER DAY);
        R(W) = 300 ;

TABLE  UTCOST  UNIT TRANSPORT COST (DOLLARS PER CASE)
*              FROM CANNERY  C  TO WAREHOUSE  W
               NEW YORK       CHICAGO        KANSAS
SEATTLE            2.5            1.7            1.8
SAN-DIEGO          2.5            1.8            1.4

VARIABLES  X  SHIPMENTS (CASES OF TINS PER DAY);

EQUATIONS  SUPPLY  AVAILABILITY CONSTRAINT (CASES OF TINS PER DAY)
           DEMAND  REQUIREMENT CONSTRAINT (CASES OF TINS PER DAY)
           COST    COST ACCOUNTING EQUATION (DOLLARS PER DAY) ;

*   AVAILABILITY CONSTRAINT IMPOSED ON EACH CANNERY
    SUPPLY(C)..
        SUM(W, X(C,W))          =L=          A(C);
*   - - - - - - - - - - - -             - - - - - - - - -
*   TOTAL SHIPMENTS LEAVING             AVAILABILITY AT
*   CANNERY  C                          CANNERY  C
*   - - - - - - - - - - - -             - - - - - - - - -

*   REQUIREMENT CONSTRAINT IMPOSED BY EACH WAREHOUSE
    DEMAND(W)..
        SUM(C, X(C,W))          =G=          R(W);
*   - - - - - - - - - - - - - -         - - - - - - - - -
*   TOTAL SHIPMENTS ARRIVING            REQUIREMENT AT
*   AT WAREHOUSE  W                     WAREHOUSE  W
*   - - - - - - - - - - - - - -         - - - - - - - - -

*   COST ACCOUNTING EQUATION REFLECTING TRANSPORT COST
    COST..

*          SUM((C,W), UTCOST(C,W) * X(C,W)) =E=  TRCOST;
*          - - - - - - - - - - - - - - - - - -
*                 TOTAL TRANSPORT COST
*          - - - - - - - - - - - - - - - - - -

MODEL  CANNERY  THE CANNERY TRANSPORTATION MODEL / ALL / ;

SOLVE  CANNERY  USING  LP  MINIMIZING  TRCOST ;

DISPLAY  X.AL, SUPPLY.MC ;
```

A model in GAMS is the selection of a subset of the symbolic equations. In the above example all equations are included in the model. Once a model is defined, a particular algorithm must be chosen. In this case linear programming (LP) is selected to minimize the variable TRCOST in the model CANNERY. Display statements can be used to get selected pieces of data. Here we have asked for the activity levels associated with the variables (X.AL), and the shadow prices (marginal costs) associated with the availability constraints (SUPPLY.MC). Note that throughout the model description, each statement has started with a key word, and terminated with a semicolon.

This section was written to give the reader a quick overview of several important aspects of the language in GAMS. The cannery example does not portray some of the complexities associated with the representation of large-scale models. That is why a more extensive description of the notation in GAMS is developed in the next section.

4. A MINIMAL VERSION OF THE LANGUAGE IN GAMS

Most problems associated with model building can be reduced to a basic question involving communication. How can one communicate data and its associated complex mathematical structures when the human mind is limited in its power to grasp and comprehend many issues simultaneously. The only tool available to us is our power of abstraction which aids us in understanding the complexity of real world phenomena. It allows us to define partitionings, mappings, nestings, and short-hand notation. The language in GAMS is essentially a short-hand notation which takes advantage of any partitionings, mappings and nestings. In this chapter we will examine the syntactic and some of the semantic rules that govern the notation. We have organized the material by subsections, each describing an important part of the language.

4.1 Sets and Set Mappings

A simple (one-dimensional) set in GAMS is a finite collection of labels. These sets play an important role in the indexing of algebraic statements. The cannery example in section 3 contains two such simple sets (namely C and W), and both their syntax and use are illustrated there. Several one-dimensional sets can be related to each other in the sense that there is a correspondence between them. As an example consider the correspondence between countries and regions. Depending on one's viewpoint, this is a one-to-many or one-to-one correspondence. To each country corresponds a specific set of regions, while each region corresponds to one specific country only. As we shall see, these correspondences play an important role in GAMS as they can be used to control the domain of definition of assignment statements and symbolic equations.

The syntax for set correspondences is much like the one for single sets. Consider the following illustration

```
SET  CR  COUNTRY-REGION CORRESPONDENCE /

     INDONESIA.N-SUMATRA
     INDONESIA.N-JAVA
     MALAYSIA.W-MALASIA
     .

     .                              /;
```

or,

```
SET  CR  COUNTRY-REGION CORRESPONDENCE /
     INDONESIA.(N-SUMATRA, E-JAVA), MALAYSIA.W-MALAYSIA, .../;
```

Note that the period is used as an operator to relate the elements of the different sets, and that the order of the elements in the correspondence is fixed (in this case country first, region second). In order to reduce unnecessary repetition, the parentheses can be used when several elements in one set correspond to a single element of the other set. There can be any number of sets in a correspondence. The following few lines illustrate a 3-dimensional set mapping.

```
SET  RZD  REGION ZONE DISTRICT MAPPING /

     NORTH.IRRIGATED.(W-NORTH, C-NORTH, E-NORTH)
     CENTRAL.(IRRIGATED.(NW-UPPER, NE-UPPER)
              RAINFED.(S-UPPER, W-LOWER, E-LOWER))
              .
              .                                   /;
```

There are ways to change the information contents of sets and set mappings. This can be done via algebraic assignment statements, which require all sets to be indexed. Assume that a set R of regions has been defined, and that a copy of this set is desired. Then one can write the following GAMS statements.

```
SET  R  COPY OF SET R ;  RR(R) = R(R) ;
```

The next example is a redefinition of RR on the basis of the above set correspondence RZD. Assume that the new set RR should contain all regions that are not rain-fed. The instruction SUM, already mentioned in the previous section, denotes a union instead of a summation when applied to sets.

```
RR(R) = R(R) - SUM(D, RZD(R, 'TROPICAL', D)) ;
```

Note that the 3-dimensional correspondence RZD requires 3 driving indices. Since the middle index is invariant, we have used the quotes to indicate a specific element rather than the entire set.

4.2 Data Tables

Tabular arrangements of data are a very convenient way to describe multi-dimensional parameters. The unit cost table in section 3 is an example of a

2-dimentional parameter. The following table illustrates a 4-dimensional parameter, where 3 dimensions are captured in the row descriptions, while the fourth dimension is contained in the column label.

```
        TABLE  L  LABOR COEFFICIENTS IN HOURS PER RAI
*                 BY REGION, CROP ROTATION, TECHNOLOGY AND MONTH
                                    JANUARY   FEBRUARY   MARCH   APRIL

NORTH-UPP.SUGARCANE.TRAD-BUFF          2          2         2      12
NORTH-UPP.SUGARCANE.MOD-TRACT          1          2         2      10
             .
             .

+                                     MAY       JUNE      JULY   AUGUST

NORTH-UPP.SUGARCANE.TRAD-BUFF         12         35        30      45
NORTH-UPP.SUGARCANE.MOD-TRACT         12         30        30      40
             .
             .
```

Note that we have specified the units for the entire table in the table heading. As it stands at the moment, unit analysis has to be done by the model builder, although one of our goals is to make automatic unit analysis an integral part of the data base system in GAMS. The order of the sets used in the row and column descriptions in the table statement must be maintained in later references to the parameter. For the above example this will be L(R,C,T,M) where R, C, T and M refer to the simple sets. Note that all columns could not fit on one line. Any table, however, can be continued by using a plus operator at the beginning of each new set of column headings.

4.3 Assignment and Equation Statements

Most of the syntax used in assignment statements and equations are the same, although it is straightforward to detect if a GAMS statement is an assignment or an equation.

An assignment statement in GAMS is an instruction to perform some data manipulation and store the result. It can be compared to a FORTRAN statement where the result of the operations performed is stored under the name that appears on the left side of the equal sign. As an example consider the parameter DIST(I,J) indicating the distance from location I to location J, where the elements in the sets I and J are identical. Assume that initially only the lower triangular part of DIST was specified in a TABLE statement, and that we are interested in specifying the entire matrix. We can write the following sentence

```
DIST(I,J) = DIST(I,J) + DIST(J,I) ;
```

The right-hand side is defined for each 2-tuple of the Cartesian product of the sets I and J. A copy of DIST(I,J) is stored in a temporary work array, and the entries in DIST(I,J) are replaced with the results from the additions for all pairs (I,J) in a parallel fashion. Note that all values of DIST(I,J) that were not defined in the TABLE statement are assumed to be zero. An alternative but equivalent GAMS statement for the above replacement is as follows.

```
DIST(I,J) = MAX(DIST(I,J), DIST(J,I)) ;
```

Here the MAX operator selects the largest of the two values inside the parentheses.

An equation in GAMS is a symbolic representation of one or more constraints to be used as part of a simultaneous system of equations, or an optimization model. It always begins with the equation name, possibly indexed, followed by two dots (periods). We again refer to the equations in the Cannery example of section 3. In the next section we will develop additional examples of equations and assignment statements while describing the role of the conditional operator used in the language.

4.4 The $ Operator

Partitioning large models by using driving indices provides an elegant short-hand notation. Complexities, however, are introduced when there are restrictions imposed on the partitionings. As these complexities arise continually in large-scale models, we have strived for an elegant and effective way to incorporate them in a model statement.

Let us begin with an example. Define the sets R and D as regions and districts respectively. Assume that for each district in a region we know the level of income YD(R,D), and that we want to determine the regional income YR(R) for each of the regions. Writing the assignment statement

```
YR(R) = SUM(D, YD(R,D)) ;
```

is meaningless as not every district is contained in each region. We need to use, therefore, the relationship between the sets R and D. Let RD be the set correspondence between these two sets. Then we can write the following assignment statement

```
YR(R) = SUM(D$RD(R,D), YD(R,D)) ;
```

Here the dollar sign is used as a conditional operator. For each specific region R it restricts the sum to be over those elements of D for which the correspondence RD(R,D) is defined.

Let A be a name or an expression in GAMS, and let B be a name or a true-false expression. Then the phrase A $ B is a conditional statement in GAMS where the name A is considered or the expression A is evaluated if and only if the name B is defined or the expression B is true.

When the dollar operator is used in an assignment statement, it can appear both on the right and on the left of the equal sign. When it appears on the left, it controls the domain over which the assignment is defined. Whenever the condition following the name on the left is not true the existing data values contained under that name remain unaffected. If on the other hand that same condition is applied to the right of the equal sign, the existing values contained in the name on the left will be set to zero whenever the condition is not true.

In order to illustrate the conjunctive use of the dollar operator and logical phrases constrained in an assignment statement, consider the next example. Let the sets P, I and M denote processes, plants and machines respectively. The parameter K(M,I) denotes the number of units of available capacity of machine M in plant I, while the parameter B(M,P) describes the required number of units of capacity of machine M per unit level of process P. We want to define a zero-one parameter, PPOSS(P,I), indicating which processes P need to be considered for plant I. We can write the following set of logical relations always resulting in either a zero or one.

```
PPOSS(P,I) = SUM(M $ (K(M,I) EQ 0), B(M,P) NE 0) EQ 0 ;
```

Here the expression B(M,P) NE 0 will contain a value 1 if process P is dependent on machine M, and 0 otherwise. These values are summed over all machines M that are not available in plant I. If the resulting sum is zero for process P then the process is not dependent on unavailable machines, and should therefore be considered. Note that PPOSS is one in this case. If the resulting sum is not 0, the process is dependent on at least one unavailable machine, and should therefore not be considered. The parameter PPOSS is set to zero in this case.

When the dollar operator appears in an equation statement, it is used to control the generation of equations and/or variables. As an illustration let CAP be an equation name referring to capacity constraints, and let Z be a variable name referring to levels of process operation. Using the notation of the previous paragraph, we can write the following symbolic equation.

```
CAP(M,I) $ (K(M,I) GT 0)..
    SUM(P $ PPOSS(P,I), B(M,P) * Z(P,I))  =L=  K(M,I) ;
```

In this example the system will generate an equation for a specific pair of machines and plants only when the capacity of that machine in that plant is strictly positive. Similarly, only those variables that refer to processes which can be operated at a positive level will be generated.

The last page of this paper contains a simplified version of the grammar in GAMS. The grammar is a set of rules that determines which constructs are permitted in the language. It describes the syntax. It is also the most compact and complete way to describe the modeling language in GAMS.

Equations and Assignments

```
<equation>           ::=  <left part> .. <expression> <type> <expression> ;
<assignment>         ::=  <left part> = <expression> ;
<left part>          ::=  <left> | <left> $ <primary>
<left>               ::=  ident | ident ( <index list> )
<index list>         ::=  <index expression> | <index list> , <index expression>
<index expression>   ::=  <simple index> | <simple index> <lag operator> <expression>
<simple index>       ::=  ident | 'index'
<primary>            ::=  <variable> | number | SUM ( <con control> , <expression> ) |
                          function ( <expression list> ) | ( <expression> )
<variable>           ::=  ident | ident ( <index list> )
<expression list>    ::=  <expression> | <expression list> , <expression>
<con control>        ::=  <control> | <control> $ <primary>
<control>            ::=  ( <ident list> ) | ident
<expression>         ::=  <term> | <unary operator> <term> |
                           <expression> <binary operator> <term>
<term>               ::=  <primary> | <term> $ <primary>
<unary operator>     ::=  - | + | NOT
<binary operator>    ::=  - | + | * | / | ** |EQ|GT|GE|LT|LE|NE|AND|OR|XOR
<type>               ::=  =L= | =G= | =E=
<lag operator>       ::=  + | - | ++ | --
```

Set Definitions

```
<set definition>     ::=  SET <declaration list> ; | CONSTANT SET <declaration list> ;
<declaration list>   ::=  <declaration> | <declaration list> , <declaration>
<declaration>        ::=  <set name> | <set name> / <value list> /
<set name>           ::=  ident | ident text
<value list>         ::=  <element text> | <value list> , <element text> | ALL
<element text>       ::=  <element> | <element> text
<element>            ::=  <simple element> | <element> . <simple element> |
                          <element> . ( <element list> ) | ( <element list> )
<element list>       ::=  <element> | <element list> , <element>
<simple element>     ::=  index | 'index'
```

Table Definitions

```
<table definition>   ::=  TABLE  <table name> <table body> ;
<table body>         ::=  eol <columns> <table values> |
                             <table body> eol + <columns> <tables values>
<columns>            ::=  <element>| <columns> <element>
<table values>       ::=  <row> | <table values> <row>
<row>                ::=  eol <element> | eol <element> <row values>
<row values>         ::=  <signed number> | <row values> <signed number>
<table name>         ::=  <identifier> | <identifier> <text>
```

SOFTWARE DESIGN FOR ALGORITHMS OF HIERARCHICAL OPTIMIZATION

A. Kalliauer

Österreichische
Elektrizitätswirtschafts-
Aktiengesellschaft
1010 Vienna, Austria

This paper deals with a software concept, which was developed during realisation of an optimization algorithm for solving a class of large-scale problems with a linear constraint system showing a nested structure with coupling variables. Special attention shall be drawn to the interactions between the software concept to be designed and choice and realisation of the mathematical method. The principal ideas, on which the software concept is based, seem to be applicable to a fairly general class of algorithms in Hierarchical Optimization.

1. Introduction:

In various applications - as for example in economical planning - we often have to deal with large-scale dynamical models. For the major part of problems these models are structured (e.g. models of decentralised planning). The problem can be taken as a system of interconnected subproblems (e.g. corresponding to subregions). Furthermore long-time planning models can be multiple-subdivided with respect to time. (In a planning model concerning a year's period weekly and daily subintervalls could be defined.) Thus we get a nested-structured model.

We obtained our results working on the optimal operation control for the Austrian hydro storage system for power production over the year's period. One of the main purposes is the evaluation of new storage projects in future power systems. This planning purpose required the development of a general and abstract model forcing us to write well designed computer programs to develop a good design of model and algorithm.

Optimization methods advantageously using such hierarchical structures are called decomposition techniques (see [1], [2]). Hierarchical Optimization - shown on a two-level system - means definition of a master problem and subproblems according to the structure. Respective algorithms generally work in the following way: In an iterative process at first all the subproblems are optimized giving locally optimal solutions, then optimization steps in the master problem are made towards the global optimum using the information (e.g. dual variables) derived from the local suboptima. This in turn gives better starting points for the local optimizations. The process is terminated with the optimum for the total problem being reached. In the case of a multilevel or nested-structured system a master problem itself can represent a subproblem in a higher level system. The hierarchical optimization results in a nested process of subproblem solutions.

The structure of the problem appears in the structure of the constraint system (see Fig. 1 and Fig. 2). Applying decomposition techniques the working matrices, which result from the constraint system, maintain the same sparseness of the structure.

2. Basic aspects for realisation:

Software design cannot be made independent of the choice of the method. Thus, one of the main aspects is to base software design on a clear and transparent optimization concept. In the case of a nested structure it is usefull to apply the same optimization concept for both the subproblems and the master problem. Thus each subproblem uses working matrices of the same kind. (The working matrices of a subproblem are composed of vectors derived from the constraints local to this subproblem or to one of the interior subproblems. Local constraints are constraints of a subproblem, which do not correspond to some interior subproblem.)

Now we should try to choose a method or transform a method such that the working matrix for the master problem consists of the working matrices of the subordinated subproblems and that the matrices of the subproblems can be taken out of the working matrix of the master problem. In this way updating of the working matrices for one subproblem can be performed together with updating for the remaining sub- and master-problems. In this case initialisation work can be omitted when changing from one subproblem to another. (The additional expenditure to keep all matrices ready for use can be neglected for very large structures, because this refers only to a small part of the problem.) Concerning the storage requirement we need at maximum the requirement for the matrix corresponding to the total problem when solved with a compactification method.

In our realisation we used a partitioning concept solving problems with coupling variables (see [3], [4]). To illustrate this concept we consider a simple unstructured problem

$$\max_{x,y} f(x,y)$$

$$\text{s.t.} \quad A\,x + Ey \le b.$$

We define a subproblem as follows:
Optimize the system for x to a given fixed vector y:

$$Z(y) = \max_{x} f(x,y)$$

$$\text{s.t.} \quad A\,x \le (b - Ey).$$

The master problem consists in changing the "coupling" vector y.

Thus, we have defined a chain-like structure with a master problem and one subproblem. This concept can be used in the same sense for nested-structured block-angular systems. As solution algorithm for master problem and subproblems we use a primal method based on the concept of feasible direction algorithms ([5]). Since during the iteration process constraints are activated and inactivated, the working matrices have to be updated.

In our simple example we have applied the partitioning concept to an unstructured problem. That means we are able to use this concept neglecting whether or not a structure exists. This comes up to working with linear dependent constraint vectors, which has to be handled anyway.

This circumstance gives rise to special application of the partitioning concept column by column neglecting whether or not there exists a structure. This very essential transformation of the algorithm we have called "successive" partitioning. Here one constraint vector maximally is assigned to each subproblem. Advancing from one subsystem to the next means to consider the next row of the integrated working matrix system. Thus we have no difference between working off one row after the other in one submatrix and proceeding from one subproblem to the next. We will call this concept "row by row" algorithm.

Although column by column partitioning means some more work to be carried out at the local matrices (which can be neglected for very sparse systems), this concept results in a uniform algorithm, which allows us to design a clear software also for very generally structured hierarchical systems. Now the software elements will be considered to be designed as a basis for such an algorithm.

3. Software elements:

Software elements are developed to facilitate writing an algorithm for a general hierarchical system without paying too much attention to the complexity of the realisation problem. Although software design is not independent of the method, we should try to distinguish as clear as possible between the level of algorithm routines and the software level.

The main components of the software elements are divided into 3 distinct groups:

(a) structure handling,
(b) data management,
(c) mathematical vector operations.

ad (a): Instead of the row index k in a usual matrix we assign to each row a "row-identifier" k, which consists of the information about the subsystem and the local position of the row. Structure handling comprises the routines to manipulate the row identifier k, when proceeding in the structure. The order of vectors is defined by the partitioning procedure and begins at the innermost system. Change of the row identifier can be made in the positive and negative senses corresponding to "k = k + 1" and "k = k - 1". The following sets of rows have to be considered: all vectors of all the subproblems interior to a given subproblem - subtree - or all vectors on a branch in the subsystem tree from a subnode to a goal-node (e.g. the uppermost system).

Example of routines

k = ROW-IDENTIFIER (subsystem-no, position-no)
k = NEXT-ROW-IN-SUBTREE (vertex, k, sense, end)
k = NEXT-ROW-ON-BRANCH (goal-node, k, sense, end)
(sense = $\pm$ 1, end ... end criterion: TRUE or FALSE)

ad (b): For large systems the working matrices of all subproblems often cannot be stored in core storage. Data management includes the routines for storage and preparation of all the row vectors needed. In the case where the locally used data exceeds the storage capacity, the software switches automatically to a virtual storage concept.

A routine, which brings in a row vector q_k from a matrix Q and puts it into vector v, can be considered as

GET-ROW-FROM-Q (k,v).

ad (c): The vectors of the matrices used are stored and handled in a compactified form to save storage requirements. "Vector" means the unity of information about the structure and addresses and of the data itself.

To these vectors (e.g. rows of working matrices, gradient, feasible direction vector) operations are applied during the algorithm routines as for example inner products, linear combinations, assignments. The software elements enable the user to deal no longer with the interior structure of the vectors.

Example of routines:

Assignment a = b	ASSIGN (a,b)
inner product c = a . b	INPROD (a,b,c)
linear combination y = a + c . b of vectors a and b	LINCOM (y,a,b,c)

4. Design of the algorithm:

Taking these software elements as a basis we are able to concentrate on the algorithm itself. The nature of the algorithm is given by the solution strategy common to all the subproblems: in our case the feasible direction method.

The main routines, which handle the structured matrices, are:

- the determination of the feasible direction vector and dual variables,
- the determination of the maximal steplength,
- the updating of the working matrices in the case of activation of a constraint and
- the updating in the case of an inactivation.

All these routines proceed row by row in the working matrices. This will be illustrated be means of our algorithm:

Let Q denote the (structured) matrix of active constraints q_k (k = 1, ..., m). The resulting working matrices for the successive partitioning algorithm let be denoted by $\tilde{Q}$ (derived from Q) and by $\tilde{\Gamma}$ (coefficient matrix).

The rows of $\tilde{Q}$ can be represented as

$$\tilde{q}_k = q_k - \sum_{j=1}^{k-1} \tilde{\gamma}_{kj} \tilde{q}_j$$

$$(\text{or } Q = \tilde{\Gamma} \tilde{Q}).$$

Updating of $\tilde{Q}$ and $\tilde{\Gamma}$ gives new row vectors $\hat{\tilde{q}}_k$ and $\hat{\tilde{\gamma}}_k$ for the new matrices $\hat{\tilde{Q}}$ and $\hat{\tilde{\Gamma}}$. Roughly speaking, this can be performed in the following way:

$$\hat{\tilde{q}}_k = \tilde{q}_k + \beta_1 \cdot \text{qhelp}_k$$

(β_1: coefficient built up by an inner product,
qhelp_k: help vector),

$$\text{qhelp}_k = \text{qhelp}_{k-1} + \beta_2 \cdot \tilde{q}_{k-1}$$

A similar transformation holds for coefficient vector $\hat{\tilde{\gamma}}_k$. The determination of the feasible direction vector involves linear combinations of the gradient and of the vectors $\tilde{q}_k$.

These routines hold for the structured as well as for the unstructured case. They can be developed the same way as for an unstructured problem neglecting the real structure of the problem, neglecting the dimensionality of the data to be handled and neglecting the interior construction of the vectors.

The iteration steps for one subproblem resemble to those of a conventional method. The only part, which is specific for the structured case, is a routine determining the subproblem to be optimized next ("iteration control"). This routine gets activated as soon as some local iteration criterion is satisfied (e.g. whenever a suboptimum point is reached or whenever the local increase in the objective function slows down or whenever some maximum number of iteration steps for the actual subproblem is exceeded.

Such an iteration control enables us to run different optimization strategies for large-scale systems. Allowing only iteration steps for the uppermost problem the algorithm behaves as a compactification method. If we perform at each subproblem a full suboptimization and advance to the higher level problem not before all the interior subproblems are locally optimized, we work in the original sense of a decomposition method. But with this iteration control modul we can examine many other iteration strategies, which will accelerate the whole algorithm, between the two strategies mentioned before.

5. Conclusion

The paper has shown some realisation aspects of methods for nested structured systems. A good software design allows to develop the algorithm as in a laboratory-like environment (data in core storage, non-compactified vectors, unstructured problem). The designer of an algorithm can work and change on his ideas, independent of the software problems at large-scale problems. On the other hand the construction of the appropriate software elements can be made without knowledge of the mathematical theory of a sophisticated algorithm. Such a concept used for similar problems may allow realisation of a greater number of methods.

6. References

[1] Geoffrion, A.M.:
Elements of Large-scale Mathematical Programming, Part I and II.
Management Sci., Vol. 16, No. 11 (1970)

[2] Wismer, D.A.:
Optimization Methods for Large-scale Systems.
Mc Graw Hill Comp., N.Y., 1970

[3] Kalliauer, A.:
Compactification and Decomposition Methods for NLP Problems with Nested Structures in Hydro Power System Planning.
9. Int. Symposium on Math. Programming (Budapest, 1976)

[4] Kalliauer, A.:
A Computational Method for Hierarchical Optimization of Complex Systems.
IFIP-Working Conference on Modelling and Optimization of Complex Systems (Novosibirsk, 1978)

[5] Zoutendijk, G.:
Mathematical Programming Methods. North-Holland, 1976

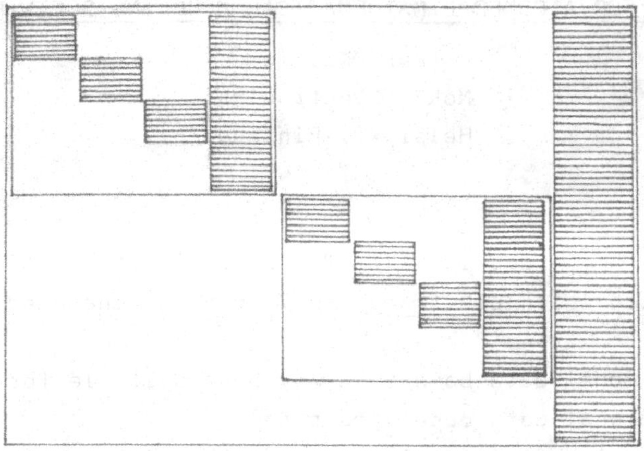

Fig. 1: Example of a constraint matrix with a three-level structure with coupling variables

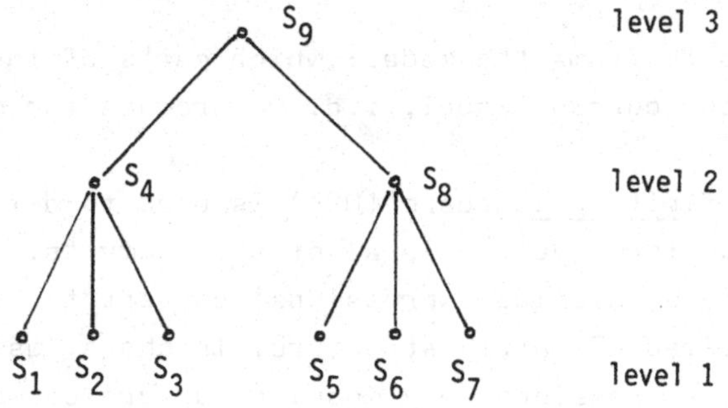

Fig. 2: Structure of matrix in Fig. 1

OUTLINES FOR A GENERAL MATHEMATICAL MODELING SOFTWARE

Kari Kallio
Nokia Electronics
Helsinki, Finland

ABSTRACT

The paper deals with the development of a general optimization modeling system with which arbitrary models can be generated. This consists of

1. organizing the model data base in a way best suitable for modeling purposes (relational data base structure)
2. developing a model definition language, with which model construction can be easily expressed and built (based on predicate calculus).

The model construction process is best illustrated by an example. Suppose that we want to build a linear programming production model and that we have the basic data available in the computer in some form. The first step is to define, what kind of model is to be generated (LP production model in this case), which assigns those basic data tables that are required.
The needed tables are then picked to a model data base with relational structure.
The second step is to frame the model: which parts of the model data base are used in the current model, i.e. to produce the matrix rows and columns.
Then, the model definition language (MDL) is used to define the model structure and to perform the retrieval of the elements. So, the MDL statements (predicate calculus expressions) convert the data base structure to the aimed LP matrix structure. In short: using compact predicate calculus expressions, any model structure can be defined. It is only these expressions that are given to the computer, which takes care of the rest: retrieves all the needed data from model data base and sets up the matrix structure in a form ready for solution algorithm.

CONCEPTS AND NOTATIONS

A relation can be presented as a table in which each row represents a tuple. Each tuple may occur only once in the table.
In the tabular presentation of a relation, it is customary to name the table, and to name each column, e.g.

MATRIX

ROW	COLUMN	VALUE
RPROD01	CPROC14	.274
RPROD23	CPROC05	.365
RPROD23	CPROC07	.532
.	.	.
.	.	.

A column or a set of columns whose values uniquely identify a row of a relation is called a key of the relation. In the MATRIX relation the key consists of the ROW and COLUMN columns. A relation can have more than one key and then it is customary to designate one of the keys as the primary key. A relation represented as a table is usually denoted by the tablename which is followed by the columnnames enclosed in parentheses, where the primary key columns are underlined e.g.

MATRIX(<u>ROW</u>,<u>COLUMN</u>,VALUE).

A projection of a relation is a relation which has been derived from the first one by dropping columns and removing resulting multiple occurrences of the same row. In this paper projections are denoted by inserting asterisks for the dropped columns, e.g.

MATRIX(ROW,*,*)

is a unary relation of the matrix row names. We also use the same name for a relation and for an element inclusion predicate i.e.

R(x) is TRUE if $x \in R$ and
FALSE otherwise,

and also

R(x,*) is TRUE if for some c $(x,c) \in R$ and
FALSE otherwise.

"For some" is used for existential and "for all" for universial quantifiers and "not" is used for negation. Character strings are given between apostrophes ('). "&" is used as concatenation operator for joining character strings e.g. 'X' & 'Y' = 'XY'.
Next, we apply the above concepts to an LP case model.

LP MATRIX AS A RELATION

Here we represent in a relational form the information found in most LP package input files. The representation is general, not oriented towards any special LP package, but is in no way unique. The rhs and ranges vectors are considered as columns and the column upper-, lower and fixed-bound vectors as rows. Our relations are

ROWS(ROW,ROW-TYPE)
COLUMNS(COL,COL-TYPE)
MATRIX(ROW,COL,VAL)

where ROW - matrix row names (character string)
COL - matrix column names (character string)
VAL - matrix element values (number)
COL-TYPE - matrix column types (character string)
ROW-TYPE - matrix row types (character string).

For normal matrices - generalized upper bounding, separable programming and possible other special operating modes are not considered - possible row types are

'EQ','LE','GE','FR' - in the usual meaning
'BUP' - upper bound row
'BLO' - lower bound row
'BFX' - fixed bound row

and the possible column types are

'CV' - continuous column
'BV' - binary column
'IV' - integer column
'RHS' - rhs column
'RNG' - ranges column.

The extension of the possible types to special operating modes is not difficult.

THE CASE MODEL

Our case model is a profit maximizing production model where

- products and semiproducts are made from semiproducts and raw materials,
- the production uses a number of limited processing capabilities, which can to some extent be increased by overtime working and
- markets for products and availability of raw materials are limited.

The structure of the model is given in figure 1. The basic relations - input data tables - are

RECEIPT(PROD,USED-MAT,MAT-USE)
PROD-PROC(PROD,PROC,PROC-USE)
PRODUCT(PROD,PRICE,MARKET-LIM)
RAWMAT(RMT,COST,AVAIL-LIM)

PROCESS(PROC,COST-NORM,CAP-NORM,COST-OVRT,CAP-OVRT),

where RECEIPT	- table of raw material and semiproduct usages for products
PROD-PROC	- table of process usages for products
PRODUCT	- product vector with price and market limit attributes
RAWMAT	- raw material vector with cost and availability attributes
PROCESS	- process vector with cost and capacity attributes
PROD	- product code
PROC	- process code
RMT	- raw material code
USED-MAT	- product or raw material code
MAT-USE	- usage of material USED-MAT in production of product PROD (unit/unit)
PROC-USE	- usage of process PROC in production of product PROD (unit/unit)
PRICE	- unit selling price of a product
MARKET-LIM	- market upper limit for selling a product
COST	- unit cost of raw material
AVAIL-LIM	- upper limit of raw material availability

COST-NORM, COST-OVRT, CAP-NORM and CAP-OVRT are normal and overtime costs and capacities of a process.

In our LP matrix the general variable naming rule is: the ordinary variable names will begin with 'R' for rows and 'C' for columns, which are followed by material or process codes - we suppose a unique coding.

A process and a product each need two columns in the matrix, which must be identified by the name coding. The normal process usage and total production columns are denoted by appending '1' at the end of the basic name. The overtime process usage and the sold production columns are denoted by appending '2' at the end of the basic name. For example the column "amount of product P1234 sold out" is coded as "CP12342", which can also be represented as 'C' & 'P1234' & '2'.

THE LP MATRIX DEFINITION

The LP matrix is defined by a number of predicate definitions because of notational reasons. According to our earlier convention, we may think of a predicate as denoting a relation with the same name.

We suppose, that the model user supplies a list - a unary relation - P(PROD) of products to be included in the model. Our first task is to enlargen P to list PRODS(PROD), which includes also all semiproducts used in the production.

```
(1) PRODS(x) = P(x) or ( RECEIPT(x,*,*) and for some y
                         ( PRODS(y) and RECEIPT(y,x,*) ) )
```

Because of the finite length of the production path, the recursive definition of PRODS can be evaluated in a finite number of steps. Also lists of raw materials and processes in production of PRODS are needed.

```
(2) RAWS(x) = for some y ( PRODS(y) and not PRODS(x) and
                           RECEIPT(y,x,*) )
(3) PROCS(x) = for some y ( PRODS(y) and PROD-PROC(y,x,*) )
```

Raw material rows x total production columns

```
(4) SUB1('R'&x, 'C'&y&'1',v) = PRODS(y) and RAWS(x) and
                               RECEIPT(y,x,v)
```

Processing rows x total production columns

```
(5) SUB2('R'&x,'C'&y&'1',v) = PRODS(y) and PROCS(x) and
                              PROD-PROC(y,x,v)
```

Production rows x total production columns

```
(6) SUB3('R'&x,'C'&y&'1',v) = PRODS(x) and PRODS(y) and
                              (RECEIPT(y,x,v)
                              or ( x=y and v=-1 ))
```

Production rows x sold production columns

```
(7) SUB4('R'&x,'C'&y&'2',v) = v=1 and ( PRODS(x) and x=y)
```

All diagonal -1 matrices

```
(8) SUB5('R'&x,'C'&y,v) = v=-1 and
                          ( (RAWS(x) and y=x)
                       or ( PROCS(x) and
                            ( y=x&'1' or y=x&'2' ) ) )
```

Object vector

```
(9) PROF('C'&x,v) = ( RAWS(x) and RAWMAT(x,-v,*) )
                  or for some y
                     ( PROCS(y) and
                        ( x=y&'1' and PROCESS(y,-v,*,*,*) )
                   or   ( x=y&'2' and PROCESS(y,*,*,-v,*) ) )
                   or for some y
                     ( x=y&'2' and
                       ( PRODS(y) and PRODUCT(y,v,*) ) )
```

Upper bound vector BND is defined analogiously to PROF. We decide that defaulf value for rhs element is 0 and default type for column type is 'CV' and we do not need the COLUMNS relation at all.

The main matrix definition

```
(10) MATRIX(r,c,v) = SUB1(r,c,v) or SUB2(r,c,v) or SUB3(r,c,v) or
                     SUB4(r,c,v) or SUB5(r,c,v)
                     or ( r='PROFIT' and PROF(c,v) )
                     or ( r='UPPERBND' and BND(c,v) )
```

The row information

```
(11) ROWS(r,t) = t='EQ' and for some y
                 ( r='R'&y and (PRODS(y) or PROCS(y) or
                 RAWS(y) ) )
             or
             ( r='PROFIT' and t='FR' )
             or
             ( r='UPPERBND' and t='BUP' )
```

COMPUTER IMPLEMENTATION

Basically the significance of the presented method of matrix generation does not entirely depend on if it is implemented or not as a programming language. As a notation only it can serve as an exact and concise documentation of a model and as an interface between the modeller and the matrix generator programmer.

A model structuring can be made by submatrix definitions, which further can be mapped one-to-one to generator program modules.

Some of the existing relational data manipulation languages are almost capable of performing the desired tasks - extended by the concatenation operator and with their own syntaces. Efficiency is another question because the relational systems have not been optimized for manipulation of large sparse matrices.

It the implementation is not based on any existing high level relational system, the basic tools needed are:

- an intelligent optimizer to transform the relational definitions to efficient access paths and operation sequences and
- an efficient binary-ternary relational storage system for vectors and large sparse matrices.

In addition we should be able to define and reference virtual matrices i.e. only a definition or a pointer structure is physically stored; values are stored in data tables or in actual - physical - submatrices.

The matrix definition could be evaluated either by an interpretive system or could be translated to executable code. In the second case we could speak of a matrix generator generator.

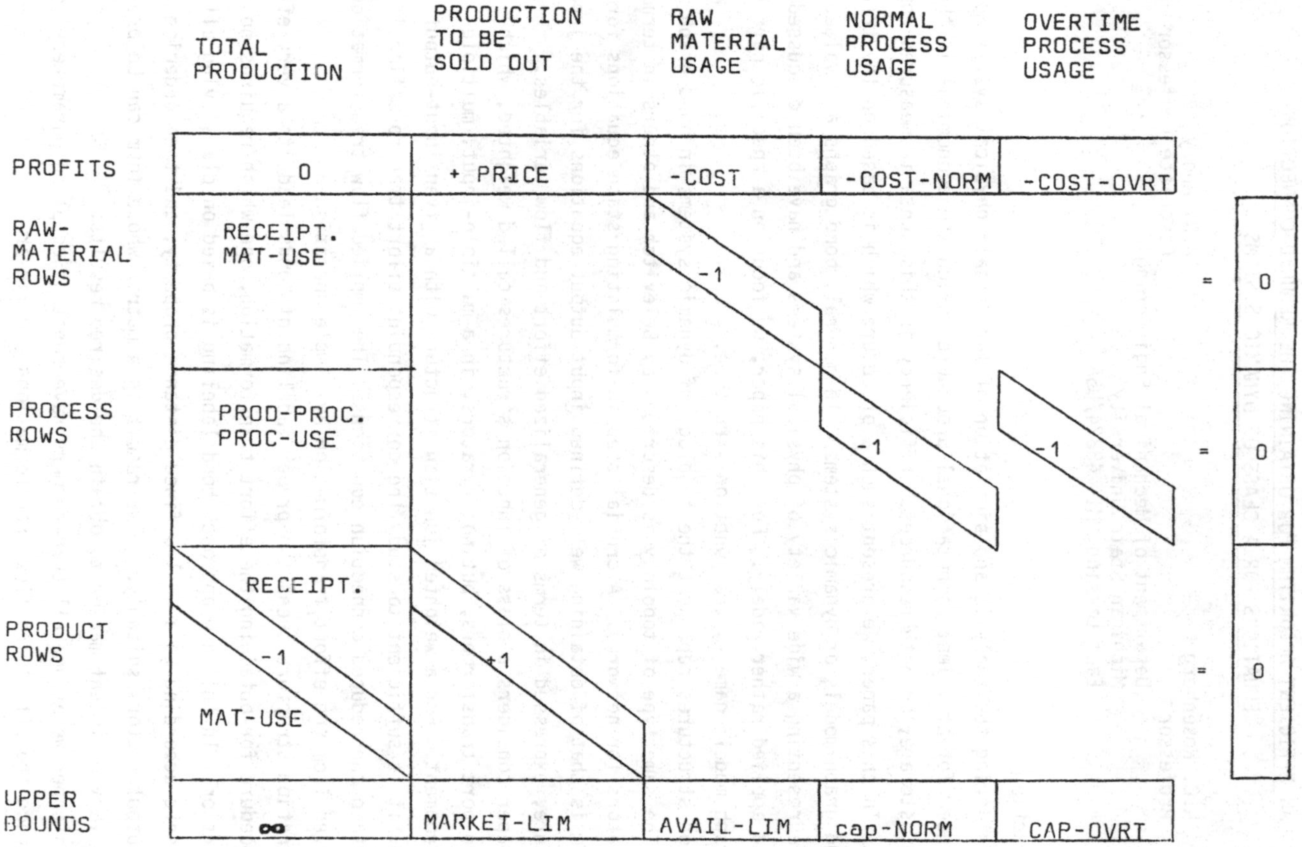

FIGURE 1: THE CASE MODEL

AN EFFICIENT ALGORITHM FOR OBTAINING THE REDUCED CONNECTION EQUATIONS FOR A CLASS OF DYNAMIC SYSTEMS

R.C. Rosenberg
Professor

A.N. Andry, Jr.
Associate Professor

Department of Mechanical Engineering
Michigan State University
East Lansing, MI 48824/USA

1. Introduction

In considering the analysis and simulation of large-scale physical system models, there is a need for efficient computational procedures which are economical in their use of time, storage, or both resources. Efficiency in this sense is measured comparatively. In this paper, we present such a procedure which is based on the treatment of bond graph models of dynamic systems. In general, bond graphs are valuable tools for representing a wide variety of physical systems and have been discussed, studied, and applied rather widely. For this paper, we focus on a specific part of the bond graph model; namely, the junction structure.

Junction structures represent the topology of dynamic systems in bond graph notation. The same type of topology is described by Belevitch and others in terms of wiring operators for networks. A crucial step in formulating state equations for a given system is that of obtaining well-defined input-output equations for the junction structure, expressed in terms of generalized effort and flow variables.

This paper considers a class of junction structures called weighted, which includes two-port transformers, but not gyrators, in a multiple-input, multiple-output environment. For a weighted junction structure with a given input-output orientation, it is sufficient to study the corresponding effort transformation in order to obtain the reduced connection equations. The implied flow transformation can be derived from the effort transformation by simple operations.

For junction structure trees the proper labeling of bonds leads to a very efficient procedure for obtaining the effort transformation, one which requires no matrix inversion. In this new approach bond labeling is based on classifying all bonds into sets according to a well-defined distance property, and then ordering the input and output vectors suitably. The result is a matrix whose form can be processed in a very efficient manner to obtain the desired results.

In this paper we assume all transformers have constant coupling parameters (or moduli). Hence the effort and flow transformations are linear.

2. Weighted Junction Structures

A weighted junction structure is an open bond graph composed of the elements {0, 1, TF} as defined previously. Associated with a given weighted junction structure (WJS) are two (linear) transformations; namely, an effort transformation, T_e, and a flow transformation, T_f.

As an example consider the WJS of figure 1(a). One TF, one 0, and one 1 comprise the graph. The bonds and ports are labeled for convenience; and power reference directions are shown by the half-arrows.

The equations in fig. 1(b), (c), and (d) correspond to the properties of the 1-junction, TF element, and 0-junction, respectively, for the power orientations shown.

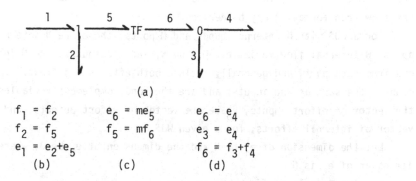

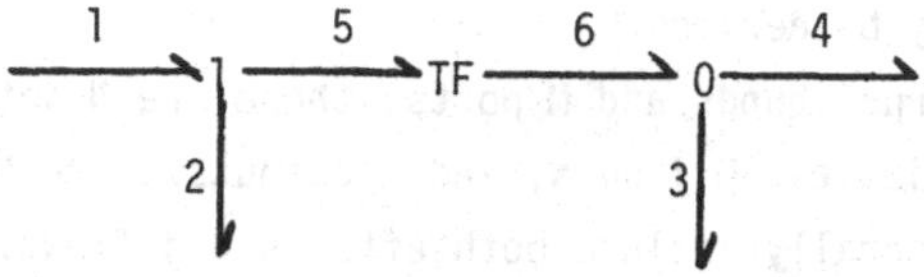

(a)

(b)	(c)	(d)
$f_1 = f_2$	$e_6 = me_5$	$e_6 = e_4$
$f_2 = f_5$	$f_5 = mf_6$	$e_3 = e_4$
$e_1 = e_2 + e_5$		$f_6 = f_3 + f_4$

Figure 1. An example WJS.
(a) the graph
(b) 1-junction relations
(c) TF relations
(d) 0-junction relations

The objective is to obtain a characterization of the WJS at its ports. The resulting characterization can be written in two parts, one relating effort variables at the ports, the other relating flow port variables. One such expression for the example is given by equation [1]. The transformation

$$\begin{bmatrix} f_1 \\ f_2 \\ e_3 \\ e_4 \end{bmatrix} = \left[\begin{array}{cc|cc} 0 & 0 & m & m \\ 0 & 0 & m & m \\ \hline m & -m & 0 & 0 \\ m & -m & 0 & 0 \end{array}\right] \begin{bmatrix} e_1 \\ e_2 \\ f_3 \\ f_4 \end{bmatrix} \qquad [1]$$

contains T_e and T_f, shown as the nonzero subarrays in the equation. Observe that the algebraic sum of port powers is zero.

As is well known, the effort transformation always is decoupled from the flow transformation for WJS graphs, because the elements {0, 1, TF} only take efforts and flows into flows. Since WJS graphs are power-conserving, it suffices to study either T_e or T_f. From one the other may be derived by adjusting for power orientations.

In the remainder of this paper we shall concentrate on T_e, the effort transformation, without loss of generality.

3. Formulation of T_e for Tree Junction Structures

In this section we describe a new procedure for formulating the system connection equations (i.e., the junction structure equations), which permits an efficient reduction of the equations to input-output form. We shall concentrate on the effort transformation, since, for a WJS, that provides the necessary information from which the flow transformation may be derived.

For a WJS with N internal bonds and M ports, there are N internal effort variables, N internal flow variables, M inputs, and M outputs. The M inputs are chosen one from each port, and generally include both efforts and flows. The M outputs are ordered the same as the inputs, and are the bond complement variables. Let $\underline{e}_i$ be the vector of effort inputs, $\underline{e}_o$ be the vector of effort outputs, and $\underline{e}_t$ be the vector of internal efforts, for a given WJS.

Let the dimension of $\underline{e}_i$ be K and the dimension of $\underline{e}_o$ be J, where K+J=M. The dimension of $\underline{e}_t$ is N.

For the example of figure 2, we have

$$\underline{e}_i = (e_2, e_4, e_5)$$
$$\underline{e}_o = (e_1, e_3, e_6)$$
and $\underline{e}_t = (e_7, e_8, e_9, e_{10})$.

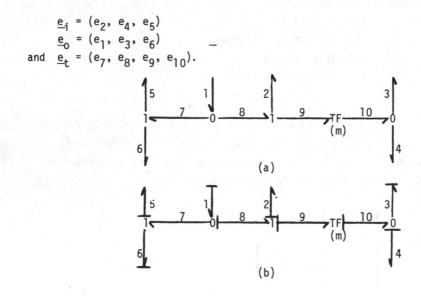

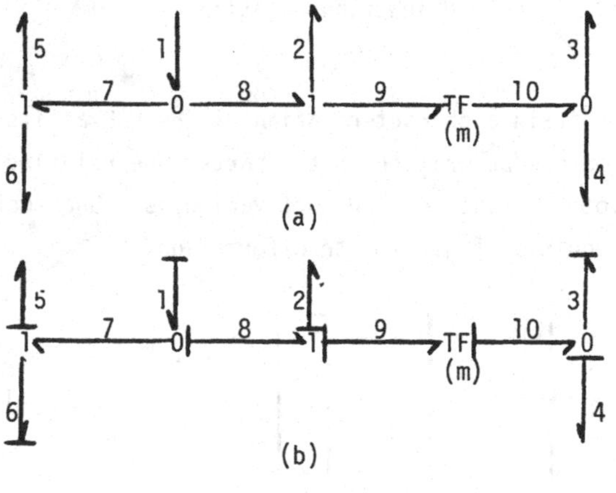

Figure 2. A WJS example.
(a) graph without causality
(b) graph with causality

The system connection equations are derived from the 0, 1, and TF nodes. There are fourteen such constraints in all. The equations can be summarized in the form

$$\begin{bmatrix} \underline{e}_o \\ \underline{e}_t \end{bmatrix} = \begin{bmatrix} S_{11} & S_{12} \\ S_{21} & S_{22} \end{bmatrix} \begin{bmatrix} \underline{e}_i \\ \underline{e}_t \end{bmatrix} \qquad [2]$$

where the subarrays S_{ij} have the appropriate dimensions.

For this example we obtain

$$\begin{bmatrix} e_1 \\ e_3 \\ e_6 \\ e_7 \\ e_8 \\ e_9 \\ e_{10} \end{bmatrix} = \begin{bmatrix} 0 & 0 & 0 & \quad 0 & 1 & 0 & 0 \\ 0 & 1 & 0 & \quad 0 & 0 & 0 & 0 \\ 0 & 0 & -1 & \quad 1 & 0 & 0 & 0 \\ 0 & 0 & 0 & \quad 0 & 1 & 0 & 0 \\ 1 & 0 & 0 & \quad 0 & 0 & 1 & 0 \\ 0 & 0 & 0 & \quad 0 & 0 & 0 & m \\ 0 & 1 & 0 & \quad 0 & 0 & 0 & 0 \end{bmatrix} \begin{bmatrix} e_2 \\ e_4 \\ e_5 \\ e_7 \\ e_8 \\ e_9 \\ e_{10} \end{bmatrix} \qquad [3]$$

Formal reduction of eq. [2] leads to the results

$$\underline{e}_t = (I-S_{22})^{-1} S_{21}\, \underline{e}_i \qquad [4]$$

and

$$\underline{e}_o = [S_{11}+S_{12}(I-S_{22})^{-1}S_{21}]\, \underline{e}_i, \qquad [5]$$

or

$$\underline{e}_o = T_e \cdot \underline{e}_i, \quad \text{for convenience.}$$

For the specific case of eq. [3], we find

$$\begin{bmatrix} e_1 \\ e_3 \\ e_6 \end{bmatrix} = \begin{bmatrix} 1 & m & 0 \\ 0 & 1 & 0 \\ 1 & m & -1 \end{bmatrix} \begin{bmatrix} e_2 \\ e_4 \\ e_5 \end{bmatrix} \qquad [6]$$

after suitable matrix algebra is completed. If the formal matrix approach is used the array $(I-S_{22})^{-1}$ must be found. Typically S_{22} is large and sparse, and the computation can be expensive, both in storage and time costs if the inversion is carried out.

4. Main Result

We present the following procedure which makes the inversion of the array $(I-S_{22})$ unnecessary for WJS trees.

(1) Assign causality to the junction structure according to the standard procedure.

(2) Assign to each bond an index based on its causal distance from the input set. Take the input set to be index 0.

(3) Order all bonds by their index derived in (2) above. Order of bonds within

a set (same indices) is not important.

(4) Write the S_{ij} arrays in terms of the ordered bond sets.

The result is an S array which is lower triangular, with subarrays defined by similar index sets of bonds. S can be reduced to T_e form (i.e., input-output form) without the need to invert any matrices.

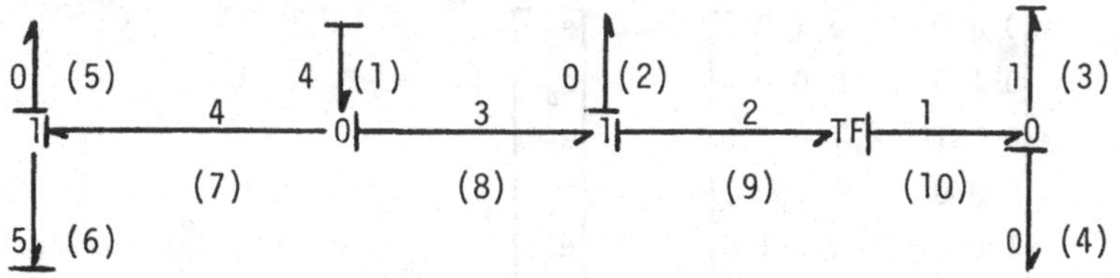

Figure 3. Causal distances for the WJS of figure 2.

Figure 3 shows the bonds of figure 2(b) labeled by causal distance. Original bond numbers are in parentheses. The input set (e_2, e_4, e_5) is labeled 0. The set of bonds whose efforts can be found in terms of set 0 is labeled 1 (i.e., e_3, e_{10}). The final set is 5, and includes only bond 6. Writing the effort equations for the WJS according to the procedure, we obtain

$$\begin{bmatrix} e_3^* \\ e_{10} \\ \hline e_9 \\ \hline e_8 \\ \hline e_1^* \\ \hline e_7 \\ \hline e_6^* \end{bmatrix} = \begin{bmatrix} 0 & 1 & 0 & & & & & & \\ 0 & 1 & 0 & & & & & & \\ 0 & 0 & 0 & 0 & m & & & & \\ 1 & 0 & 0 & 0 & 0 & 1 & & & \\ 0 & 0 & 0 & 0 & 0 & 0 & 1 & & \\ 0 & 0 & 0 & 0 & 0 & 0 & 1 & & \\ 0 & 0 & -1 & 0 & 0 & 0 & 0 & 0 & 0 \end{bmatrix} \begin{bmatrix} e_2 \\ e_4 \\ e_5 \\ e_3 \\ e_{10} \\ e_9 \\ e_8 \\ e_1 \\ e_7 \end{bmatrix} \qquad [7]$$

Reduction of the lower triangular S array can be accomplished from the top down in subarrays. The result for T_e is the same as given by eq. [6]. Output variables have been marked by an asterisk for convenience.

5. Conclusions

We summarize the results of this paper by stating that the proper labeling of bonds for a weighted junction structure tree leads to a system connection matrix that can be reduced to input-output form with maximum efficiency and without the need for matrix inversion. The procedure for labeling sorts bonds into equivalent sets according to their causal distance from the input set. For weighted junction structures only the effort transformation was considered, since the flow transformation can be derived from power orientation information, given the effort transformation.

Implementation of this procedure on a computer would represent a significant improvement over the current, formal matrix method now used by ENPORT-4. Extension of the procedure to include arbitrary junction structures (i.e., with loops and with gyrator nodes) would be valuable, and is being studied. Finally, the extension of the approach to include modulated (i.e., nonlinear) TF and GY nodes would generate improvements in the processing of nonlinear models.

Bibliography:

1. V. Belevitch, Classical Network Theory, Holden Day, San Francisco, 1968.
2. D. Karnopp and R. Rosenberg, System Dynamics: A Unified Approach, John Wiley, New York, 1975.

Characteristics of Incremental Assignment Method

Keiji Yajima
Computation Center
Institute of JUSE
(Japanese Union of
Scientists and Engineers)
Tokyo Japan

1. Introduction

The incremental assignment method had been brought out in a study of highway construction planning. The origin and destination survey which deals with a set of demand quantity transfered from one place to another offered a basis for the scientific investigation. The quantity that served for a forecast of future demand had been assigned as much as possible for the existing transportation routes in the analysis and it arose a rather difficult problem.

The assignment problem can be solved by a method in which each transportation demand quantity alotted to the shortest path that takes minimum required time. The consequence of application of this method seemed to disagree with the actual transportation flow. From the exprience we recognize the shortest path would alter in accordance with the degree of traffic congestion. The assignment technique pursued this important fact and got into such step that demand is assigned without consideration of road capacity and then required time is revised in consideration of both flow quantity and capacity limit. This method seems to us splendid but also differed slightly from the real path selection procedure.

The incremental assignment method then came out incorporated with following steps:

1) select the origin and destination points at random,
2) find out the shortest path between origin and destination,
3) assign some percentage of load to that path,
4) revise the passage time in accordance with the actual load assigned to that path,
5) repeat 1 - 4 untill the total demands are satisfied.

The incremental assignment method was recommended to simulate the automobile flow in urban system.

2. Multi-commodity trasportation problem and incremental assignment method

In our preliminary research we had to deal with the total quantity of various kinds of commodities incorporated with the existing railway transportation network. A multi-commodity problem can be handled with the large scale linear programming technique. The incremental assignment method as a handy tool, on the other hand, has raised following problems:

1) the randomness of origin and destination selections could have some effects on the solution,
2) the optimum size of increment is not obvious,
3) piecewise linear cost function (Fig.1,[1]) is not realistic, as the capacity limit is unbounded with regard to the number of vehicles.

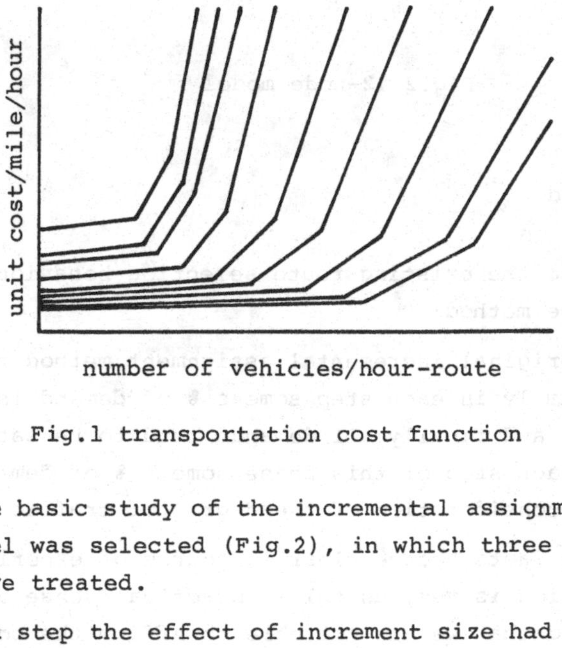

Fig.1 transportation cost function

Aiming the basic study of the incremental assignment method a small size model was selected (Fig.2), in which three commodities and twelve nodes are treated.

As a first step the effect of increment size had been observed and it became clear that the total transportation cost happens to fluctuate more seriously depending on the scale of increment size. Generally speaking, from the theoretical point of view the incremental assignemnt method in an original form appeared unsatisfactory in the sense that we could afford an example in which there is no possibility to attain optimum solution by the method.

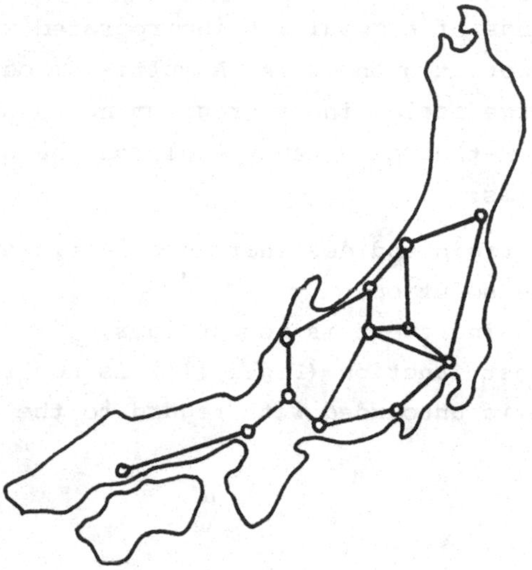

Fig.2 12-node model

3. Two-phase method

To approximate the existing route selection behaviour we adopted following two-phase method:

1) phase I. The original incremental assignment method cooresponds to this phase, namely in each step some ε % of demand is assigned to optimum route, and finally all demands tend to be satisfied.
2) phase II. In each step of this phase some ε % of demand is reassigned in consideration of the degree of traffic congestion.

The following facts became clear in course of experiments that the feasible solution is very useful in practical sense on the other hand we are assured that we are not able to affirm the convergence to the optimum solution. So we stepped into the two-phase method with a penalty function and further a penalty function with modified capacity limit.

4. The charactaristics of two-phase method

Introducing a simple example (Fig.3), we discuss the characteristics of incremental assignment method in detail. Provided that there are two kinds of materials M and N incorporated with railway transportation metwork, these two materials are to be transfered from point A to points B and C. More exactly 20 tons of material M should be transfered from A to B per day and the same quantity of N from A to C is required per day. To transport material M from point A to B could have either the path APB or AQB and to transport material N there are two possible paths APC and AQC. The capacity limits are assigned to AP and AQ as 20 tons per day for both routes and the rest of routes is assumed not to have limits.

Let us assume the transportation costs 5 dollars for routes AP ans PC, 10 for AQ, 15 for QC (Fig.3).

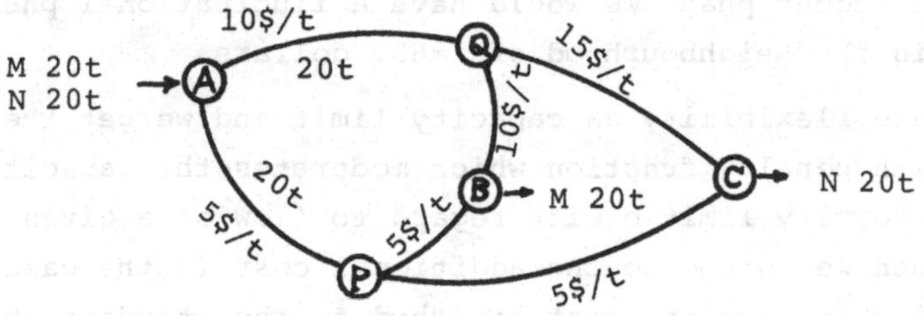

Fig.3 5-node model

If we analyse this problem by linear programming technique, we would easily get the solution. Let us represent non-negative variables x_1, x_2, y_1, y_2 such that

x_1 : transportation quantity of commodity M via path APB

x_2 : transportation quantity of commodity M via path AQB

y_1 : transportation quantity of commodity N via path APC

y_2 : transportation quantity of commodity N via path AQC

the we get as final solution

$$x_1 = 0, \quad x_2 = 20, \quad y_1 = 20, \quad y_2 = 0, \tag{1}$$

and total cost is 600 dollars.

We are not able to attain this solution by using the incremental assignment method, as transportation cost of route AP is less than that of route AQ so both commodities M and N are assigned to route AP until the capacity limit is attained. Using random assignment technique it is expected naturally that the approximately same quantities that means 10 tons for both commodities M and N would be alloted to route AP. Getting the capacity limit in the allocation procedure, the route AQ is selected and we get the final result such as

$$x_1 = 10, \quad x_2 = 10, \quad y_1 = 10, \quad y_2 = 10, \qquad (2)$$

so the total cost would fluctuate around the exact value of f=650 which corresponds to the solution (1).

We could not expect the oprimum solution even for the two-phase incremental assignment method. Approaching to the feasible solution, in next phase the total amount of transportation quantities is reduced and those cutted-off quantities are assigned again to the optimum routes. In this second phase we would have a fluctuational phenomena of total cost in the heighbourhood of f=650 dollars.

We introduce flexibility on capacity limit and we get the two-phase method with penalty function which moderates the capacity limit. If there is a capacity limit c with regard to flow of a given route that is $s \leqq c$, then we introduce the additional cost in the case of $s>c$ $\exp(p(s-c))$ which is an extra cost attached to the standard charge. A value p is chosen properly and we would compute the optimum cost for a given p. After that we increase the value of p.

Let $p=1$, $\varepsilon=0.01$ and as initial values we take $x_1=10$, $y_2=10$ then after hundreds number of iterations the solution is in the meighbourhood of $x_1=1$, $x_2=19$, $y_1=20$, $y_2=0$.

As a second example we take $p=2$, $\varepsilon=0.01$ and the same initial values for x_1 and y_2 then we get the solution such as $x_1=0.5$, $x_2=19.5$, $y_1=19.9$, $y_2=0.1$.

Stepping forward this direction if we take $p=20$, $\varepsilon=0.01$ then we are forced getting into miserable oscillation.

Taking into consideration of $z=x_1-y_2$ and its stability, the penalty function $\exp(p(s-c))$ had been modified to a $\exp(p(s-c))$ which indicates that we would have an alteration of capacity limit c to c' and we get exp(p(s-c')). This is called capacity modification method and it contributes convergence acceleration in the sense of that z attains quickly final value.

5. Applications

Applying the two-phase method with penalty function we solved the multi-commodity problem in national railway corporation quite successfully([5],[6]).

We also treated the highway construction plan in which there exist 27 nodes, 19 kinds of commodityes, three kinds of branches between two adjacent modes, namely, a sea route, a railway and a road ([3],[4]). With small computing cost we could evaluate the effects of plan. For example some destricts were designated to take consideration of unbalance among the production, consumption and transportation capacity.

In urban system there are several kinds of industries and between each industry there exists a degree of dependence, for example simply either the preference of adjacence or disadjacence. From the given dependence matrix the distribution pattern could be gotten in fixed block([10]).

In the design phase of communication metwork the failure time circuit assignment problem plays important role, and the incremental assignment method has been applied ([8]).

6. Conclution

As the characteristics of this method, it could be summarized : (a) the algorithm reflects the existing procedure, (b) intermediate results are also valuable, (c) large scale network can be treated by the small size computer.

References

[1] B. V. Martin and M. L. Manheim, A research program for compatison of traffic assignment techniques, HRb Record. 88 (Jan. 1964).

[2] B. V. Martin, A computer program for traffic assignment research, MIT Research Report R 64-41 (Dec. 1964).

[3] Principles and techniques in planning of transportation networks, Ministry of Transportation, Planning Board, Japan, Report No. 15, 1969, Report No. 32, 1970 (in Japanese).

[4] Survey report on the effects of national highway network construction, Ministry of Construction, Bureau of Road, Japan, 1970 (in Japanese).

[5] S. Moriguti, M. Iri, A. Nagaya, A sequential method in multi-commodity transportation problem, Conference Report 1-1-7, pp. 21-22, Operations Research Society of Japan, 1970 (in Japanese).

[6] S. Moriguti, M. Iri, Y. Tukamoto, Flow capacity correction method in multi-commodity transportation problem, Conference Report, 1-1-8, pp. 23-24, Operations Research Society of Japan, 1970 (in Japanese).

[7] System analysis of new urban system in which population concentrates effectively, JUSE, 1971 (in Japanese).

[8] Y. Ishizaki, N. Yoshida. S. Sasabe and Y. Ishiyama, Multi-commodity flow approach to assignment of circuit in case of fialure in a communication network, IX International Conference on Mathematical Programming, Budapest, Hungary, 1976.

[9] A special issue on incremental assignment method, Communications of the Operations Research Society of Japan, Vol. 22, No. 12, pp. 688-719, 1977 (in Japanese).

[9.1] K. Okudaira, History and applications, pp. 688-694.

[9.2] M. Iri, Incremental assignment method from the view point of mathematical programming, pp. 695-701.

[9.3] S. Moriguti, Oscillation and convergence in incremental assignment method, pp. 702-710.

[9.4] S. Goto, Applications of incremental assignment method to the design of communication network, pp. 711-719.

[10] K. Okudaira, Introduction on urban system analysis, Shokokukan-sha, 1976 (in Japanese).

STOCHASTIC MODELLING OF SOCIO-ECONOMIC SYSTEMS

Mirosław Bereziński, Jèrzy Hołubiec

Polish Academy of Sciences

Systems Research Institute

Newelska 6, 01-447 Warszawa, POLAND

1. INTRODUCTION

It is a well known fact that the social and economic phenomena and processes are stochastic ones and objective in nature. Their intrinsic internal mechanism is expressed in the form of numerous socio-economic laws which are nothing other than a manifestation of the operation of the law of large numbers. This fact obliges us to treat social and economic variables as stochastic ones and to use the stochastic methods in the field of mathematical modelling of socio-economic systems. Despite strong trends in economics and related sciences in recent years to use basic results of stochastic mathematics for social and economic phenomena and processes modelling the models of socio-economic systems are at least nonlinear and dynamic but still deterministic. However, real socio-economic systems are nor deterministic but stochastic ones and that is why the deterministic economics must be considered as dealing with the mathematical expectations of the random variables, that really characterize the empirical socio-economic systems[2] [10], Although this way of introducing of stochasticity into models of socio-economic systems is very common, it is not unique one. The present economics is faced with the necessity to build models considering social and economic variables as being represented by stochastic processes. In this paper an attempt is made to trace some selected problems and motivations of queueing networks approach to socio-economic systems modelling.

2. QUEUEING NETWORK AS A MODEL OF NATIONAL ECONOMY

Assume that the whole national economy is divided into a set of n distinct sectors interconnected to each other by a set of links. The links reflect real interdependencies and intersector flows. Each sector produces a specific product and consumes parts of output production of the remaining sectors. Let X_i /i = 1,2,...,n/ be the output of the i-th sector per unit of time and let x_{ij} /i, j = 1,2,...,n/ denote that part of output of the i-th sector which goes to the j-th sector. Symbol x_{ii} stands for a part of the output of the i-th sector which remains in the sector. Usually, not all of the production of the given sector is consumed by other sectors or is used up in the same sector. In general, there remains a certain surplus, i.e. the final product of the given sector. Let us denote the final product of the i-th sector by x_i/ i= 1,2,...,n/. It is well known [8], [9], that the balance of national economy, according to the system proposed by W. Leontief [9], can be represented by the expanded matrix of the balance of production as shown in Table 1. From the mathematical point of view the middle part of the table is nothing other than a square matrix of intersector flows and reflects cooperation between the sectors. Introducing the notion of technical coefficients of production defined by the known formula

$$a_{ij} = \frac{x_{ij}}{X_j} \qquad /i,\ j = 1,2,\dots,n/$$

we obtain that the cooperation of the sectors can be reflected in the form of the matrix of the technical coefficients of production $[A = a_{ij}]$ where i, j = 1,2,....,n. Now it is important

Table 1

Scheme of Leontief's Balance of National Economy

Total outputs	Intersector flows	Final outputs
X_1	$x_{11}x_{12} \cdots x_{1n}$	x_1
X_2	$x_{21}x_{22} \cdots x_{2n}$	x_2
$\vdots$	$\vdots \ \vdots \quad \vdots$	$\vdots$
X_n	$x_{n1}x_{n2} \cdots x_{nn}$	x_n

to notice that the coefficients a_{ij} which indicates how many units of output of the i-th sector are needed to produce one unit of output of the j-th sector, can be viewed as probabilities that a part x_{ij} of the total production X_1 of the i-th sector goes to the j-th sector. Assume that we consider an open economy and let a_{oi} be the probability that an external flow enters sector i, i = 1,2,....,n. Similarly, let $a_{i,n+1} = 1 - \sum_{j=1}^{n} a_{ij}$ be the probability that a part of production of the i-th sector leaves the system, i.e. the probability that the final of the i-th sector is x_i. Let γ_i be a rate of the external flow going to the i-th sector and λ_i let be the total input rate to the i-th sector. One can obtain that λ_i is the solution to the linear equation [5].

$$\lambda_j = \gamma_j + \sum_{i=1}^{n} \lambda_i a_{ij}$$

or $$\lambda = \gamma + \lambda A \qquad /1/$$

where $\lambda = [\lambda_1, \lambda_2, \ldots, \lambda_n]$, $\gamma = [\gamma_1, \gamma_2, \ldots, \gamma_n]$. Consider the network of sectors of Fig. 1.

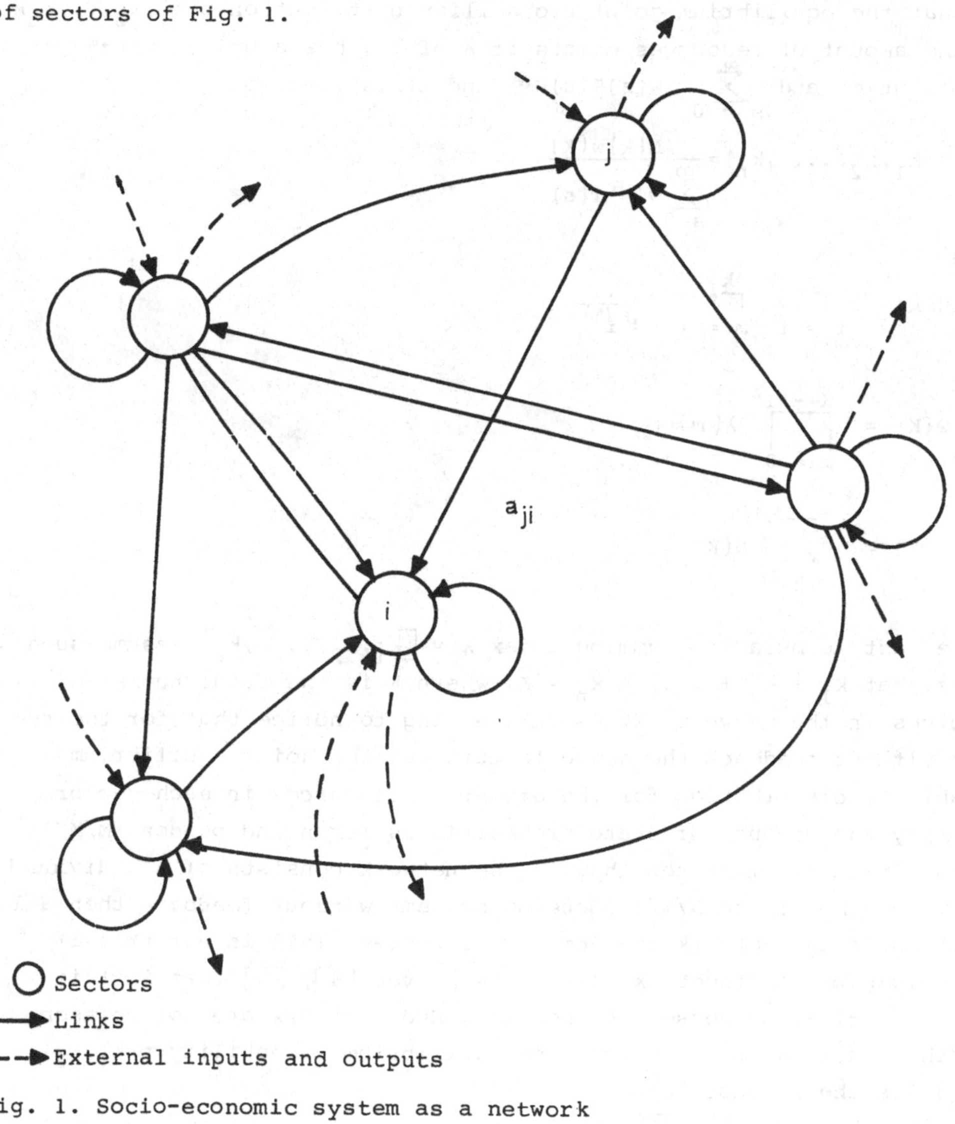

Fig. 1. Socio-economic system as a network

Assume that each sector operates as a M/M/1 queueing system. The streams of resources enter the network and are directed to sectors. Let k_i be the amount of resources present at sector i and assume k_i to be non negative integers. Vector $k = [k_1, k_2, \ldots, k_n]$ will denote the network state. Assume the service times at sector i to be statistically independent and exponentially distributed with mean $1/\mu_i$. If the process of the circulation of resources in the network is assumed to be a first order Markov chain and the process of the external arrivals is a Poisson one with the parameter $\hat{\gamma}$ then the most fundamental results for the open queueing network is given by the Jackson's theorem. It states [4] [6] that the equilibrium joint probability distribution $p(k_1, k_2, \ldots, k_n)$ for the amount of resources exists if λ of /1/ has a unique non-negative solution and $\sum_{s=0}^{\infty} w(s)T(s) < \infty$ and it is given by.

$$p(k_1, k_2, \ldots, k_n) = \frac{N(k)w(K)}{\sum_{s=0}^{\infty} w(s)T(s)}$$

where

$$N(k) = \prod_{i=1}^{n} \prod_{m=1}^{k_i} \frac{\lambda i}{\mu_i(m)}$$

$$w(K) = \prod_{m=0}^{K-1} \hat{\lambda}(m)$$

$$T(K) = \sum_{k} N(K)$$

In the last formula the summing index $k = [k_1, k_2, \ldots, k_n]$ assume such values that $k_1 + k_2 + \ldots + k_n = K$, where K is the total number of resources in the network. It is interesting to notice that for the network without feedback the above formula for the joint equilibrium probability distribution for the amount of resources in each sector is simply the product of state probabilities for n independent M/M/1 queues. One can remark too that if the network consists of n individual sectors assumed to be M/M/1 queueing systems without feedback then all the flows in the network are Poisson processes. This is not true if the sectors posses feedback. It can be proved [4], [6] that in this case the arrival processes to the individual sectors are not Poisson. Newertheless, also in this case the equilibrium probability $p(k_1, k_2, \ldots, k_n)$ has the product form

$$p(k_1, k_2, \ldots, k_n) = \prod_{i=1}^{n} p_i(k_i)$$

where $p_i(k_i)$ is the equilibrium probability of finding k_i resources in the i-th sector considered to be an M/M/1 or M/M/m_i queueing system [4], [6], Symbol m_i denotes the number of servers in the i-th sector.

As we have seen, one of the basic assumptions in the queueing network approach to the modelling of national economy system is that the routing process is first-order Markov chain. Then, the dynamics of the network is described by the Chapman-Kolmogorov equation, which we present in the form given by E. Gelenbe and R. Muntz [4]

$$\frac{d}{dt} a(k,t) = - (\hat{\lambda}(K) + \sum_{i=1}^{n} \mu_i(k_i)(1 - a_{ii})\ a(k,t) + \sum_{i=1}^{n} \hat{\lambda}(K-1) a_{oi}\ a(c(k,i),t) + \sum_{i=1}^{n} \mu_i(k_i + 1) a_{i,n+1} a(b(k,i),t) + \sum_{i=1}^{n} \sum_{\substack{j=1 \\ i \neq j}}^{n} \mu_i(k_i - 1) a_{ij}\ a(d(k,i,j),t)$$

where $b(k,i)$ and $c(k,i)$ are vectors identical to vector k except for the element k_i which is repleced by $k_i + 1$ and $k_i - 1$ respectively. Vector $d(k,i,j)$ is also identical to k except for k_i which is replaced by $k_i + 1$ and k_j which is replaced by $k_j - 1$.

This model has been applied to present the evolution of the production system in Poland. First of all the Markov property in the intersector flows matrix, corresponding to the production system, has been stated. This property is extremely strong [2]. Next, the production system was decomposed into two subsystems: industry and the rest of the production. Markov chain as a model of evolution of this two-sector production system has been found [2]. A part of numerical results present Fig. 2. It explains

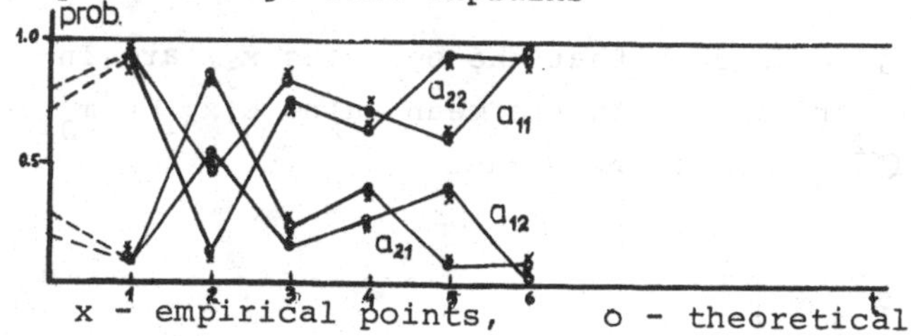

Fig. 2. Transition probabilities matrices.

changes of transition probabilities matrix in the successive years. The other results, especially concerning to the production system equilibrium studying one can find in [2].

3. PRODUCTION SECTOR AS A QUEUEING SYSTEM

In the modelling of national economy it is very important to posses adequate models of its production sectors. If one consider the production system to be a queueing network, it is necessary to treat the production sectors as queueing systems. It is known that the most elementary queueing system is M/M/1 one and the most general is G/G/m system [6]. The first one assumes that the input to the sector is Poisson process and that the production process is characterized by the probability distribution of service time which is assumed to be exponential one. It is assumed too that the whole sector is viewed as an integral whole. As opposed to the M/M/1 queueing system, the model G/G/m assumes that both input process and service time distribution are quite general and that the sector includes m internal subsystems. There is a lot of theoretical results concerning queueing systems and queueing networks modelling – see for example [3], [4] , [6] . Some possibilities of their application in the domain of socio-economic systems modelling, as well as the first numerical results are given in [2]. In this paper we should like to pay attention on the one of the most interesting and most important problems in socio-economic modelling, i. e. on the modelling of production function. It is clear that from the mathematical statistics point of view a production function is nothing other than a regression function. One of the most known production functions is generalized Cobb-Douglas production function

$$X_i = C \prod_{j=1}^{n} x_{ji}^{\alpha_{ji}} \qquad /2/$$

where $\sum_{j=1}^{n} \alpha_{ji} \leq 1$. Note [2], that if we assume the service time distribution in the sector j to be such that the outpusts x_{ji} are independent log-normal random variables with the mean value $E(x_{ji}) = m_{ji}$ and the variance $D^2(x_{ji}) = \sigma^2_{ji}$ and if the series

$$\sum_{j=1}^{n} \alpha_{ji} m_{ji} \quad , \quad \sum_{j=1}^{n} \alpha^2_{ji} \sigma^2_{ji}$$

converge, where α_{ji} are non-negative constrants, then the random variable /2/ has the log-normal distribution with $E(X_i) = B + \sum_{j=1}^{n} \alpha_{ji} m_{ji}$ and $D^2(X_i) = \sum_{j=1}^{n} \alpha_{ji}^2 \sigma_{ji}^2$. Here B is a constant such that $C = e^B$

Thus we obtain a probabilistic interpretation of a generalized Cobb-Douglas production function. Other results on the stochastic version of Cobb-Douglas production function are included in [2] .

4. CONCLUDING REMARKS.

In the paper some general remarks and suggestions on the possibility of application of queueing networks approach to stochastic modelling of socio-economic systems are presented. As far as we know it is a first attempt of considering the socio-economic system to be an open queueing network. It seems that this approach can be very fruitful especially in investigations of the equilibrium states of economy. The queueing network approach include in a natural way the basic socio-economic variables, considering them as random variables with given or unknown probability distribution. What more, the first results [2]with the aggregation and decomposition of such networks suggests that it is possible to build stochastic model of socio-economic system as a hierarchical queueing network, which changes its states according to Chapman-Kolmogorov stochastic differential equation [2].

REFERENCES

1 Adomian G., On the Modelling and Analysis of Nonlinear Stochastic Systems, Invited Lecture at the Second Int. Conf. on Mathemat. Modelling, St Louis, Missouri, July 1979 r.

2 Bereziński M., Queueing network approach to the modelling of production system, Internal Raport of the Systems Research Institute, Warsaw 1979 /in Polish/.

3 Burke P.J., The output of a queueing system, Operat. Res., 4, 699-704, 1956.

4 Gelenbe E., Muntz R.R., Exast and Approximate Solutions to Probabilistic Models of Computer System Behaviour, Part I, Acta Informatica 7, 35-60, 1976.

5 Gnedenko B., The Theory of Probability, Mir Publishers, Moscow 1976.

6 Kleinrock L., Queueing Systems, Vol. I, II, John Wiley, New York 1977.

7 Labetoulle j., Pujolle G., Soula Ch., Distribution of the Flows in a general Jacson Network, IRIA, Research Report N° 341, 1979.

8 Lange O., Introduction to Econometrics, Polish Scientific Publishers, Warsaw 1978.

9 Leontief W., The Structure of American Economy 1919-1939, Oxford Univ. Press, Oxford, England 1951.

10 Tintner G., Some Aspects of Stochastic Economics, Stochastics, Vol. 1, pp. 71-86, 1973.

OPTIMAL ALLOCATION OF A SEISMOGRAPHIC NETWORK BY NONLINEAR PROGRAMMING

Bruno Betrò

Istituto di Matematica - Università di Milano - Milano - Italy

ABSTRACT. The optimal allocation of a seismic network as an application of D-optimal design is considered together with the resulting nonlinear programming problem. Some examples of optimal allocation are given for a network in central Italy.

INTRODUCTION. In every country in which a certain level of seismic activity is present the problem arises of deeply studying those portions of the crust of the Earth which can give origin to earthquakes, in order to gain the best possible information about the actual seismic risk levels. A basic instrument for such study is the correct identification of the spatial parameters of the seismic source or in other words the so called hypocentre. This latter has to be located starting from seismic waves arrivals recordered at a certain number of seismographic stations. Thus the location is affected by errors, induced by fluctuations in the seismographic equipments, errors in the readings of seismograms, imperfect knowledge about the portion of earth crossed by the seismic waves. An analysis of the main problems connected to the hypocentre location has been recently carried out in Betrò et al. (1979). A factor which strongly influences the accuracy in the hypocentre location is the geometry of the network of seismographic stations with respect to the hypocentre.

Thus it is important to give criteria and algorithms for distributing a set of seismic stations so that the error in the location of earthquakes is kept as low as possible. The problem has as yet received scarse attention in the literature, with the exception of the papers by A. Kijko (see Kijko (1978) which also contains a list of references to preceedings papers), Archetti and Betrò (1978), considering the case of finding the minimum number of stations ensuring uniform coverage of a seismic area, Archetti and Betrò (1979), also discussing optimality criteria for a fixed number of stations.

In this paper the problem of the optimal spatial distribution of a seismic network (composed by a fixed number of stations) is considered as a problem of optimal design of experiments in a nonlinear regression scheme. In particular the basic features of the so called D-optimal design, when applied to the design of seismic networks, and the resulting nonlinear constrained optimization problem will be examined. Some examples of application are given for a seismic network in central Italy.

Formulation of the hypocentre location problem as a nonlinear regression problem.

The problem of the hypocentre location, when the seismic source is appropriately described by a point in space and an instant in time, can be described mathematically by a nonlinear regression scheme. The approach was firstly introduced by Geiger (1910). We'll denote by $\underline{x}_o$ the vector of the three spatial hypocentre coordinates, by t_o the origin time of the seismic event, by $\underline{x}_i$, $i=1,\dots,n$ the positions of n seismic stations, by t_i the firs arrival time of seismic waves observed at the ith station. The time t_i will be thought as affected by a a random normal error ε_i with mean 0 and variance σ_i^2. Thus the model can be introduced

(1) $\quad t_i = t_o + f(\underline{x}_i, \underline{x}_o) + \varepsilon_i$

where the function f is the so called travel time function from an hypocentre placed at $\underline{x}_o$ to the station placed at $\underline{x}_i$.

The simplest form of f is when the waves propagate in an homogeneous and isotropic medium with constant velocity v. In this case we have

(2) $\quad t_i = t_o + \frac{1}{v}\,\mathrm{dist}(x_i, x_o) + \varepsilon_i$

where dist(.,.) is the usual euclidean distance.

The unknown parameters t_o, $\underline{x}_o$ can be obtained by the least square method. Assuming independence of the errors ε_i, the least squares estimates $\hat{t}_o, \hat{\underline{x}}_o$ are obtained solving the problem

$$\min_{t,\underline{x}} \sum_1^n (t_i - t - f_i(\underline{x}, \underline{x}_i))^2/\sigma_i^2 \;.$$

As well known by least square theory, under wide conditions, $\hat{t}_o, \hat{\underline{x}}_o$ tend to t_o, $\underline{x}_o$ in probability when n tends to infinity. Moreover $\hat{t}_o$, $\hat{x}_o$ is asymptotically normal with mean t_o, $\underline{x}_o$ and covariation matrix

$(J^T\Sigma^{-1}J)^{-1}$, with

(3) $$J = \left[\frac{\partial \underline{f}(\underline{x}_o)}{\partial \underline{x}} \middle| \underline{1}\right]$$

where $\frac{\partial \underline{f}}{\partial \underline{x}}$ is the Jacobian matrix of the vector function $\underline{f}(\underline{x}) = (f(\underline{x}_1,\underline{x}),\ldots,f(\underline{x}_n,\underline{x}))$, $\underline{1}$ is an n-vector with unit components and

$$\Sigma = \begin{bmatrix} \sigma_1^2 & & 0 \\ & \ddots & \\ 0 & & \sigma_n^2 \end{bmatrix}.$$

Thus, setting $\underline{h}_o = (t_o, \underline{x}_o)$, $\underline{h} = (t,\underline{x})$ we have that the ellipsoid

(4) $$(\underline{h}-\hat{\underline{h}}_o)^T(J^T\Sigma^{-1}J)(\underline{h}-\hat{\underline{h}}_o) \le \chi^2_{p/4}$$

where $\chi_{p/4}$ is the quantile of order p of the χ^2 distribution with 4 degrees of freedom, is approximately a confidence region at level p for $\underline{h}_o$.

Actually a precise estimation of $\hat{t}_o$ is of scarce practical relevance and hence, in place of the ellipsoid (4), it is convenient to consider the three-dimensional ellipsoid

(5) $$(\underline{x}-\hat{\underline{x}}_o)\ S\ (\underline{x}-\hat{\underline{x}}_o) \le \chi^2_{p/3}$$

where

$$S = J_{11}^T\Sigma^{-1}J_{11} - J_{11}^T\Sigma^{-1}\underline{1}\,\underline{1}^T\Sigma^{-1}J_{11}/\underline{1}^T\Sigma^{-1}\underline{1}$$

with $J = \begin{bmatrix} J_{11} & \underline{1} \\ \underline{1}^T & 1 \end{bmatrix}$, J given by (3) and $\chi^2_{p/3}$ p quantile of the χ^2 distribution with 3 degrees of freedom. Ellipsoid (5) is an approximate confidence region for $\underline{x}_o$ at the level p. Its shape heavily depends on the allocation of the seismic stations relative to the hypocentre position. Under some circumstances, it may be even unbounded along some direction, which corresponds to the singularity of S. This fact occurs if and only if one of the following conditions holds

(6a) $J_{11}\underline{z} = \underline{0}$ for some $\underline{z} \neq 0$;

(6b) $J_{11}\underline{z} = \underline{1}$ for some $\underline{z}$.

Indeed, if z is such that $J\underline{z}=0$ or $J\underline{z}=\underline{1}$ then $S\underline{z}$ is easily seen to vanish. Conversely, if $\underline{z}\neq 0$, $S\underline{z}=0$ and $J_{11}\underline{z}\neq 0$, $\underline{z}$ can be assumed such that $\underline{1}^T\Sigma^{-1}J_{11}\underline{z} = \underline{1}^T\Sigma^{-1}\underline{1}$, as $\underline{1}^T\Sigma^{-1}J_{11}\underline{z}$ is not allowed to vanish, and hence

$$(J_{11}\underline{z}-\underline{1})^T\Sigma^{-1}(J_{11}\underline{z}-\underline{1}) =$$
$$= \underline{z}^TJ_{11}^T\Sigma^{-1}J_{11}\underline{z} - \underline{z}^TJ_{11}^T\Sigma^{-1}\underline{1} - \underline{1}^T\Sigma^{-1}J_{11}\underline{z} + \underline{1}^T\Sigma^{-1}\underline{1} =$$

$$= \underline{z}^T J_{11}^T \Sigma^{-1} J_{11} \underline{z} - \underline{z}^T J_{11} \Sigma^{-1} \underline{1} =$$

$$= \underline{z}^T J_{11}^T \Sigma^{-1} J_{11} \underline{z} - \underline{z}^T J_{11}^T \Sigma^{-1} \underline{1} \frac{\underline{1}^T \Sigma^{-1} J_{11} \underline{z}}{\underline{1}^T \Sigma^{-1} \underline{1}} =$$

$$= \underline{z}^T \left(J_{11}^T \Sigma^{-1} J_{11} - J_{11}^T \Sigma^{-1} \underline{1} \frac{\underline{1}^T \Sigma^{-1} J_{11}}{\underline{1}^T \Sigma^{-1} \underline{1}} \right) \underline{z} = \underline{z}^T S \underline{z} = \underline{0}$$

which implies $J_{11}\underline{z} = \underline{1}$.

Optimality criteria for a seismic network.

If the asymptotic properties can be assumed to hold approximately for finite n, the probabilistic meaning of relation (5) suggests that optimality criteria for a seismic network should be based on optimization of some functional related to the matrix $S = S(\underline{x}_o; \underline{x}_1, \ldots, \underline{x}_n)$, aiming at a reduction of the "dimensions" of the ellipsoid.

Various types of such criteria have been analyzed in the framework of the so called optimal design of experiments (Fedorov (1972)). The most popular of such criteria is the D-optimal design criterion, consisting in the maximization of the determinant of the matrix of the coefficients of the confidence ellipsoid, which corresponds to minimization of the volume of the ellipsoid. Its feasibility for the design of seismic networks has been analyzed in Kijko (1978) and in Archetti and Betrò (1979). They suggest the introduction of a probability measure $P(d\underline{x}_o)$ reflecting a seismic relevance a priori assigned to points $\underline{x}_o$ of a certain three-dimensional region K and the averaging of determinant with respect to P. Consequently the optimal allocation of a network can be reduced to the problem

$$\max_{\underline{x}_1, \ldots, \underline{x}_n} \int_K \det(S(\underline{x}_o; \underline{x}_1, \ldots, \underline{x}_n)) P(d\underline{x}_o).$$

More meaningfully from the geometrical point of view, as the volume of (5) is proportional to 1/ det(S) we prefer to consider the problem

$$(7) \qquad \min_{\underline{x}_1, \ldots, \underline{x}_n} \int_K \frac{P(d\underline{x}_o)}{\sqrt{\det(S(\underline{x}_o; \underline{x}_1, \ldots, \underline{x}_n))}} .$$

In real situations it is not likely that the stations can be set anywhere; tipically seismic stations cannot be set over the sea.

Thus a certain number of constraints, expressed by a set C in the space of $\underline{x}_1, \ldots, \underline{x}_n$ has to be added to problem (7), obtaining finally the constrained optimization problem

$$(7') \quad \min_{\underline{x}_1, \ldots, \underline{x}_n \in C} \bar{V}(\underline{x}_1, \ldots, \underline{x}_n) ,$$

$$\bar{V} \quad (\underline{x}_1, \ldots, \underline{x}_n) = \int_K \frac{P\ (d\ \underline{x}_o)}{\sqrt{\det\ (S(\underline{x}_o, \underline{x}_1, \ldots, \underline{x}_n)}} \ .$$

Optimal allocation of a station for the network of Ancona.

Criterion (7') was used to attempt some optimization for the seismic network existing around the town of Ancona, on the Adriatic coast of Italy (fig. 1).

The network consists of six stations (black dots in the figure) all placed at null depth. In the region earthquakes occur mainly under the sea and hence the coast represents a serions constraint for the allocation of the seismic stations.

Seismic waves were assumed to propagate with constant velocity v = 5 km / sec and reading errors of first arrival times to have a standard deviation $\sigma = 0.1$.

Three different points were assumed to represent possible seismic sources : they are indicated in fig. 1 by $\underline{x}_o^1, \underline{x}_o^2, \underline{x}_o^3$.

Referring to the coordinate system of the map completed with the depth expressed in kilometers, their coordinates are

$$\underline{x}_o^1 = (21,\ 17,\ 1);\ \underline{x}_o^2 = (6,\ 25,\ 5);\ \underline{x}_o^3 = (16,\ 27,\ 1)$$

We'll discuss here the simple case, particularly attractive for the possibilities of graphical displaying, in which optimization is performed with respect to the position of one station. Five stations were fixed to their present position, that is in fig. 1 MB, PL, MS, MT, CF, and the matrix S considered in dependence only a sixth station $\underline{x}_6$. $\underline{x}_6$ was constrained to lie at null depth within the region C bounded by the coast and the borders of the figure.

These three latter constraints aim at taking into account the phisical fact that beyond a certain distance small events are no longer registe

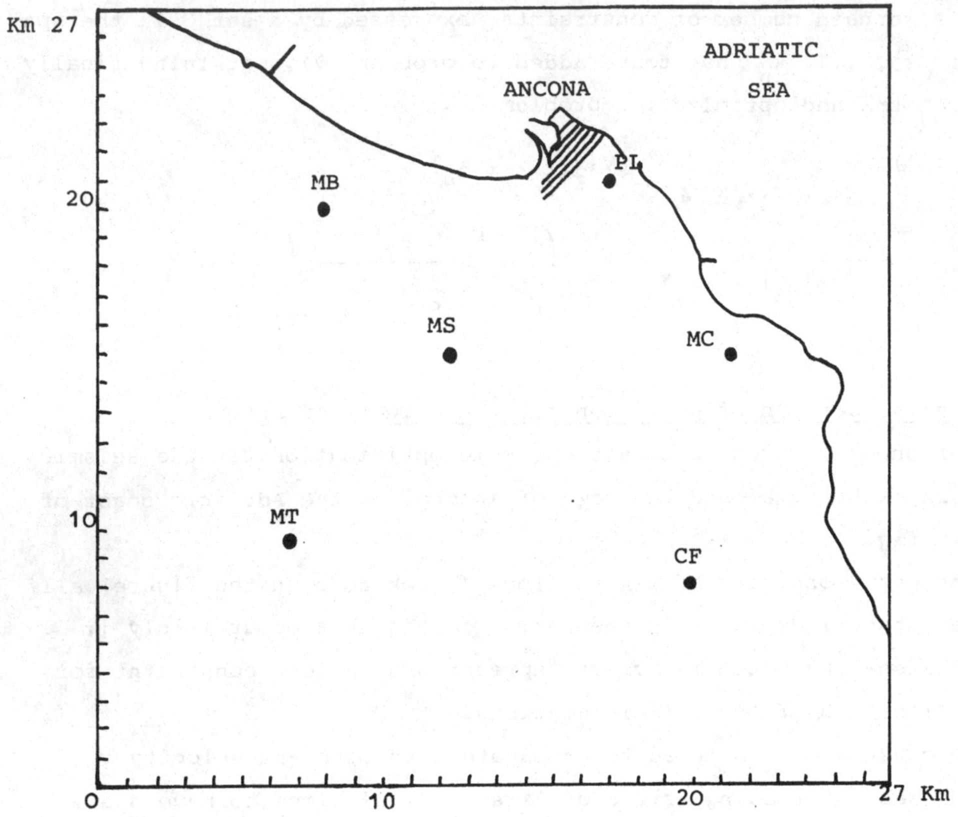

Fig. 1 - The seismografic network near Ancona

red by the stations.

According to these assumptions, problem (7') is reduced to the problem

$$(7'') \quad \min_{\underline{x}_6 \in C} \bar{V}(\underline{x}_6), \quad \bar{V}(\underline{x}_6) = \sum_1^3 \frac{p_i}{\sqrt{\det S(\underline{x}_o^i; \underline{x}_6)}}, \quad \sum_1^3 p_i = 1 .$$

Four different situations were considered for the weights p_i

(8a) $p_1 = 1$, $p_2 = p_3 = 0$

(8b) $p_1 = p_3 = 0$, $p_2 = 1$

(8c) $p_1 = p_2 = 0$, $p_3 = 1$

(8d) $p_1 = p_2 = p_3 = 1/3$.

Figg. 2-3-4-5 display some level contours of function $\bar{V}$ ($\underline{x}_6$) given by (7") in correspondence of conditions (8a) - (8d), and a third degree spline representation of the coast. Such representation has the smoothness required by the most effective nonlinear programming routines.

The usefulness of the optimization approach is clearly supported by the numbers: sensible reductions of the volume of the confidence ellipsoid can be gained minimizing $\bar{V}$ ($\underline{x}_6$) and the optimal location of $\underline{x}_6$ is not obvions.

A relevant feature of the optimization problem is the presence of several local minima for the objective function, due to the constraints. Level contours reveal approximately the location of local minima and hence a local routine is sufficient for solving problem (7"). This is not the case of course of the more general problem (7'), for which global seeking methods should be used (Towards Global Optimization (1975) - (1978)).

Minimization of $\bar{V}$ ($\underline{x}_6$) was performed by means of subroutine OPRQP of the OPTIMA package (N.O.C. (1976)),designed for efficiently finding local solutions to nonlinear programming problems.The routine was started from different points lying in the region of attraction of different minima. Table I stresses the good performance of OPRQP.

<u>CONCLUSIONS</u>. The feasibility of the optimization of the allocation of a seismc network by nonlinear programming should have been proved, at least if criterion (7') is accepted as good. The criterion may fail if region (5) cannot be regarded as a confidence region for $\underline{x}_o$, because of a number of stations not sufficiently high to ensure that the approximation given by asymptotic properties of $\hat{t}_o$, $\hat{\underline{x}}_o$ is good enough. Another paper is being prepared investigating such question.

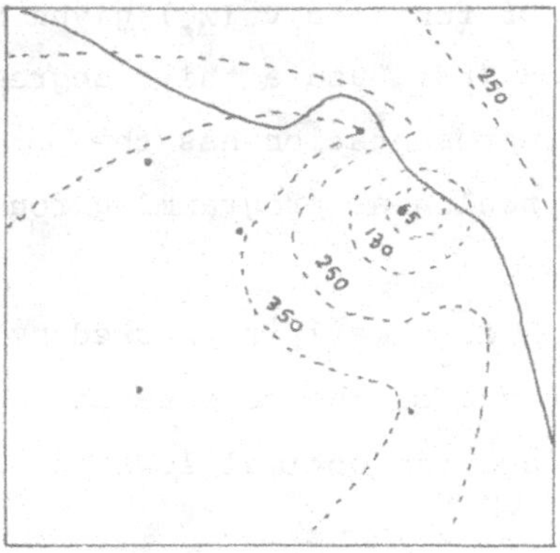

Fig.2 - Level contours of $\bar{V}(\underline{x}_6)$ with condition (8a) (Km^3)

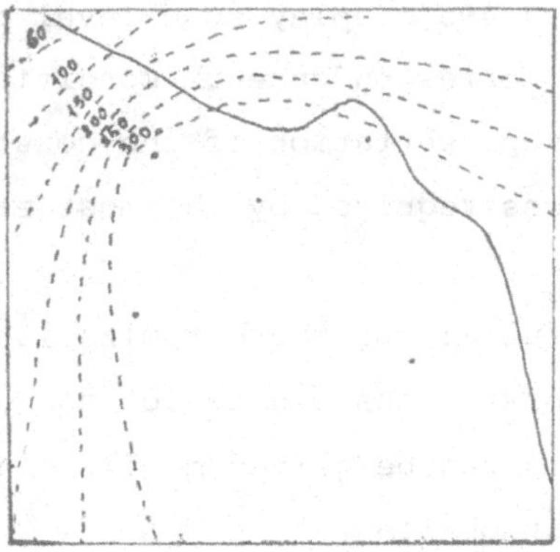

Fig.3 - Level contours of $\bar{V}(\underline{x}_6)$ with condition (8b) (Km^3)

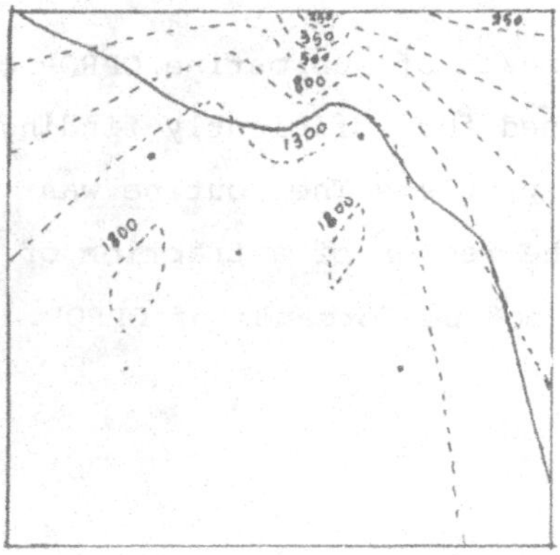

Fig.4 - Level contours of $\bar{V}(\underline{x}_6)$ with condition (8c) (Km^3)

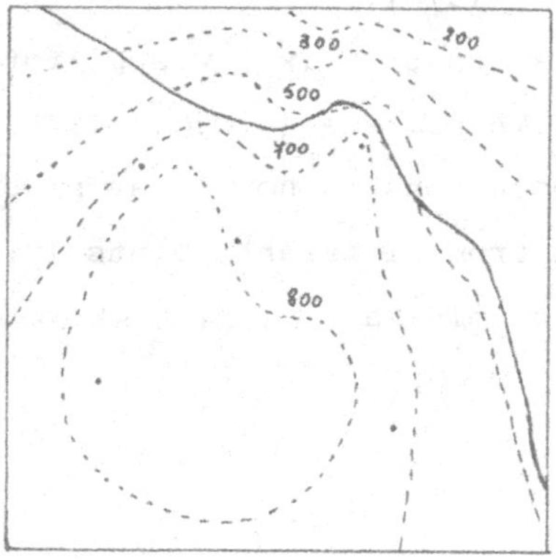

Fig.5 - Level contours of $\bar{V}(\underline{x}_6)$ with condition (8d) (Km^3)

p_1 p_2 p_3	Starting point	Initial value	Final point	Final value	Function evalu tions + gradient evalutions
1 0 0	(21.00,17.00)	29.29	(20.91,16.93)	29.04	5 + 5
	(1.00,20.00)	314.91	(-0.03,28.26)	219.60	65 + 9
	(26.00,10.00)	197.04	(27.00, 5.78)	184.51	45 + 23
	(16.00,25.00)	207.52	(17.32,22.95)	265.78	29 + 24
0 1 0	(1.00,25.00)	67.68	(0.00,28.21)	51.37	37 + 8
	(16.00,25.00)	201.68	(26.33,22.97)	263.03	17 + 17
0 0 1	(16.00,27.00)	190.03	(15.25,22.71)	871.48	22 + 21
	(1.00,20.00)	868.59	(0.00,28.21)	376.40	25 + 8
	(26.00.10.00)	869.88	(24.38,15.32)	784.77	255 + 101
$\frac{1}{3}$ $\frac{1}{3}$ $\frac{1}{3}$	(6.00-15.00)	846.67	(0.01,28.22)	215.70	82 + 30
	(16.00,25.00)	268.14	(25.52,22.87)	478.93	26 + 24
	(23.00,15.00)	483.21	(23.29,16.69)	429.45	66 + 43

Table I - Performance of OPRQP

REFERENCES.

Archetti F. and Betrò B. (1978), the Multiple Covering Problem and its Application to the Optimal Dimensioning of a Large Scale Seismic Network, Lecture Notes in Control and Information Sciences, Vol. 7, Springer Verlag.

Archetti F. and Betrò B. (1979), Optimization Problems Arising in the Design and Exploitation of Seismographic Networks, in Numerical Methods for Dynamical Systems, Dixon L.C.W. and Szegö G.P. eds, North Holland.

Betrò B., Di Natale M., Stucchi M., Zonno G. (1979), Some Considerations on the Hypocentre Location Problem, to appear.

Fedorov V.V. (1972), Theory of Optimal Experiments, Academic Press, New York.

Geiger L. (1910), Herdbestimmung bei Erdbeben aus den Ankungtszeiten, K. Gesell. Wiss. Goett., 4, 331 - 349.

Kijko A. (1978), Methods of the Optimal Planning of Regional Seismic Networks, Publ. Inst. Geophys. Pol. Acad. Sc., A - 7 (119).

N.O.C. (1976), OPTIMA, The Hatfield Polytechnic, Hatfield, Hertfordshire.

Towards Global Optimization (1975), Dixon L.C.W. and Szegö G.P. eds, Vol I, North Holland.

Towards Global Optimization (1978), Dixon L.C.W. and Szegö G.P. eds, vol. II, North Holland.

STOCHASTIC APPROACH TO THE TWO-LEVEL OPTIMIZATION OF THE COMPLEX OF OPERATIONS

Z.Bubnicki
Technical University of Wroclaw
Institute of Engineering Cybernetics
Control Systems Group, Poland

M.Staroswiecki, A.Lebrun
University of Lille
Centre d'Automatique
France

1.Introduction

The paper deals with the problem of time-optimal resources and tasks allocation in the complex of static operations with random parameters. The direct approach to this very specific stochastic optimization problem [1,2] is usually very complicated bacause of the great computational difficulties. The decomposition consists here in dividing the whole complex into subcomplex which leads to the two-level allocation system (Fig.1).

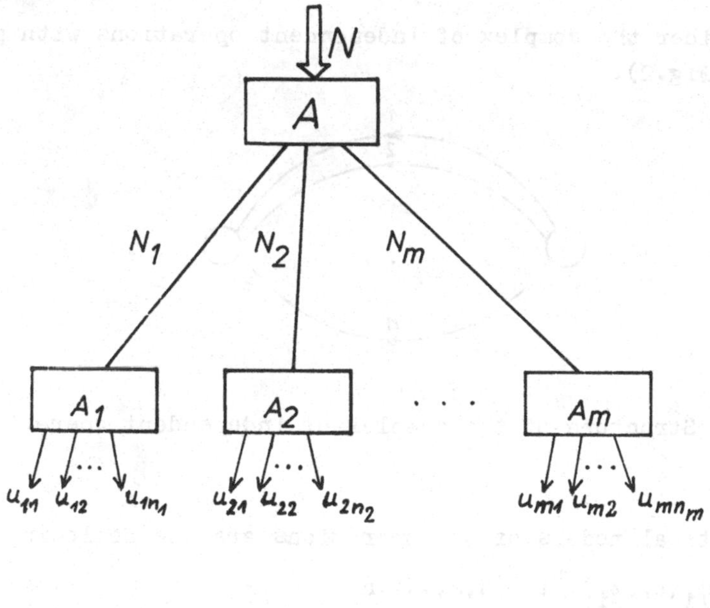

Fig.1. Two-level allocation system. A - allocation in subcomplexes, $A_1,\dots,A_m$ - allocation is seperate operations in each subcomplex.

On the upper level the global amount of resources (or the global seize of tasks) N is distributed among the subcomplexes and on the lower level the amount N_l ($l = 1,2,\dots,m$) in each subcomplex is distributed among the seperate operations according to the decision variables u_{lj} (the amount of resources or the size of tasks for j-th operation i l-th subcomplex).

Different versions of this approach were presented in [3,4,5] with the expected execution time of the complex as an optimization criterion. In this paper the problem of finding the solution maximizing the probability that the execution time is less than a given number is formulated and then the decomposition and two-level optimization with this form of a criterion is proposed for the complex of independent operations. The simple example of the complex with four operations and with rectangular probability densities of the random parameters is given.

2. Stochastic optimization

Let us consider the complex of independent operations with parallel structure (Fig.2).

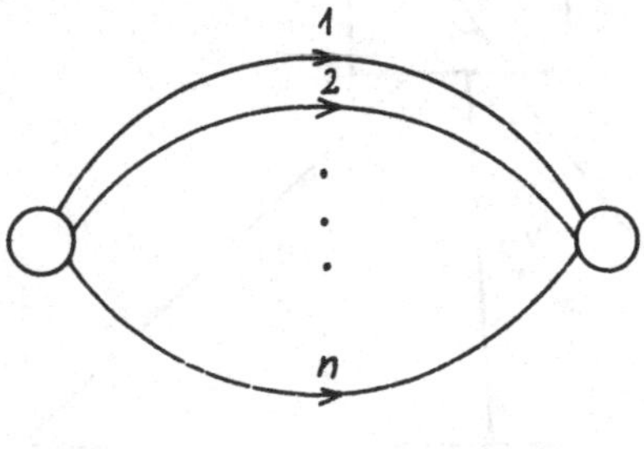

Fig.2. Structure of the complex of independent operations.

The mathematical models of the operations are the following

$$T_i = \varphi_i(u_i, z_i), \quad i = 1,2,\dots,n \tag{1}$$

where

T_i - execution time,

u_i - decision variable (the amount of resources or the size of tasks),

z_i - random parameter, i.e. realization of the random variable $\underline{z}_i$; $\underline{z}_i$ are assumed to be independent for the different i with the known probability distribution $F_{i,z}(z)$. Knowing (1) and $F_{i,z}(z)$ we can determine probability distribution for $\underline{T}_i$

$$F_i(\tau,u_i) = P(\underline{T}_i < \tau) \qquad (2)$$

or we can consider (2) as the initial form of the stochastic model of the operation.

The following forms of the criteria of stochastic optimization can be introduced

$$Q_1 = E(\underline{T}) = \int_0^\infty \tau \, d\left[\prod_{i=1}^{n} F_i(\tau,u_i)\right] \triangleq G_1(u_1,u_2,\ldots,u_n), \qquad (3)$$

$$Q_2 = \max_i E(\underline{T}_i) \triangleq G_2(u_1,u_2,\ldots,u_n), \qquad (4)$$

$$Q_3 = P(\underline{T} < \alpha) = \prod_{i=1}^{n} F_i(\alpha,u_i) \triangleq G_3(u_1,u_2,\ldots,u_n), \qquad (5)$$

where

$\underline{T} = \max_i \underline{T}_i$, $\underline{T}_i = \varphi_i(u_i,\underline{z}_i)$, α- given parameter.

The problem is to find $u_1^*,u_2^*,\ldots,u_n^*$ minimizing Q_1 or Q_2 or maximizing Q_3 with constraints

$$\sum_{i=1}^{n} u_i = N, \; u_i \geqslant 0 \; (i = 1,2,\ldots,n), \qquad (6)$$

where N is the global amount of resources of the global seize of tasks.

The determination of the function G_1 and the extremalization of G_1,G_2 or G_3 with constraints (6) is very difficult from the computational point of view and usually requires the application of the special numerical algorithms.

Stochastic optimization problem can be extended for the complex of dependent operations with a structure described by the graph more complicated than that on Fig.1. Then the "decomposition in time" may be applied which reduces the problem to the stochastic optimization of the separate parts of the complex, each of them has the parallel structure as was described in [6].

3. Decomposition and two-level optimization

The decomposition of the stochastic optimization problem with the criterion Q_1 was presented in [3,4,5]. It leads to the following formula

$$\min_{\substack{u_1,\dots,u_n \\ \sum_{i=1}^{n} u_i = N}} E\left[\max_{i=1,\dots,n} \varphi_i(u_i,\underline{z}_i)\right] =$$

$$= \min_{\substack{N_1,\dots,N_m \\ \sum_{l=1}^{m} N_l = N}} \left\{\max_{l}\left[\min_{\substack{u_{l1},\dots,u_{ln_l} \\ \sum_{j=1}^{n_l} u_{lj} = N_l}} E\left[\max_{j=1,\dots,n_l} \varphi_{lj}(u_{lj},\underline{z}_{lj})\right]\right]\right\} \quad (7)$$

where z_{lj} is the random parameter of the j-th operation in the l-th subcomplex, n_l is the number of the operations in the l-th subcomplex and m is the number of subcomplexes.

Another approaches may be obtained by putting E after the first "min" or after the first "max" or after the second "max" in the right-hand side of (7). In general, none of these approaches gives the correct result, i.e. the same as the result of the direct optimization. In other words, for each approach the equality (7) is not satisfied.

The criterion Q_2 is easy to obtain as the function of $u_1,\dots,u_n$ and the optimization problem is reduced to the known problem for the deterministic case, but on the other hand it has no practical interpretation. The criterion Q_3 proposed in this paper is easy to obtain as the function of $u_1,\dots,u_n$ and the corresponding optimization problem is easy to decompose and to obtain the correct results, the same as the results of the direct optimization.

Using the known properties of the distribution functions $F_i(\alpha,u_i)$ which are increasing functions of u_i for resources and decreasing functions of u_i for tasks - it is easy to prove the following.

Theorem

$$\max_{\substack{u_1,\dots,u_n \\ \sum_{i=1}^{n} u_i = N}} \prod_{i=1}^{n} F_i(\alpha,u_i) = \max_{\substack{N_1,\dots,N_m \\ \sum_{l=1}^{m} N_l = N}} \prod_{l=1}^{m}\left[\max_{\substack{u_{l1},\dots,u_{ln_l} \\ \sum_{j=1}^{n_l} u_{lj} = N_l}} \prod_{j=1}^{n_l} F_{lj}(\alpha,u_{lj})\right],$$

where $F_{lj}(\alpha,u_{lj})$ is the probability distribution for $\underline{T}_{lj}$, and the results u^*_{lj} of the optimization via decomposition are the same as the solutions u^*_i of the direct optimization problem.

The decomposition leads here to the two-level optimization. On the lower level we maximize

$$\prod_{j=1}^{n_l} F_{lj}(\alpha, u_{lj}) \triangleq G_3^{(l)}(u_{l1},\dots,u_{ln_l}), \quad (l=1,2,\dots,m) \tag{8}$$

according to $u_{l1},\dots,u_{ln_l}$ with constraints

$$\sum_{j=1}^{n_l} u_{lj} = N_l, \quad u_{lj} \geqslant 0, \quad (j=1,2,\dots,n_l).$$

As the results we obtain

$$Q^{(l)} = \max_{u_{l1},\dots,u_{ln_l}} \prod_{j=1}^{n_l} F_{lj}(\alpha, u_{lj}) \triangleq G^{(l)}(N_l)$$

and

$$u_{lj}^{*} = \varkappa_{lj}^{*}(N_l). \tag{9}$$

On the upper level we maximize

$$Q_3 = \prod_{l=1}^{m} G^{(l)}(N_l)$$

according to $N_1,\dots,N_l$ with constraints

$$\sum_{l=1}^{m} N_l = N, \quad N_l \geqslant 0 \quad (l = 1,2,\dots,m).$$

Substituting the results of the maximization $N_1^{*},\dots,N_m^{*}$ into (9) we obtain the optimum values of the decision variables u_{lj}^{*} ($j = 1,\dots,n_l$; $l = 1,\dots,m$).

4. Example

Consider the simple example of four operations with the probability distributions

$$F_i(\tau, u_i) = \begin{cases} 0 & \tau \leqslant 0 \\ k_i u_i \tau & 0 < \tau \leqslant \dfrac{1}{k_i\, u_i} \\ 1 & \tau > \dfrac{1}{k_i u_i} \end{cases},$$

$i = 1,2,3,4$. Let $u_1 = u_{11}$, $u_2 = u_{12}$, $u_3 = u_{21}$, $u_4 = u_{22}$, $k_1 = k_{11}$,

$k_2 = k_{12}$, $k_3 = k_{21}$, $k_4 = k_{22}$. Let us start with the first subcomplex and assume

$$k_{12}u_{12} > k_{11}u_{11}. \tag{10}$$

Then, according to (8)

$$G_3^{(1)}(u_{11},u_{12}) = F_{11}(\alpha,u_{11})F_{12}(\alpha,u_{12}) = \begin{cases} k_{11}k_{12}u_{11}u_{12}\alpha^2 & \alpha \leqslant \frac{1}{k_{11}u_{11}} \\ k_{12}u_{12}\alpha & \frac{1}{k_{11}u_{11}} < \alpha \leqslant \frac{1}{k_{12}u_{12}} \\ 1 & \alpha > \frac{1}{k_{12}u_{12}} \end{cases}$$

Now, two cases may be considered.

I. Maximization of

$$G_3^{(1)}(u_{11},u_{12}) = k_{11}k_{12}u_{11}u_{12}\alpha^2$$

with constraints (10), $\alpha k_{11}u_{11} \leqslant 1$, $u_{11}+u_{12} = N_1$, $u_{11} \geqslant 0$, $u_{12} \geqslant 0$. The solution exists for

$$\frac{1}{k_{11}} + \frac{1}{k_{12}} > \alpha N_1 \tag{11}$$

and the results are the following:

(i) For

$$k_{12} < k_{11} \quad \text{and} \quad k_{11}\alpha N < 2 \tag{12}$$

$$u_{11}^* = u_{12}^* = \frac{N_1}{2}, \quad Q^{(1)} = k_{11}k_{12}\left(\frac{\alpha N_1}{2}\right)^2.$$

(ii) For

$$k_{12} > k_{11} \tag{13}$$

$$u_{11}^* = \frac{k_{12}N_1}{k_{11}+k_{12}}, \quad u_{12}^* = \frac{k_{11}N_1}{k_{11}+k_{12}}, \quad Q^{(1)} = \left(\frac{k_{11}k_{12}\alpha N_1}{k_{11}+k_{12}}\right)^2. \tag{14}$$

(iii) For

$$k_{11}\alpha N_1 > 2 \tag{15}$$

$$u_{11}^* = \frac{1}{k_{11}\alpha}, \quad u_{12}^* = N_1 - \frac{1}{k_{11}\alpha}, \quad Q^{(1)} = k_{12}\alpha\left(N_1 - \frac{1}{k_{11}\alpha}\right).$$

II. Maximization of

$$G_3^{(1)}(u_{11},u_{12}) = k_{12}u_{12}\alpha$$

with constraints (10), $\alpha k_{11}u_{11} > 1$, $\alpha k_{12}u_{12} \leqslant 1$, $u_{11} + u_{12} = N_1$,

$u_{11} \geqslant 0$, $u_{12} \geqslant 0$.

The solution exists for

$$k_{11} \alpha N_1 > 1 \tag{16}$$

and is the following

$$u^*_{11} = N_1 - \beta, \quad u^*_{12} = \beta, \quad Q^{(1)} = k_{12} \alpha \beta$$

where

$$\beta = \min\left(\frac{k_{11}N_1}{k_{11}+k_{12}}, \; N_1 - \frac{1}{\alpha k_{11}}, \; \frac{1}{\alpha k_{12}}\right).$$

For given N_1, k_{11}, k_{12}, α it is necessary to prove the conditions (11) and (16), to chose the proper case (13),(14),(15) if the condition (11) is satisfied and if both conditions (11),(16) are satisfied - to choose the solution from the cases I and II for the greater value of $Q^{(1)}$. The considerations and results for the second subcomplex are analogous. For finding the optimum values N^*_1, N^*_2 it is necessary to consider all the possible cases with the maximization of $G^{(1)}(N_1) \cdot G^{(2)}(N_2)$. The situation is mach simpler if for two subcomplexes (11) is satisfied and(16) is not satisfied for any $N_1 < N$, i.e. if

$$\frac{1}{k_{11}} + \frac{1}{k_{12}} > \alpha N, \quad \frac{1}{k_{21}} + \frac{1}{k_{22}} > \alpha N, \tag{17}$$

and $k_{11} \alpha N < 1$, $k_{21} \alpha N < 1$. (18)

Then, for the case (13), i.e. for

$$k_{12} > k_{11}, \quad k_{22} > k_{21} \tag{19}$$

we obtain the solution (14) and consequently

$$Q_3 = Q^{(1)}Q^{(2)} = \left(\frac{k_{11}k_{12} \alpha N_1}{k_{11}+k_{12}}\right)^2 \left(\frac{k_{21}k_{22} \alpha N_2}{k_{21}+k_{22}}\right)^2.$$

Maximization of Q_3 with constraints $N_1 + N_2 = N$, $N_{1,2} \geqslant 0$ gives

$$N^*_1 = N^*_2 = \frac{N}{2}$$

and finally

$$u^*_{11} = \frac{k_{12}N}{2(k_{11}+k_{12})}, \quad u^*_{12} = \frac{k_{11}N}{2(k_{11}+k_{12})}, \quad u^*_{21} = \frac{k_{22}N}{2(k_{21}+k_{22})}, \quad u^*_{22} = \frac{k_{21}N}{2(k_{21}+k_{22})},$$

E.g. for $k_{11} = \frac{1}{4}$, $k_{12} = 1$, $k_{21} = \frac{1}{3}$, $k_{22} = 2$, $N = 2$, $\alpha = 1$ conditions (17),(18),(19), are satisfied and the numerical results are the following

$$N_1^* = N_2^* = 1,\ u_{11}^* = 0{,}80,\ u_{12}^* = 0.20,\ u_{21}^* = 0.85,\ u_{22}^* = 0.15.$$

5. Some generalization - Polyoptimization problem [7]

The stochastic optimization of the complex of independent operations may be considered as a special case of the following polyoptimization problem:

$$\underline{y}_i = \Phi_i(x_1,x_2,\dots,x_n;\underline{z}_i),\ i = 1,2,\dots,n,$$

$$Q_1 = \underset{\underline{z}_1,\dots,\underline{z}_n}{E}\left[\max_i \Phi_i(x_1,x_2,\dots,x_n;\underline{z}_i)\right],$$

$$Q_2 = \max_i \underset{\underline{z}_i}{E}\left[\Phi_i(x_1,x_2,\dots,x_n;\underline{z}_i)\right],$$

$$Q_3 = P\left[\max_i \Phi_i(x_1,x_2,\dots,x_n;\underline{z}_i) < \alpha\right]$$

where

x_i - decision of i-th decision maker,

y_i - cost (or loss) of i-th decision maker,

F_i - cost (or loss) function,

z_i - random parameter; z_i and z_j stochastically independent for $i \neq j$. The problem is to find $x_1^*,x_2^*,\dots,x_n^*$ as to minimize Q_1 or Q_2 or to maximize Q_3 (three different versions of the stochastic optimization). The deterministic version of this polyoptimization problem was called "social approach" and was considered in [8]. This is a generalization of the former stochastic optimization of the complex of operations in which

$$\Phi_i(x_1,x_2,\dots,x_i;z_i) = \varphi_i(x_i,z_i);\ y_i = T_i.$$

If

$$F_i(\gamma,x_1,x_2,\dots,x_n) = P(\underline{y}_i < \gamma)$$

is the probability distribution function for $\underline{y}_i$, then the probability distribution function for

$$\underline{y} \triangleq \max_i \Phi_i(x_1,x_2,\dots,x_n;\underline{z}_i)$$

is the following

$$F(\gamma;x_1,x_2,\dots,x_n) = P(\underline{y} < \gamma) = \prod_{i=1}^{n} F_i(\gamma;x_1,x_2,\dots,x_n),$$

and

$$Q_3 = \prod_{i=1}^{n} F_i(\alpha;x_1,x_2,\ldots,x_n).$$

Optimization problem is now much more complicated and cannot be decomposed as in a special case for the complex of operations.

6. Final remarks

The suggested approach to the stochastic optimization of the complex of operations may be considered as some basis for the numerical methods and algorithms which are much simpler for the criterion Q_3 then for Q_1. The criterion Q_3 may be also applied for the wide class of complexes of dependent operations. As was mentioned above, the "decomposition in time" leads to the sequential optimization of the complexes with parallel structure in the successive time intervals. The applications of the "decomposition in space" suggested in this paper leads to three-level optimization: the optimal allocations for the seperate operations and for the subcomplexes in each time interval on the first and the second level, respectively and the optimal coordination between the different time intervals on the third level. The application of the decomposition methods for dependent operations and for the presented generalization (polyoptimization problem) requires the further investigations.

References

1 Bubnicki Z.: Optimal control of the complex of operations with random parameters. Podstawy Sterowania. Tom 1, z.1, 1977.

2 Bubnicki Z., Markowski J.: Probabilistic problems of the time--optimal control of independent operations, Viena 1977, Pergamon Press.

3 Bubnicki Z.: O pewnych problemach czasowo-optymalnego sterowania kompleksem operacji w warunkach probabilistycznych. Podstawy Sterowania, vol.2, No 3, 1972.

4 Bubnicki Z.: Two-level optimization and control of the complex of operations. Papers of IFAC/IFORS/IIASA Workshop on Systems Analysis Applications to Complex Programs, Pergamon Press 1977, Bielsko-Biała p.p. 1407-1412.

5 Bubnicki Z.: Two-level optimization and control of the complex of operations. VI IFAC Congres, Pergamon Press, Helsinki, 1978.

6 Bubnicki Z.: Time-optimal control of dependent operations with random parameters. Systems Science, Vol.3, No 3, 1977, pp.227-236.

7 Bubnicki Z., Staroswiecki M.: Optimization of the complex of operations as some example of a special polyoptimization problem. Paper presented on International Conference Systems Science V, Wroclaw, Poland, 1977.

8 Staroswiecki M.: Contribution á l´analyse et á la commande de systémes complexes á critéres multiples. These d´etat es science. Université de Lille, Centre d´Automatique, 1978.

SOME RESULTS ON TIMED PETRI-NETS.

Ph.CHRETIENNE.
Université Paris 6
4 place Jussieu, Paris 5ième
FRANCE

I. INTRODUCTION.

Petri-nets have been found an adequate tool to describe the state transitions of rather complicated systems (as asynchronous systems). Many coordination problems have been modeled successfully with them.

However, these models need more information in order to study some quantitative aspects as utilisation rates , delays .....which are of main interest for a practical point of view.

So, we are interested in more sophisticated models called Timed Petri-Nets (TPN) in which the time dimension is introduced. In this paper, we first give a formal and rigorous definition of the execution of a TPN ; then, we give some general results on what we call "a program" : finally, we extend Ramachandani previous results on strongly periodic event graphs to general Petri-nets.

II. DEFINITIONS-NOTATIONS-RULES OF THE GAME.

Petri graph.

A Petri graph is an oriented bipartite graph $G = (T_U P, V)$ with $V \subset T.P_U P.T$.

$T \neq \emptyset$ is the set of transitions (or tasks)

$P \neq \emptyset$ is the set of places (associated with state components of the system)

Petri-net.

A Petri-net is a pair $R = (G,M)$ where G is a Petri- graph and M a mapping

$$M : P \longrightarrow N \quad .$$

This mapping is called the initial marking (or state) of R .

M(p) can also be called the number of tokens of place p .

An example.

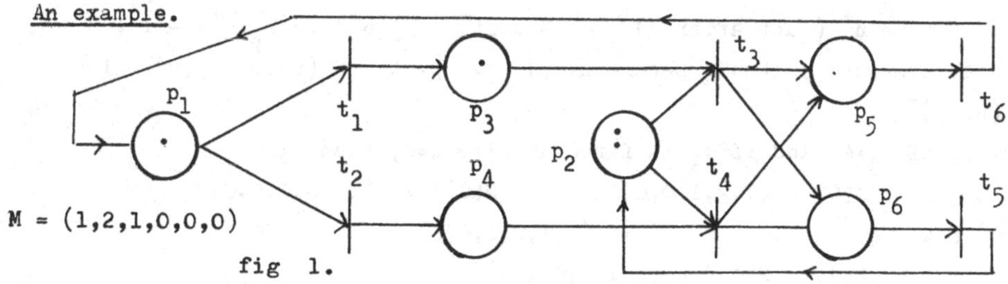

fig 1.

Some notations.

Consider place p, we call p^+ the set of the output transitions of p and p^- the set of the input transitions of p. Consider transition t, we call t^+ (resp t^-) the set of the output (resp input) transitions of t.

We shall not consider Petri-nets with loops (a place p belonging to t^+ and t^-), so in this case, we can associate to a Petri-net an incident matrix E where

$$e_{ij} = \begin{cases} -1 & \text{if } t_j \in p_i^+ \\ +1 & \text{if } t_j \in p_i^- \\ 0 & \text{elsewhere} \end{cases} .$$

Rules of the game.

Firable transition : a transition is said to be firable if

$$\forall p \in t^- \quad M(p) > 0 \quad ;$$

The firing of transition t will change the marking in the following way :

$$\forall p \in t^- \quad M(p) := M(p) - 1$$

$$\forall p \in t^+ \quad M(p) := M(p) + 1 \quad .$$

A sequence $s = t_{i_1} t_{i_2} \ldots\ldots\ldots t_{i_q}$ of transitions is feasible for an initial marking M if each transition of s can be fired in the s order .

If n(s) is the column vector whose t-component is the number of occurences of t in s, then we have the state equation :

$$\underset{\text{new state}}{M'} = \underset{\text{old state}}{M} + E.n(s) \quad .$$

Timed Petri-net (TPN).

A timed Petri-net is a Petri-net whose transitions have a strictly positive duration ; we define

$$t \in T \longrightarrow d_t \in N^+ \text{ (set of positive integers)} \quad .$$

The time dimension must be introduced in the notation of the state of the system, so we note M (u) the marking of the TPN at time u .

The firing of transition t at time u is feasible if :

$$\forall p \in t^- \quad M_p(u) > 0 \quad .$$

The firing of transition t has the following effects :

at time u^+ (just after u) $\quad \forall p \in t^- \quad M_p(u^+) := M_p(u^+) - 1$;

at time $(u+d_t)^-$ (just before $u+d_t$) $\quad \forall p \in t^+ \quad M_p(v) := M_p(v) + 1$

where $v=(u+d_t)^-$.

On figure 1, if $d_3=4$, and if t_3 is fired at time u=0, then :

$$M(0) = (1,2,1,0,0,0) \qquad M(0^+) = (1,1,0,0,0,0)$$

for $0^+ \leqslant u \leqslant 4^-$ $\quad M(u) = (1,1,0,0,0,0)$

$$M(4) = M(4^-) = (1,1,0,0,1,1) \quad .$$

A transition t, fired at time u, is said to be active from time u^+ to time $(u+d_t)^-$.

In order to construct the state equation, we have to define two staired functions associated with each transition $t \in T$:

$D_t(u)$ = number of initialised firings of t in the closed interval $[0 \quad u]$

$F_t(u)$ = number of achieved firings of t in $[0 \quad u]$.

With these two functions, the state equation is given for each place p, by :

$$M_p(u) = M_p(0) + \sum_{p^-} F_t(u) - \sum_{p^+} D_t(u) .$$

III. CONCEPT OF PROGRAM.

The notion of sequence is not sufficient for TPN essentially because parallelism is not allowed in the execution of tasks (transitions). So, we have defined the more general concept of program .

A program is a mapping G $\quad G : T \longrightarrow \{D_t\}$ where

D_t is a staired function $\quad D_t \quad u \geqslant 0 \longrightarrow N$ with the following shape :

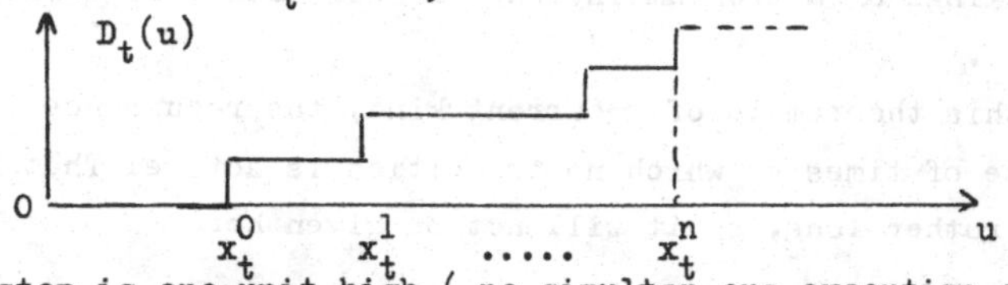

fig 2.

- each vertical step is one unit high (no simultaneous execution of the same task);
- each horizontal step is an integer longer than d_t.

A useful notion is that of a subprogram $G(0,u_0)$ of program G ; if G is the program defined by the set D_t , then $G(0,u_0)$ is the program defined by the set of functions $D'_t(u)$, $t \in T$ where

$$D'_t(u) = D_t(u) \quad \text{if} \quad u \leqslant u_0$$

$$D'_t(u) = D_t(u_0) \quad \text{if} \quad u > u_0 .$$

The concept of program or subprogram is still too large for realistic applications, so , we have classed programs in smaller and more adequate subsets.

Feasible program.

A program is feasible if there exists an initial marking M(0) such that :

$$\forall u \geqslant 0 \qquad 0 \leqslant M(u) \leqslant m \quad ;$$

the marking of each place is positive and bounded .

Finite program.

A program is finite if $\quad \forall t \in T \quad \exists K_t \in N \; / \; \forall u \geqslant 0 \quad D_t(u) < K_t$;

the D_t function will remain constant from time $u = u_t$.

Complete program.

A program is complete if $\quad \forall t \in T \quad D_t \neq \dot{0}$ (the null function) .

Periodic program.

A program is periodic if :

$$\forall t \in T \quad \exists r_t \in N \; / \; \forall n \geq 1 \quad x_t^{(n)} - x_t^{(n-1)} = r_t \; ;$$

r_t is the firing period of transition t.

IV. GENERAL RESULTS ON PROGRAMS.

Consider a TPN and the associated PN for an initial marking M(0). The sets of marking reachable from M(0) in the TPN and in the PN are linked by the two propositions :

P_1 : Each marking of the PN can be reached by a program in the TPN ;

P_2 : If, at time u, no transition is active, there exists a sequence s of the PN such that : $M(0) \xrightarrow{s} M(u^-)$.

This last proposition says that, in most of the cases, the sequence of the successive markings of a program includes a subsequence of markings of the associated PN .

The proof of this theorem is of recurrent kind, the recurrence index being that of the sequence of times at which no transition is active. That proof is not difficult but rather long, so it will not be given here .

The second general result is the following :

Theorem 2. A finite program is feasible .

Here also, the proof is of recurrent kind, based on the sequence v_n of times where at least one transition is initialised. From the proof, we can actually construct one initial marking of the finite program.

The third result deals with the initial minimum marking that can be allocated to a program in order to remain feasible. This critical marking is calculated by a simple formula :

Theorem 3. $$B_p(v_{n+1}^+) = \mathrm{Max}'\,(B_p(v_n^+) \, , - a_p(v_{n+1}^+))$$

where :

$B_p(v_0^+) = 0$;

$B_p(u)$ is the initial minimum marking of place p for the subprogram G(0,u) of program G ;

$$a_p(u) = \sum_{p^=} F_t(u) - \sum_{p^+} D_t(u) \quad .$$

The proof of this formula is not difficult . A corollary of theorem 3 is rather useful because it gives a way to know if a program is feasible : a program G is feasible if and only if all the functions $a_p(u)$ defined on R^+ are upper and lower bounded.

Figure 2) below gives an example of the behaviour of the sequence $B_p(v_n^+)$ for a very simple program ; we consider the TPN of fig 1) with the following data :

t	1	2	3	4	5	6
d_t	4	3	4	2	2	2

firing times for t_1 : 0,6,10 ; for t_2 : 2,5,8 ; for t_3 : 0,5,9
for t_4 : 0,2,4 ,6,8,10 ; for t_5 : 6,8,10
for t_6 : 0,2,4 .

We consider place p_1 and we draw the function $a_1(u)$; we have :

$$-a_1(u) = -F_5(u) + (D_1(u) + D_2(u)) .$$

The function $B_1(u)$ is drawn in dotted line .

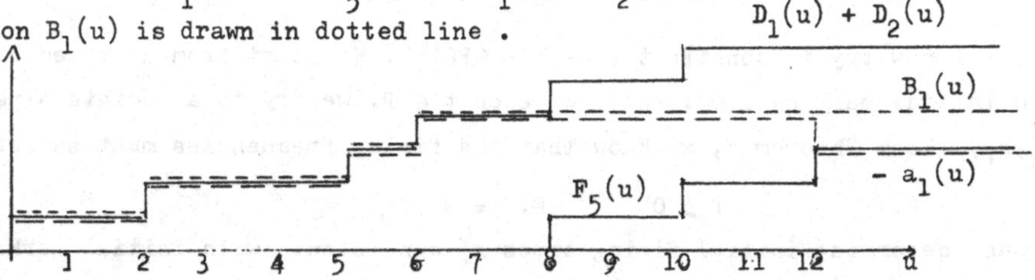

fig 3.

V. COMPLETE PERIODIC PROGRAMS.

We now focus our attention on infinite programs which are periodic and complete. First, we shall briefly recall RAMACHANDANI results (1973) on the particular case of PN called event graphs. Then, w'll show some new results on general PN.

A complete periodic program (CPP) is entirely defined by $(x_t^0, r_t, d_t)\ t \in T$. We assume that the durations d_t are known integers and we must construct a CPP (determine positive values for the x_t^0 and r_t) which is feasible .
The first important result is the following :

Theorem 4. The necessary and sufficient condition for a CPP to be feasible is : $\forall\, p \in P \quad \sum_{p^+} f_t = \sum_{p^-} f_t$ where $f_t = 1/r_t$ is the firing frequency of transition t .

Proof. If G is a CPP, we have defined the time functions

$$a_p(u) = \sum_{p^-} F_t(u) - \sum_{p^+} D_t(u) .$$

If time u_0 is such that $u_0 \geqslant \max_T(x_t^0 + d_t)$ (each transition has already be executed once) and if we note $R_p = \underset{\dot{p}}{\mathrm{lcm}} (r_t)$ where $\dot{p} = p^- \cup p^+$

(lcm = least common multiple) , then we study the variation of the marking of place p between time u and $u+R_p$; that is to say : $a_p(u+R_p) - a_p(u)$.

For each t of $\dot{p}$, we have : $R_p = k_t . r_t$ (by the definition of R_p) ; so we can write :

$$\forall\, t \in p^+ \quad D_t(u+R_p) - D_t(u) = k_t = R_p . f_t \ ;$$

$$\forall\, t \in p^- \quad F_t(u+R_p) - F_t(u) = k_t = R_p . f_t \ .$$

From that, we deduce :

$$a_p(u+R_p) - a_p(u) = R_p\left(\sum_{p^=} f_t - \sum_{p^+} f_t \right) .$$

If we want the CPP to be feasible, the marking of place p must remain bounded and positive for all $u \geqslant 0$; so, it is obviously necessary that :

$$\sum_{p^=} f_t - \sum_{p^+} f_t = 0 .$$

The sufficient aspect of the condition is a direct consequence of the corollary of Theorem 3 ; if $\forall p \in P \quad \sum_{p^-} f_t - \sum_{p^+} f_t = 0$, then the functions $a_p(u)$ are bounded ; so, the CPP is feasible .

We shall now try to construct feasible CPP's . We start from a timed Petri-net with initial marking M(0), and, for each $t \in T$, we try to associate a pair (x_t^0, r_t). From Theorem 4, we know that the firing frequencies must satisfy :

$$\text{(i)} \qquad f \geqslant 0 \qquad E.f = 0 .$$

We must determine initial firing times x_t^0 consistent with initial marking M(0).

We present RAMACHANDANI results on strongly connected event-graphs [1].

<u>Event graph</u>.

An event graph is a Petri-net for which :

$$\forall p \in P \quad \text{Card}(p^+) = \text{Card}(p^-)$$

(we note Card(E) the number of elements of the set E) .

In fact, an event graph is an oriented graph whose vertices are called transitions and edges are called places.

We also assume that the graph is strongly connected.

The incidence matrix E (see page 2) of an event graph is exactly the incidence matrix of the corresponding oriented graph (edges.nodes). So, the previous condition (i) is equivalent (because the graph is strongly connected) to :

$$\forall t \in T \quad f_t = f_0 \quad (f_0 > 0) \quad ; \text{ all the transitions}$$

have the same firing frequency.

If $M_p(0)$ is the initial marking of place p, $M_p(u)$ will remain positive if and only if :

$\forall n$, the firing number n of p^+ occurs after the end of the firing number $n-M_p(0)$ of p^- .

That last obvious condition can be written :

$$\forall n \geqslant M_p(0) \quad x^0_{p^+} + (n-1).r_0 \geqslant x^0_{p^-} + (n-1).r_0 + d_{p^-} ;$$

and, for each place p, we have :

$$\forall p \in P \qquad x^0_{p^+} - x^0_{p^-} \geqslant d_{p^-} - M_p(0).r_0 .$$

This linear system is a classical system of potential inequalities [2] we frequently meet in scheduling problems; it is known that it has a solution if and only if the length of each cycle path of the graph is <u>not positive</u> when edge (p^-,p^+) is valued by : $d_{p^-} - M_p(0).r_0$.

With that last condition, we can get the minimal period r_0^{min} which is given by the formula :

$$r_0^{min} = \underset{c}{\mathrm{MAX}} \left(\frac{\sum_{p\in c} d_{p^-}}{\sum_{p\in c} M_p(0)} \right)$$

where c is a cycle path of the graph . Searching r_0^{min} is greatly simplified when one remarks that it is necessary and sufficient to get the non positivity constraint on a <u>basis of cycle paths</u> of the graph .

Then, the values x_t^0 are the lengths of the longest paths from an origin to the node associated with transition t.

We now come to the presentation of results of the same kind on <u>general timed Petri-nets</u>. We start from any strongly connected Petri-net and an initial marking $M(0)$. We must determine firing frequencies $f_t, t\in T$ and initial firing times x_t^0 so that the corresponding CPP is feasible .

We know that , with a solution of

$$f > 0 \qquad E.f = 0 \ ,$$

w'll get a periodic marking, but we must choose x_t^0 values for the marking to remain positive, that is to say :

$$\text{(ii)} \qquad \forall\, u \geq 0 \quad \forall\, p\in P \quad M_p(u) = M_p(0) + \sum_{p^-} F_t(u) - \sum_{p^+} D_t(u) \quad 0 \ .$$

In fact w'llget a system of sufficient conditions on the unknowns to satisfy condition (ii). Consider transition t whose firing period is r_t and initial firing time x_t^0, the structure of the functions $D_t(u)$ and $F_t(u)$ is the following

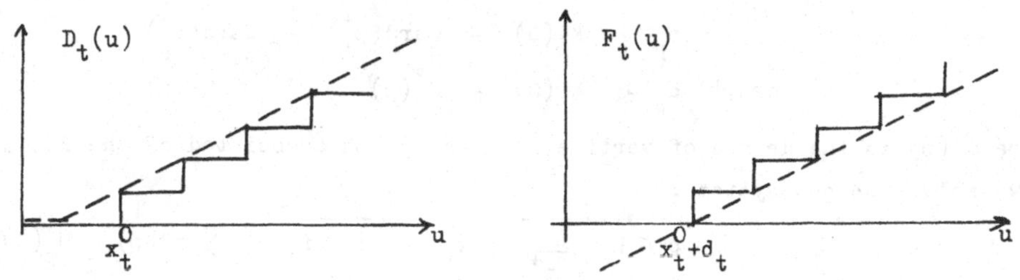

fig 4.

It is very easy to bound these functions as we se on figure 4, by the following inequalities :

$$\text{(iii)} \qquad \forall\, u \geq 0 \qquad F_t(u) \geq f_t.(u-(x_t^0+d_t)) \quad ;$$

$$\text{(iv)} \qquad \forall\, u \geq x_t^0 - r_t \qquad D_t(u) \leq f_t.(u-(x_t^0-r_t)) \quad .$$

If we add the new constraint (that is not a strong one as w'll see) :

$$\text{(v)} \qquad \forall\, t\in T \qquad x_t^0 \leq r_t$$

the two inequalities (iii) and (iv) are satisfied for all $u \geq 0$.

From the inequalities (iii),(iv) and (v) , we get :

$$\forall u \geq 0 \quad M_p(u) \geq M_p(0) + \sum_{p^-} f_t.(u-(x_t^0+d_t)) - \sum_{p^+} f_t.(u-(x_t^0-r_t))$$

and, with relation (i), that inequality simplifies a lot :

$$\forall u \geq 0 \quad M_p(u) \geq M_p(0) + \sum_{p^-} f_t(x_t^0+d_t) - \sum_{p^+} f_t(x_t^0-r_t) \quad .$$

The "good thing" is that the bounding term does not contain the time variable u; we remenber that already occurs in event graphs (see page 6), but there we had a necessary and sufficient condition and here, the condition is only sufficient. If we note b_p the known quantity : $M_p(0) + \sum_{p^-} f_t.d_t - \sum_{p^+} f_t.r_t$,

we remark that $f_t.r_t = 1$ for every t , so we have :

$$b_p = M_p(0) + \sum_{p^-} f_t.d_t - \mathrm{Card}(p^+) \quad .$$

From these results, we shall get the wanted condition :

$$\forall u \geq 0 \quad M_p(u) \geq 0 \quad ,$$

by solving the <u>linear system of inequalities</u> :

$$\text{(vi)} \qquad \forall p \in P \quad \sum_{p^+} f_t.x_t^0 - \sum_{p^-} f_t.x_t^0 \geq b_p \quad ;$$

$$\text{(vii)} \qquad \forall t \in T \quad x_t^0 \leq r_t \quad .$$

We had assumed that for each transition t : $r_t \geq d_t$, that restriction can obviously be written $f_t.d_t \leq 1$; first we remark that this condition is easy to cbtain because the linear system (i) is homogenous ; then we can get a little bit restrictive sufficient solution to our problem if we bound b_p by :

$$b_p \leq M_p(0) - \mathrm{Card}(p^+) - \mathrm{Card}(p^-)$$

$$\text{or} \quad b_p \leq M_p(0) - d^\circ(p)$$

where $d^\circ(p)$ is the degree of vertice p in the graph associated of the PN, and if we solve the new system :

$$\forall p \in P \quad \sum_{p^+} f_t.x_t^0 - \sum_{p^-} f_t.x_t^0 \geq d^\circ(p) - M_p(0) \quad ;$$

$$\forall t \in T \quad x_t^0 \leq r_t \quad .$$

Some remarks can be made about the linear system (vi).
If x_t^0 is a solution of (vi) , then $x'^0_t = x_t^0 + a$ (constant) is also a solution. System (vi) looks like a <u>generalised system of potential inequalities</u>, but it has not the main property of classical such systems :

if x_t^0 , $t \in T$ is a solution of (vi) and y_t^0, $t \in T$ an other solution of (vi), then z_t^0 defined by :

$$z_t^0 = \mathrm{Min}\ (\ x_t^0, y_t^0\ )$$

is not always a solution of (vi) .

The final remark we shall make about the linear system (vi) ,(vii) can be very useful if the initial marking can be choosen :

$$\text{if} \quad \forall\ p \in P \qquad M_p(0) \geqslant d^\circ(p)$$

then, for <u>any vector f</u> solution of (i),

$$x_t^0 \ = \ 0,\ t \in T$$

is a solution of (vi) and (vii) . For that particular solution, all transitions can be fired for the first time at time 0 .

---

[1] RAMACHANDANI .C. "Analysis of asynchronous concurrent systems by timed Petri-nets" Ph.D. Thesis,E.E. DEPT, MIT, 1973.

[2] ROY.B. "Algèbre moderne et théorie des graphes" Tome 2, DUNOD 1970 .

NON EQUILIBRIUM COMPUTER NETWORK DISTRIBUTION

by

P. Hammad
University of Aix Marseille
3, Av. R. Schuman
13100, Aix en Provence

and

J.M. Raviart
University of Valenciennes
59326, Valenciennes (France)

ABSTRACT:

This paper is concerned with the computation of non equilibrium open queueing network distributions by using diffusion technique, with an application to a model of a packet switching computer network.

The main result is the computation of the transient period time duration for one queue and also for an open queueing network. It is important when you want to make some measures or simulation. It is foreseen to generalize the results to a closed queueing network and to extend the computations to variable arrival rates and service rates.

I - INTRODUCTION.

The theory of networks of queues is developing rapidly under the impact of problems which have been raised by the mathematical modelling of multiprogrammed computer systems. An explicit information about the behaviour of a queue is often wanted ; for example when measuring, you need to determine the length of the observation period ; also when simulating, you need the sampling rate. So, because of the non stationarity of the workload of computer systems, the use rate and the throughput undergo some fluctuations, thus it is interesting to estimate the transient time.

An analytical solution of a queue can be obtained directly for some service time distributions and interarrival time distributions (see Jackson's and BCMP theorems |3|) or by approximation : iterative techniques |2|, diffusion method (|4|,|6|,|7|). The diffusion method, chosen here, is a continuous approximation of the queue length and its probability distribution is then described by a diffusion equation which has to be solved with appropriate boundary conditions.

But the published results do not take into account either the system transient period evolution and duration, or interarrival time distribution change with time.For a simple M/M/1 queue, the transient probability distribution is complex. The use of diffusion approximation allows to answer some of those questions because of the relatively short computation of the probability distribution.

This paper is about open queueing networks. Indeed in such a network it is possible to isolate any particular station inside the network : it is enough to be able to compute the mean and the variance of the inter arrivals. It is for that reason that at first,only one queue GI/G/1(*) is studied and then the general network.

In a first part, starting from the well known diffusion equation (Kolmogorov equation) and its analytical solution, the evolution of a single queue is studied at any instant t for some initial distributions (Normal, exponential and bimodal), by the classical methods of sampling and numeric computation. Then in a second part an open queueing network is considered with an application to a model of a packet switching computer network with 16 queues : it is the subnetwork CIGALE of the French computer network CYCLADES |7|.

(*) GI/G/1 : General Independent interarrival time distribution/General service time distribution/a single server.

II - TRANSIENT BEHAVIOUR OF A SINGLE QUEUE.

The usual assumptions for the diffusion approximation (*) of a queue GI/G/1 are taken here (unbounded capacity, FIFO...).

Let $N(t)$ be the number of customers in a station at time t : it is the difference between the arrival number and departure number from the original instant. $N(t)$ is approximated by a continuous path process $X(t)$ and the variations $dX(t)$ should be approximatively normally distributed with mean $\beta\, dt$ and variance $\alpha\, dt$ with :

$$(1)\quad \beta = \lambda - \mu\ ,\quad \alpha = \lambda.C + \mu.K$$

where λ is the arrival rate, μ the processing rate, C the squarred coefficient of variation of the interarrival time and K the squarred coefficient of variation for the service time.

The probability density function $f(x,t)dx = \Pr[x \leq X(t) < x + dx]$ satisfies, for $X(t) \in \mathbb{R}^+$, the diffusion equation

$$(2)\quad \frac{\partial f}{\partial t} = \frac{\alpha}{2}\frac{\partial^2 f}{\partial x^2} - \beta\frac{\partial f}{\partial x}\ .$$

The choice of the boundary conditions does not influence the inquired results and then we shall take reflecting boundaries (|6|,|7|,|9|) instead of absorbing boundaries (|1|,|3|,|4|) for simplifying reasons.

The general solution of (2) is

$$(3)\quad f(x,t) = \int_0^\infty f_o(x_o)\, g(x,x_o,t)\, dx_o$$

where f_o is the distribution $f(x,t)$ at the initial time and where the fundamental solution $g(x,x_o,t)$ of (2) is given by (|5|) :

$$(4)\quad g(x,x_o,t) = \frac{1}{\sqrt{2\pi\alpha t}}\left[e^{-\frac{(x-x_o)^2}{2\alpha t}} + e^{-\frac{(x+x_o)^2}{2\alpha t}}\right] e^{\frac{\beta}{\alpha}(x-x_o) - \frac{\beta^2 t}{2\alpha}}$$

$$- \frac{2\beta}{\alpha\sqrt{\pi}}\, e^{\frac{2\beta x}{\alpha}} \int_{\frac{x+x_o+\beta t}{\sqrt{2\pi\alpha\, t}}}^{\infty} e^{-y^2}\, dx$$

The steady state solution of (2) is given by

$$(5)\quad f_\infty(x) = \frac{2\,|\beta|}{\alpha}\, e^{-2\frac{\beta}{\alpha}x}$$

(*) For more precisions see |1|,|3|,|4|,|6|,|7|,|9|.

The distribution f(x,t) is studied for some initial distributions : normal, exponential ana bimodal. The figure 1 shows the evolution in two particular cases. It is easy to verify here and on other cases that the initial distribution does not influence the evolution of the transient period. Here its duration is between 10 and 60 units of time.

t		0	0.1	0.25	0.63	1	2.51	6.3	10	25.1	63	100
(a)	m	5	4.93	4.84	4.65	4.45	3.87	3.04	2.6	1.83	1.83	1.8
	σ^2	5	5.14	5.44	6.1	6.53	7.1	6.36	5.4	3.5	3.5	3.5
(b)	m	5	4.95	4.87	4.67	4.49	3.80	2.53	1.8	0.91	0.93	0.97
	σ^2	2	2.08	2.24	2.66	3.11	4.7	4.78	3.45	1.20	1.01	1.06
(c)	m	5	4.52	3.97	3.16	2.71	1.91	1.85	1.85	1.94	1.95	1.95
	σ^2	2	4.27	6.68	7.54	6.84	4.71	3.81	4	4.23	4.2	4.2

TABLE 1. VALUES OF MEAN AND VARIANCE.
a) f_o bimodal $\alpha = 1$, $\beta = -0.5$; b) f_o gaussian $\alpha = 1$, $\beta = -0.5$;
c) f_o gaussian $\alpha = 20$, $\beta = -5$

From the distribution f(x,t) the mean and the variance of the queue length are computed at each sampling time for many values of the couple of parameters α and β and for different initial means and variances : it has been noticed that the initial values did not influence the evolution time. Some of the computations are given on figures 2 and 3 and table 1. The mean evolution with time is monotonous ; it is the same for the variance if f_o is exponential ; on the other hand if f_o is gaussian the variance has a maximum. By comparing the numeric results and the figures, it can be verified that

$$T = \frac{k}{|\beta|} , \quad t_o = \frac{k'}{\alpha} , \quad t_M = \frac{k''}{|\beta|} ,$$

where T is the steady state time, t_o is the initial evolution time, t_M is the maximum variance time, k, k', k' are constants, independent of the initial distribution f_o and of α,β ; their values are about :

$$k \simeq 15 \ , \ k'' = 3 \ , \ k' = 0.1.$$

The main result here lays on T which is the time of the steady state.It can alsc be considered as the duration of the evolution period (k"<< k). Later on, we use μ and $u^{(*)}$ instead of β and it results

$$T = \frac{k}{\mu(1-u)} \quad \text{with } k \simeq 15 \ .$$

(*) u is the utilization factor (λ/μ)

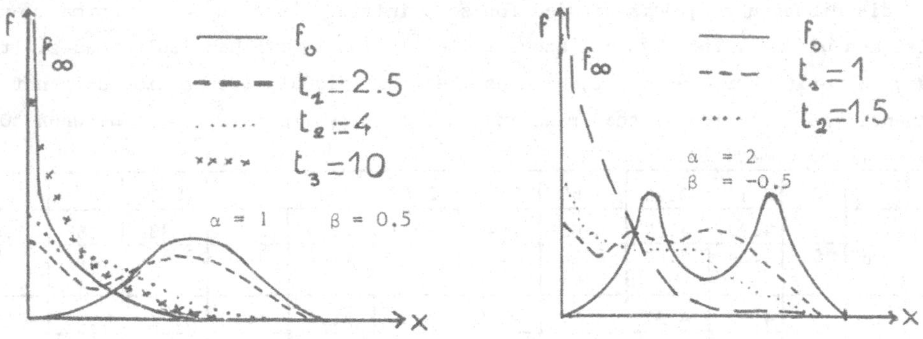

FIGURE 1. EVOLUTION OF THE DISTRIBUTION a) f_o gaussian b) f_o bimodal

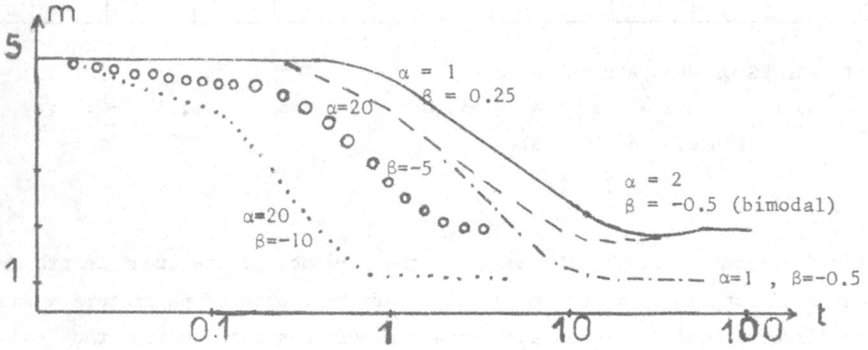

FIGURE 2. EVOLUTION OF THE LENGTH MEAN

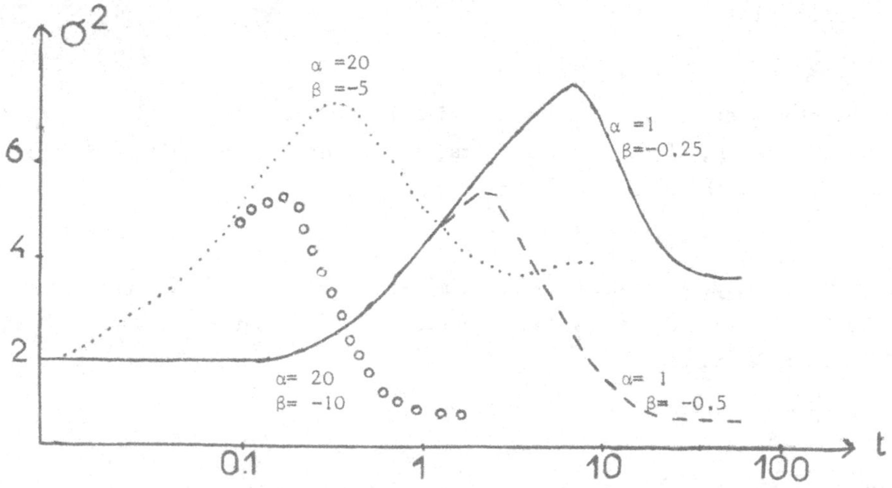

FIBURE 3. EVOLUTION OF THE LENGTH VARIANCE

As an example for an $E_2/M/1$. Erlang 2/Exponential/1 server with $\mu = 100$ per hour and $u = 0.9$, the equilibrium is obtained after one hour and a half.

The transient period time only depends upon the service rate and the utilization factor u ; for a fixed μ the more the queue is used ($u < 1$), the longer T is.

III - APPLICATION TO AN OPEN QUEUEING NETWORK : CIGALE

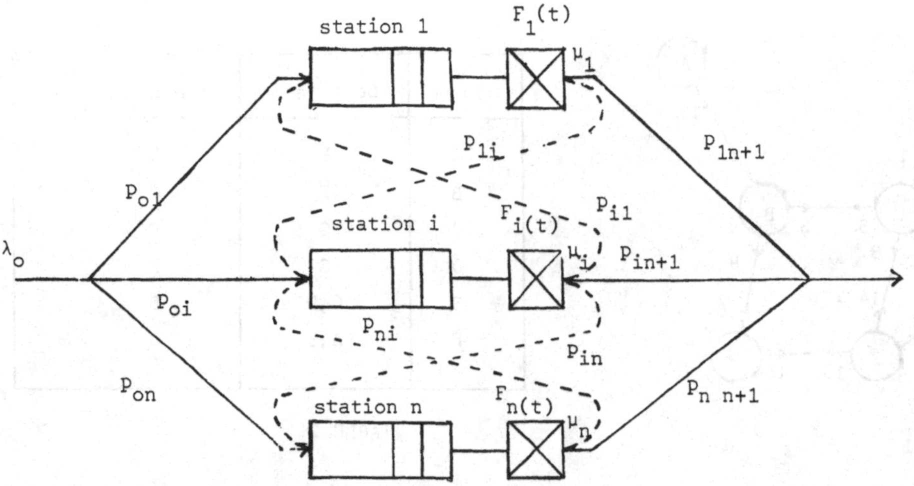

FIGURE 4.

We consider the general queueing network of figure 4 |4|, composed of FIFO stations, each composed of one server with general service time distribution. The external arrivals constitute a renewal process of rate λ_o ; the variance of the interarrival time is C_o. The transition of customers from station i to j is defined by the probability matrix $P = |p_{ij}|$. The service times at station i have a distribution function $F_i(t)$ with rate μ_i and variance of interservice time K_i. The arrival rate λ_i in the queue i is then |1|

$$(6)\ \lambda_i = \lambda_o e_i ,$$

where the expected number of visits e_i , that a customer of the network will do to the station i, is the solution of the system of equations

$$(7)\ e_i = p_{oi} + \sum_1^n e_j p_{ji} , \quad i = 1, 2, \ldots, n .$$

Then we need to determine the squarred coefficient C_i of variation of the interarrival time to queue i : they are expressed relatively to the squarred coefficients L_i of variation of the interoutput time :

$$(8)\ C_i = (1/\lambda_i) \sum_{j=0}^{n} [(L_j - 1) p_{ji} + 1] \ \lambda_j p_{ji}$$

where L_i is the solution of the system of equations.

$$(9) \qquad L_j = u_i^2(K_i+1) + (1-u_i)\left[2u_i + 1 + \lambda_i^{-1} \sum_{j=0}^{n} \left[(C_j-1)p_{ji}+1\right] \lambda_j p_{ij}\right] - 1$$

with $u_i = \lambda_i/\mu_i$.

The preceding results are used for the analysis of the packet switching sub-network CIGALE of the computer network CYCLADES |8|. The CIGALE topology is shown on figure 5 (|1|,|4|).

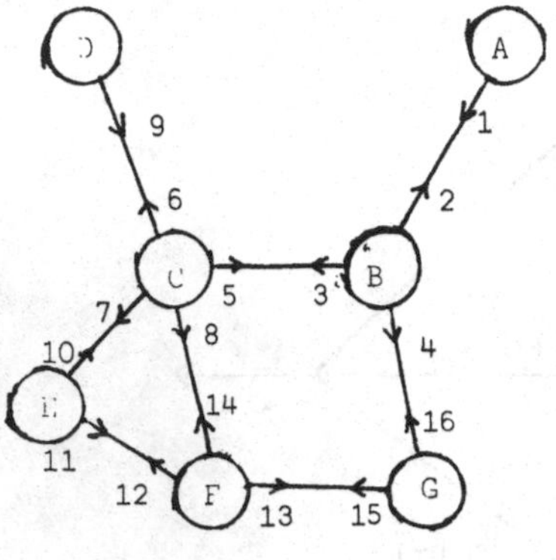

FIGURE 5.

Source	Destination	Via Node
A	F	C
B	F	C
C	G	F
D	G	F
G	C,D	B
F	A,B	C

TABLE 2

Nodes A to G are minicomputers and the numbers on the arcs connecting the nodes refer to output queues (16 queues). The packets come from the outside of the network to any node i and are assumed to go to any other node j with equal probability. Packet routing is fixed by the shortest line transit time and by the table 2 if there is any "ambiguity" . The service time is 0.4 seconds for a packet on all lines, except for the first, second, third and fifth lines which are faster (0.3 seconds for one packet).

Packet routing is made independent from the customers by using a probabilistic routing. The probabilities p_{ij} must then be computed. Every line of the network is transmitting packets with known sources and destinations and p_{ij} is the ratio of the stream going through the queue i towards the queue j to the total stream of the queue i. For example $p_{11} = p_{12} = p_{15} = \ldots = p_{1n} = 0$, $p_{13} = 2/3$, $p_{14} = 1/6$, $p_{1\,17} = 1/6$ and so on...

The arrivals from outside have been considered as exponential ($C_o = 1$) and the services as constant ($K_i = 0$). The table 3 gives the computed results of the mean length L and of the time T for each queue with an arrival rate λ_o increasing from 10 to 15 (where there is a saturation for the queue 3). Here T and L go up with λ_o but the times T are not very high because the network is not very much saturated.

In the table 4 the service times have been modified for the queues 3, 5, 8, 9 from 0.3 s to 0.2 s by packet and then the network can absorb a larger number of packets from outside. It is seen that the network transient period time ($\max_i T_i$) can become very high (for example 1260 s or 21 mn for the queue 6) when some queues have an utilization factor u_i very close to 1. So, the mean queue length increases greatly.

		λ_o = 10		λ_o = 11		λ_o = 13		λ_o = 15	
i	μ_i	T	L	T	L	T	L	T	L
1	3.33	7.87	0.354	8.5	0.446	10.2	0.629	12.6	0.900
2	3.33	7.87	0.340	8.5	0.394	10.2	0.531	12.6	0.714
3	3.33	15.75	0.966	21	1.335	63.1	4.34	-	-
4	2.5	7.41	0.115	7.59	0.13	7.97	0.161	8.4	0.194
5	3.33	10.50	0.564	12.1	0.688	17.5	1.115	31.5	2.174
6	2.5	12.6	0.505	14.1	0.610	18.8	0.924	28	1.515
7	2.5	10.5	0.353	11.3	0.412	13.5	0.564	16.8	0.775
8	2.5	15.75	0.731	18.8	0.932	30.7	1.713	84	5.05
9	2.5	14	0.617	16.1	0.846	23.3	1.444	42	2.99
10	2.5	9.7	0.257	10.3	0.360	11.9	0.490	14	0.666
11	2.5	7.4	0.116	7.6	0.132	7.97	0.164	8.4	0.199
12	2.5	7.4	0.112	7.6	0.126	7.97	0.155	8.4	0.188
13	2.5	12	0.435	13.3	0.516	17.14	0.734	24	1.085
14	2.5	9.7	0.297	10.3	0.360	11.9	0.490	14	0.666
15	2.5	7.4	0.116	7.6	0.132	7.97	0.164	8.4	0.199
16	2.5	9.7	0.297	10.3	0.360	11.9	0.490	14	0.666

TABLE 3

		λ_o =15		λ_o =17		λ_o =19		λ_o =21	
i	μ_i	T	L	T	L	T	L	T	L
1	3.33	12.6	0.900	16.6	1.34	24.2	2.19	45	4.50
2	3.33	12.6	0.762	16.6	1.09	24.2	1.68	45	3.24
3	5.0	10.5	0.914	15.7	1.44	31.5	2.91	-	-
4	2.5	8.4	0.194	8.87	0.232	9.4	0.273	10	0.318
5	5.0	7	0.548	8.5	0.726	10.9	0.989	15	1.43
6	2.5	28	1.57	54.8	3.36	12.60	82.6	-	-
7	2.5	16.8	0.807	22.1	1.17	32.3	1.85	60	3.65
8	5.0	5.6	0.394	6.3	0.495	7.3	0.62	8.57	0.785
9	5.0	5.2	0.374	5.8	0.47	6.56	0.59	7.5	0.75
10	2.5	14	0.666	17	0.918	21.7	1.31	30	1.99
11	2.5	8.4	0.199	8.9	0.239	9.4	0.28	10	0.33
12	2.5	8.4	0.188	8.9	0.223	9.4	0.26	10	0.30
13	2.5	24	1.23	40	2.23	120	7.1	-	-
14	2.5	14	0.666	17	0.918	21.7	1.31	30	1.99
15	2.5	8.4	0.199	8.9	0.239	9.4	0.283	10	0.33
16	2.5	14	0.666	17	0.918	21.7	1.31	30	1.99

TABLE 4

REFERENCES.

|1| Badel M., Zonzon M. : Validation d'un modèle à processus de diffusion pour un réseau de file d'attente général : Application à Cyclades. Rapport IRIA n°209 (Décembre 1976).

|2| Chandy M., Herzog W., Woo L. : Parametric analysis of queueing networks. IBM J. Res. Devel. vol. 19, p. 36-42 (1975).

|3| Gelenbe E., Muntz R.R. : Probabilistic models of computer systems Part I (exact results). Acta Informatica vol. 7, p. 35-60 (1976).

|4| Gelenbe E., Pujolle G. : The behaviour of a single queue in a general queueing network. Acta Informatica vol. 7, p. 123-160 (1976)

|5| Hammad P. : Le rôle de l'énergie libre pour des systèmes régis par l'équation de Fökker-Planck. Annales de l'Institut Henri Poincarré, Vol.25,2(1976)

|6| Kobayashi H.: Application of the diffusion approximation queueing networks Part I and II. JACM vol. 2 p. 316-328, p. 459-469 (1974).

|7| Newell G.F. : Queues with time independent arrival rates : I the transient through saturation J. Appl. Prov. vol. 5 p.436-451 (1968).

|8| Pouzin L. : CIGALE the Packet Switching Machine of the Cyclades Computer Network. Proc. IFIP Congress 74, Stockolm, North Holland p. 155-159 (August 1974).

|9| Reiser M., Kobayashi A. : Accuracy of the diffusion approximation for some queueing system. IBM J. Res. Devel. Vol. 18 p. 110-124 (1974).

DYNAMIC PROGRAMMING OF STOCHASTIC ACTIVITY NETWORKS WITH CYCLES

E. Höpfinger
University of Ulm
Mathematik VII
FRG

1. INTRODUCTION. Network techniques are used in the planning and control of projects which consist of a large number of interrelated tasks, or activities. An activity network usually represents two particular aspects of a project

(i) the precedence relationship among the activities, and

(ii) the duration of an activity.

The basic precedence relationship is represented by a directed graph the directed edges of which represent activities. The cycles of the graph indicate that activities can be carried out several times. Contrary to the traditional networks the precedence relationship and the durations of the activities occur at random. The stochastic structure is obtained by the assumption that for distinct vertices of the directed graph exactly one of the emanating activities will be carried out. Such models are called generalized activity networks (e.g. Elmaghraby 1977) or GERT networks (e.g. Neumann 1975), or stochastic activity networks. Particularly simple networks are present if exactly one of the activities emanating from a vertex is carried out for all vertices except for the sinks. Under special assumptions for the transition probabilities these networks are equivalent to semi-Markov decision processes for which a theory and procedures are well established (Nicolai 1977). The associated stochastic process is given by $(X_n, T_n)_{n \in \mathbb{N}}$ where the random variable X_n denotes the n-th realized vertex and T_n the time of the realization of X_n. The decisions influence the transition probabilities and entail costs dependent on the occurrence times of the vertices. Optimal decisions are those minimizing the costs. However the problem differs from the most general semi-Markov decision processes in that it is substantially non-stationary and usually allows no optimal stationary policy. Nevertheless standard techniques of dynamic programming can be applied (Nicolai 1977).

In the case of more general stochastic networks these results cannot be used directly. Already for acyclic networks one has to consider a stochastic process $(X_n, T_n)_{n=1,2...}$ where X_n denotes a subset of vertices of V which occur simultaneously at time T_n. X_n can contain more than one vertex. More restrictive is the fact that $(X_n, T_n)_{n=1,2,...}$ generally is not a Markov process, i.e., more information about the history $(X_1, T_1, a_1'; ...; X_n, T_n)$ is needed for the transition probabilities than the last state and the action a_n' taken. This problem was treated by Höpfinger (1978) for general acyclic networks and subsets of the histories containing the relevant information were identified. In this paper we establish an adequate modification of these subsets called "monitoring states" such that the present stochastic activity networks can be conceived as dynamic programming problems. Two optimization procedures whose key ingredients are "N-state contraction" and "monotonicity" properties are outlined based on Denardo (1967).

2. THE PRECEDENCE RELATIONSHIP. It is assumed that the precedence relationship among the activities can be represented by a finite directed graph (V,E) where the activities are represented by the directed edges. Each directed edge is unanimously determined by a pair $(i,j)\in V\times V$ where i denotes the initial vertex and j the final vertex; hence the set of directed edges E is identified with a subset of the Cartesian product $V\times V$ of the vertices.

Let $P(j) = \{i \mid (i,j) \in E\}$ designate the set of predecessors of a vertex j and $S(i) = \{j \mid (i,j) \in E\}$ the set of successors of i. If $P(j) = \emptyset$, then j is called a source, and if $S(i) = \emptyset$, then i is said to be a sink. The sets of sources and of sinks are assumed to be non-void, respectively.

The directed graph (V,E) contains finitely many subgraphs (V_ν, E_ν) $(\nu = 1,\dots,n_N)$ which are disjoint: $V_\mu \cap V_\nu = \emptyset$, $E_\mu \cap E_\nu = \emptyset$ $(\mu \neq \nu)$ and are acyclic. Each subgraph (V_ν, E_ν) is assumed to have exactly one source r_ν and one sink s_ν, and $|S(s_\nu)| = 1$. The vertices of (V,E) are subdivided into six classes according to their entrance or their exit. Each vertex can be considered a "receiver" and a "source" or "emitter". Three kinds of receiving vertices are introduced, W, U, O,

1) A vertex $i \in W$ will be realized at the completion time of any activity leading into it. All vertices of $V\setminus(V_1 \cup \dots \cup V_{n_N})$ and the sources r_ν $(\nu = 1,\dots, n_N)$ form W.

2) A vertex $i \in U \cap V_\nu$ will be realized if r_ν has been realized and all edges leading into i are realized by the same realization of r_ν.

3) A vertex $i \in O \cap V_\nu$ will be realized if r_ν has been realized and one edge leading into i is realized via the realization of r_ν.

The exact definition of the properties of a vertex $i \in U \cup O = (V_1 \cup \dots \cup V_{n_N})\setminus \{r_1,\dots, r_{n_N}\}$ will be given below. It should be noted that a simple definition is only possible for acyclic networks. Two types of "emitter" vertices are defined; the deterministic and the stochastic.

1) $i \in S$ is called stochastic if exactly one activity emanating from i is executed if i is realized. i is supposed to be the only predecessor of all its successors.

2) $i \in D$ is called deterministic if all emanating activities must be undertaken.

It is assumed that $S \supseteq V\setminus(V_1 \cup \ldots \cup V_{n_N})$ or $D \subseteq V_1 \cup \ldots \cup V_{n_N}$. If the source r_ν of an acyclic subgraph is realized it is required that for each possible choice of an emanating edge from a stochastic vertex a directed walk of realized edges and vertices exists which terminates with the sink s_ν. In the following, let (V,E) have one source and at least one sink and let each vertex of $V\setminus(V_1 \cup \ldots \cup V_{n_N})$ be reachable from a source and a sink of (V,A), i.e., there exist connecting walks.

3. THE ASSOCIATED STOCHASTIC PROCESS. With each pair $(i,\tau) \in V \times \mathbb{R}_+$ is associated a set $\Delta_{i\tau} \neq \emptyset$ of admissible actions. Let $i \in V$ be realized at time τ and let $a \in \Delta_{i\tau}$ be the chosen action. For each emanating activity (i,j) a distribution function $F_{ij}^a(\tau,.)$ for the duration of (i,j) is given and a probability $p_{ij}^a(\tau)$ of execution of (i,j) if $i \in S$. The activities last almost always longer than zero $F_{ij}^a(\tau,o) = o$.

Let $(\Omega, \mathfrak{A})$ be a measurable space and let for each directed edge (i,j) and each stochastic vertex $k \in S$ exist "independent" sequences of stochastic processes $(\nu \in \mathbb{N})$: $D_{i,j}^\nu := \{D_{i,j}^\nu(\tau): \Omega \to \mathbb{R}_+, \mathfrak{A} - \mathcal{B}_+$ measurable, $\tau \in \mathbb{R}_+\}$, $B_k^\nu := \{B_k^\nu(\tau): \Omega \to S(k)$, $\mathfrak{A} - \mathfrak{P}(S(k))$ measurable, $\tau \in \mathbb{R}_+\}$. $D_{i,j}^\nu(\tau)$ is regarded as the duration of the activity (i,j) if vertex i is realized as the ν-th vertex at time τ, and $B_k^\nu(\tau)$ as the chosen activity emanating from k if k is realized as the ν-th vertex at time τ. $\mathcal{B}_+$ denotes the set of Borel measurable subsets of the non-negative reals. $\mathfrak{P}(S(k))$ is the power set of $S(k)$. It is assumed that $(\tau,\omega) \to D_{ij}^\nu(\tau,\omega)(\nu \in \mathbb{N})$ are $\mathcal{B}_+ \times \mathfrak{A} - \mathcal{B}_+$ measurable and $(\tau,\omega) \to B_k^\nu(\tau,\omega)$ are $\mathcal{B}_+ \times \mathfrak{A} - \mathfrak{P}(S(k))$ measurable.

Each map $\gamma \in \times_{i\in V,\ \tau \in \mathbb{R}_+} \Delta_{i\tau}$ is assumed to define a probability measure $P^\gamma: \mathfrak{A} \to \mathbb{R}$ such that $P^\gamma(D_{ij}^\nu(\tau) \le t) = F_{ij}^{\gamma(i,\tau)}(\tau,t)$ and $P^\gamma(B_k^\nu(\tau) = e) = p_{ke}^{\gamma(i,\tau)}(\tau)$. For each map γ the required independence of the stochastic processes B_k^μ, D_i^ν, $(\mu,\nu \in \mathbb{N})$ is equivalent to: Let $\{(k_1,\mu_1), \ldots, (k_m,\mu_m), (i_{m+1}, j_{m+1}, \nu_{m+1}), \ldots, (i_n, j_n, \nu_n)\}$ be a finite set of indices and $M_1, \ldots, M_n$ finite sets of times.

Then $P^\gamma(B_{k_i}^{\mu_i}(\tau) = e_\tau\ (\tau \in M_i),\ D_{i_e j_e}^{\nu_e}(\tau) \le d_e\ (\tau \in M_e),\ i = 1, \ldots, m;$

$$e = m+1, \ldots, n) = \prod_{i=1}^{m} P^\gamma(B_{k_i}^{\mu_i}(\tau) = e_\tau\ (\tau \in M_i)) \prod_{e=m+1}^{n} P^\gamma(D_{i_e j_e}^{\nu_e}(\tau) \le d_e\ (\tau \in M_e))$$

where $e_\tau \in S(k_i)$ and $d_e \in \mathbb{R}$ are arbitrarily chosen.

Remark: One can generate such stochastic processes if the duration is constructed from a function $f: \mathbb{R}_+ \times \Omega \times \Delta \to \mathbb{R}_+$, Δ a set of actions, where f is measurable on the product space and where a probability measure P on $\mathfrak{A}$ is given. Analogously B_i^ν is derived from a function g on $\mathbb{R}_+ \times \Omega \times \Delta$. Then the transformation by these functions into an appropriate infinite Cartesian product of the reals generates an example for the model described above.

The realization of the project or the stochastic network is described by a finite or infinite sequence $(X_n, T_n)_{n = 1,2,\ldots}$ of states where X_n denotes the set of vertices realized at the point of time T_n. For formal reasons a new vertex $|V| + 1$ and directed edges $(i, |V|+1)$ are introduced for all sinks i of (V,N) and $i = |V|+1$. Then $D^{\nu}_{i,|V|+1} = 1$ for the additional edges. The augmented directed graph is denoted by (V^+, E^+). Then the realization is given by an infinite sequence $(X_n, T_n)_{n \in \mathbb{N}}$ where $X_n \subseteq V^+$ and $T_1 < T_2 < \ldots$. Given the processes B^{μ}_k and D^{ν}_{ij} the process (X_n, T_n) is defined as follows:

Let $1 \in V^+$ denote the only source of (V^+, E^+). Then $X_1 = \{1\}$ and $T_1 = o$.

If $1 \in S$ so $X_2 = \{B^1_1 (o)\}$ and $T_2 = D^1_{1X_2}(o)$ by abuse of notation. Now assume that $X_2, X_3, \ldots, X_n$ only contains vertices of $V^+ \setminus (V_1 \cup \ldots \cup V_n)$.

Then $X_{n+1} = \{B^n_{X_n} (T_n)\}$ and $T_{n+1} = T_n + D^n_{X_n, X_{n+1}} (T_n)$ where by abuse of notation X_n, X_{n+1} represent their single element, too.

Let $X_n = \{r_\nu\}$ for an acyclic subnetwork (V_ν, X_ν). (X_n, T_n) causes the realization of a subset W_n of vertices of V_n which is determined implicitly: $r_\nu \in W_n$; the vertex $j \in (V_\nu \setminus \{r_\nu\}) \cap U$ belongs to W_n if $P(j) \subseteq W_n$ and $B^n_{ij} (T^i) = j$ holds for each $i \in P(j) \cap S$. T^i denotes the realization time of i which will be determined below. The vertex $j \in (V_\nu \setminus \{r_\nu\}) \cap O$ belongs to W_n if $P(j) \cap D \cap W_n \neq \emptyset$ or there is an $i \in P(j) \cap S \cap W_n$ such that $B^n_{ij} (T^i) = j$. The realization times T^j of $j \in W_n$ are determined by $T^{r_\nu} = T_n$.

If $j \in (W_n \cap U) \setminus \{r_\nu\}$ then $T^j = \max (T^i + D^n_{ij} (T^i) \mid i \in P(j))$, if $j \in (W_n \cap O) \setminus \{r_\nu\}$ then $T^j = \min (T^i + D^n_{ij} (T^i) \mid i \in P(j) \cap D \cap W_n$ or $(i \in P(j) \cap S \cap W_n$ and $B^n_i (T^i) = j))$.

Then a finite sequence $(X_e, T_e)'_{e = n+1, \ldots, m}$ exists such that $X_{n+1} \cup \ldots \cup X_m = W_n$, $T_{n+1} < T_{n+2} < \ldots$, and, T_e denotes the realization time of each vertex of X_e.

As soon as $s_\nu \in X_\mu$ we drop the remaining sequence $(X_e, T_e)'$ with $e = \mu+1, \ldots, m$ since this one is not relevant for the general process. Hence $(X_e, T_e) = (X_e, T_e)'$ $(e = n, n+1, \ldots, \mu)$ and $X_{\mu+1} = S(s_\nu)$, $T_{\mu+1} = T_\mu + D^{\mu}_{s_\nu, f} (T_\mu)$ with f successor of s_ν. The process (X_n, T_n) continues in the same way as described above.

4. THE STOCHASTIC DECISION MODEL. As previously stated as soon as a vertex i is realized at time τ or equivalently $i \in X_n$ and $\tau = T_n$ for a state an action $a \in \Delta_{i\tau}$ can be chosen. Then the cost $c(i, \tau, a)$ has to be expected. The value of the function $c: \times_{i \in V^+, \tau \in \mathbb{R}_+} \Delta_{i\tau} \to \mathbb{R}$ is composed of the expected cost of the duration of the executed emanating activities, the immediate cost of the action a chosen, and the cost caused by the realization of i at time t (e.g. penalties for infringing deadlines). However $c(|V|+1, \tau) = o$ $(\tau \in \mathbb{R}_+)$. Given state (X_n, T_n) a family $a'_n = (a_i)_{i \in X_n}$ has to be chosen and the cost $\sum_{i \in X_n} c(i, \tau, a_i)$ is to be expected. c is assumed to be bounded.

As soon as several activities may be executed simultaneously the process $(X_n, T_n)_{n \in \mathbb{N}}$

is no longer Markov as is obvious.
Fortunately not the whole history $(X_1, T_1, a_1'; \ldots; X_{n-1}, T_{n-1}, a_{n-1}'; X_n, T_n)$ and the actions a_n' are relevant for the transition to the next state (X_{n+1}, T_{n+1}) but only a subset which we will call "monitoring state". If $X_n \subseteq (V^+ \setminus (V_1 \cup \ldots \cup V_{n_N})) \cup \{r_1, \ldots r_{n_N}\}$ the monitoring state of the process coincides with (X_n, T_n). Now assume $X_n \subseteq (V_\nu \setminus \{r_\nu\})$ for a $\nu \in \{1, \ldots, n_N\}$. Let n' be the largest index such that $r_\nu \in X_{n'}$ and $n' < n$. Then the transition to (X_{n+1}, T_{n+1}) depends on the vertices i of $X_{n'} \cup \ldots \cup X_n$ such that $S(i) \nsubseteq X_{n'} \cup \ldots \cup X_n$ for $i \in D$ and $S(i) \cap (X_{n'} \cup \ldots \cup X_n) = \emptyset$ for $i \in S$. Let $W_{n,1}'$ denote this set of vertices. Then the set of possibly realized vertices at the next step is a subset of the set of successors of $W_{n,1}'$ diminished by the already realized successors of the deterministic vertices in this set: $W_{n,2}' = (X_{n'} \cup \ldots X_n) \cap S\,(W_{n,1}' \cap D)$. Furthermore, one must note the already realized vertices of O which may possibly be reached by walks starting from $W_{n,1}'$. Let their set be denoted by $W_{n,3}' = (X_{n'} \cup \ldots \cup X_n) \cap O \cap \{j \in V_\nu \mid \exists \text{ path from } W_{n,1}' \text{ to } j\}$. Then the associated monitoring state is given by $(X_n, T_n; W_{n,1}, T_{n,1}, a_{n,1}; W_{n,2}; W_{n,3})$ where $W_{n,1} \subseteq W_{n,1}' \setminus X_n$ is any set including all vertices one successor of which can be realized in case of favorable choices of emanating activities at stochastic vertices. Usually one will not use the minimal set $W_{n,1}$ because of the expensive search techniques. $T_{n,1}$ is the family of times of realizations of the vertices of $W_{n,1}$, $a_{n,1}$ the family of associated actions. $W_{n,2} \supseteq (X_{n'} \cup \ldots \cup X_n) \cap S(W_{n,1} \cap D)$ may contain vertices of $S\,(W_{n,1} \cap D)$ the realization of which is impossible. $W_{n,3}$ is any set of realized vertices of O including all vertices which can be reached by a "realized path".

The set of vertices which may be realized at the next step is given by
$X_{n+1}^o := ((O \cap P(X_n \cup W_{n,1})) \setminus W_{2,n}) \cup \{j \in U \setminus W_{n,2} \mid P(j) \subseteq X_n + W_{n,1}\}$. An obvious but lengthy formula determines the probability that exactly a given subset of X_{n+1}^o will be realized and at which time each of these vertices will occur. Especially X_{n+1} is a proper or improper subset of X_{n+1}^o. Since X_{n+1}^o is finite only finitely many X_{n+1} are possible. By construction it is clear that there exist only finitely many sequences $X_{n'}, \ldots, X_n$ such that $r_\nu \in X_{n'}$, $s_\nu \in X_n$ which are part of a stochastic process $(X_e, T_e)_{e=n', \ldots, n}$. A set $X_n \subseteq V_\nu$ can only be realized once again if the process enters (V_ν, E_ν) anew.

Let Ξ denote the set of all monitoring states. If $x = (X_n, T_n; \ldots)$ for some n then the set of actions is $\Delta_x = \times_{i \in X_n} \Delta_{i\,T_n}$. The policy space Δ is defined as the Cartesian product of the action spaces, i.e., $\Delta = \times_{x \in \Xi} \Delta_x$. An element $\delta \in \Delta$ is called policy. Exept for monitoring states $(X_n, T_n, \ldots)$ with $s_\nu \in X_n$ the cost incurred at $x = (X_n, T_n, \ldots)$ for action $\delta_x = (a_i)_{i \in X_n}$ equals $c\,(x, \delta_x) = \sum_{i \in X_n} c(i, T_n, a_i)$.

If $s_\nu \varepsilon X_n$ then a stochastic process $(X_n, T_n)''$ may continue inside (V_ν, E_ν) for some steps. It is assumed that for this process cost-minimizing actions are chosen such that $c(x, \delta_x)$ for $x = (X_n, T_n, .)$, $s_\nu \varepsilon X_n$ denotes minimal expected summed up costs relative to $(X_n, T_n)''$.

Let $\mathbb{B}$ be the collection of all bounded functions from Ξ to the reals $\mathbb{R}$, i.e., $b \varepsilon \mathbb{B}$ if and only if $b: \Xi \to \mathbb{R}$, $b(|V|+1, \tau) = o$ $(\tau \varepsilon \mathbb{R}_+)$ and $\sup \{|b(x)|, x \varepsilon \Xi) < \infty$. For each $u \varepsilon \mathbb{B}$ and each policy δ a return function is defined by

$$[H_\delta (u)] (x): = c(x, \delta_x) + \inf_{\substack{b \text{ integrable} \\ b \geq u}} \int_\Xi \mu (dy| x, \delta_x)\, b(y)$$

where $\mu(.| x, \delta_x)$ is the probability measure over Ξ determined by x and action δ_x. $b \geq u$ if this relation holds for each x. H_δ maps $\mathbb{B}$ into $\mathbb{B}$ and satisfies monotonicity: $H_\delta (u) \geq H_\delta (v)$ if $u \geq v$. In the case that H_δ has a unique fixed-point v_δ, i.e., $H_\delta (v_\delta) = v_\delta$, v_δ is called the return function of the policy δ. Given that H_δ has a unique fixed-point for each δ then a major task is the determination of the optimum return function c^* defined by $c^*(x) = \inf \{v_\delta (x) \mid \delta \varepsilon \Delta\}$ and an optimal policy δ^* such that $c^* = v_{\delta^*}$. Based on the paper of Denardo (1967) existence of c^* will be proved.

5. EXISTENCE OF THE OPTIMUM RETURN FUNCTION. Let R denote the maximal number of monitoring states $(X_n, T_n; \ldots)_{n=1, \ldots,}$ such that they can follow each other as realized states of process and that no set X_n will appear a second time.

Assumption: There exists a real number $\eta > o$ such that for each policy $\delta \varepsilon \Delta$ and for each vertex $i \varepsilon W \cap O = V \setminus (V_1 \cup \ldots . \cup V_{n_N})$ a path $[i_o, i_1 \ldots, i_\mu]$, $i_o = i$, $i_\mu = |V|+1$, exists such that $\eta < p_{i_\nu, i_{\nu+1}}^{\delta(i_\nu \tau)} (\tau)$ for $\nu = o, \ldots, \mu$ and $\tau \varepsilon \mathbb{R}_+$.

By a lengthy but straightforward proof one obtains

Theorem: H_δ is a R-stage contraction on $\mathbb{B}$ such that

$$\rho(H_\delta (u), H_\delta (v)) \leq \rho (u, v), \quad \rho (H_\delta^R (u), H_\delta^R (v)) \leq (1 - \eta^R) \rho (w, v) \quad (u,v \varepsilon \mathbb{B}).$$

The metric ρ is defined by $\rho (u,v) = \sup (|u (x) - v (x)| \mid x \varepsilon \Xi)$. Let $L : \mathbb{B} \to \mathbb{B}$ be defined by $(Lv) (x): = \inf (h(x,a,v) \mid a \varepsilon \Delta_x)$ where $h (x,a,v): = [H_\delta (v)](x)$ for $\delta_x = a$.

According to Theorem 4 (Denardo 1967) H_δ has a unique fixed-point v_δ, the optimum return function c^* is the unique fixed-point of L, and $\rho (L^R v, c^*) \leq (1 - \eta^R) \rho (v, c^*)$ if $v \geq c^*$. The existence of optimal policies depends on continuity and measurability requirements (Hinderer, chapter 17, 1970).

This result can be exploited for the determination of the optimum return function, because $\rho (L^m v, c^*) \to o$ for $m \to \infty$ and each $v \varepsilon \mathbb{B}$, $v \geq c^*$. Choose any $v \varepsilon \mathbb{B}$, such that $v \geq c^*$. After that determine $Lv =: v'$ and $K = \sup \left(\left| \frac{v'(x) - v(v)}{v'(x)} \right|, x \in \Xi \right)$. If $K < \varepsilon^*$ for a prescribed $\varepsilon^* > o$ then stop. Otherwise calculate Lv' and continue.

Based on Theorem 4 by Denardo (1967) one can apply a policy improvement algorithm: Choose any $\delta \in \Delta$. Determine v_δ. Then calculate Lv_δ and a policy γ (if possible) such that $H_\gamma(v_\delta) = Lv_\delta$. If $K \leqq \varepsilon^*$ then determine v_γ and stop. Otherwise start from the beginning with γ instead of δ. It should be noted that γ exists, if each Δ_x is finite. Otherwise continuity and measurability conditions have to be satisfied.

REFERENCES:

Denardo, E.V., Contraction Mappings in the Theory Underlying Dynamic Programming, SIAM Review, Vol. 9, No. 2 (1967).

Elmaghraby, S., Activity Networks; John Wiley & Sons, New York/Sydney/Toronto 1977.

Hinderer, K., Foundations of Non-stationary Dynamic Programming with Discrete Time Parameter, Lecture Notes in Operations Research and Mathematical Systems, Vol. 33, Springer Verlag, Berlin/Heidelberg/New York, 1970.

Höpfinger, E., Foundations of Time Analysis and Optimization of Acycle Activity Networks, Habilitationsschrift, University of Karlsruhe, 1978.

Neumann, K., Operations Research Verfahren III, Carl Hanser Verlag, München/Wien, 1975.

Nicolai, W., Zeit- und Kostenanalyse von Projekten mit Hilfe von Netzwerken und zugeordneten stochastischen Prozessen, doctoral dissertation, University of Karlsruhe, 1979.

A NECESSARY CONDITION FOR THE ELIMINATION OF CRANE INTERFERENCE*

R.W. Lieberman and I.B. Türksen

Department of Industrial Engineering
University of Toronto, Toronto
Ontario, M5S 1A4, Canada

*Supported by the Natural Sciences and Engineering Research Council of Canada

ABSTRACT

Copper smelters and steel mills utilize large overhead cranes to transport material from one location to another within the plant. The instructions "move material from source A to sink B" are termed "jobs". Because all cranes share the same track and cannot pass each other, the set of jobs may impose a set of conflicting demands on the cranes. Two cranes assigned to specified jobs may be mutually blocked in their transit from source to sink. This is known as crane interference. The completion of one of the jobs must of necessity delay the completion of the other. Delays due to crane interference may be minimized by an appropriate assignment of cranes to jobs.

In this paper, a model is presented in which the scheduling problem reduces to that of constructing batches of jobs which can be assigned to cranes so that crane interference is eliminated. A necessary but not sufficient condition for such job batches can be determined by deriving the partition number. The partition number is defined to be the minimum cardinality of an ordered partition of an n-tuple. An efficient algorithm ($O(n^2)$) is described to determine a minimum ordered partition.

This is followed by a graph-theoretic representation of the problem in which it is shown that the determination of the partition number is equivalent to finding the stability number of a transitive 1-graph. The complement graph is identified as a permutation graph.

Job batches are constructed using a heuristic procedure. If the necessary condition is not satisfied, then the batching strategy will not completely eliminate crane interference. There may exist another strategy which will eliminate interference. Therefore, other strategies should be examined before the use of the heuristic procedure.

I INTRODUCTION

In some industrial processes, overhead cranes are required to move materials to and from various locations. The cranes all share a common track and consequently are unable to pass one another. Cranes engaged in some activity at a specific location may block other cranes from beginning or completing service on waiting jobs causing delays in throughput. Prime examples of where such problems occur are copper smelters and steel mills. Typically, the cranes travel along a single gantry running between various facilities requiring crane service such as reverberatory, converter and anode furnaces. The process requires that material be transferred from one facility (the source location) to some other facility (the sink location). The specification of the source and sink locations is termed a job.

If n jobs require service at time zero, a scheduling problem exists as to the assignment of the $m \geq 2$ cranes to the n jobs so as to minimize the delays in processing due to crane interference. A trade-off exists between the number of "potential" parallel servers (m) and the number of "potential" occurrences of crane interference. Certainly, no interference can occur with exactly one crane. However, the total completion time or makespan could be halved if two cranes were used in parallel so that no interference occurs. As the number of servers increases, so does the likelihood of crane interference.

In order to utilize $m \geq 2$ cranes most effectively, the m cranes must be assigned to the jobs so that no interference exists, that is the crane system is made to behave as an m-parallel server (mps) system. This may or may not be possible depending on the structure of the jobs themselves. A set of jobs for which no such assignment is possible is said to possess a positive congestion. If a zero interference assignment is possible, the job set has a zero congestion.

One possible way to process jobs is to assign all m cranes to m jobs $(m \leq n)$, and when the jobs are completed, assign all m cranes again to jobs and continue in this manner until all the n jobs are completed. Such a strategy is called batching since in effect, the n jobs are partitioned into $\lceil \frac{n}{m} \rceil$ sub-sets or batches ($\lceil x \rceil$ is the least integer greater than or equal to x) and the jobs in a batch are processed simultaneously and independently from jobs in another batch. This strategy will yield an interference-free solution provided that the job set has the proper structure.

In section II, a model is described which reduces the scheduling problem to one of finding a proper partition of the job set so that the batching strategy yields a solution with no interference. A necessary condition is presented for the existence of such a partition. In section III, an efficient algorithm ($O(n^2)$) is described which determines the parameter required for the necessary condition. A graphical representation is shown in section IV in which the problem is equivalent to determining the stability number of a transitive 1-graph. Finally, in section V, some examples are discussed in order to illustrate the ideas presented.

II THE MODEL

A crane system is defined to be a set of m cranes $C = \{C_1, C_2, \ldots, C_m\}$, a set of N job locations $L = \{1, 2, \ldots, N\}$ and a set of n jobs $J = \{J_1, J_2, \ldots, J_n\}$. Both the cranes and the locations are numbered sequentially from left to right and together compose the crane aisle (See Figure 1).

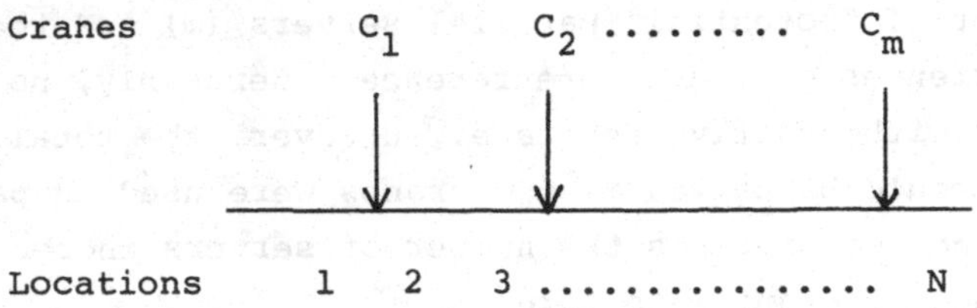

FIGURE 1

A job J_i is an ordered pair (ℓ_{i1}, ℓ_{i2}), ℓ_{i1} is the source location and ℓ_{i2} is the sink location, $\ell_{i1}, \ell_{i2} \in L$, $\ell_{i1} \neq \ell_{i2}$. The model assumes for simplicity that each location is either a source or a sink, but not both. The set of sources is denoted as $L' \subset L$ and the set of sinks as $L'' \subset L$ and $L' \cup L'' = L$, $L' \cap L'' = \emptyset$ where $\emptyset$ is the null set. We assume that the number of sources is equal to the number of sinks, i.e. $|L'| = |L''|$. The model is static in that it assumes all jobs to be ready for processing at time 0. The cranes are identical and crane travel times are taken as zero. Job processing times are equal; τ time units at the source and τ units at the sink, a total of 2τ. Once a crane begins to process a job, it cannot begin a new job until the current job is finished (no pre-emptions). Once a crane begins processing on the source location of a job, it is said to be committed to process the sink location.

The system state at time k consists of the present crane positions and the future crane commitments. The state variable is:

$$X(k) = (x_1', x_2', \ldots, x_m' | x_1(k), x_2(k), \ldots, x_m(k)).$$

$x_i(k)$ represents the position of crane C_i at time k:

$$x_i(k) = \begin{cases} 0 & \text{if crane } C_i \text{ is idle at time k} \\ \ell \in L & \text{if crane } C_i \text{ is busy at location } \ell \\ & \text{at time k} \end{cases}$$

x_i' is the future commitment of crane i:

$$x_i' = \begin{cases} 0 & \text{if crane } C_i \text{ is uncommited} \\ \ell \in L'' & \text{if crane } C_i \text{ is committed to service} \\ & \text{location } \ell \in L''. \end{cases}$$

It is assumed that any number of cranes can be positioned while idle at the ends of the crane aisle and between any locations.

The completion time of job J_i is the time that the sink location of job J_i completes service and is denoted F_i.

<u>Definition 1</u>

The state X(k) is <u>feasible</u> iff the following condition holds:

$$x_i(k) < x_j(k) \quad \text{iff} \quad i < j,\ x_i(k),\ x_j(k) \neq 0$$

This condition expresses the fact that cranes cannot pass each other.

A solution to the crane scheduling problem is a specification for each job location ℓ_{ij} of

(a) the start time σ_{ij}, $i = 1,2,\ldots,n$, $j = 1,2$, and

(b) the crane assigned C_p, $p = 1, 2, \ldots, m$.

A solution or <u>schedule</u> is completely described by a listing of all system states X(k), k = 1, 2, ..., M where M is the makespan. A schedule is feasible iff the system states X(k) are feasible for all k. The set of feasible solutions is S_F. Each job set J functionally determines the set S_F.

Associated with each feasible schedule $s \in S_F$ is a makespan M(s). A minimum feasible solution $s^* \in S_F$ is one in which

$$M(s^*) \leq M(s) \qquad \forall s \in S_F.$$

A lower bound on makespan for the crane system can easily be derived by removing the single-track constraint. In effect, the system now behaves as an m-parallel server (mps) system.

<u>Lemma 1</u>

A lower bound on makespan for the crane system is $M^* = 2\left\lceil \frac{n}{m} \right\rceil \tau$.

<u>Proof</u>

With no interference and m identical processors, the system can process up to m jobs in parallel, each batch of at most m jobs consuming 2τ time units. The minimum number of batches is $\left\lceil \frac{n}{m} \right\rceil$. Hence

$M^* = 2\left\lceil \frac{n}{m} \right\rceil \tau.$ Q.E.D.

Because the processing times of all locations are all equal, a batching strategy, one that processes jobs simultaneously in batches, seems like a fruitful approach to minimize interference since it would allow locations started together to finish together. What is desired is to construct batches of jobs in such a way that by starting jobs of a given batch simultaneously, the jobs will finish simultaneously.

To see how such batches should be constructed, consider a batch of m jobs, say $B = \{J_{i_j} | J_{i_j} \in J,\ j = 1,2,\ldots,m\}$. The batch B consists of a sub-set of sources $L'(B)$ and a sub-set of sinks $L''(B)$ that is $L'(B) = \{\text{source locations in } B\}$ and $L''(B) = \{\text{sink locations in } B\}$. The set of sources $L'(B)$ can always be started together by assigning crane C_1 to the smallest numbered source, C_2 to the next largest, and so on assigning C_m to the largest. However, each of the m cranes is now committed to process the sink location associated with its respective source. The sink locations can only be started together if they form an increasing sequence.

Define the operator ∇ to be the <u>source numerical order operator</u>: $\nabla J = (J_{i_1}, J_{i_2}, \ldots, J_{i_n})$ such that $\ell'_{i_j} < \ell'_{i_{j'}}$ iff $i_j < i_{j'}$, where $\ell'_{i_j} \in J_{i_j} \cap L'$. In other words, the operator forms an ordered set of jobs from an unordered set of jobs, the order being on the <u>source</u> locations in increasing location number order. For example, consider the batch $B = \{(5,8), (2,4), (6,7), (1,3)\}$ with $m = 4$. $\nabla B = ((1,3), (2,4), (5,8), (6,7))$. The sets $L'(\nabla B) = (1,2,5,6)$ and $L''(\nabla B) = (3,4,8,7)$ are defined to be the ordered sets of sources and sinks respectively induced by ∇B. It can be seen from the state vector that if

$$X(0) = (3,4,8,7|1,2,5,6) = (L''(\nabla B)|L'(\nabla B)),$$

the locations of $L''(\nabla B)$ cannot be processed together because of interference between cranes C_3 and C_4 at locations 8 and 7 respectively (feasibility is violated). However, if job (5,8) were replaced by job (10,9), then $X(0) = (L''(\nabla B)|L'(\nabla B) = (3,4,7,9|1,2,6,10)$ yields an interference-free processing of batch B in time 2τ. The key then, is to construct batches of jobs so that the $L''(\nabla B)$'s are monotonically increasing sequences or "as close as possible" to a monotonically increasing sequence.

A measure of how close a given sequence is to a monotonically increasing sequence is the partition number denoted Q. It is the minimum number of ordered subsets in a partition of the sequence such that each subset is monotonically increasing. Coming back to the example, $L''(\nabla B)$

= (3,4,8,7), it can be seen that the sinks could be processed in 2 sub-batches, namely (3,4,8,0) and (7,0,0,0) or (3,8,0,0) and (4,7,0,0). Of course, they could be processed in 3 or 4 sub-batches: (3,0,0,0), (4,8,0,0), (7,0,0,0) or (3,0,0,0), (4,0,0,0), (0,8,0,0) and (0,0,0,7). But 2 sub-batches is the minimum number of sub-batches that can be formed to complete the jobs, and therefore takes the least time. So $Q(L''(\nabla B)) = 2$. Since $L'(\nabla B)$ is always a monotonically increasing sequence, $Q(L'(\nabla B)) = 1$.

Let $P_n = (p_1, p_2, \ldots, p_n)$ be an n-tuple of unique integers and let $\Gamma_k = (\gamma_{k1}, \gamma_{k2}, \ldots, \gamma_{kq_k})$ be ordered subsets of P_n such that

(1) $\bigcup_{\forall k} \Gamma_k = P_n$

(2) $\bigcap_{\forall k} \Gamma_k = \emptyset$

The partition $\Gamma = \{\Gamma_k\}$ is an ordered partition if the following conditions hold:

(a) $\gamma_{ki} < \gamma_{kj}$ iff $i < j$, $\forall k$.

(b) $\gamma_{ki} = p_r$ and $\gamma_{kj} = p_s$
and $i < j$ iff $r < s$, $p_r, p_s \in P_n$.

The partition Γ_{min} is a minimum ordered partition (MOP) iff $|\Gamma_{min}| \leq |\Gamma|, \forall \Gamma$. The partition number of P_n is $Q(P_n) = |\Gamma_{min}|$.

To determine whether batches of jobs B_j, can be constructed so that $L''(\nabla B_j)$ are monotonically increasing, it is necessary to examine ∇J, the set of ordered jobs. Obviously, if $Q(L''(\nabla J))$ is greater than the number of batches required for an mps solution, interference must occur. This forms a necessary condition for an interference-free solution.

Theorem 1

For a given set of jobs J, if $Q(L''(\nabla J)) > \left\lceil \frac{n}{m} \right\rceil$, then the makespan under the batching strategy is greater than the lower bound, i.e., $M > M^*$.

Proof

$Q(L''(\nabla J))$ is the cardinality of a MOP for the n-tuple $L''(\nabla J)$. This is the minimum number of batches B which can be constructed from J such that each ordered set $L''(\nabla B)$ forms a monotonically increasing sequence. If this number exceeds $\left\lceil \frac{n}{m} \right\rceil$, certainly more than $\left\lceil \frac{n}{m} \right\rceil$ batches are required. Therefore the makespan using the batching strategy is $M > 2\left\lceil \frac{n}{m} \right\rceil \tau$. Q.E.D.

A necessary condition, then, for a non-interference solution under batching is $Q(L''(\nabla J)) \leq \left\lceil \frac{n}{m} \right\rceil$.

III AN ALGORITHM TO CONSTRUCT A MOP

An algorithm is now presented which constructs s MOP from a given n-tuple P_n. Once the MOP is constructed, $Q(P_n)$, the partition number is determined. The complexity of the algorithm is $O(n^2)$. The MOP's construction proceeds as follows: the elements of P_n are scanned from left to right in one pass. The first element p_1 is placed in Γ_1 as the first element, that is $\gamma_{11} = p_1$. Each element p_i, $i \geq 2$ is examined and is either placed in an existing subset Γ_j or becomes the first element of a new subset Γ_k where $k = \max(j) + 1$. At any stage, if ℓ_j is the most recent element to be placed into subset Γ_j and there are currently k subsets, then p_i becomes the first element of a new subset Γ_{k+1} if $p_i < \ell_k$. Otherwise, it joins the subset Γ_j for which $\ell_j < p_i < \ell_{j-1}$ where $\ell_o > \max_{1 \leq i \leq n} p_i$. The algorithm is now stated in Pidgeon Algol as in [1] (pp. 33-39).

Algorithm MOP

Input: An n-tuple $P_n = (p_1, p_2, \ldots, p_n)$
Output: A MOP $\Gamma(P_n) = \{\Gamma_k | k=1,2,\ldots,Q(P_n)\}$
where $\Gamma_k = (\gamma_{k1}, \gamma_{k2}, \ldots, \gamma_{k\ell_k})$.

```
begin
    k ← 1; ℓ1 ← 1; j ← 1; γ11 ← p1;
    for i ← 2 until n do
        begin
            if pi < γkℓk then begin
                                  k ← k + 1;
                                  j ← k;
                                  ℓj ← 1
                              end
            else if k = 1 then goto new;
                 else for j ← k step -1 until 2 do
            begin
                 if γjℓj < pi < γj-1ℓj-1 then
                          begin
new:                          ℓj ← ℓj + 1;
                              goto set
                          end;
                 else if j = 2 then
                          begin
```

```
                    j ← 1;
                    goto new
                  end
        end;
set:    b_{jℓ_j} ← p_i
end
```

As an example, consider $P_5 = (3,2,4,1,5)$.

(1) $\gamma_{11} = p_1 = 3$ $\quad\quad \Gamma_1 = (3)$

(2) $p_2 = 2 < \gamma_{11}$ $\quad\quad \therefore\ \gamma_{21} = 2,\ \Gamma_2 = (2)$

(3) $p_3 = 4 \geq \gamma_{21}$

$p_3 \geq \gamma_{11}$ $\quad\quad \therefore\ \gamma_{12} = 4\ \ \Gamma_1 = (3,4)$

(4) $p_4 = 1 < \gamma_{21}$ $\quad\quad \therefore\ \gamma_{31} = 1,\ \Gamma_3 = (1)$

(5) $p_5 = 5 \geq \gamma_{31}$

$p_5 \geq \gamma_{21}$

$p_4 \geq \gamma_{12}$ $\quad\quad \therefore\ \gamma_{13} = 5$ and $\Gamma_1 = (3,4,5)$.

Hence $\Gamma = \{\Gamma_1, \Gamma_2, \Gamma_3\}$ where $\Gamma_1 = (3,4,5)$

$\Gamma_2 = (2)$

$\Gamma_3 = (1)$

and therefore $Q(P_5) = 3$.

In a worst case analysis, the number of elementary operations is $f(n) \leq \frac{1}{8}(n^2 + 8n + 7)$. Therefore, the complexity of the algorithm is $O(n^2)$.

Theorem 2

The partition constructed by algorithm MOP is a minimum ordered partition.

Proof

By construction, suppose we have H ordered subsets:

$$\Gamma = \left\{ \begin{array}{l} \Gamma_1 = (\gamma_{11}, \gamma_{12}, \ldots, \gamma_{1q_1}) \\ \Gamma_2 = (\gamma_{21}, \gamma_{22}, \ldots, \gamma_{2q_2}) \\ \vdots \\ \Gamma_H = (\gamma_{H1}, \gamma_{H2}, \ldots, \gamma_{H\gamma_H}, \ldots, \gamma_{Hq_H}). \end{array} \right\}$$

By construction, in each subset Γ_k

$\gamma_{ki} < \gamma_{kj}$ iff $i < j$. Also, since the scan is from left to right in P_n, for $\gamma_{ki} = p_s$ and $\gamma_{kj} = p_t$ then

$$i < j \ \text{ iff } \ s < t.$$

Therefore the partition Γ is an ordered partition. Also, for any element in Γ_H, say $\gamma_{Hr_H} = p_s$, an element must exist in Γ_{H-1}, say $\gamma_{H-1,r_{H-1}}$ such that $\gamma_{Hr_H} < \gamma_{H-1,r_{H-1}}$ and for $\gamma_{Hr_h} = p_{i_H}$ and $\gamma_{H-1,r_{H-1}} = p_{i_{H-1}}$, $i_H > i_{H-1}$. But for this element $\gamma_{H-1,r_{H-1}}$, a similar element can be found in $\Gamma_{H-2,\gamma_{H-2}}$. Thus a monotone decreasing sequence exists $\gamma_{1r_1} > \gamma_{2r_2} > \ldots > \gamma_{Hr_H}$ which corresponds to $p_{i_1} > p_{i_2} > \ldots > p_{i_H}$ where $i_1 < i_2 < \ldots < i_H$. It is therefore clear that any ordered partition must have at least H subsets, and thus H is a minimum. Therefore, the partition Γ is a minimum ordered partition. Q.E.D.

IV GRAPHICAL REPRESENTATION

The partitioning of an n-tuple into a MOP can be modelled by defining a digraph $G = (X,U)$ where the set of vertices $X = \{x_i | x_i \in P_n\}$ and the set of arcs $U = \{(x_i, x_j) | x_i < x_j, i < j, x_i, x_j \in X\}$. For example, the digraph of $P_5 = (2,4,3,1,5)$ is $G(X,U)$ where $X = \{1,2,3,4,5\}$ and $U = \{(2,4), (2,3), (2,5), (4,5), (3,5), (1,5)\}$.

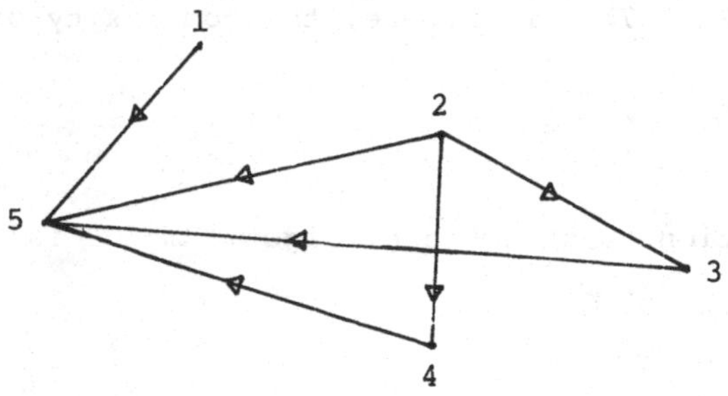

In terms of the graph, an ordered partition is simply a set of vertex-disjoint paths which cover the vertices of G. A MOP is a path cover of minimum cardinality. The cardinality is known as the path-to-vertex covering number [2]. $Q(P_n)$ then is actually the path-to-vertex covering number of $G(X,U)$.

Since a directed edge from x_i to x_j implies $x_i < x_j$, the graph is acyclic. Furthermore, since the partial orderings $x_i < x_j$ and $i < j$ are transitive, the graph is transitive. The graph is a 1-graph since there is at most one arc between any pair of vertices.

Dilworth's Theorem ([3]p. 300)

If G is a transitive 1-graph and if M is a family of paths that partitions its vertex set, then $\min_M |M| = \alpha(G)$ where $\alpha(G)$ is the stability number of the graph G.

Hence the determination of $Q(P_n)$ reduces to finding the cardinality of a maximum stable set in G(X,U).

The algorithm presented in section III is identical to the one proposed by Even, Lampel, and Pnueli in [4] for finding a minimum chromchromatic decomposition in a transitive digraph. Furthermore, the complement undirected graph of G(X,U) is defined as $G(X,\bar{U})$ where $\bar{U} = \{(x_i,x_j) | i < j \text{ and } x_i > x_j\}$ which is by definition a permutation graph.

V EXAMPLES AND DISCUSSION

To illustrate the preceding ideas, consider:

Example 1

m, the number of cranes is 4.

N, the number of job locations is 20.

n, the number of jobs is 10.

J, the job set is $\{J_1, J_2, \ldots, J_{10}\}$

$= \{(11,3), (14,19), (1,8), (15,7), (13,5), (18,16)$
$(9,17), (4,2), (10,6), (20,12)\}$.

The lower bound on makespan is:

$$M^* = 2\left\lceil\frac{n}{m}\right\rceil\tau = 2\left\lceil\frac{10}{4}\right\rceil\tau = 6\tau.$$

First, the source numerical order operator is applied to yield the set $L''(\nabla J)$:

∇J	$L'(\nabla J)$	$L''(\nabla J)$
J_3	1	8
J_8	4	2
J_7	9	17
J_9	10	6
J_1	11	3
J_5	13	5
J_2	14	19
J_4	15	7
J_6	18	16
J_{10}	20	12

Algorithm MOP is then applied to find $Q(L''(\nabla J))$:

$$\Gamma = \{\Gamma_1, \Gamma_2, \Gamma_3\}$$
$$= \{(8,17,19), (2,6,7,16), (3,5,12)\}$$

Therefore $Q(L''(\nabla J)) = 3 \le \left\lceil \frac{10}{4} \right\rceil = 3$

Since the necessary condition is met, it may be possible to construct 3 batches ofjobs, $\{B_i | i = 1,2,3\}$ with at most four jobs per batch such that $Q(L''(\nabla B_i)) = 1$. For this example, the MOP Γ is already of this form, so that

$B_1 = \{J_3, J_7, J_2\}$

$B_2 = \{J_8, J_9, J_4, J_6\}$

$B_3 = \{J_1, J_5, J_{10}\}$

The system can now process each batch simultaneously, the solution being:

$X(0) = (8,17,19,0|1,9,14,0)$

$X(\tau) = (0,0,0,0|8,17,19,0)$

$X(2\tau) = (2,6,7,16|4,10,15,18)$

$X(3\tau) = (0,0,0,0|2,6,7,16)$

$X(4\tau) = (3,5,12,0|11,13,20,0)$

$X(5\tau) = (0,0,0,0|3,5,12,0)$

$X(6\tau) = (0,0,0,0|0,0,0,0)$

Example 2

Consider Example 1 with $m = 3$. Now $M^* = 8\tau$. To eliminate all interference, we must construct four batches, each batch containing three jobs at most. Although the MOP Γ is not in this form, it is a trivial task to construct such a set of batches from the MOP. Simply create a new subset Γ_4, and put the extra element from Γ_2 into Γ_4. This defines the batch set:

$B_1 = \{J_3, J_7, J_2\}$

$B_2 = \{J_1, J_5, J_{10}\}$

$B_3 = \{J_8, J_9, J_4\}$

$B_4 = \{J_6\}$.

It may not always be straightforward to construct the required set of batches from the MOP if the necessary condition $Q(L''(\nabla J)) \le \left\lceil \frac{n}{m} \right\rceil$ is met, in fact, such a batch set may not exist. In this situation, a heuristic procedure may be applied whose time complexity is polynomial [5].

Example 3

Consider Example 1 with a new job set J such that $L''(\nabla J) = (8,4,6,2,9,3,15,13,12,10)$. Applying Algorithm MOP yields

$$\Gamma = \left\{ \begin{array}{l} \Gamma_1 = (8,9,15) \\ \Gamma_2 = (4,6,13) \\ \Gamma_3 = (2,3,12) \\ \Gamma_4 = (10) \end{array} \right\}$$

Since $Q(L''(\nabla J)) = 4 > 3 = \left\lceil \frac{n}{m} \right\rceil$, 3 batches each with no more than four jobs cannot be constructed, and $M > M^*$ for any set of batches constructed from the job set J. In this case, some other strategy other than batching may yield a schedule whose makespan is M*. For example, at any time k, some cranes can be servicing the sinks of some jobs while others are servicing sources of others. Such a strategy has been developed for systems in which m = 2 and is termed a meshing strategy [5]. It has been demonstrated that under certain conditions, a meshing strategy will yield a schedule whose makespan M is M*, even if $Q(L''(\nabla J)) > \lceil \frac{n}{2} \rceil$. Therefore, before searching for a "good" set of batches by some heuristic procedure, alternate strategies should be examined first.

REFERENCES

[1] Aho, Alfred V.; Hopcroft, John E.; Ullman, Jeffrey D.: The design and analysis of computer algorithms. Reading, Mass.: Addison-Wesley (1975).

[2] Boesch, F.T.; Chen, S.; McHugh, J.A.M.: On covering the points of a graph with point disjoint paths. Proceedings of the Capital Conference on Graph Theory and Combinatorics at the George Washington University. Graphs and combinatorics. Berlin: Springer-Valag (1973).

[3] Berge, Claude: Graphs and hypergraphs. Amsterdam: North-Holland Publishing Company (1973).

[4] Even, S.; Lampel, A.; Pnueli, A.: Permutation graphs and transitive graphs. JACM, Vol 19, No. 3, 400-410 (July, 1972).

[5] Lieberman, R.W.: Scheduling under interference constraints. Ph.D. thesis, Dept of I.E., Univ. of Toronto (1979).

OPTIMAL CONSTRUCTIONS OF PROJECT NETWORKS

(EXTENDED ABSTRACT)

Maciej M. Sysło
Institute of Computer Science
University of Wrocław
Pl. Grunwaldzki 2/4
50-384 Wrocław, Poland

PERT, CPM and other techniques which can be applied to the planning and scheduling involve the construction of *networks*. There are two types of networks which represent a *project*, i.e., the activities and their precedence relations, namely, the *activity* network and the *event* network. An activity network is a digraph D in which the nodes correspond one-to-one with the given activities and there is an arc (u,v) in D if activity u precedes activity v. There exists a unique activity network without redundant arcs for each project. In an event network E which corresponds to an activity network D, the given activities are represented by a subset of arcs of E and the precedence relations are preserved. In general, *dummy* activities (arcs of E) are introduced to satisfy the last requirement. Since there is an infinite number of different sized event networks for each project, the problem is to find for a set of activities and their precedence relations an event network with the minimum number of dummy activities. The motivation behind this problem is to minimize the time of the analysis of a network which is proportional to the number of arcs, including those which correspond to dummy activities.

Krishnamoorthy and Deo proved in [1] that the problem of finding the minimum number of dummy activities in the event network which corresponds to a given set of activities and their precedence relations is NP-complete. In [3], the precedence relations for which there exists an event network without dummy activities are characterized and it is proved that the question whether a given precedence relations require dummy activities in the event network can be answered in polynomial time. We proved also that the latter problem for not necessarily acircuit digraphs can be also solved in polynomial time.

Paper [3] contains an example of a network which shows that the number of nodes and the number of arcs in an event network cannot be minimized simultaneously. In other words, it is shown that it is possible to decrease the number of dummy activities in an event network by increasing the number of nodes.

The problem of finding an event network for a given activity network is closely related to the construction of a digraph from its *line digraph*. It is easy to notice that if an activity network is a line digraph or if it is transitively equivalent to a line digraph then there exists the corresponding event network without dummy activities.

If an activity network D is not a line digraph then we can transform D into a line digraph D' in which the precedence relations are preserved. In fact, all methods for constructing an event network find such a line digraph D'. Paper [2] contains two algorithms for finding D' from a given activity network D by a sequence of subdivisions of arcs of D and a sequence of subdivisions of arcs of D which form the complete bipartite subdigraphs, resp. It is proved that these algorithms are polynomial and optimal in the class of methods which subdivide arcs of D. In general, only approximate solutions are obtained by the method in [2].

Some well-known algorithms for finding the event network with the minimum number of dummy activities are reviewed in [3] and a new approach is proposed which gives rise to a new approximate algorithm and can be applied to produce an optimal branch--and-bound method.

References

[1] M.S. Krishnamoorthy, N. Deo, *Complexity of the minimum-dummy-activities problem in a PERT network*, Networks 9 (1979), in press.

[2] M.M. Systo, *Optimal constructions of reversible digraphs*, Report Nr N-55, Institute of Computer Science, University of Wrocław, Wrocław 1979.

[3] M.M. Systo, *Optimal constructions of event-node networks*, Report Nr N-61, Institute of Computer Science, University of Wrocław, Wrocław 1979.

ENUMERATION TECHNIQUES IN DIRECTED HYPERGRAPHS

Cyriel VAN NUFFELEN
UFSIA, Antwerpen, Belgium

ABSTRACT

In this note we extend the concept of adjacency matrix, formerly only used for graphs, to a weighted directed hypergraph which is a generalization of the usual non-directed hypergraph.
This ables us to enumerate chains in this directed hypergraph and so we obtain a generalization of the well known theorem to calculate chains in a graph.

1. HYPERGRAPHS

1.1. Let $X = \{x_1, x_2, \dots, x_n\}$ be a finite set of vertices and $\mathcal{E} = \{E_1, E_2, \dots, E_r\}$ be a family of subsets of X, called edges. Then $\mathcal{E}$ constitutes a hypergraph on X, denoted $H = (X, \mathcal{E})$ if: $\forall i;\ E_i \neq \emptyset$ and $\bigcup_i E_i = X$.

1.2. In each edge E_i, we form with the vertices of E_i, ordered m-tuples ($2 \leqslant m \leqslant \# E_i$). All ordered m-tuples so formed are labelled throughout in the hypergraph and are denoted as the set $D = \{y_1, y_2, \dots, y_t\}$.

Let $d : D \to \mathbb{R}$ be a function over these ordered m-tuples (weight 0 corresponds to deletion). Then we call the triplet $H = (X, \mathcal{E}, d)$ a weighted directed hypergraph.

1.3. It is easy to see that with a proper choice of the weight d, one can describe the common used kinds of hypergraphs. For example; the case $d \equiv 0$ corresponds to deletion of the ordering and we obtain the usual non-directed hypergraph of C. BERGE [2]. Now consider one particular ordering of all the vertices in each edge E_i, and the corresponding weights are 1; let all other weights be 0, then we have the definition of directed hypergraphs formerly used by Ph. VINCKE [5]. Also G. ALIA & P. MAESTRINI [1] and A. GERMA [3] used certain kinds of directed hypergraphs.

2. SUITES

2.1. A vertex x_j is said to be "reachable" from the vertex x_i if there exists an m-tuple y with $d(y) \neq 0$ and x_i strictly "preceeds" x_j in y.

2.2. We say x_j is "k-reachable" from x_i if there exist k different ways to reach x_j from x_i (different by the m-tuples).

2.3. A "suite" of length $n \geqslant 1$ is a sequence of vertices and m-tuples: $x_1, y_1, y_2, \ldots, y_n, x_{n+1}$ in which for all k; x_{k+1} is reachable from x_k.
Two suites are different if they differ by their vertices, their m-tuples or by the order in the suite.

2.4. By a proper choice of the weights, a suite reduces to the usual chain or walk in an undirected or directed graph.
It is also possible to redefine "reachable" in the sense that for example x_i and x_j are respectively the first and last vertex in the m-tuple, this to obtain other kinds of suites.

3. ADJACENCY MATRIX

3.1. The "adjacency matrix" $A = [a_{ij}]$ of a weighted directed hypergraph $H = (X, \mathcal{E}, d)$ with n vertices, is a n x n-matrix in which: $a_{ij} = k$ iff x_j is k-reachable from x_i and 0 otherwise.

3.2. Theorem 1

Consider a weighted directed hypergraph H and its adjacency matrix A. Then the element (i,j) of A^n equals the number of suites from x_i to x_j of length n in H.

Proof

The theorem is true for n = 1 and suppose it holds for the value n - 1.
Let $A = [a_{ij}]$, $A^{n-1} = [b_{ij}]$ and $A^n = [p_{ij}]$.
The general element p_{ij} of A^n equals:

$$p_{ij} = \sum_k b_{ik} a_{kj}.$$

Now $b_{ik} a_{kj}$ is the number of suites with length n of the form:

$$x_i, \ldots, x_k, y, x_j$$

because the number of different m-tuples y equals a_{kj} by

definition.
So, p_{ij} gives the total number of suites from x_i to x_j of length n.

3.3. If the weighted directed hypergraph reduces to a graph (directed or not), then the adjacency matrix becomes the usual adjacency matrix of a graph and theorem 1 reduces to the well known theorem for calculation of the number of walks in a graph.

4. ENUMERATION OF CHAINS

4.1. A "chain" is a suite in which all the vertices and m-tuples are different (except for the first and last vertex).

4.2. In order to enumerate this chains we introduce a second form of adjacency matrix.
The adjacency matrix $A = \left[a_{ij}\right]$ of a weighted directed hypergraph H with n vertices is the n x n-matrix in which:
$a_{ij} = h_i \hat{+} k_i \hat{+} l_i \dots$ iff x_j is reachable from x_i by the m-tuples $y_h, y_k, y_l, \dots$ and 0 otherwise.
The meaning of the separation mark $\hat{+}$ will be explained in 4.3.

4.3. The elements of the form: 0, h_i, k_j, l_p, in the adjacency matrix A will be denoted as the set M.
On this set M we apply two operations, namely $\hat{+}$ and $\hat{x}$, which satisfy the following properties:
a1. $\hat{x}$ is closed in M and defined as follows:
$k_i \hat{x} l_j \dots = k_i \hat{x} l_j \dots$ iff 1) k, l, ... are all different and differ from zero; 2) all indices i, j,... are different and 0 otherwise.
a2. $\hat{x}$ is associative.
a3. $\hat{x}$ is not commutative.
b1. $\hat{+}$ is closed in M.
b2. $\hat{+}$ is associative.
b3. $\hat{+}$ is commutative.
c1. $\hat{x}$ is distributive with respect to $\hat{+}$.
It is our intention to calculate the powers of the adjacency matrix A from 4.2. and therefore we note

$A = [a_{ij}]$, $A^{n-1} = [b_{ij}]$ and $A^n = [p_{ij}]$.

The element p_{ij} from A^n is defined as: $p_{ij} = \hat{\sum_k} a_{ik} \hat{\times} b_{kj}$.

To be practical we could in a somewhat simple-minded way say that we are calculating such as we do with the natural numbers. The most important exception is the rule a1. We call the operation a1. a "double Latin multiplication".

4.4. Theorem 2

Let $H = (X, \mathcal{E}, d)$ be a weighted directed hypergraph with adjacency matrix A (second form). Consider $A^n = [p_{ij}]$ (double Latin multiplication), then p_{ij} gives an enumeration of the different chains of length n from x_i to x_j.

Proof

Indeed, if we compare the definitions of 3.1. and 4.2., then by theorem 1 it is clear that A^n gives us suites of length n from x_i to x_j. All the vertices and m-tuples in these suites will be different by the double Latin multiplication of 4.3.

4.5. If the hypergraph reduces to a graph, the enumeration method of theorem 2, reduces to a method of A. KAUFMAN and Y. MALGRANGE [4].

REFERENCES

1. G. ALIA & P. MAESTRINI, A procedure to determine optimal partitions of weighted hypergraphs through a network-flow analogy, Calcolo, 13(1976), 191-211.

2. G. BERGE, Graphes et Hypergraphes, Dunod, Paris, 1970.

3. A. GERMA, Decomposition of the edges of a complete t-uniform directed hypergraph, Colloquia Mathematica Societates Ianos Bolyai; Combinatorics, 18(1976), 393-399.

4. A. KAUFMAN and Y. MALGRANGE, Recherche des chemins et circuits hamiltoniens d'un graphe, Revue Française de Rech. Opérat., 26(1963), 61-73.

5. Ph. VINCKE, Hypergraphes orientés, Cahiers du C.E.R.O., 17(1975), 407-416.

OPTIMAL DISPATCHING CONTROL OF BUS LINES

A. Adamski
Institute of Computer Science
and Control Engineering
Stanislaw Staszic University
of Mining and Mettallurgy

Al.Mickiewicza 30-059 Kraków, POLAND

INTRODUCTION.

In the near future, an improved public transport system is likely to be the only solution for reducing traffic congestion and chaos in most cities. Let us try to justify this conclusion recalling the obvious traffic problems of our cities. Increase in the social and economical activity of urban population leads to the rapid rise in vehicle traffic demand / intolerable increase of traffic volume and heavy congestion /. Such situation has caused devaluation of primal functions and features of road networks which are :

1.- to provide safe traffic conditions for vehicles and pedestrians,

2.- to provide possibilities of efficient communication between different places in urban area,

3.- compatibility with other urban systems and human environment,

through the increase of the number of accidents /e.g 50 thousand people are killed annually in U.S , 20 thousand in Japan /, decrease of efficiency / first of all in centrall parts of cities / and expansion of harmful influences upon human environment /air pollution and street noise /. In such a situation a question arises what are the possibilities of efficient solutions / if any exist / of these most serious social problems, what are remedial and preventive measures against the deepening uncontrollable traffic chaos in cities. At the first sight it seems that the problem can be radically solved by the development of existing, and creation of new transport systems. Infrastructure improvement by building more and better roads is the most effective way to gain travelling safety and efficiency but it encounters serious limits. First of all it leeds to enormous invest -

ment costs which several times exceed usual financial resources of cities / e.g. in Japan they spend about 6 billion per annum on this purpose and even this sum cannot follow the rapid increase of traffic demand /. Another important constraint results from the need of compatibility with other urban systems and human environment / e.g. urban area utilization, preservation of the character of the town e.t.c /. That is why the investments are concentrated on the critical parts of the street network. New conceptions of transport systems try to join the advantages of individual transport / large route elasticity, good door-to-door service and high accessibility for the user, travel comfort / with the good points of mass transport / programmable traffic, high capacity per unit area, lower influence on environment, lower transportation costs /. In this way the demand-activated dial-a-bus systems / Bustaxi,Retax, Rufbus / are created. They are a mixture of mass transportation systems along with comfort and some door-to-door convenience for increased accessibility for the users. Another proposal of a demand-activated system with increased capacity in high traffic density conditions with maintaining safety are the so-called Personal Rapid Transit Systems / PRT / which join features of private motor vehicles with programmed traffic / small automatic driving self-propelled vehicles operating on dedicated roadbed or tracks /. In such a system a small computer must be installed in each vehicle / whose prise exceeds many times the price of the vehicle / and a large integrated computer system for centralized control. New transportation systems, because of their limited range / in dial-a-bus systems necessity solution on-line real time large scale assignment and routing optimization problems / and enormous costs / PRT systems / do not look very promissing for the nearest future. A rational and cheap solution to the traffic problems can be found in most effective use of the existing road networks. Such solutions have a compromising character. This way act the road traffic control / monitoring / systems which, basing on continuously actuated information establish and realized appropriate control strategies in urban areas. In these large scale on-line real time systems very high reliability and large real-time processing capability are met. Most important is the development of the public transport management systems, considering that in public transport we can realize a rich repertoire of control strategies / origin-destination points and trip routes, are known /, high capacity per unit area, lower operational costs / fuel consumption / and first of all the possibility of changing traffic structure in order to decrease congestion in central parts of cities,utilizing the existing infrastructure. Public transport in central areas can be made an efficient competitor for the individual transport only provided an improved level of service /e.g. accessibility, reliability, speed, comfort, fares (3)/. In the public transport control systems these requirements can be realized in one multilevel control system (2) (3) / Fig.1 and table 1. / by means of on-line dispatching and priority control of the buse line operation, by modification and synchronization of scheduling of different bus lines, optimization of the bus network geometry and location of stops (1÷3). In the paper the above problems, with special emphasis on dispatching control are discussed.

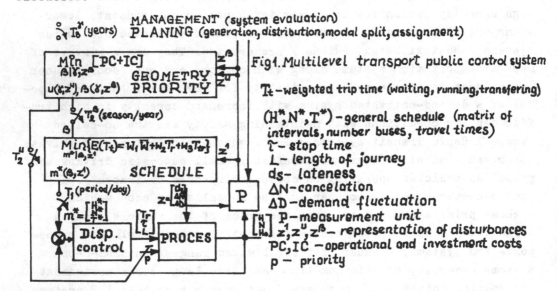

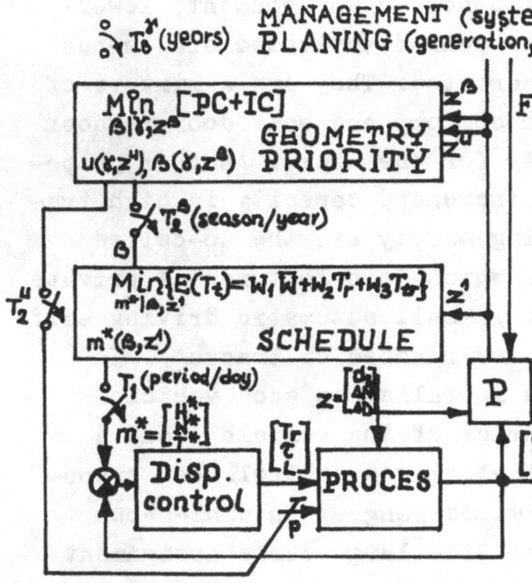

Fig1. Multilevel transport public control system

T_t – weighted trip time (waiting, running, transfering)

(H^*, N^*, T^*) – general schedule (matrix of intervals, number buses, travel times)

τ – stop time

L – length of journey

d_s – lateness

ΔN – cancelation

ΔD – demand fluctuation

P – measurement unit

z^1, z^u, z^β – representation of disturbances

PC, IC – operational and investment costs

p – priority

Table 1.

Objective	Optimization problem	Comment
Stabilization -schedule -demand	1. LQ-problem 2. realized by information transp. service system and state constraints (next section)	Dispatching control Psychological aspects
Schedule creation -frequencies f or number of buses n.	1. $PO_{\min}\; n \quad S(n)=\sum_{i,j}\left(\Theta_{ij}/\left(\frac{m\cdot x_{ij}}{v_l}+\sum_l n_l o^l_{ij}\right)\right) \mid D\leq S$ or $P[D\leq S]=\alpha$ 2. $PO_{\min}\; f \quad S(f)=f\cdot\left(T_r+T_l+c\sum_j \max[E(n^j_b)B, E(n^j_a \mid f)A]\right) \mid D\leq S$ 3. $PO_{\min}\; n \quad S(n)=c_1L_q+c_2L_s+c_3L_o+c_4n+c_5\lambda P_{str} \mid D\leq S$	OPTVBUS (8): Analytical user-oriented model (Gamma distr. intervals assumption) ($D\leq S$ means demand≤supply) (18) Analytical operator-oriented model (Bernouli distr. alightings) Simulation ($G/G_b/n$) queue model.
Network geometry and Priority control	$PO_{\min}\; u \quad S(u)=\sum_{i,j}\left(c_{ij}l_{ij}x_{ij}(u)+u_{ij}p_{ij}\right) \mid D\leq S$ Bus Transyt, Simulation models	fixed cost route selection problem. evaluation of the priority schemes by simulation.

DISPATCHING CONTROL

On the transportation route with fixed schedule served by more than one vehicle we can observe unstability in taking schedule /pairing effect /(4)(6)(13). This phenomenon results from the fact that the time required for a vehicle to load passengers is an increasing function of the number of passengers boarding and alighting. Now let some vehicle arrive somewhat behind or ahead of schedule for any reason. It must load more or less respectively than the normal number of passengers and will consequently depart even further behind or ahead of schedule at each succeding load point on its route. The next following vehicle will in turn, have fewer / more then the normal number of passengers and will, if uncontrolled get ahead of or behind schedule. The net effect of any disturbance is that vehicle tend to form pairs. The purpose of direct control in the multilevel control system Fig.1 is stabilization of the bus trajectories around schedule trajectories created in the optimization level and consequently counteraction to the bus pairing effect on the bus route. In Kraków bus pairing because of the high variation of travel and stop times and arrival intensities have essential impact on the bus line operation.

Assume :

- random passenger arrival pattern at the stops /e.g. headways are short, less than 10 minutes /.The passengers arrive acording to a Poisson process with time dependent intensity / stepwise linearized intensity levels in separate subperiods time of day, estimated separately for each stop along the route(17). Variability in intervals between subsequent buses is neglected.
- alighting models are given in the form of a probability transition matrix between bus stops along the route $P=[p_{ij}]$ or two simple alighting models estimated basing on the measurements of the number of boarding passengers n_{bk} and alighting passengers n_{ak} on the consecutive bus route stops $k=1,\ldots,n$ .The first model assumes independence between alighting passengers and that a stop attracts the whole load remaining from the preceding stops with equal force. All categories of load in the bus z_k^{si} / for si variant of service / can be reconstructed by backtracking "total bus" (17). Analytical expression for two variants of service is presented below :

$$z_{k-1}^{s1} = z_k^{s1} + n_{a,k-1}^{s1} - \frac{\mu}{1+\mu} n_{b,k-1}; \quad z_{k-1}^{s2} = z_k^{s2} + n_{a,k-1}^{s2} - \frac{1}{1+\mu} n_{b,k-1}$$

where μ - the alighting proportion. In another model the passengers are divided into groups depending on boarding stops /marked by sub - script l/. The number of passengers from each group alighting at a k-th stop $n_{a,kl}$ is proportional to their number in the bus.(3)

$$n_{a,kl} = n_{ak} \frac{n_{bl} - \sum_{m=l}^{k-1} n_{a,ml}}{\sum_{i=1}^{k-1} n_{bi} - n_{ai}}$$

-existence of an Automatic Bus Location and Identyfication system / ABL-system /(2)(3).

-passenger board first arriving bus and time of the i-th stop is determined by number of boarding passengers and in linearized form is given by $\tau = C + B \cdot n_b$.

For the city area with relatively homogeneous traffic conditions / constant passenger arrival λ_p and boarding rate λ_b/ an analytical expression (1) for the stop time of the j-th bus on the n-th bus stop is derived, as a function of trip parameters Fig. 4.

$$\tau_{jn} = Cr^n \sum_{k=1}^{j-1} \binom{n+k-2}{n-1} s^{k-1} + \binom{n+j-2}{n-1} r^{n-1} s^{j-1} \tau_{11} - r^{n-1} \sum_{k=1}^{j-1}$$

$$\binom{n+j-k-2}{n-1} s^{j-k} H^{(1)}_{k,k+1} + s^j \sum_{k=2}^{n} \binom{n-k+j-1}{n-k} r^{n-k} \tau^*_{0k} + \qquad (1)$$

$$+ \sum_{l=1}^{j} \sum_{k=2}^{n} \left[\binom{n+j-k-l-1}{n-k} r - \binom{n+j-k-l-1}{n-k-1} s \right] r^{n-k} \; T_{l,k-1} \; s^{j-l}$$

where $\tau_{11} = Cr - s\,(H^{(1)}_{01} - \tau^*_{01})$; $k_i(t) = \frac{\lambda_i(t)}{\lambda_{bi}(t)}$; $\lambda_i(t) = \lambda_p, \lambda_{bi}(t) = \lambda_b$ for $t \in [0,T]$; $r_i(t) = 1 / 1 - k_i(t)$; $s_i(t) = 1 - r_i(t)$.

H^0_{ij} / $H^{(k)}_{01}$ - intervals between scheduled departure times from the terminal or a stop on the route in the beginning of the day service.

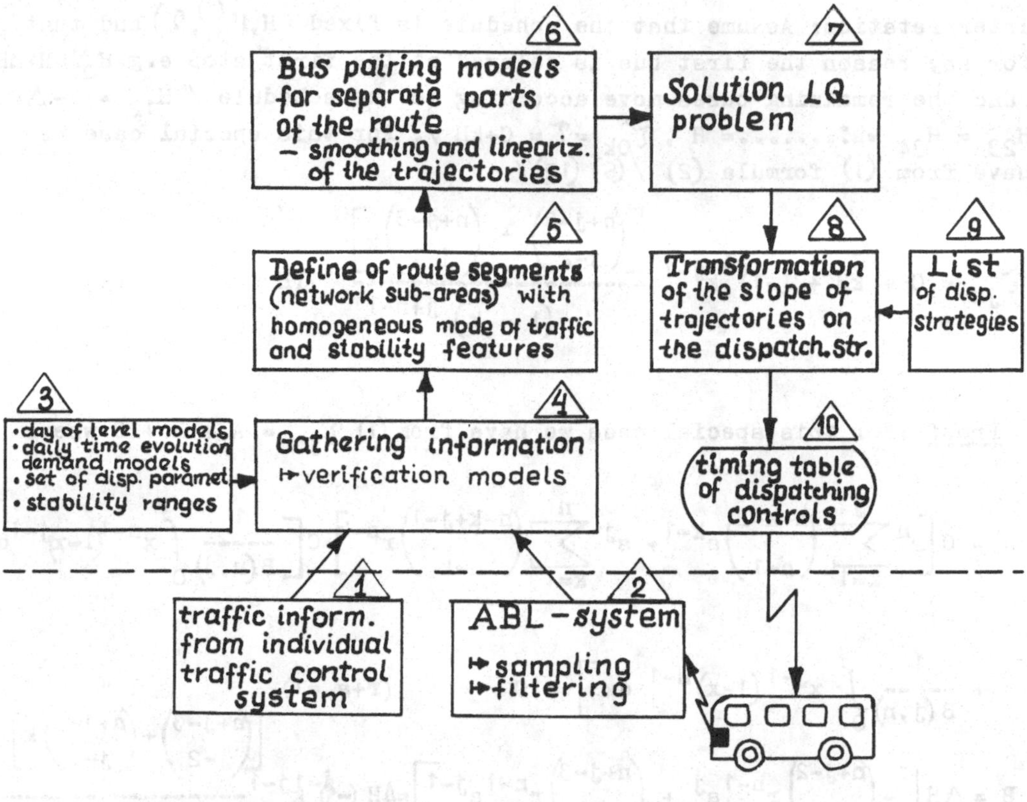

Figure 2. Idea of dispatching control a bus route

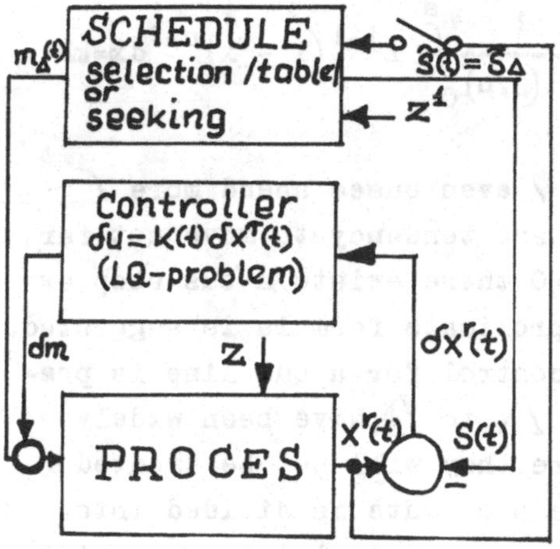

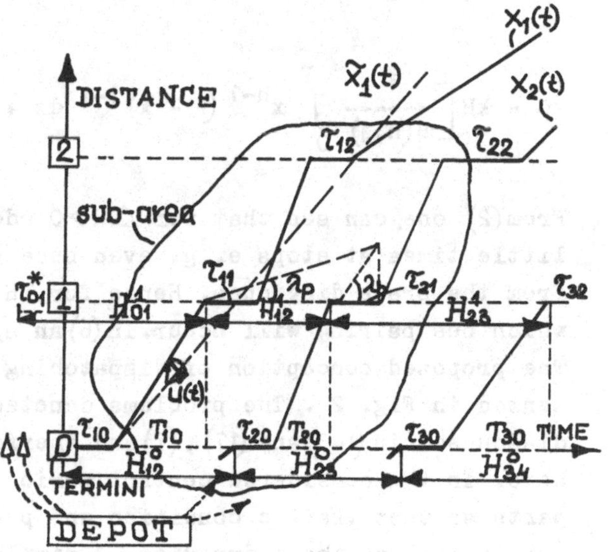

Figure 3. Dispatching control structure

Figure 4. Linear model of bus pairing effect

Interpretation: Assume that the schedule is fixed $\left(H, H_{01}^{(1)}, \tau\right)$ and that for any reason the first bus is delayed at the first stop e.g. $H_{01}^{(1)} = H + \Delta H$ and the remaining buses move according to the schedule / $H_{12} = H - \Delta H$, $H_{23} = H_{34} = \ldots\ldots = H$, $\tau_{0k}^{*} = \tau = C + kH$ /. For this special case we have from (1) formula (2) / (6) (12) /

$$\tau_{jn} = C + kH + (-1)^{j-1} k^{j-1} \frac{\left[\binom{n+j-3}{j-2} + \binom{n+j-3}{j-1} k\right]}{(1-k)^{j+n-1}} \Delta H \tag{2}$$

<u>Proof</u> : For this special case we have from (1) $\tau_{jn} = A + B + C$ where

$$A = C\left[r^{n} \sum_{k=1}^{j} \binom{n+k-2}{n-1} s^{k-1} + s^{j} \sum_{k=1}^{n} \binom{n-k+j-1}{n-k} r^{n-k}\right] = C\left[\frac{1}{B(n,j)} \int_{0}^{r} x^{n-1} (1-x)^{j-1} dx + \right.$$

$$\left. + \frac{1}{B(j,n)} \int_{0}^{s} x^{j-1} (1-x)^{n-1} dx\right] = C \qquad (r+s=1)$$

$$B = \Delta H\left[-\binom{n+j-2}{n-1} r^{n-1} s^{j} + \binom{n+j-3}{n-1} r^{n-1} s^{j-1}\right] = \Delta H\,(-1)^{j-1} k^{j-1} \frac{\left[\binom{n+j-3}{j-2} + \binom{n+j-3}{j-1} k\right]}{(1-k)^{j+n-1}}$$

$$C = kH\left[\frac{1}{B(n,j)} \int_{0}^{r} x^{n-1} (1-x)^{j-1} dx + \frac{1}{B(j,n)} \int_{0}^{s} x^{j-1} (1-x)^{n-1} dx\right] = kH$$

From (2) one can see that for $\Delta H > 0$ odd / even buses spend more / little times at stops e. g. even buses have tendency to move not far from the prece ding ones. Hence for $\Delta H > 0$ there exists a bus stop at which bus pairing will occur. In (6) an approximate formula is suggested. The proposed conception of dispatching control for a bus line is presented in Fig. 2 . The problems denoted △1 to △4 have been widely disscussed in (1-3) and (17) , (7) and therfore they will not be treated here. In the considered control period a bus route is divided into parts so that traffic condition and passenger arrival rates are similar as well as their dynamics. A simple example can be a partition into a central area part and a suburban part. These parts are treated as homogeneuous and basing on (1), smoothing of vehicle trajectories can be done Fig. 4 in respective zones. For trajectories determined in that way the following linear-quadratic problem is solved.Fig 3

$$PO_{min}\ u(t)\ S(u) = \frac{1}{2}\left[\delta x^T(T)\ Q\ \delta x(T)\right] + \frac{1}{2}\int_0^T \left(\delta\tilde{x}^T(t)\, X\, \delta\tilde{x}(t) + u^T(t)\, U\, u(t)\right) \quad (3)$$

on the trajectories $\dot{\tilde{x}}(t) = A\,\tilde{x}(t) + B\,u(t)\ \ t\in[0,T]$; $X,Q \geqslant 0,\ U > 0$.

Randomness inherent in the operation of transport systems reduces quality of passenger service far below its theoretical level, changing its behavior. Dispatching control must reduce the adverse effects of operational randomness and stabilize the trip demand / passenger behavior /. Specifity of the problem consists in the uncertainty arising from the fact that waiting time has different "disutility" for different passengers who face varying consequences of reaching destination early or late,moreover the realization of the control in the system changes the waiting time distribution which can change passengers behavior. Consequently the performance index of the control must include the costs connected with waiting times /especially important is avoidance of long interruption in the service/, trip times and "disutility" of users /depending on the difference between the real and expected /scheduled/ times of arrival of passengers to the points of destination /. Demand stabilization will be based on constraining instantaneous states at selected points of the routes $N(x(t_i), t_i)=0$ $i=1,\dots,n$ /or selected intervals on the routes e.g. route interval running through a working area /, and will be stronger in the case of fixed time trips. Constraints on the state variables can be also introduced as chance constraints /probability of exceeding some given values by instantaneous states/. The terms of the performance index represent the following requirements: the end of the trajectory should hit with a given accuracy /layover time/ the terminal and fulfil instantenous state constraints at selected points of the route, the trajectory should tract the time schedule / $\delta x(t) = x(t) - s(t)$ / and the control which changes the slope of the trajectory should not exceed a certain value / details with numerical example are include in (1)/. As a result of solving the problem (3) we obtain functions determining the time evolution of the slopes of bus trajectories. Implementation of these controls into practice is based on a hierarchical list of admissible dispatching strategies, acceptable from the passenger point of view (1)(3)(5)(10)(11).Specific variants of dispatching strategies have different parameter lists which describe precisely the conditions in which they can be applied. Typical strategies applied in practice are: punctuality, acceleration/deceleration, delaying, reserve, substitution, overtaking, curtailment. A formal description of dispatching strategies by step functions is proposed in (3) another approach by logical functions, can be found in (16). /for example the curtailment strategy has two variants: terminal and route ones, and these have 4 parameters /location, relative position before/behind turn, level of the bus load //. Transformation of the slope of trajectories into dispatching strategies gives timing table of dispatching controls for control period. Second layer controller in Fig.3 realizes repetitive control (9) provided the state of the process and an estimate of the disturbances are given. First of all basing on the actual information on availability of buses, levels of cancellation it selects /from the off-line prepared table of variants of the schedule/ or dynamically generates /dynamical programing problem; assingment of the buses to working trips basing on the position and availability of all buses currently running on the route/ actual modification of the schedule. For the schedule determined in this way a nonlinear optimization problem is solved /stop time is a nonlinear function of the number of boarding and alighting passengers and bus load/ at discrete moments of time t_j. Ilustrative example : Let us assume a hypothetical radially oriented bus route as in (18). Basing on the socio-economic data for each area, data for determining bus running and dead times, trip length p.d.f for passengers boarding at n-th stop / g_n has normal distribution/, demand estimate /modal split model of the logit-type/ it was calculated frequency as an optimal solution time minimizing problem $f_{opt} \approx 8$ bus/h e.g. schedule headway $H \approx 8$ min. The route has been divided into three zones depending on passenger demand and for such a partition we have solved an LQ problem (4)

$$\dot{x}(t)=\begin{bmatrix}1&0&0\\0&2&0\\0&0&1\end{bmatrix}x(t)+\begin{bmatrix}b_1&0&0\\0&b_2&0\\0&0&b_3\end{bmatrix}u(t)\ ;\ \begin{matrix}x(t_0)=(6,12,8)\\T=30';\ \Delta=2'\\\sum_{i=1}^{3}x_i^2(T)=6\end{matrix}\ ;\ S(u)=\sum_{i=1}^{3}\left(\frac{T_i(T)-x_i(T)}{\Delta}\right)^2+\frac{1}{2}\int_0^T\sum_{i=1}^{3}(u_i(t)-x_i(t))^2dt$$

analytical solutions are:

$$u_i(t_i,b_i)=\frac{A_i+B_ie^{(2T+t)(G_i)}+E_ie^{(T-t)(G_i)}+F_ie^{(2T-t)(G_i)}}{C_i+D_ie^{2TG_i}}$$

where $A_i, B_i, C_i, \ldots, G_i$ - are functions of b_i.

In the example admisible strategies are :punctuality, acceleration, deceleration, overtaking. A simulation program for dispatching control gives /time simulation 200 (min)./ for first bus /No. 8/ delaying on the termini 1.8 min.,for second /No.9/ transfer to the reserve /an reserve bus is send into the route/ ,third bus no changes.

Conclusions : The proposed dispatching control should be verified on a real bus line. Some forms control can be realized without an ABL system by conventional means, for example control from terminals /by boundary conditions/ or by traffic lights.

LAYER	PERIOD	OBJECTIVE	OPTIMIZATION PROBLEM	COMMENT
1	cycle	Stabilization 1.↦ schedule 2.↦ demand	1.LQ problem /see next section/ 2. realized by information transport service system /open loop system stabilizing reaction passengers throug - hout the trip	Dispatching control Psychological aspects
2	period/day	Schedule creation Min.journey time and operational costs.	Optybus 1.POmin n $S(n)\sum_{ij} \frac{\bullet_{ij}}{\frac{m\,x_{ij}}{Y_{L_n}}+\sum_l n_l\left(\frac{x_{ij}^L}{Y_L}-\frac{x_{ij}}{Y_L}\right)}$ | demand $\leqslant$ supply for passenger and vehicles 2.POmin f $S(f)=f\cdot(T_r+T_l+c\sum_{j=1} \max[E(n_b)B, E(n_d)\|f)A])$ | demand $\leqslant$ supply for each bus and stop 3. POmin n $S(n)=c_1L_q+c_2L_s+c_3L_o+c_4n+c_5\lambda P_{str}$ | demand $\leqslant$ supply	Analitical user-oriented model based on Gamma distr. intervals assumption /3/ n - number of buses f - frequencies /bus/n/ Analitical operator-oriented model based on Bernoulli distr.dlighting assumption /Edwards model/ Simulation model: $/G/G_b/n/$ quene system
3	seasons/year	1.Bus network geometry /route and stops/ 2.Priority control public transport	1.POmin $U^x K \sum c_{ij} l_{ij} x_{ij}(U)+u_{ij}p_{ij}$ | $\sum x_{ij}(U)\leqslant d_{ij}$ ↦distribution,assignment,modal split problem solution for several of priority variants ↦Bus Transyt /TARL Rep.L253 D.I.robertson/	fixed cost route selection problem. Simulation models

TABLE 2

REFERENCES

1. Adamski,A., "Optimal control of the bus line" Proc. 2-nd conference of the Institute of Transport vol.1 pp. 1-9, Warszawa 1978
2. Adamski,A., Rudnicki,A., "Bus transport system as a control plant" Proc. VII Krajowej Konferencji Automatyki vol.1 pp.722-730 Rzeszów77
3. Adamski,A., Grega,W.,Werewka,J., "Automatic Traffic Control System for Kraków" Rep.77/1 TRACOS group Inst. of Comp. Sci. Aut. Cont. 1977
4. Barnett,A.,"On controlling randomness in transit operations" Trans.Sci, 8 / 2 / 102, 1974.
5. Bly,P.H., Jackson,R., "Evaluation of bus control strategies by simulation" Rep. LR 637 TRRL, Crowthorne.
6. Chapman,R.A., Mitchel,J.F., "Modeling Tendency of Buses to Form Pairs" Trans. Sci, 12 / 2 / 165-175, 1978
7. Chapman,R.A., Gault,H.E., Jenkins,I.A., "Factors affecting the operation of bus routes" Res. Rep. No.23 TORG Univ.of Newcastle 1976
8. Erlander,S., "Accessibility, Entropy and the Distribution and Assignment of Traffic" LiTH-MAT-R-76-1 Linkóping Inst. of Tech. 1976
9. Findeisen,W., "Multilevel Control Systems" /in Polish/ PWN, 1974
10. Finnamore,A.J., Bly,P.H., "Bus control systems, some evidence on their effectiveness" AFCET Control and Transportation Systems 1974
11. Jackson,R.L., "Evaluation by simulation of control strategies for a high frequency bus service" Rep. 807 TRRL, Crowthorne 1977.
12. Newell,G.F., Potts,R.B., "Maintaining a bus schedule" Proc.2-nd conf. Australian Road Research Board, 2 /1/ 388-393,1964
13. Newell,G.F., "Control of Pairing of Vehicles on Public Transportation Route, Two Vehicle, One Control Point"Trans. Sci. 8 /3/ 1974
14. Osuna,E.E., Newell,G.F., "Control Strategies for an Idealized Public Transportation System" Trans. Sci. 6 /1/, 52-72 1972.
15. Scheele,S., "A Mathematical Programming Algorithm For Optimal Bus Frequencies" Dissert. No 12 Linkóping University 1977
16. Rudnicki,A., " Classification of the control actions in Public Transport" Proc. 2-nd conf. of the Institute of Transport 1 166 1978
17. Scalia-Tomba,G.P., Anderson,P.A., "Statistical analysis of an urban route" Rep. LiTH-MAT-R-78-11 Linkóping Institute of Technology 1978
18. Edwards,J.L., Determining transit vehicle requirements on local city bus routes: a different approach" Traff. Eng. Cont. 1979.

A STRATEGIC APPROACH TO AIR TRAFFIC CONTROL

L. Bianco, M. Cini, L. Grippo
Centro di Studio dei Sistemi di Controllo e
Calcolo Automatici - Consiglio Nazionale del
le Ricerche - Via Eudossiana 18,00184 Roma
Italy

INTRODUCTION

Present Air Traffic Control (ATC) systems are essentially based on short term, safety oriented interventions, which do not take enough into account the cost of aircraft operations.

On the other hand, today, the impact of these costs on economy is such as to justify the introduction of a planning control function capable of improving the overall efficiency of the system. In recent years, several studies have been performed in this direction and the concept of "strategic control" has been introduced [1-7]. Since this term has been given in the literature different interpretations, we introduce first a multilevel model of the ATC functions, in which the strategic control actions are clearly defined.

The remaining part of the paper is devoted to the "strategic on-line control" of individual flights and, in particular, the case of speed control on preassigned routes is considered. For this case, a real time algorithm based on a combination of dynamic programming and branch and bound techniques is proposed and numerical results are reported.

2. MULTILEVEL MODEL OF AIR TRAFFIC CONTROL

The Air Traffic Control problem is a typical large-scale problem characterized by: high dimensions, strong interactions among the component subsystems, fast dynamics, different decision makers, multiple (and often conflicting) objectives, stochastic disturbances, etc. Therefore, a realistic approach to the ATC problem involves a decomposition of the system and/or its control into smaller parts, in order to simplify the final design and implementation. In particular, we consider here a decomposition in terms of different control actions.

A first distinction is that between *off-line* and *on-line control*. Off-line control is a planning activity, based only on a long-term forecast, and carried out before the aircraft enter the system. On-line control is an activity, based on the observation of the current system state and performed in a time interval short enough to allow repeated interventions on the observed traffic.

A second distinction concerns *flow-control* and *control of individual flights*. Finally, a further decomposition criterion is referred to the time horizon and the spatial extension of the control action, which can be splitted into *tactical* and *strategic control*. Tactical control is a real-time action satisfying short term requests and interesting a limited sector of the airspace. Strategic control is a medium or long-term planning activity, aimed at managing the traffic flows and/or the individual flights according to an overall traffic optimization plan for the entire controlled airspace.

All the preceding control functions are represented in the multilevel model shown in fig. 1, where each level is characterized by a time horizon and a specific objective. Further details on this model can be found in [7].

3. STRATEGIC CONTROL OF FLIGHTS: SPEED CONTROL ON PREASSIGNED ROUTES

In [7] we considered the on-line strategic control problem, as the problem of determining optimal flight plans in a given region on the basis of successive observations of the current system state. A discretized representation of the flight plans was employed, by defining each path as a series of waypoints, waypoints altitutes and times for passing each waypoint. Then, for each planning time interval, the control problem was formulated as a mathematical programming problem, in which the objective function measures the deviations from the nominal flight plans and the constraints take into account aircraft performance capabilities, structure of the airspace, safety separation standards and traffic limitations prescribed by the flow control function.

In the present paper we consider the particular case in which the altitude levels at the given waypoints are fixed at the nominal values and control is achieved by varying the flight times. Moreover, the flight plans of already treated and cleared aircraft are assumed as constraints and the last aircraft entered the system is provided with an optimal conflict-free path.

On the basis of these assumptions, the mathematical model can be formulated as follows.

Let $K = \{1,2,\dots,k,\dots,N\}$ be the ordered index set of the given waypoints for the aircraft under consideration and denote by t_k the transit time at the waypoint k. Then, the constraints on the aircraft performance capabilities can be expressed by

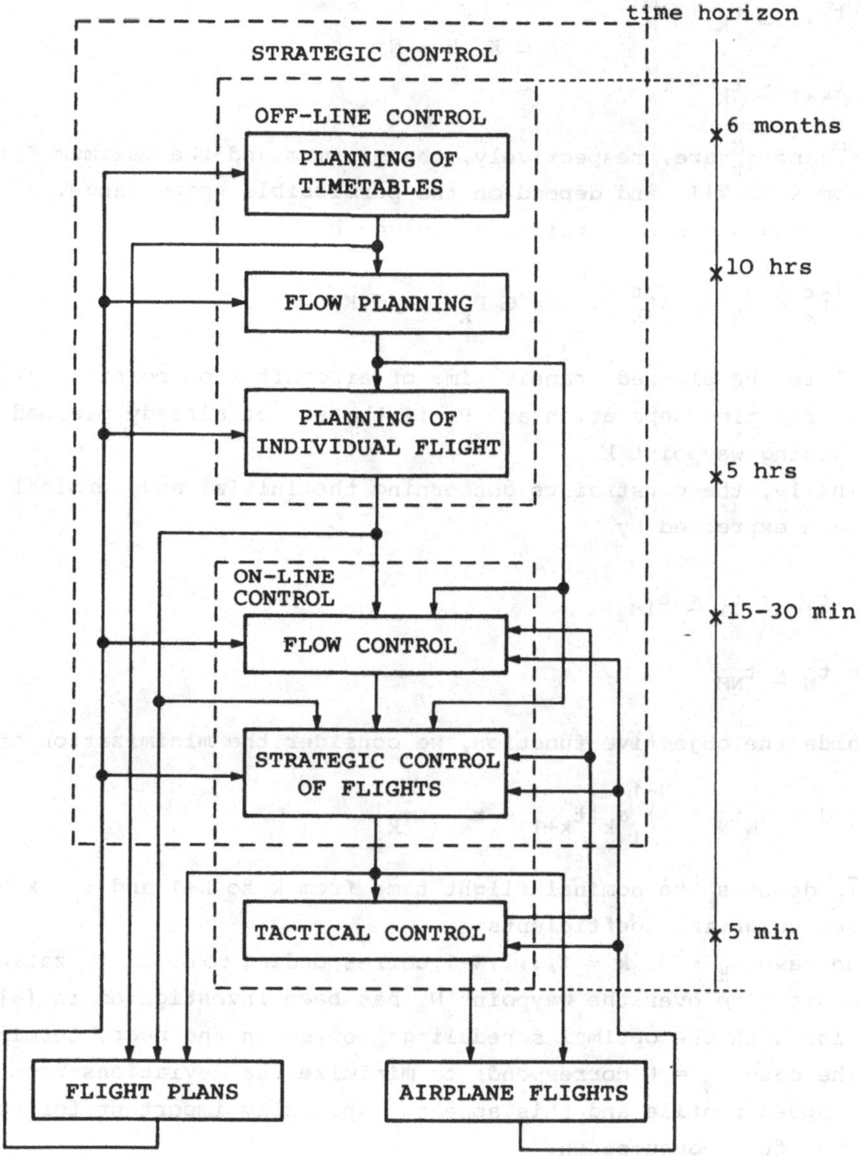

Fig.1- MULTILEVEL MODEL OF THE AIR TRAFFIC CONTROL.

$$t_{k+1} \geq t_k + \tau_k^m \qquad k \in K,\ k < N \tag{1}$$
$$t_{k+1} \leq t_k + \tau_k^M$$

where τ_k^m and τ_k^M are, respectively, the minimum and the maximum flight time from k to k+1 and depend on the permissible speed range.

The separation constraints are given by;

$$\left|\hat{t}_k^r - t_k\right| \geq \Delta t_k^r \quad , \quad r \in P_k \ , \ k \in K \tag{2}$$

where $\hat{t}_k^r$ is the planned transit time of aircraft r on point k, Δt_k^r is the required time separation and P_k is the set of already planned aircraft passing waypoint k.

Finally, the constraints concerning the initial and terminal waypoints are expressed by

$$t_{1m} \leq t_1 \leq t_{1M} \tag{3}$$

$$t_N \leq t_{NM} \tag{4}$$

As regards the objective function, we consider the minimization of

$$J = \alpha_N t_N + \sum_{k=1}^{N-1} \alpha_k \left| t_{k+1} - t_k - \bar{\tau}_k \right| \tag{5}$$

where $\bar{\tau}_k$ denotes the nominal flight time from k to k+1 and α_k, $k \in K$ are given weighting coefficients.

The case $\alpha_k = 0$, $k = 1,\dots,N-1$, corresponding to the minimization of the transit time over the waypoint N, has been investigated in [8] in connection with the optimal scheduling problem in the near terminal area. The case $\alpha_N = 0$ corresponds to minimize the deviations from the nominal speed profile and this appears especially important for reducing the fuel consumption.

The problem of minimizing (5) under the constraints (1)-(4) can be formulated as a mixed integer linear programming problem, the combinatorial aspect depending on the constraints (2).

4. SOLUTION ALGORITHM

Preliminarly, in order to simplify the description of the solution algorithm let us introduce the following definitions.

DEFINITION 1. The set of *permissible time windows* at waypoint k is the set of time intervals

$$W_k^\ell = \left[a_k^\ell\ ,\ b_k^\ell\right] \quad , \quad \ell = 1,\dots,n_k$$

with

$$a_k^{\ell+1} > b_k^\ell$$

such that any $t_k \in \bigcup_{\ell=1}^{n_k} W_k^\ell$ satisfies the constraints (2)-(4).

DEFINITION 2. For a given time interval $[c_k,d_k]$ at waypoint k, the *reflection* $R\{[c_k,d_k]\}$ of $[c_k,d_k]$ on the waypoint k-1 is the interval:

$$R\{[c_k,d_k]\} = \left[c_k - \tau_{k-1}^M\ ,\ d_k - \tau_{k-1}^m\right]$$

In particular, for $c_k = d_k = t_k$ we write $R\{t_k\} = [t_k - \tau_{k-1}^M, t_k - \tau_{k-1}^m]$.

DEFINITION 3. A sequence of time intervals U_k, $k = j,\dots,N$ is said *admissible* between j and N if

$$U_k \subseteq R\{U_{k+1}\} \quad , \quad k = j,\dots,N-1$$

and, $\forall k \in \{j,\dots,N\}$, $\exists \ell \in \{1,\dots,n_k\}$ such that $U_k \subseteq W_k^\ell$.

It can be remarked that, employing Definitions 1 and 2, the constraints (1)-(4) can be rewritten as

$$t_k \in R\{t_{k+1}\} \quad , \quad k = 1,\dots,N-1$$

$$t_k \in \bigcup_{\ell=1}^{n_k} W_k^\ell \quad , \quad k = 1,\dots,N$$

Note further that, if $\{U_k\}$ is an admissible sequence between 1 and N, there exists at least one speed profile on the preassigned trajectory satisfying all the problem constraints.

The solution algorithm proposed here utilizes a hybrid approach combining the dynamic programming and the branch and bound techniques. Dynamic programming allows to transform the mixed problem into a discrete problem, which is then solved by an enumerative procedure. A relevant feature of the method is that dynamic programming is employed without discretizing the continuous variables t_k, and this permits to obtain the exact optimal solution.

The algorithm is based on the generation of a set of trees, which associate with each waypoint k a set of nodes $\{U_k^i\}$ at level k, where

$U_k^i = [c_k^i, d_k^i]$ represents a suitable time interval contained in $\bigcup_{\ell=1}^{n_k} W_k^\ell$.

A path on a tree, from level k to level N, will be a sequence of admissible intervals in the sense of Definition 3. Moreover, it will be shown that with each node can be associated a function of the form:

$$f_k^i(t_k) = \gamma_k^i |t_k - u_k^i| + \delta_k^i \tag{6}$$

such that, $\forall t_k \in U_k^i$, $f_k^i(t_k)$ is the minimum cost to reach level N, starting from level k at time t_k and following the corresponding path on the tree.

At level N, which corresponds to waypoint N, we set:

$$U_N^i = W_N^i \, , i = 1,\ldots,n_N \tag{7}$$

and

$$\gamma_N^i = \alpha_N \, , \; u_N^i = 0 \, , \; \delta_N^i = 0 \, , \tag{8}$$

so that

$$f_N^i(t_N) = \alpha_N t_N \, . \tag{9}$$

In order to clarify how the algorithm operates, we prove first the following propositions.

PROPOSITION 1. Let $U_{k+1}^i = [c_{k+1}^i, d_{k+1}^i]$ be a time interval at level k+1, $k \leq N-1$, with the associated function $f_{k+1}^i(t_{k+1})$ of the form (6), then, for a given $t_k \in R\{U_{k+1}^i\}$:

$$\begin{aligned}\min_{\substack{t_{k+1} \in U_{k+1}^i \\ t_k+\tau_k^m \leq t_{k+1} \leq t_k+\tau_k^M}} & \left\{\gamma_{k+1}^i |t_{k+1}-u_{k+1}^i| + \delta_{k+1}^i + \alpha_k |t_{k+1}-t_k-\bar{\tau}_k|\right\} = \\ & = \gamma_{k+1}^i |\hat{t}_{k+1}-u_{k+1}^i| + \delta_{k+1}^i + \alpha_k |\hat{t}_{k+1}-t_k-\bar{\tau}_k|\end{aligned} \tag{10}$$

where, $\hat{t}_{k+1}$ is the solution of

$$\min_{\substack{t_{k+1} \in U_{k+1}^i \\ t_k+\tau_k^m \leq t_{k+1} \leq t_k+\tau_k^M}} |t_{k+1}-t_k-\bar{\tau}_k| \qquad \text{for} \quad \gamma_{k+1}^i \leq \alpha_k \tag{11}$$

or the solution of

$$\min_{\substack{t_{k+1} \in U_{k+1}^i \\ t_k+\tau_k^m \leq t_{k+1} \leq t_k+\tau_k^M}} |t_{k+1} - u_{k+1}^i| \qquad \text{for} \quad \gamma_{k+1}^i \geq \alpha_k \tag{12}$$

PROOF. The proof follows immediately from the form of the cost function (10).

PROPOSITION 2. Let $U^i_{k+1} = [c^i_{k+1}, d^i_{k+1}]$ be a time interval at level k+1, $k \leq N-1$, with the associated cost function $f^i_{k+1}(t_{k+1})$, then:

i) there exist, at level k, a partition of $R\{U^i_{k+1}\}$ into sub-intervals S^{ih}_k, $h = 1,\ldots,p$, and functions

$$f^{ih}_k(t_k) = \gamma^{ih}_k |t_k - u^{ih}_k| + \delta^{ih}_k \tag{13}$$

such that, $\forall t_k \in S^{ih}_k$:

$$f^{ih}_k(t_k) = \min_{\substack{t_{k+1} \in U^i_{k+1} \\ t_k+\tau^m_k \leq t_{k+1} \leq t_k+\tau^M_k}} \{f^i_{k+1}(t_{k+1}) + \alpha_k |t_{k+1} - t_k - \bar{\tau}_k|\} \tag{14}$$

ii) The subintervals S^{ih}_k, the coefficients γ^{ih}_k, u^{ih}_k, δ^{ih}_k and the solution $\hat{t}_{k+1}$ of (14) are those reported in Table 1.

PROOF. We prove the proposition for the case $\gamma^i_{k+1} \leq \alpha_k$ (case A of Table 1).

Let $t_k \in R\{U^i_{k+1}\}$, then by Prop. 1 we have that the solution $\hat{t}_{k+1}$ of (14) is the solution of (11). Now, it can be easily verified that:

for $t_k \in S^{i1}_k \triangleq [c^i_{k+1} - \tau^M_k, c^i_{k+1} - \bar{\tau}_k]$, $\hat{t}_{k+1} = c^i_{k+1}$;

for $t_k \in S^{i2}_k \triangleq [c^i_{k+1} - \bar{\tau}_k, d^i_{k+1} - \bar{\tau}_k]$, $\hat{t}_{k+1} = t_k + \bar{\tau}_k$;

for $t_k \in S^{i3}_k \triangleq [d^i_{k+1} - \bar{\tau}_k, d^i_{k+1} - \tau^m_k]$, $\hat{t}_{k+1} = d^i_{k+1}$.

By substituting the value of $\hat{t}_{k+1}$ into (10), we can define the functions $f^{ih}_k(t_k)$, h = 1,2,3 of the form (13), where:

$$\gamma^{i1}_k = \alpha_k, \quad u^{i1}_k = c^i_{k+1} - \bar{\tau}_k, \quad \delta^{i1}_k = \delta^i_{k+1} + \gamma^i_{k+1} |c^i_{k+1} - u^i_{k+1}|$$

$$\gamma^{i2}_k = \gamma^i_{k+1}, \quad u^{i2}_k = u^i_{k+1} - \bar{\tau}_k, \quad \delta^{i2}_k = \delta^i_{k+1}$$

$$\gamma^{i3}_k = \alpha_k, \quad u^{i3}_k = d^i_{k+1} - \bar{\tau}_k, \quad \delta^{i3}_k = \delta^i_{k+1} + \gamma^i_{k+1} |d^i_{k+1} - u^i_{k+1}|.$$

A similar proof can be given for the case $\gamma^i_{k+1} \geq \alpha_k$ (case B of Table 1). ■

Now given node U^i_{k+1} at level k+1, the set of nodes at level k connected with U^i_{k+1} is the set

$$\left\{U^j_k : U^j_k = S^{ih}_k \cap W^\ell_k, \; U^j_k \neq \emptyset, \; h = 1,\ldots,p, \; \ell = 1,\ldots,n_k\right\}$$

where S^{ih}_k, $h = 1,\ldots,p$ is the appropriate partition of $R\{U^i_{k+1}\}$ introduced in Prop. 2. With each node U^j_k it is associated the cost function $f^j_k(t_k) = f^{ih}_k(t_k)$ defined in Prop. 2.

By employing repeadly this procedure it is possible to construct a set of trees, whose root nodes are the permissible time windows at level N. An example is shown in fig. 2.

Case	Subcase	Partition of $\mathcal{R}\{U^i_{k+1}\}$	$\hat{t}_{k+1}$	γ^{ih}_k	u^{ih}_k	δ^{ih}_k
Case A: $\gamma^i_{k+1} \le \alpha_k$		$S^{i1}_k=[c^i_{k+1}-\tau^M_k, c^i_{k+1}-\bar{\tau}_k]$	c^i_{k+1}	α_k	$c^i_{k+1}-\bar{\tau}_k$	$\delta^i_{k+1}+\gamma^i_{k+1}\lvert c^i_{k+1}-u^i_{k+1}\rvert$
		$S^{i2}_k=[c^i_{k+1}-\bar{\tau}_k, d^i_{k+1}-\bar{\tau}_k]$	$t_k+\bar{\tau}_k$	γ^i_{k+1}	$u^i_{k+1}-\bar{\tau}_k$	δ^i_{k+1}
		$S^{i3}_k=[d^i_{k+1}-\bar{\tau}_k, d^i_{k+1}-\tau^m_k]$	d^i_{k+1}	α_k	$d^i_{k+1}-\bar{\tau}_k$	$\delta^i_{k+1}+\gamma^i_{k+1}\lvert d^i_{k+1}-u^i_{k+1}\rvert$
Case B: $\gamma^i_{k+1} \ge \alpha_k$	B1: $c^i_{k+1} \ge u^i_{k+1}$	$S^{i1}_k=[c^i_{k+1}-\tau^M_k, c^i_{k+1}-\tau^m_k]$	c^i_{k+1}	α_k	$c^i_{k+1}-\bar{\tau}_k$	$\delta^i_{k+1}+\gamma^i_{k+1}\lvert c^i_{k+1}-u^i_{k+1}\rvert$
		$S^{i2}_k=[c^i_{k+1}-\tau^m_k, d^i_{k+1}-\tau^m_k]$	$t_k+\tau^m_k$	γ^i_{k+1}	$u^i_{k+1}-\tau^m_k$	$\delta^i_{k+1}+\alpha_k\lvert\tau^m_k-\bar{\tau}_k\rvert$
	B2: $d^i_{k+1} \le u^i_{k+1}$	$S^{i1}_k=[c^i_{k+1}-\tau^M_k, d^i_{k+1}-\tau^M_k]$	$t_k+\tau^M_k$	γ^i_{k+1}	$u^i_{k+1}-\tau^M_k$	$\delta^i_{k+1}+\alpha_k\lvert\tau^M_k-\bar{\tau}_k\rvert$
		$S^{i2}_k=[d^i_{k+1}-\tau^M_k, d^i_{k+1}-\tau^m_k]$	d^i_{k+1}	α_k	$d^i_{k+1}-\bar{\tau}_k$	$\delta^i_{k+1}+\gamma^i_{k+1}\lvert d^i_{k+1}-u^i_{k+1}\rvert$
	B3: $c^i_{k+1} \le u^i_{k+1} \le d^i_{k+1}$	$S^{i1}_k=[c^i_{k+1}-\tau^M_k, u^i_{k+1}-\tau^M_k]$	$t_k+\tau^M_k$	γ^i_{k+1}	$u^i_{k+1}-\tau^M_k$	$\delta^i_{k+1}+\alpha_k\lvert\tau^M_k-\bar{\tau}_k\rvert$
		$S^{i2}_k=[u^i_{k+1}-\tau^M_k, u^i_{k+1}-\tau^m_k]$	u^i_{k+1}	α_k	$u^i_{k+1}-\bar{\tau}_k$	δ^i_{k+1}
		$S^{i3}_k=[u^i_{k+1}-\tau^m_k, d^i_{k+1}-\tau^m_k]$	$t_k+\tau^m_k$	γ^i_{k+1}	$u^i_{k+1}-\tau^m_k$	$\delta^i_{k+1}+\alpha_k\lvert\tau^m_k-\bar{\tau}_k\rvert$

TABLE 1.

The optimal solution of problem (1)-(5) can be obtained, in principle, by generating the complete set of trees and performing a forward iteration from level 1. In fact, optimal value $t^*_1 \in U^{i^*}_1$ at level 1 is defined by

$$f^{i^*}_1(t^*_1) = \min_{i=1,\dots m_1} \left[\min_{t_1 \in U^i_1} f^i_1(t_1) \right]$$

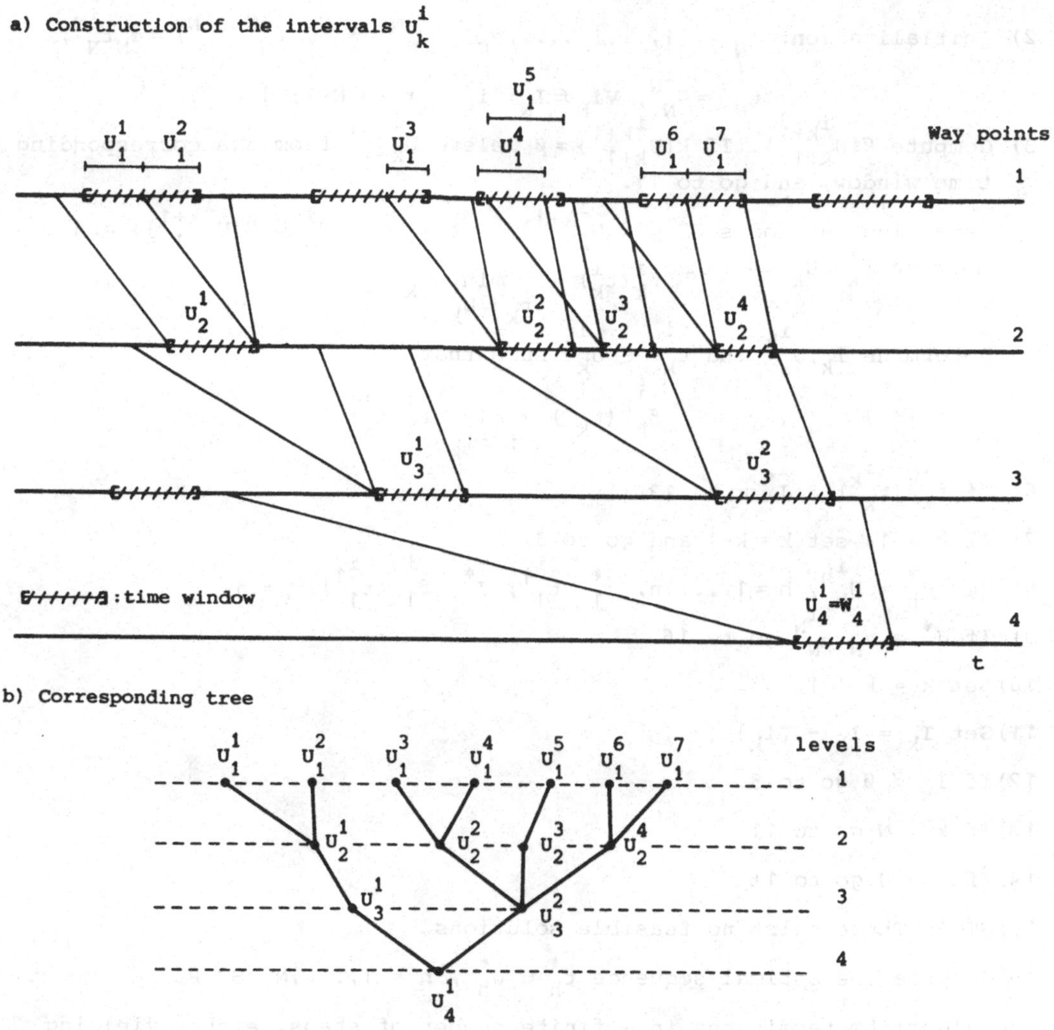

Fig.2 - GENERATION OF A TREE OF ADMISSIBLE INTERVALS.

where U_1^i , $i = 1,\ldots,m_1$ represent the nodes at level 1. Then, the optimal value t^*_{k+1} is obtained from t^*_k by employing the expressions of $\hat{t}_{k+1}$ reported in Table 1.

However, in order to avoid enumeration of all nodes, we propose the branch and bound algorithm described below.

1) Input data: number of levels N; time windows $W_k^{\ell_k}$, weighting coefficients α_k and parameters τ_k^m, $\bar{\tau}_k$, τ_k^M, $\forall k \in \{1,2,\ldots,N\}$; upper bound of the optimal cost J^*.

2) Initialization: $I_N = \{1,\ldots,i_N,\ldots,n_N\}$; $U_N^{i_N} = W_N^{\ell_N}$, $f_N^{i_N}(t_N^{i_N}) = \alpha_N t_N^{i_N}$,

$$t_N^{i_N} = c_N^{i_N},\ \forall i_N \in I_N;\ i_N = 1;\ k = N-1;\ j = 0.$$

3) Compute $R\{U_{k+1}^{i_{k+1}}\}$. If $R\{U_{k+1}^{i_{k+1}}\} = \emptyset$ delete $U_{k+1}^{i_{k+1}}$ from the corresponding time window, and go to 11.

4) Determine the nodes $U_k^i \subseteq R\{U_{k+1}^{i_{k+1}}\}$, set $I_k = \{i : U_k^i \subseteq R\{U_{k+1}^{i_{k+1}}\}\}$ and compute $t_k^i \in U_k^i$ such that $f_k^i(t_k^i) = \min_{t_k \in U_k^i} f_k^i(t_k)$

5) Determine $i_k, U_k^{i_k}$ and $t_k^{i_k} \in U_k^{i_k}$ such that

$$f_k^{i_k}(t_k^{i_k}) = \min_{i \in I_k} f_k^i(t_k^i)$$

6) If $f_k^{i_k}(t_k^{i_k}) \geq J^*$ go to 13.

7) If $k > 1$, set $k = k-1$ and go to 3.

8) Set $U_h^* = U_h^{i_h}$, $h = 1,\ldots,n$, $t_1^* = t_1^{i_1}$, $J^* = f_1^{i_1}(t_1^{i_1})$, $j = j+1$

9) If $J^* = \alpha_N c_N^{i_N}$ go to 16.

10) Set $k = k + 1$

11) Set $I_k = I_k - \{i_k\}$

12) If $I_k \neq \emptyset$ go to 5.

13) If $k < N$ go to 11

14) If $j > 0$ go to 16

15) STOP. There exist no feasible solutions

16) Compute the optimal sequence $t_h^* \in U_h^*$, $h = 1,\ldots,N$. STOP.

The algorithm terminates in a finite number of steps, either yielding the optimal solution, or indicating that the admissible set is empty. An admissible solution is produced whenever step 8 is reached.

5. COMPUTATIONAL RESULTS AND CONCLUSIONS

The algorithm has been implemented on a computer Univac 1110/22 and different series of test problems were performed in correspondence of 5,10 and 15 levels.

For each case the parameters $\bar{\tau}_k, \tau_k^M, \tau_k^m, \alpha_k, a_k^\ell, b_k^\ell$ were randomply generated, assuming a uniform distribution of their values on pre-established intervals.

The results obtained are summarized in table 2.

Number of levels	5	10	15
first admissible solution average computation time (sec.)	10^{-2}	6.8×10^{-2}	1.3
average global computation time (sec.)	2.5×10^{-2}	1.8	64

TABLE 2.

It is clear that the algorithm can be surely employed in real time when the number of waypoints is less than 15. When the waypoints are 15 the computation time is still acceptable but the behaviour of the algorithm begins to be critical.

However, it must be remarked that the test problems worked out correspond to particular congestion situations that rarely occur in the real ATC operation environment.

In conclusion,on the basis of the results obtained, the proposed approach seems to be satisfactory for solving the speed control problem on preassigned routes.

Thus, the extension of this algorithm to the general "strategic on-line control of flights" problem will be the subject of a future research.

REFERENCES

[1] A.BENOIT et al.:*Study of Automatic Conflict Detection and Resolution in Air-Traffic Control Planning*. The 7th ICAS Congress, Rome, 14-18 Sept. 1970.

[2] *Fourth Generation Air Traffic Control Study*, Trasp. Systems Center, Cambridge, USA, June 1972.

[3] S.RATCLIFFE, H.GENT: *The Quantitative Description of a Traffic Control Process*, Journal of Navigation, Vol. 27, No. 3, July 1974.

[4] R.L.ERWIN et al.: *Strategic control Algorithm Development*. Boeing Commercial Airplane Company, Aug. 1974.

[5] A.BENOIT et al.: *The Introduction of Accurate Aircraft Trajectory Predictions in Air Traffic Control*. Plans and Developments for Air Traffic Systems. AGARD-CP-188, Feb. 1976.

[6] A.BENOIT et al.: *An evolutionary application of advanced flight path prediction capability to ATC*. Proc. Intern. Conference on Electronic Systems and Navigation Aids.,Paris,Nov.14-18, 1977.

[7] L.BIANCO,M.CINI,C.GRIPPO:*Pianificazione e Controllo Strategico del Traffico Aereo*.Rapporto CSSCCA,R.78-28, December 1978.

[8] L.TOBIAS: *Automated Aircraft Scheduling Methods in the Near Terminal Area*. J. Aircraft, vol. 9, n. 8, August 1972.

EDP PROJECT AND COMPUTER EQUIPMENT SELECTION BY THE USE OF LINEAR PROGRAMMING

George E. Haramis
Olympic Airways / EDP Dept.
Athens Airport - West Terminal
Athens, Greece

Abstract

Criteria for selecting Management Information Systems for implementation by the EDP Center are discussed with particular emphasis on the operational, technical and economical aspects. Integer Linear Programming /net present value method/ and General Linear Programming methods are presented in order to optimize project and computer equipment selection respectively.

1. EDP Project Selection

The problem of selecting Management Information Systems /MIS/ which will be implemented, is in general a problem of financial selection, and as such it can be classified in one of the two general categories of selection problems.

Firstly, in the category of simple consequence problems which have solutions that are simply acceptable. Secondly, in the category of optimization problems with acceptable solutions, but which at the same time permit the finding of the optimum solution among the solutions, i.e., the solution which maximizes the output using a particular method in every case.

Furthermore, in order that the solution to the problem is effective, it must be distinguished for its completeness and at the same time unquestionably accepted by those who will implement it.

The literature on the subject of project selection per se is very rich; on the contrary it is relatively poor on the subject of EDP project selection. Limiting ourselves to the latter, the most important criteria for their selection refer to matters which deal with the operation of the Electronic Data Processing /EDP/ Center; these criteria are:

- The cost of developing the new system, i.e., the costs of system's analysis, design and implementation.

- The incremental cost of running the EDP Center due to the operation of the new system.
- The available computer time, i.e., the time which has not been allocated to the operation of other systems.
- The time-period within a 24-hours day, week or month, during which the computer is available for the execution of the new system.
- The capital return time for the capital which will be invested for system´s development.

Certainly apart from these criteria, the safest and most economical one for the selection of the systems which will be computerized is the amount of money the Corporation saves becouse of the operation of a new system.

System selection in general is essentially based on decision making methods; although system selection is considered as relatively simple, in the case of data processing it becomes exceedingly complicated becouse of the nature of the data processing environment. In such an environment, the corporation´s executives /users/ who cooperate with the EDP Center seek, each one for himself not only one but more than one systems or services, while showing no interest for other departments´ needs. This not good image is completed by the lack of understanding and recognition of the EDP Center personnel´s efforts by the coroporate executives.

2. Long-range Planning and M I S Development

The problem of selecting the systems which are going to be implemented by the EDP Center within a specified period of time, coincides with the long-range corporate planning for organizing and operating the corporation´s information basc.

This planning is completed through the realization of a master plan, which must take into account:

a. The corporate policy /goals, objectives, etc/.

b. The research for locating the systems which are going to be implemented.

c. The study for determining whether the development of the new system is possible /feasibility study/

Corporate policy is expressed through:

- The coroporate objectives which, for the EDP Center in particular, refer to the development of information organization.
- The strategy which must be followed by the corporation. As far as data processing is concerned, this strategy refers to the development of

systems for decision making.
- Assessment and evaluation methods which, especially for systems analysis and design, are not limited only to matters of quantity, quality, precision, security and time-charts, but also determine modifications within the system, aimed towards improwing decisions effectiveness.

The research for locating these systems, which are going to implemented was aimed towards office organization systems during the last decade; however, now it searches for information systems which shall provide data for decision making; this research is based on:
- Identifying the most significant regions in the corporation, where decision making take place.
- Determining in which of these regions always exist difficulties without satisfactory solutions.
- Obtaining and processing the related information, and determining the corresponding essential decisions that need to be taken in order to solve these problems.

The study of whether the development of a /new/ system is possible, is being effected from the operational, the technical and the economical viewpoint.
The study of the possible outcome that the new system will have in relation to the eventual difficulties in its operation /operational feasibility study/, refers to examining whether system operation fulfills the objectives of its implementation.
In relation to the technical problems involved /technical feasibility study/, the study refers to a comparison between the presently available technical capabilities /equipment and personnel qualifications/, and these which are necessary for the system's implementation.
Finally, the economic problems /economic feasibility study/ are difficult to determine, since they refer to many factors which cannot be evaluated such as system's life-duration, and which lie beyond assesments on reducing rosts or increasing profits.

3. Integer Linear Programming and Investment Decision for EDP Systems Development

The final selection of the systems which are going to be implemented among the systems which were provisionally selected /as it has been previously stated/, is based on the net present value method. The problem is an allocation problem on a specified sum from the data processing center's budget among n "competitive" investments to the implementation of systems.

It is also a problem of integer linear programming, since the system to be implemented either shall be realized in its whole, or its realization will be postponed.

If we define as:

X_j the variables which indicate the systems to be implemented, and which must be integers either 0 /postponement of realization/ or 1 /realization/

$$0 \leqslant X_j \leqslant 1$$

V_j the net present value of the system j, for: j=1,2,3,...,n where n= the number of the systems to be implemented, and

T_i i=1,2,3,...,m where m= the number of relative time-periods,

K_i the present value of the available capital during time-period i. /K_m shall be the largest sum which can be allocated by the data processing center during the year m/.

K_{ij} the present value of the capital needed during time-period i for the system j,

then:

the problem is to find the n positive integer or zero values of the variables X_1, X_2, X_3,...,X_n, which satisfy the equations:

$$K_{11}X_1 + K_{12}X_2 + K_{13}X_3 + \dots + K_{1v}X_v \leqslant K_1$$
$$K_{21}X_1 + K_{22}X_2 + K_{23}X_3 + \dots + K_{2v}X_v \leqslant K_2$$
$$K_{31}X_1 + K_{32}X_2 + K_{33}X_3 + \dots + K_{3v}X_v \leqslant K_3$$
$$\vdots$$
$$K_{m1}X_1 + K_{m2}X_2 + K_{m3}X_3 + \dots + K_{mv}X_v \leqslant K_m$$

and which maximize the economic or objective function

$$Z = V_1X_1 + V_2X_2 + V_3X_3 + \dots V_nX_n$$

4. Computer Equipment Selection

It is useless to elaborate here on the resuts of an incorrect selection of the data processing center equipment. At the process of determining the electronic computer system which shall satisfy the corporation's needs, it is very important to identify the characteristics of the Central Processing Unit and the peripherals.

In addition, special attention must be given to the following points:

- Instruction execution speed by the central processing unit.
- The possibility of extending the computer system's capabilitics, that is to increase the memory capacity of the central processing unit, or to increase the number of magnetic disk units, magnetic tapes or to increase their speed, etc.
- The assurance for a trouble-free operation of the computer, the probability of mechanical failures and the time required for their repair.
- The possibility of carrying-out the work in another nearby data processing center, in case the computer is down.
- The possibility of carrying-out the work by a larger computer of the same family, in case the data processing center must expand.
- The ability to apply multi-programming, tele-processing, multi-processing techniques.
- The number or ready programs and applications offered by the computer manufacturer.
- The number of computer programming languages and their capabilities.
- The cost of buying or renting and operating the computer along with the cost of buying or renting the hardware and software; the cost of applications design, the cost of application programs implementation, the cost of training analysts-programers-operators, the cost of installing the computer, the cost of air-conditioning equipment, the rent of machine room, etc.

In addition to these, it is necessary to investigate the following:

- Optimizing the operating system limitations with respect to the main sections of the computer memory.
- Determining the optimum combination between the number and speed of computer peripherals.
- The optimum combination of channels and the units connected through them should be calculated, their optimum utilization should be defined.

5. Linear Programming and Investment Decisions for D.P. Equipment Selection.

In general, the steps for data processing equipment selection are:

- Determining the computer configuration.
- Estimating the computer times for instruction execution.
- Defining selection criteria.
- Setting specifications for the manufacturers.

- Evaluating offers and choosing the supplier.
- Negociating the contract's terms.

Always the problem of corporate investment on technical equipment has been important. To alleviate this problem, several related methods have been developed, most of which are based on the net present value of the "suitable" equipment.
In the case of computers, the sense of the word "suitable" was given in the previous chapter.
Having these considerations in mind regarding the net present value method and denoting with X_j the variables which determine "suitable" data processing equipment, we are trying to find the n real positive or zero values $X_1, X_2, X_3, \ldots\ldots X_n$ in order to select the optimum data processing equipment which is also economical.
Hence, the problem appears to be a General Linear Programming Problem formulated as follows:

Maximization $Z = \sum_{j=1}^{j=n} V_j X_j$ subject to the constraints:

$$\sum_{j=1}^{j=n} K_{ij} X_j \leq K_i$$

for $i = 1,2,3,\ldots\ldots, m$

$j = 1,2,3,\ldots\ldots, n$

and $X_1 \geq 0$, $X_2 \geq 0$, $X_3 \geq 0, \ldots\ldots\ldots, X_n \geq 0$

where :

m = the number of relative time periods,

n = the number of kinds of data processing equipment,

K_i= the present value of the capital available during time-period i,

K_{ij}= the present value of the capital required during time-period i, for equipment j.

V_j= net present value of the equipment j /computer system or part of a computer/.

REFERENCES

1. M. Blackman, The Design of Real Time Applications, John Wiley and Sons Ltd, 1975.
2. Control Data Institute, Project Management, 1975.
3. Due-Clower, Intermediate Economic Analysis, R. D. Irwin, Inc., 5th ed., 1968.
4. G. Haramis, A Contribution to Systems Analysis and Design Methodology, Ed. Hellenic Organization of Systems Sciences, 1974.
5. W. Greenwood, Decision Theory and Information Systems, South Western Publishing Co, 1969.
6. A. Lazaris, Economic Analysis-Economic Programming, 1965.
7. C. Mao, Quantitative Analysis of Financial Decisions, The Mcmillan Co, 1969.
8. P. Steriotis, Linear Programming, Athens School of Economics /Business Administration Institute/.
9. Zimmerman-Sovereign, Quantitative Models for Production Management, Prentice Hall, Inc., 1974.

IMPACT OF FINANCING ON OPTIMAL R & D RESOURCE ALLOCATION

S. H. Hung*, J. C. Hung+, and L. P. Anderson#
The University of Tennessee
Knoxville, Tennessee 37916
U. S. A.

I. INTRODUCTION

The growing volatility of the state of society, technology, energy sources, and market conditions, coupled with the competitiveness of the business world, accelerates the growing importance of research and development (R & D) for business, industry and government as well as educational institutions. To cope with new problems, to meet new demands, to prosper, and, most important, to survive, innovations are needed which can only be achieved through effective R & D.

In the late summer of 1979, a question was raised: should the U.S. Government provide the funding necessary to keep the financially troubled Chrysler Corporation alive as a U.S. and, in fact, worldwide manufacturing firm. It appeared that Chrysler was the victim of fast changing market conditions. The crisis could have been averted had the company maintained effective R & D programs including economic forecasting, prediction of gasoline supply, exploration for the use of alternative energy sources, study of market changes, and development of more fuel-efficient engines. An effective R & D program not only results in marketable new products but also provides management with information for sound decision-making.

On the national scene, the present high inflation rate is at least partially caused by increased oil prices and growing shortages. The inflation rate could have been kept low had the U.S. Government better directed the national R & D effort during the past decade. It is inconceivable that a nation which can put men on the moon cannot develop the use of alternative energy sources in time.

It is our view that, for any organization, R & D efforts may not always assure success, but lack of the same will almost assure failure. It is easy to stress the importance of R & D for an organization but this leads to two decision-making problems for management:

Problem 1. The selection of R & D projects and determination of their funding levels.

---

* Graduate Student, Department of Finance
+ Professor, Department of Electrical Engineering
# Professor, Department of Finance

Problem 2. The determination of the optimal level of financing for the organization's R & D.

There are always more potential R & D projects that an organization can afford to undertake. Because of the many factors involved, some of them quite intangible, project selection is a complex decision-making process. A method for selecting projects and determining funding levels has been proposed by Greenblott and Hung.[1] This method will be incorporated with new developments in this paper.

An organization's R & D is made possible by its financial investment. R & D expenditures, like other costs, reduce the net income of the organization. Consequently, there is a tendency to minimize such expenditures if the process is viewed myopically. Because of the uncertainty of the expected returns, the times lags between expenditures and returns, the activities of competitors, and life cycles of new products, the decision for R & D expenditures is especially difficult to make. Furthermore, recognizing the cost of money and its varying nature in the market, an R & D expenditure which is feasible in one year may not be so in another. While insufficient R & D is detrimental for an organization, excessive R & D may result in insolvency. This leads to Problem 2, namely, the determination of the optimal level of financing for the organization's R & D. A viable solution to this problem, which is not yet available, will be the main concern of this paper.

The determination of the optimal R & D financing level requires the consideration of financial aspects of the situation. Among them are the estimated R & D return in monetary units, money market conditions, the firm's desired rate of return, and the firm's value to investors. A prerequisite to the estimation of the R & D return is the knowledge of selected projects, which is the result of the project selection process in Problem 1. Methods for project evaluation strictly from the monetary point of view have been proposed, and have been used in the financial world for those projects whose monetary returns can be estimated.[2] These methods will serve as guidelines for the present discussion. To find a solution for Problem 2, a bridge needs to be built between these methods and project goals. This is due to the fact that some of the project goals are not monetary.

In the following section, a resource allocation method will be reviewed concisely, followed by a summary of various methods of investment evaluation. Then, R & D as an investment will be discussed. Based on these discussions, two methods for the determination of optimal financing levels for R & D will be proposed.

II. A RESOURCE ALLOCATION METHOD

R & D resource allocation consists of both project selection and funding level determination. Greenblott and Hung have proposed an analytical method for R & D resource allocation. [1] Their method does not require knowledge of the monetary return of each project, which is the most difficult data to obtain. Instead, a utility function is used as a quantitative measure for all projects. This amounts to a measure of the relative importance of each project. The utility function approach has the advantage of being able to take non-monetary factors into consideration. The method is useful in three ways. First, it establishes a priority ranking for all proposed projects according to their relative importance. Second, for a fixed total R & D budget, the method selects projects and assigns their individual funding levels for the maximum total utility. Third, the method can supplement the conventional "intuitive decision making" by highlighting important factors.

In this method, a project, say project k, has a utility function composed of five factors

$$U_k(l_k) = v_k\, p_k(l_k)\, m_k\, (l_k)\, f_k\, (l_k)\, q_k\, (l_k) \tag{1}$$

where v_k is the impact value, p_k is the probability of success, m_k is the manpower factor, f_k is the facility factor, q_k is the timeliness factor, and l_k is the funding level. The five factors are elaborated below.

<u>Impact Value to the Company</u>. The impact value of a proposed project is evaluated with the following three considerations: (1), Impact of the proposed project on the organization's mission and company goal; (2), Technical readiness contributed by the project, and (3), Nontechnical objectives of the project.

<u>Probability of Success</u>. It is convenient and useful to define success probability in terms of the estimated level of progress of the R & D activity and it is assumed that the probability of success is a function of the funding level.

<u>Manpower Factor</u>. This factor indicates the availability of manpower for the proposed project. It reflects the degree of impact on the project due to the manpower problems such as head count and competence. The manpower factor also depends on the funding level.

<u>Facility Factor</u>. Similar to the manpower factor, the facility factor reflects the degree of the impact on the project due to space and capital equipment limitation. This factor also depends on the funding level.

<u>Timeliness Factor</u>. This factor estimates the market's acceptability of the proposed product at any time and is a function of the funding level since timeliness depends on the probable completion date, which normally depends on the funding level. In general, an accurate timeliness-versus-time curve is usually not possible; therefore, a reasonable, simplified approximation based on analytical and intuitive estimation should be used.

For n proposed projects the total R & D utility is given by

$$U = \sum_{k=1}^{n} U_k (l_k) \tag{2}$$

Optimal resource allocation is achieved by adjusting l_k for maximum U with the constraint

$$\sum_{k=1}^{n} l_k = L \tag{3}$$

where L is the fixed budget level.

III. INVESTMENT EVALUATION METHODS

In the business world, there are five commonly used methods of investment evaluation: payback, average rate of return, net present value, internal rate of return, and profitability index.[2,3] These all serve a common purpose in measuring the monetary return on investment.

Payback Period. This method evaluates the payback period T and compares it with some predetermined standard. It indirectly measures the return on investment through measurement of time. The procedure does not consider the time value of money nor the income generated after the payback period.

Average Rate of Return. The average rate of return $\bar{r}$ is defined as the ratio of the average annual after-tax profit P to the average investment in that period. Assuming straight line depreciation the average investment is one half the total investment I_T. Thus

$$\bar{r} = 2P/I_T \tag{4}$$

Like the payback period method, this method also does not consider the time value of money. In fact, the payback period is related to the average rate of return via:

$$T = 2/\bar{r} \tag{5}$$

Net Present Value (NPV). The net present value, N, is defined as

$$N = \sum_{t=o}^{n} E_t/(1+k)^t \tag{6}$$

where E_t is the estimated cash-flow for period t, and k is a prescribed rate of return. Positive E_t indicates an inflow of cash; negative E_t indicates an outflow. The investment proposal is acceptable if NPV is non-negative. This method of evaluation is a discounted cash-flow approach which considers the time value of money.

Internal Rate of Return (IRR). The method of internal rate of return determines the rate of return r from the estimated cash flow through the relationship

$$\sum_{t=o}^{n} E_t/(1+r)^t = 0 \qquad (7)$$

where E_t, as before, represents the estimated cash-flow in period t. The investment proposal is acceptable if r is equal to or greater than a prescribed rate of return. Like the NPV method, the IRR method is also a discounted cash-flow approach which considers the time value of money.

Profitability Index (PI). Using the notation already defined, the profitability index J is defined as

$$J = (1/E_o) \sum_{t=1}^{n} E_t/(1+k)^t \qquad (8)$$

As long as J is equal to or greater than 1, the investment proposal is acceptable. It should be pointed out, however, that NPV and PI methods are two forms of the same concept. In fact,

$$J = (N/E_o) - 1 \qquad (9)$$

Each of these methods has its merits and weaknesses. An astute investor would supplement one method with others for a better evaluation. Reference 2 contains a thorough and more advanced discussion on this subject.

IV. R & D AS AN INVESTMENT

Investment decision-making requires consideration of the return on investment which is defined to be the monetary return, according to financial language. So far, the determination of return on investment for R & D has not been discussed. Indeed, this is a difficult task because some of the R & D projects may not contribute directly to products for sale.

For example, the R & D projects in a manufacturing firm can be classified into types as shown in Figure 1. In the figure, the product and process oriented R & D projects are of the type whose return on investment can often be estimated with sufficient accuracy. On the other hand, the non-product oriented R & D projects, which include basic research and management information analyses, are of a type where the return on investment for each project can seldom be accurately estimated. R & D projects of this type must be viewed as "insurance" investments that are needed under competitive and fast changing market conditions. Their return on investment cannot easily be estimated individually, but may, sometimes, be estimated collectively.

Figure 2 is a simplified diagram showing income and expenditure flows between various parts of a manufacturing firm. The diagram helps one to visualize the investment nature of R & D programs. Solid lines indicate income and expense flows while dashed lines represent the output of the R & D effort.

V. OPTIMAL FINANCING LEVEL

The investment evaluation methods presented in Section III are used for deciding whether a proposed investment is acceptable or not. They are not intended to generate the information used in determining the best investment level. This section is devoted to methods for determining the optimal financing level for R & D.

As mentioned before, R & D funding decision making must consider the return on investment. For those R & D activities that are not directed toward specific products for sale, their returns are especially difficult to estimate. In the ensueing discussion, two approaches will be proposed: one assumes that the return on R & D can be readily estimated while the other assumes that part of the R & D return can hardly be estimated. The former will be called the "return-on-investment" approach, and the latter, the "equivalent return-on-investment" approach. Each approach may use more than one method, depending on the criterion chosed for optimization.

1. Return-On-Investment Approach.

When the return for proposed R & D can be estimated, any one of the investment evaluation methods described in Section III can be used, but only NPV, IRR and the PI method can be extended for optimal investment decision-making.

Payback Period. Payback period is a function of investment level and will be denoted by T(L) where L is the investment level. A higher investment level certainly accelerates R & D results but does not necessarily shorten the payback period, since the return during the payback period is the net income after all expenses, including taxes and interests. Using the payback method the optimization criterion is to adjust L such that

$$T(L) = \text{minimum} \tag{10}$$

Average Rate of Return. Average rate of return R(L) is also a function of investment level L. A higher investment level in general leads to higher returns but does not always lead to a higher rate of return. Here the optimal investment level is obtained by choosing L such that

$$\bar{r}((L) = \text{maximum} \tag{11}$$

Figure 3 depicts the general form of average return as a function of investment level. Notice the nature of the diminishing return at a high investment level. The slope of the line tangent to the return curve at point A is the maximum average rate of return, and the corresponding investment L_A is the optimal investment level.

Net Present Value. The net present value should depend on L because the cash-flow per period depends on L. Hence we can write Eq. (6) as

$$N(L) = \sum_{t=1}^{n} E_t/(1+k)^t - L \tag{12}$$

where -L has been substituted for E_o. For optimal investment level, L is chosen so that

$$N(L) = \text{maximum} \tag{13}$$

Internal Rate of Return. Optimal financing based on the method of internal rate of return can be obtained based on Eq. (7) which can be written as

$$\sum_{t=1}^{n} E_t(L)/[1+r(L)]^t - L = 0 \tag{14}$$

with $E_o = -L$. Since the expected cash-flow $E_t(L)$ depends on L, the solution for r, which is the internal rate of return, also depends on L. Optimal investment level L is so adjusted that

$$r(L) = \text{maximum} \tag{15}$$

Profitability Index. Eq. (8) can be rewritten as

$$J(L) = 1/L \sum_{t=1}^{n} E_t(L) \tag{16}$$

For optimality, the profitability index J(L) is maximized with respect to L.

2. Equivalent Return-On-Investment Approach.

For R & D involving projects whose returns are difficult to estimate, such as non-product oriented projects, a method is proposed below for determining a utility-on-investment which can then be employed for investment level optimization.

It has been shown in Section II that given a group of proposed projects it is always possible to assign to each a utility function $U_k(l_k)$ which represents the relative importance of each project. The total utility of R & D is given by Eq. (2).

Assume the following scenario. An organization initially budgets L_0 dollars for R & D, but would like to know: 1) how to allocate L_0 to n proposed projects; and 2) if L_0 is the optimal investment level. It is assumed that not all the projects are product oriented, therefore the return on investment for the total R & D is hard to estimate. One can allocate by using the resource allocation method presented in Section II which does not require the knowledge of returns.

A by-product of the resource allocation method is a total utility function, U(L). This function is useful for determining the optimal investment level. The general shape of the utility function is shown in Figure 4, where the scale of the ordinate is arbitrary. Note that the shape of U(L) is similar to that of R(L) in Figure 3; the slope also diminishes at high values of L. In Figure 4, U_A is the point of maximum average rate of utility, with a corresponding investment level L_A. This point may be considered the optimal investment point because it offers the largest utility-on-investment. Notice that the point of maximum utility U_M is not optimal since the marginal utility near U_M is considerably lower than that at U_A.

If the initial budget is larger than L_A, such as L_0 (in Figure 4), the optimal investment is still at L_A. That is, the initial budget L_0 can be reduced by an amount of $L_0 - L_A$.

If the initial budget is less than L_A, such as L_0 in Figure 5, then the firm should consider additional funding. In order to make this financing decision and equivalent return analysis is needed. Since L_0 is the initial budget that the firm has earmarked for R & D, it is assumed that the return on this investment is non-negative. To be conservative, take the return on this investment to be zero. Define the "break-even cost of utility" as

$$B_E = L_0/U_0; \tag{17}$$

this results in a measure of dollars per utility. The purpose of this quantity is to relate the arbitrarily scaled utility values to dollar values in a conservative way. One assumes here that zero profits will result from the initial budget. Denote the line passing through points 0 and U_0 as the break-even line, B_E. Along this line $L = B_E U$. If the additional R & D investment reduces the cost of utility, the new utility point should be above the B_E-line. Let U be the utility associated with a $L > L_0$. The break-even cost for this utility would be L' obtained in a way as shown in Figure 5. The gain on the investment may be considered as

$$\Delta L = L' - L = B_E U - L \tag{18}$$

Define the "equivalent return-on-investment" as

$$R_E = \Delta L/L = B_E U/L - 1 \tag{19}$$

This can then be treated in the same fashion as the usual return-on-investment with any of the investment optimization methods previously discussed.

IV. CONCLUSION

The main results of this paper are: 1) viewing the R & D effort in an organization as a financial investment, and 2) proposing methods for optimal R & D investment. When the return-on-investment for the R & D can readily be estimated, the usual investment evaluation methods can be extended for optimization. When a direct estimation of return is difficult, a quantity called "equivalent return-on-investment" is used instead. A method for determining the equivalent return-on-investment from an arbitrarily scaled utility function is proposed, providing a way for the optimization of investment levels for R & D activities, which are not all product-directed.

REFERENCES

1. B. J. Greenblott and J. C. Hung, "A Structure for Management Decision Making," IEEE Transactions on Engineering Management, Vol. EM-17, No. 4, 1970, pp. 145-158.
2. L. P. Anderson, V. V. Miller, and D. L. Thompson, The Finance Function, Intext Educational Publishers, Scranton, Pennsylvania, 1971.
3. J. C. Van Horne, Financial Management and Policy, 4th edition, Prentice-Hall, Inc., 1977.

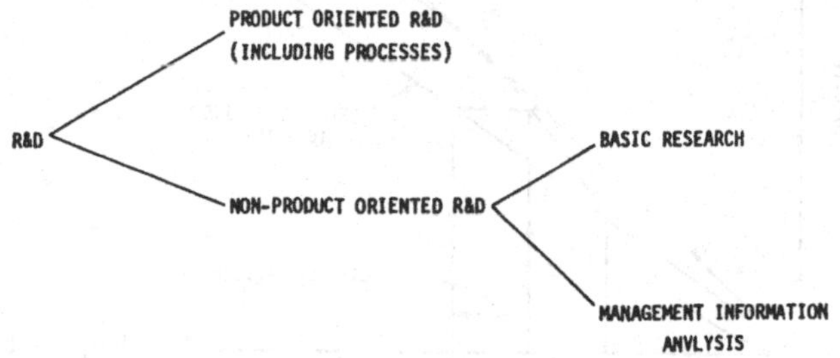

FIGURE 1

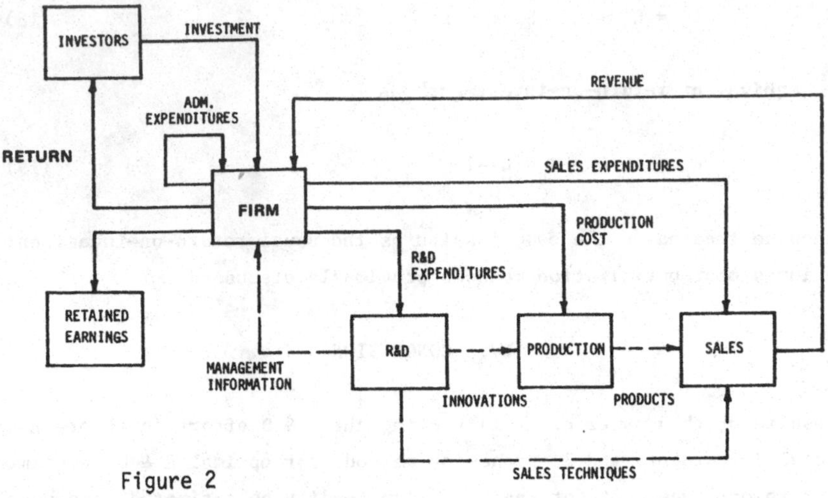

Figure 2

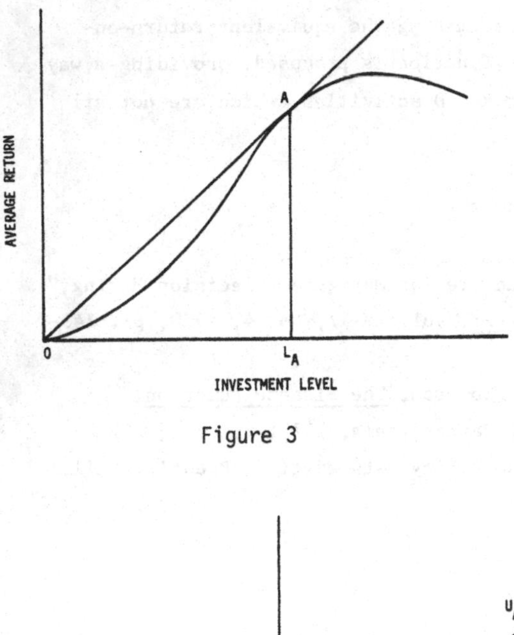

Figure 3

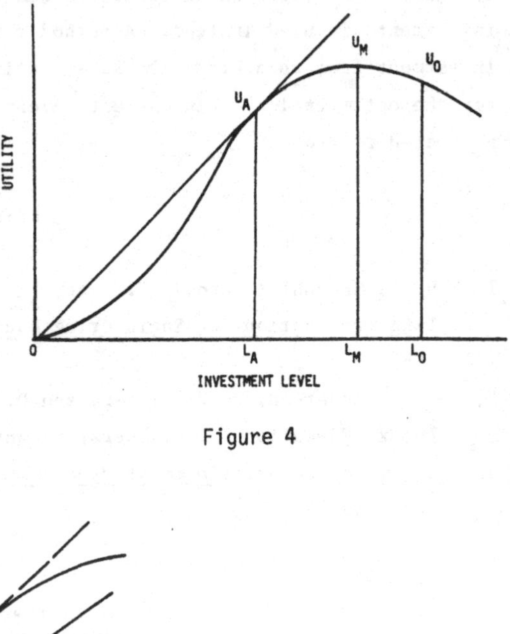

Figure 4

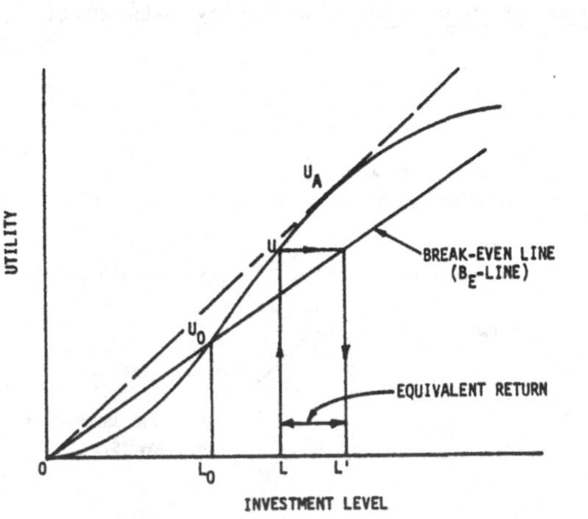

Figure 5

ON AN INEXACT TRANSPORTATION PROBLEM

Janusz Kacprzyk, Maciej Krawczak

Polish Academy of Sciences
Systems Research Institute
Newelska 6, 01-447 Warszawa, POLAND

ABSTRACT

Uncertainty and variability of some parameters, e.g. supply and demand, in many practical transportation problems often lead to difficulties in implementations. Hence, an inexact transportation problem is formulated, in which those parameters are represented by value intervals. It is shown that such a problem may be equated with an auxiliary conventional transportation problem of higher dimension. An application to the shipment of flat glass from factories to regional wholesales is presented.

1. INTRODUCTION

The presented formulation of transportation problem originated from some experience gained during the development and running of a computer system for the management of glass distribution for a group of glass works. The system concerns the shipment of glass from 9 factories producing various assortments (not each assortment in each factory) to 17 regional wholesales and covers such issues as e.g. determination of best transportation routes and means, reusable and non-reusable containers, losses, etc.

The system proved to be very useful and gave considerable savings. As to some difficulties occuring during its use, let us point out the following one. One of optimization problems solved in the system is the transportation problem. Some data needed for it, mainly the supply (production capacities) and to a lesser extent - demand, may vary considerably and in an unpredictable way. Thus, since they must be given in advance, then a great uncertainty exists.

To account for this uncertaintly and reflect it, an inexact transportation problem is formulated, in which the supply and demand are given as value intervals. For solving the problem, the inexact linear programming due to Soyster [3] may be employed. However, the problem is shown in the paper to be transformable into an auxiliary conventional transportation problem of higher dimension. Thus, it may be solved by the transportation problem package used so far. The package is very efficient and, moerover, provides many useful additional informations, e.g. for further analyses.

To illustrate the approach presented, an example of transporting some assortment of flat glass from 5 glass works to 17 regional wholesales is shown.

2. INEXACT LINEAR PROGRAMMING

The idea of inexact linear programming is due to Soyster [3] and consists in the replacement of the conventional linear programming problem by

$$\begin{aligned} &\sup(c_1x_1 + \dots + c_nx_n) \\ \text{subject to: } &x_1K_1 + \dots\dots + x_nK_n \subseteq K \\ &x_1,\dots\dots\dots,x_n \geqslant 0 \end{aligned} \tag{1}$$

where: $K_1,\dots,K_n,K \subseteq E^m$ are non-empty convex sets and "+" is set-theoretic.

In a more specific case, when the sets $K_1,\dots,K_n,K$ are intervals $[\inf K_j \div \sup K_j]$, $j = 1,\dots,n$, and $[\inf K \div \sup K]$, respectively, it may be proved [2,3] that the solution of (1) is equivalent to the solution of

$$\begin{aligned} &\sup(c_1x_1 + \dots + c_nx_n) \\ \text{subject to: } &\sum_{j=1}^{n} \sup K_jx_j < \sup K \\ &\sum_{j=1}^{n} \inf K_jx_j \geqslant \inf K \\ &x_1,\dots\dots, x_n \geqslant 0 \end{aligned} \tag{2}$$

Thus, in this formulation some uncertainty as to coefficients and resources may be reflected.

3. INEXACT TRANSPORTATION PROBLEM

As opposed to the conventional formulation of transportation problem [1]

$$\min \sum_{i=1}^{m} \sum_{j=1}^{n} c_{ij}x_{ij}$$

$$\text{subject to: } \sum_{j=1}^{n} x_{ij} = b_j \qquad (3)$$

$$\sum_{i=1}^{m} x_{ij} = a_i$$

$$x_{ij} \geqslant 0; \ \sum_j b_j = \sum_i a_i; \ i = 1,\ldots,m; \ j = 1,\ldots n;$$

where: x_{ij} - volume of commodity to be transported from supplier i to receiver j, c_{ij} - transportation costs, a_i - volume to be sent(supply) from i and b_j - volume to be supplied (demand) to j, the inexact transportation problem is formulated here as

$$\min \sum_{i=1}^{m} \sum_{j=1}^{n} c_{ij}x_{ij}$$

$$\text{subject to: } \sum_{j=1}^{n} x_{ij} \subseteq A_i \qquad (4)$$

$$\sum_{i=1}^{m} x_{ij} \subseteq B_j$$

$$x_{ij} \geqslant 0; \ i = 1,\ldots,m; \ j = 1,\ldots,n.$$

However, in the sequel it will be assumed that $A_i = [\underline{a}_i + \bar{a}_i]$ and $B_j = [\underline{b}_j + \bar{b}_j]$, where $\underline{a}_i = \inf A_i$ and $\bar{a}_i = \sup A_i$; analogously for $\underline{b}_j$ and $\bar{b}_j$. Hence, the inexact transportation problem will be meant as

$$\min \sum_{i=1}^{m} \sum_{j=1}^{n} c_{ij}x_{ij}$$

$$\text{subject to: } \sum_{j=1}^{n} x_{ij} \subseteq [\underline{a}_i \div \bar{a}_i]$$

$$\sum_{i=1}^{m} x_{ij} \subseteq [\underline{b}_j \div \bar{b}_j] \qquad (5)$$

$$x_{ij} \geqslant 0; \ i = 1,\ldots,m; \ j = 1,\ldots,n.$$

Evidently, the above problem may be transformed into an equivalent linear programming problem due to (2). However, since the available software package used so far for solving conventional transportation problems is very efficient and provides many auxiliary possibilities, e.g. printouts, analyses, then it is better to try to derive a conventional transportation problem equivalent to (5). For convenience, let us graphically represent this equivalent problem in terms of suppliers, receivers, flows of commodities (above arcs) and transportation costs (below arcs) as shown in Fig. 1.

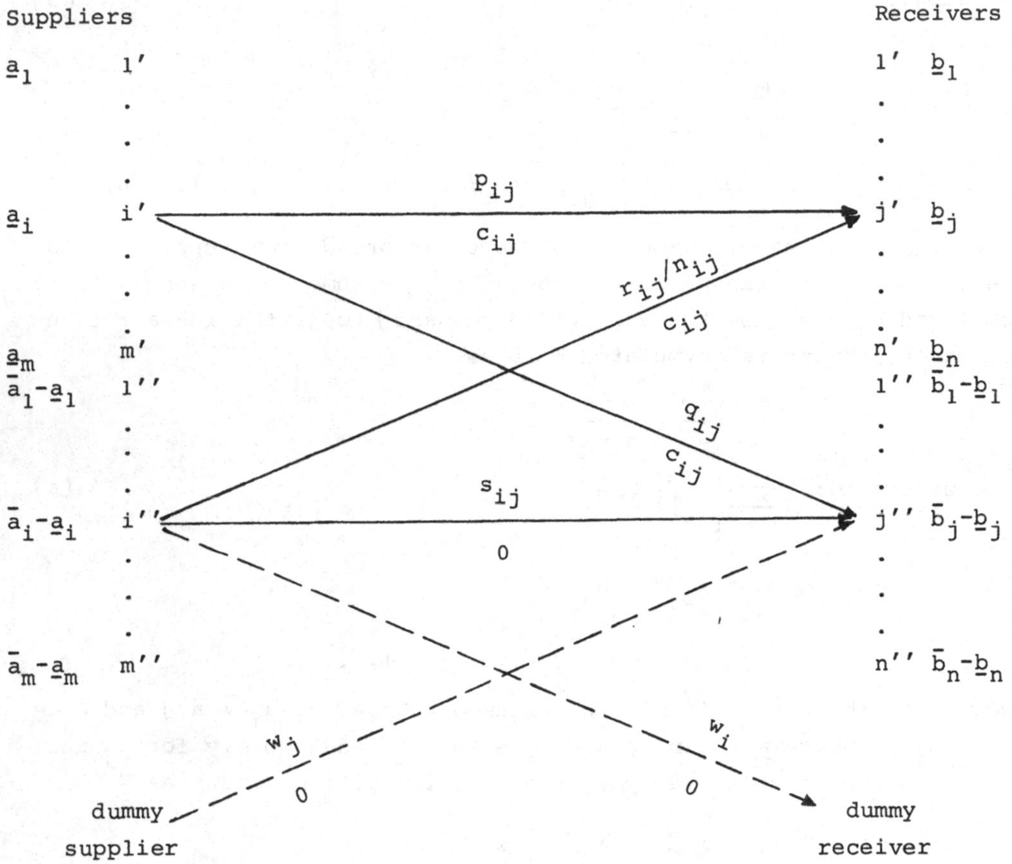

Fig. 1.

First, let us remark that each supplier i may be split into two auxiliary suppliers i' and i'' sending $\underline{a}_i$ and $\bar{a}_i - \underline{a}_i$, respectively. Analogously, each receiver j may be split into j' and j'' receiving $\underline{b}_j$ and $\bar{b}_j - \underline{b}_j$, respectively.

Now, let us denote: $\sum_{i=1}^{m} \underline{a}_i = \underline{A}$, $\sum_{i=1}^{m} \bar{a}_i = \bar{A}$, $\sum_{j=1}^{n} \underline{b}_j = \underline{B}$ $\sum_{j=1}^{n} \bar{b}_j = \bar{B}$.

Then, two situations are possible:

$$\underline{A} \leqslant \bar{B} \leqslant \bar{A} \tag{6}$$

$$\underline{B} \leqslant \bar{A} \leqslant \bar{B} \tag{7}$$

In the case (6), an overproduction $\bar{A} - \bar{B}$ exists, which is directed to a dummy receiver. In the case (7), an unsatisfied demand $\bar{B} - \bar{A}$ occurs, which is covered by a dummy supplier. Evidently, the dummy supplier and dummy receiver do not physically exist, the respective volumes of commodity remain in fact at suppliers'.

Let us now consider the case (6). The dimension of the problem is now $2m \times (2n+1)$. The volumes transported and the respective transportation costs (in parantheses) are as given in Tab.1.

	1 .... n	n + 1 ... 2n	2n+1	$\sum_j$
1 ... m	$p_{ij}(c_{ij})$	$q_{ij}(c_{ij})$	$t_i = 0$ (∞)	$\underline{a}_i$
m+1 ... 2m	$r_{ij}(c_{ij})$	$s_{ij}(0)$	$w_i(0)$	$\bar{a}_i - \underline{a}_i$
$\sum_i$	$\underline{b}_j$	$\bar{b}_j - \underline{b}_j$	$\bar{A} - \bar{B}$	$\bar{A}$

Tab.1.

The meanings of s_{ij}, w_i, and t_i are as follows: $\sum_{j=1}^{n} s_{ij}$ - volume of commodity left at the i-th supplier's, $\sum_{i=1}^{m} s_{ij}$ - volume that j-th receiver does not obtain in relation to $\bar{b}_j$, w_i - overproduction of the i-th supplier, $\sum_{i=1}^{m} w_i = \bar{A} - \bar{B}$, and t_i - artificial variables to obtain an uniformity of description.

For the case (7), the dimension of the problem is $(2m+1) \times 2n$. The volumes transported and the transportation costs are as given in Tab.2.

	1 ... n	n+1 ... 2n	$\sum_i$
1 ... m	$p_{ij}(c_{ij})$	$q_{ij}(c_{ij})$	$\underline{a}_i$
m+1 ... 2m	$n_{ij}(c_{ij})$	$s_{ij}(0)$	$\bar{a}_i - \underline{a}_i$
2m+1	$t_j = 0$ (∞)	w_j (0)	$\bar{B} - \bar{A}$
$\sum_j$	$\underline{b}_j$	$\bar{b}_j - \underline{b}_j$	$\bar{B}$

Tab.2.

The meanings of s_{ij} and t_j are similar as before, while w_j denotes the uncovered demand, i.e. to be satisfied by the dummy supplier,

$$\sum_{j=1}^{n} w_j = \bar{B} - \bar{A}.$$

Thus, the physical volumes forwarded from the i-th supplier to the j-th receiver are as follows:

a/ for (6)

$$x_{ij} = p_{ij} + q_{ij} + r_{ij} \tag{8}$$

b/ for (7)

$$x_{ij} = p_{ij} + q_{ij} + n_{ij} \tag{9}$$

4. A NUMERICAL EXAMPLE

As an example, let us show a real problem of transporting an assortment of flat glass between 5 glass works and 17 regional wholesales. Tab.3. contains both the most important data, i.e. $\underline{a}_i$'s, $\bar{a}_i$'s, $\underline{b}_j$'s, and $\bar{b}_j$'s, and the solution, i.e. the nonzero x_{ij}'s and appropriate sums.

Suppliers \ Receivers	1	2	3	4	5	6	7	8	9	10	11	12	13	14	15	16	17	$\sum_j$	$[\underline{a}_i \div \bar{a}_i]$
1	2			7						4				25		5	6	49	[47÷53]
2			25							3		51		30	7		47	163	[160÷17[illegible]]
3	6			159		19	18	30		2			52			5		291	[268÷30[illegible]]
4	12	41			4		57			2		15						131	[123÷137]
5		10			48	10		5	61	8	31			5	44	43		265	[244÷27[illegible]]
$\sum_i$	20	51	25	166	52	29	75	35	61	19	31	66	52	60	51	53	53		
$[\underline{b}_j \div \bar{b}_j]$	[18 ÷ 22]	[49 ÷ 56]	[25 ÷ 26]	[159 ÷ 170]	[48 ÷ 52]	[29 ÷ 32]	[70 ÷ 75]	[30 ÷ 35]	[59 ÷ 66]	[18 ÷ 21]	[30 ÷ 35]	[59 ÷ 66]	[47 ÷ 52]	[55 ÷ 60]	[44 ÷ 51]	[48 ÷ 53]	[47 ÷ 53]		

Tab.3

5. CONCLUDING REMARKS

The presented formulation of inexact transportation problem has some advantages both from the theoretical and practical point of view. As to the first one, the uncertainty is accounted for in a simple and efficient way and, moreover, it leads to a formulation being not qualitatively different than the conventional one. Hence, e.g. efficient tools for solving conventional problems may be employed. Moreover, for instance, the conventional transportation problem is a particular case of the present formulation, when $\underline{a}_i = \bar{a}_i$ and $\underline{b}_j = \bar{b}_j$, for all i,j.

From the practical point of view, the approach is highly appreciated by the users, i.e. the sales departmens in factories, mainly because now they are not forced to give in advance exact values of a_i's and b_j's, which they do not known in fact. Thus, they do not tend-as before-to start from tentative, in a sense random values, assuming that the problem may always be recomputed, if only more precise data would be available. Such an attitude resulted in many computational runs, which is now avoided to a large extent.

The increase of dimension is not very important, because the problem may also now be solved by the same highly efficient software package for solving the conventional transportation problems.

First experiences show also that in the case of a large imbalance between the supply and demand a proper work of inventories is very important. However, this problem is anyway crucial in this production activity.

REFERENCES

1 Dantzig G.B.: Linear Programming and Extensions, Princeton University Press, 1963.

2. Negoita C.V., Minoiu S., Stan E.: On considering imprecision in dynamic linear programming, Econ. Comput. and Econ. Studies and Res., No. 3, 1976.

3 Soyster A.L.: Convex programming with set-inclusive constraints and applications to inexact linear programming, Op. Res., Vol.21, No.5, 1973.

INTEGER PROGRAMMING AS A TOOL FOR PLANT ADJUSTMENT PROBLEM

Ignacy KALISZEWSKI, Marek LIBURA

Polish Academy of Sciences
Systems Research Institute
Newelska 6, 01-447 Warszawa, POLAND

Hanna MISIEWICZ

PROMASZ, Warszawa, POLAND

The problem of plant adjustment,i.e., the problem of choosing the most suitable technologies and plant equipment to satisfy a new production plan is being considered here. This choice must be done to minimize the total cost of adjustment considering the production plan and resources constraints.

Problems of the type stated above usually lead to mixed integer programming formulation which allows to incorporate into a model many logical relations.

The paper presents the mathematical formulation of a practical problem in its possible variants and shows the results of attemps at a solution using commercial mixed integer programming package.

1. INTRODUCTION

A planning office for machine industry factories has to solve the following problem. For a given plant a new production plan is to be introduced which differs from the old one with respect to the quantity and the list of products produced. Each product can be produced according to different technologies. The equipment of the plant considered can be partially changed by selling the existing machines and buying new ones. For each machine type the number of machines to be bought is limited.

A very restrictive constraint in this problem is the space necessary to place the machines. It is foreseen that in most cases this space will not be sufficient enough,for usually the new quantities to be produced are much larger than the last plan requirements and the new machines are more "space consuming".

Therefore the number of square metres used is to be a variable.

This variable must be penalized for taking values greater than the space we have at our disposal by some fixed and linear costs related to new plant department buildings.

Another very important constraint is the employment. All plants considered employ high qualified workers and they are not available in any number. As the labour force is for a planer a problem of special concern and besides labour costs there are some other aspects which are to be taken into account e.g. educational policy , this problem must be treated with special care.

There are some other constraints especially the ones concerning resources and energy which are to be considered.

The planning office is supposed to propose such an adjustment to the new plant requirements which minimizes the total costs.

The problem as stated above does not include time dynamic. Satisfactory from planists point of view results with a "static" formulation was a clause for its future refinements.

Until now plant adjustments have been done manually by methods of balancing and according to some heuristic rules. This approach when done by a skilful team may lead to good solutions but is extremally time consuming. Therefore it has been decided that this process should be computerized. It would make possible to prepare some software applicable to all plants of the machine industry and additionally to make use of mathematical programming to improve the quality of adjustment variants.

It was necessary then to model the problem. A number of logical conditions of the type "if ... then ..." has been formulated as integer constraints which resulted in a mixed-integer programming problem. This problem is presented in the Sect.2.

It has turned out that this model when applied to the real plants, leads to problem of rather large size. The greatest problem considered had 1113 variables and 767 constraints. The preparation of data for several problems of that size is manageable only when there exists a system for handling them. Such a system for transforming the data from the forms used in plant or company documents to the required input form of a commercial package and for carrying out all the necessary post-optimal computations has been prepared and it is briefly presented in Sect.3.

Despite of the size, the problems have been attacked by a commercial mixed integer programming code. This code incorporates different tree-search strategies which could be chosen by a user. As it had been predicted, many runs would be necessary, it was obvious that the seek-king for the most suitable strategy for problems of the considered type was of the greatest interest. The experiences in that field are presented in the Sect. 3. The same section presents the results of problem solving. Optimal integer solution in a reasonable time have been obtained only for a few problems; for all other problems acceptable sub-optimal solutions have been found.

2. MODEL OF PLANT ADJUSTMENT

Suppose that we consider a specific plant. Let i be a PRODUCT index, $i \in I$, and J_i be the set of possible TECHNOLOGIES for a product i. All machine types used by technologies J_i, $i \in I$, form the set M. For each machine type m M we denote

Z_m - number of the type m machines existing before the adjustment,

z_m - integer variable which indicates the number of retained machines of the type m existing before the adjustment,

y_m - integer variable which indicates the number of machines of the type m to be bought,

D_i - number of units of the i-th product to be manufactured,

Y_m - upper limit for y_m,

f_m - capacity of type m retained machine in the planning horizon,

g_m - capacity of type m bought machine in the planning horizon,

a^m_{ij} - capacity units number of type m machine necessary to manufacture a unit of the i-th product when j-th technology is used,

x_{ij} - continuous variable which indicates the share of the technology j in manufacturing the product i.

Then the following balance constraints should hold:

$$\sum_{i \in I} \sum_{j \in J_i} D_i a^m_{ij} x_{ij} - f_m y_m - g_m z_m \leqslant 0, \qquad m \in M \qquad /1/$$

where

$$\sum_{j \in J_i} x_{ij} = 1, \qquad i \in I \qquad /2/$$

$$x_{ij} \geqslant 0, \quad i \in I, \quad j \in J_i \qquad /3/$$

The y_m and z_m variables are obviously bounded

$$0 \leqslant y_m \leqslant Y_m, \qquad m \in M \qquad /4/$$

$$0 \leqslant z_m \leqslant Z_m, \qquad m \in M \qquad /5/$$

By u_m we denote a variable which indicates the number of m type machines to be sold. Then u_m and z_m are related as follows

$$z_m + u_m = Z_m, \qquad m \in M \qquad /6/$$

For some technological reasons it is required that the shares of technologies x_{ij} are not to be to small. Then there is a need to incorporate into our model the following implications:

"if $x_{ij} > 0$ then $x_{ij} \geqslant r_{ij}$"

where r_{ij} are appropriate share levels. Such implications are modelled by the pairs of inequalities

$$\begin{aligned} x_{ij} - \delta_{ij} &\leqslant 0 \\ -x_{ij} + r_{ij}\delta_{ij} &\leqslant 0 \end{aligned} \qquad /7/$$

where δ_{ij} are binary variables, $i \in I$, $j \in J_i$. Another requirement is not to use to much technologies for a given product manufactured. These conditions can be easily modelled using the variables δ_{ij}, namely

$$\sum_{j \in J_i} \delta_{ij} \leqslant k_i, \quad i \in I \qquad /8/$$

where k_i is for each product a given integer number.

Now we have to consider the problem of a space necessary to place the machines. A machine of the type m to be installed and operated reguires p_m square metres. The space constraint can be written as

$$\sum_{m \in M} p_i(y_i + z_i) \leqslant P \qquad /9/$$

where P is the number of square metres available. It may happen that to satisfy the production plan this constraint has to be violated. We introduce a continuous variable v (a surplus variable) which indicates the amount of square metres used above the limit P

$$\begin{aligned} \sum_{m \in M} p_i(y_i + z_i) - v &\leqslant P \\ v &\geqslant 0 \end{aligned} \qquad /10/$$

Obviously, this variable has to be equal zero when the constraint /9/ is satisfied. An additional condition is that the plant might be enlarged by at least S square metres or not changed at all. To each positive value of v, a linear cost d and fixed cost D are connected. The following relations model the "space" problem:

$/ \ldots$objective function$\ldots / + dv + D\delta'$

$$\sum_{i \in M} p_i y_i + z_i - v \leqslant P$$
$$v - M\delta' \leqslant 0$$
$$-v + S\delta' \leqslant 0 \qquad /11/$$
$$\delta' \text{ - binary}$$

where M is the largest violation of /9/ expected.

The further investigations show that when new machines are to be bought some extra fixed costs must be paid. This is due to some more sophisticated installations and servicing which are necessary for new machines. By binary variables λ_m, $m \in M$, these fixed costs, say F_m, can be introduced into the model follows

$/ \ldots$objective function$\ldots / + \ldots + F_m \lambda_m + \ldots$

$$Y_m \lambda_m - y_m \geqslant 0$$
$$-\lambda_m + y_m \geqslant 0 \qquad /12/$$
$$\lambda_m \text{ - binary}$$

Some preliminary computations have revealed an ill behaviour of the branch-and-bound method when solving the problem stated above. During the search process a large number of feasible solutions has been produced in which $y_m > 0$ and $u_m > 0$, i.e., the machines of m type are simultaneously sold and bought. To eliminate these redundant solutions the constraints (13) have been introduced which for $m \in M$ model the implication "if u_m 0 then $y_m = 0$ "

$$z_m \sigma_m - u_m \geqslant 0$$
$$y_m (1 - \sigma_m) - y_m \geqslant 0 \qquad /13/$$
$$\sigma_m \text{ - binary}$$

Finally, the model is completed by some additional resource constraints, e.g. water, electricity, internal transport etc. These constraints are modelled in usual way as linear constraints.

The objective function has been proposed by economists and it has the following form (including the penalty terms of (11)):

$$\min \sum_{i \in I} \sum_{j \in J_i} D_i c^1_{ij} x_{ij} + \sum_{m \in M} \left(c^2_m y_m + c^3_m z_m + c^4_m v_m + F_m \lambda_m \right) + D\delta' + dv$$

The model constraints plus the objective function constitute the mixed integer programming problem.

3. GENERAL DESCRIPTION OF PROBLEM ORIENTED COMPUTER SYSTEM AND COMPUTATIONAL EXPRERIENCES

The model presented in the preverious section is in fact only a part of a system developed for the solution of plant adjustment problem. The system consists of theree parts:

(i) input data analysis, preliminary computations, problem generation,

(ii) model optimizer,

(iii) output analysis, computations of the technological process parameters.

The basic data of the problem are contained in the matrices which correspond to machine types and indicate unit production time. Ddfferent sets containd also information on prices and available resources.

The aim of the first part of the system is to check the data accuracy and completness and to generate the model in the form required by a mixed-integer package. The very important points are the lower and upper limits for the model variables. This problem is treated in the system with a special care.

The computations in part three of the system are applied to the optimal or some feasible solutions found in the process of optimization Their aim is to prepare the technological process with respect to technological routes and machine characteristics for a new equipment variant chosen.

The problem optimizer is the MIP/370 IBM package. The whole system has been implemented on IBM 370/145 computer. The system has been tested first on several variants of a rather small real-life problem (plant 0 in the Table 1) which has been chosen by technology designers as a representative one. The problem after generation has 55 rows, 130 variables among them 61 integer and 24 binary variables grouped in 3 Special Ordered Sets - SOS. The first attempts to solve this problem in a reasonable time by a standard MIP tree-search strategy failed. For this reason several strategies available on so called sophisticated level of MIP were investigated and the most appropriate strategies for different phases of solution process have been determined. The MIP/370 strategy with SWI = 6, SWi = 0, i $\neq$ 1, turned out to be the best for fast finding (in about 2 min CPU) of very good (frequently optimal) solutions to the test problem. The shortest time for optimality proving appeared when the strategy with SWI = 2, SWi = 0, i $\neq$ 1, was used. It has been also observed that an optimal solution for a partial relaxation of the problem with integrality condition retained only for technology choice variables, could be computed relatively fast(for test problems in 0.1-0.2min)This relaxation followed by appropriate rounding of remaining integer variables has been used to determine very good upper bounds for objective function, which resulted in a significant acceleration of the search process.

More then 20 problems have been solved for 5 plants. For a given plant all problems have had the same structure and have differed only in parameters of the right hand sides and objective function. These different right hand sides have been mainly related to different levels of employment. Because in the frames of integer programming there is no efficient parametric programming procedure, the problems with different right hand sides have been solved separately.

The problems considered have been of rather large sizes (Table 1) and only acceptable suboptimal solutions have been obtained.

Table 1

number of plant	|I|	|M|	number of columns	number of rows	number of SOS	0-1 variables in SOS	integer variables	continuous variables
0	3	15	130	55	3	24	61	45
1	27	39	333	147	27	220	111	2
2	24	90	324	120	24	229	92	3
3	70	90	324	166	70	229	92	3
4	320	258	1113	767	320	667	348	98

Times for problems solving are given in Table 2. /all times in CPU minutes/.

Table 2.

number of plant	LP solution	upper bounds for objective functions	search for integer solutions
1	< 1	2	7 - 25
2	< 1	1 - 6	2 - 11
3	< 1	1 - 4	2 - 10
4	2 - 6	8.5 - 20	15

REFERENCES

1. Garfinkel R.S., Nemhauser G.L.: Integer Programming. New York 1972 J. Wiley a. Sons Inc.

2. IBM Mathematical Programming System Extended/370, Mixed Integer Programming/370. Program Reference Manual, IBM.

A CUTTING SEQUENCING ALGORITHM

Oli B.G. Madsen
IMSOR
THE INSTITUTE OF MATHEMATICAL STATISTICS
AND OPERATIONS RESEARCH
The Technical University of Denmark
DK 2800 Lyngby - Denmark

Abstract:

In some cutting stock problems additional constraints have to be satisfied in order to give solutions which can be used in practice. In this paper we are focusing on a specific constraint which is often met, e.g., in the glass industry. The constraint requires that pairs of corresponding pieces should be cut within a certain time interval. A too large time distance between corresponding pieces will make the identification, matching, storing and administration of pieces much more difficult.

This additional constraint may be handled directly in the cutting stock algorithm but this requires construction of a new extended type of cutting stock algorithm. In this paper other solution strategies are suggested. One of these strategies are discussed in more details. First the cutting stock problem is solved without the time constraint. When the cutting patterns are found a sequencing procedure is used for the ordering of the cutting patterns in such a way that the maximal time distance between corresponding pairs of pieces is reduced but the amount of waste remains constant. The sequencing procedure is based on algorithms for reducing the bandwidth of matrices.

The method is applied to a case from the glass industry and computational experiences and numerical results are reported.

KEY WORDS: Cutting Stock, Cutting Pattern Sequencing, Bandwidth Reduction, Glass Industry.

1. INTRODUCTION

Since the Gilmore and Gomory papers [6-9] were published at the beginning of the sixties, many papers discussing the theory and application of cutting stock problems, trim problems, depletion problems, bin packing problems, loading problems etc. have appeared. A part of this literature has been concerned with applications within the glass industry. Hahn [10] solves a glass cutting problem using the one dimensional Gilmore and Gomory method (GGM) twice. The knapsack problem is solved by dynamic programming. Dyson and Gregory [3] have also solved a glass cutting problem using the GGM. The knapsack problem is solved partly by branch and bound and partly by a heuristic method. Christensen [1] and Madsen [11] have solved a glass cutting problem by methods very similar to the methods used by Hahn.

In general there are two characteristics of glass cutting problems. First the problems are two- or three-stage guillotine cutting problems. Secondly a number of additional restrictions have to be considered, e.g. maximal number of cuts, minimal width of waste, maximal number of pieces in a cutting pattern, best pattern allocation to minimize the time elapsed producing an order, etc. The first characteristic makes the problem easier to solve. The second characteristic normally also makes the problem easier to solve apart from the pattern allocation.

This paper is mainly concerned with the problem of selecting the optimal sequence in which the cutting patterns are to be processed. Several methods to solve this problem are suggested and two of them is implemented and compared.

2. THE GLASS CUTTING PROCESS

The final rectangular glass pieces are cut from rectangular raw glass sheets (stock sheets or plates) of a given dimension. The cutting is performed in three stages. In the first operation the raw glass sheets are cut into sections (see Figure 1) using guillotine cuts i.e. cuts crossing completely from one side of the stock sheet to the other side. Then each section is divided into smaller pieces (strips) again using guillotine cuts. Finally the glass pieces may be trimmed on one side to obtain the required dimension. Several different sizes of stock sheets may be available in virtually unlimited supply.

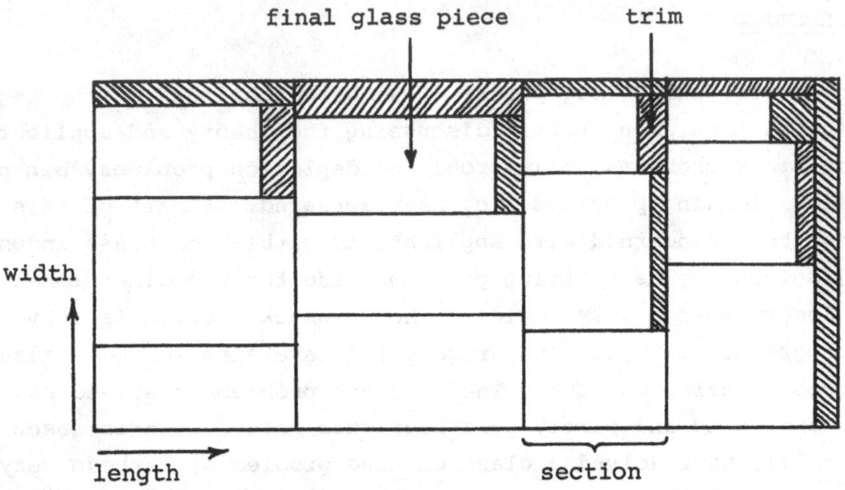

Fig. 1 Example on a cutting pattern. The hatched area is waste.

In mathematical terms the problem stated above is called 'a free two-stage guillotine cutting problem with trim (nonexact case)'. Using a week as a reasonable planning-horizon it is often necessary to plan the cutting of 350 stock sheets into 3500-4000 pieces of glass.

As mentioned in section 1 the sequence in which the cutting patterns are processed can be highly important. An order from a customer often consists of several glass pieces. From an administrative and practical point of view it is desirable to keep the glass pieces belonging to one order as close as possible. This, in fact, gives rise to two conflicting objectives: to minimize the waste and to minimize the 'distance' between corresponding glass pieces. As a special case of this cutting pattern sequencing problem (pattern allocation) one can consider the production of thermopanes. The thermopanes are composed of two or three rectangular pieces of glass of equal size (pairs of corresponding glass pieces). These pairs or triples should be cut within a certain time interval. Larger time distances will make the identification, storing, matching, and administration of glass pieces much more difficult. In the following section some methods to solve the sequencing problem will be discussed.

3. SOLUTION METHODS

3.1 THE HORIZON SHRINKING METHOD (HSM)

The cutting sequencing problem can be solved very easily by a 'quick and dirty' method which could be called horizon-shrinking. The planning horizon is divided into K smaller time periods. Then K small cutting-stock problems are solved and the average distance between corresponding glass pieces (ABCG) is observed. If the distance is too big then K is increased and the problems are solved again. This is a very simple and fast method partly because the cutting-stock problems are of smaller dimensions. Christensen [1] and Madsen [11] have solved a thermopane case with a planning-horizon of 3 days. The solution with a 3 day planning-horizon gave a reduction of the waste percentage from 12% to 5% but ABCG was 6 times as big as the solution normally used by the company. Putting $K = 12$ gave a waste percentage of 8% but the same ABCG as normally used by the company. Furthermore the computing time was reduced from 609 sec. to 120 sec. on an IBM 370/165.

3.2 THE PARALLEL METHOD (PAM)

This method can only be used in the thermopane case or similar cases. If you have a production of double thermopanes the cutting problem can be solved as a cutting-stock problem where only one of the glass pieces from a thermopane is used as data. After the cutting patterns have been determined you cut two of the stock sheets in the same way. If this cutting is done in a sequence the ABCG will be only one stock sheet. This method has not been tried yet in practice.

3.3 THE SALESMAN METHOD (TSM)

This is a two stage approach where the first stage consists of producing a minimum loss set of cutting patterns. The second stage of the method involves sequencing this set of cutting patterns so that the distance between glass pieces belonging to the same order is minimized. The sequencing problem can be formulated as a travelling salesman problem. Dyson and Gregory [3] constructed a cost matrix $C = \{c_{ij}\}$ where c_{ij} means the number of orders on raw glass sheet no. i which do not appear on sheet no. j. c_{ii} is set to an appropriate large number.

If the variable $x_{ij} = 1$ then sheet no. j shall be cut immediately after sheet no. i .

The practical difficulty with this method is the excessive amount of computer time needed and the limited number of stock sheets that could be handled. In the thermopane-case mentioned in section 2 a one week production easily consists of the cutting of 350 stock sheets into 3500-4000 pieces of glass. This means that a travelling salesman problem with 350 points has to be solved. If the production has to be planned on a daily basis then an exact algorithm could solve the problem in short time.

3.4 THE BAND WIDTH REDUCTION METHOD (BRM)

The remaining part of this paper will discuss the BR-method. As in section 3.3 a quadratic matrix will be used to indicate the connection between orders and stock sheets.

Define a symmetric matrix $A = \{a_{ij}\}$ where

$a_{ij} = 1$ means that raw glass sheet i and j have at least one corresponding pair of glass pieces in common $(i \neq j)$.

$a_{ij} = 0$ otherwise $(i \neq j)$

$a_{ii} = 0$

It is assumed that the cutting sequence of raw glass sheets corresponds to the ordering of rows and columns in A , i.e. row 3 corresponds to the raw glass sheet which is cut as no. 3 in the sequence. Interchange of column j_1 and j_2 and row j_1 and row j_2 will lead to another raw glass sheet cutting sequence but the same waste. The objective is now to exchange rows and columns in such a way that the bandwidth $\beta(A)$ of A is minimized where

$$\beta(A) = \max_i \left(i - \min_j \left(j \mid a_{ij} \neq 0\right)\right)$$

$\beta(A)$ will be equal to the maximum DBCG. It is not always possible to make $\min \beta(A)$ less than the distance required. This depends on the cutting patterns but for a fixed cutting pattern min $\beta(A)$ will lead to the best solution obtainable.

Reducing the bandwidth of matrices has been of great interest in connection with finite element methods used in the civil engineering sciences. There the Cuthill-McKee algorithm [2] is very often used to find an ordering having a small but not always minimal bandwidth. Combined with the algorithm of Gibbs, et al. [4] to construct a starting solution the Cuthill-McKee algorithm can give an efficient ordering of a 1500 by 1500 matrix in a few seconds on e.g. an IBM 370/165 computer. The execution time and the total storage requirements for the algorithm increases linear respectively as $m^{1.5}$ [5] where m is the no. of rows in A .

In order to use the algorithms a positive definit A-matrix is required. In our case A will be positive definite if $a_{ii} \equiv 1 + \sum_{j \neq i} a_{ij}$. It is then possible to use a planning horizon of one week (i.e. m = 350) and still obtain a reasonable DBCG.

The definition of A can be generalized in such a way that $a_{ij} = 1$ means that sheet i and j have at least one order in common where an order can contain e.g. all the glass pieces required by one customer.

4. COMPUTATIONAL RESULTS

The BR-method has been implemented on an IBM 370/165 computer. The algorithm is divided into two parts and consists of 534 FORTRAN statements. Part one is constructing the A-matrix based on customer-data and the cutting patterns from the cutting stock algorithm. Part two is reducing the bandwidth of the A-matrix using the Cuthill-McKee algorithm. As case a part of the data from the thermopane firm has been used. 125 stock sheets of equal size are cut into 1500 glass pieces. The total CPU-time used for part one and two was less than 6 sec. Some of the results are shown in figure 2 and 3. Figure 2 shows DBCG just after the use of the cutting stock algorithm while figure 3 shows the same results after the use of the bandwidth reduction algorithm. The main results are shown in table 1. From table 1 it can be seen that maximum DBCG has been reduced to a third and the average DBCG has been increased by 50%. Further results can be seen in Nørgaard-Nielsen [12].

	Before bandwidth reduction	After bandwidth reduction
Average DBCG	14.7	21.4
Maximum DBCG	120	44
Percentage of stock sheets with DBCG = 1 or 2	35%	10%

Table 1 Main computational results. The measures in line 1 and 2 are in numbers of stock sheets.

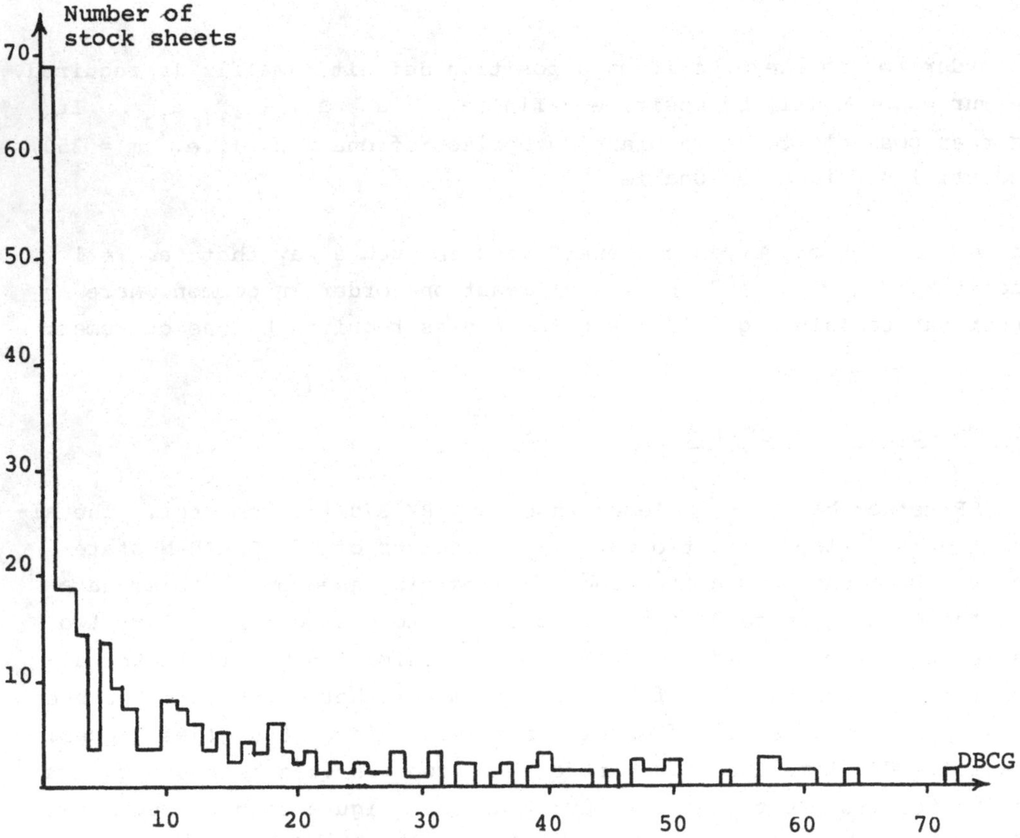

Fig. 2 The number of stock sheets as a function of DBCG (distance between corresponding pairs of glasspieces) before the use of the bandwidth reduction algorithm. DBCG is measured in stock sheets. 5 observations are not shown in the figure, namely 88, 96, 107, 107 and 120.

5. CONCLUSION AND SUGGESTIONS TO FUTURE WORK

The results from section 4 seem to indicate that the objective shall not be to minimize the maximal bandwidth, i.e. the maximal distance between corresponding pairs of glass pieces ~ the maximal time elapsed producing an order. This gives an average distance which at least in the case-study is not acceptable.

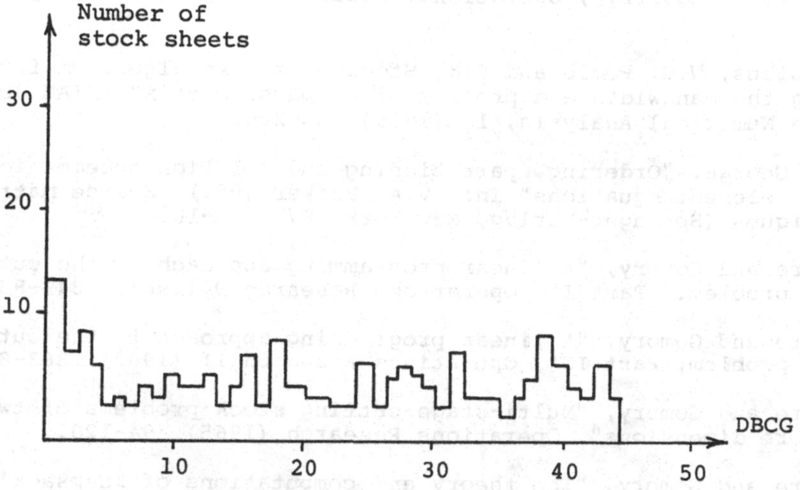

Fig. 3 The number of stock sheets as a function of DBCG (distance between corresponding pairs of glass-pieces) after the use of the bandwidth reduction algorithm. DBCG is measured in stock sheets.

An alternative way to solve the problem is either to change the objective (e.g. minimize the average distance) or to construct a more detailed cost function (as e.g. in section 3.3, the salesman method) or to make a combination of horizon shrinking and bandwidth reduction or to handle the ABCG-constraint directly in an extended cutting stock algorithm.

6. ACKNOWLEDGEMENT

I am indebted to Pia Nørgaard-Nielsen, The Danish Bank, Copenhagen, for her assistance. She has implemented the method described in section 3.4 and she has constructed the corresponding computer programs.

7. REFERENCES

[1] Bendt G. Christensen, "Planning of glass cutting" (in Danish), report from IMSOR, The Institute of Mathematical Statistics and Operations Research, The Technical University of Denmark (Copenhagen, August 1975).

[2] E. Cuthill and J. McKee, "Reducing the bandwidth of symmetric matrices", in: Proceedings of 24th Nat. Conf. Assoc. Comput. Mach. (ACM publication P-69, 1122 Ave. of the Americas, New York, 1969).

[3] R.G. Dyson and A.S. Gregory, "The cutting stock problem in the flat glass industry", Operational Research Quarterly (1974), 41-54.

[4] N.E. Gibbs, W.G. Poole and P.K. Stockmeyer, "An algorithm for reducing the bandwidth and profile of a sparse matrix", SIAM Journal on Numerical Analysis, 13 (1976), 236-250.

[5] Allan George, "Ordering, partitioning and solution schemes for finite element equations" in: V.A. Barker (ed.), Sparce matrix techniques (Springer-Verlag, New York 1977), 52-101.

[6] Gilmore and Gomory, "A linear programming approach to the cutting stock problem. Part I", Operations Research 9 (1961), 849-859.

[7] Gilmore and Gomory, "A linear programming approach to the cutting stock problem. Part II", Operations Research 11 (1963), 863-888.

[8] Gilmore and Gomory, "Multi-stage cutting stock problems of two and more dimensions", Operations Research (1965), 94-120.

[9] Gilmore and Gomory, "The theory and computations of knapsack functions", Operations Research 14 (1966), 1045-1074.

[10] Susan G. Hahn, "On the optimal cutting of defective sheets", Operations Research 16 (1968), 1100-1114.

[11] Oli B.G. Madsen, "Glass cutting in a small firm", Mathematical Programming 17 (1979), 85-90.

[12] Pia Nørgaard-Nielsen, "Optimal cutting sequences in connection with glass cutting in the production of thermopanes" (in Danish), unpublished report from IMSOR, The Institute of Mathematical Statistics and Operations Research, The Technical University of Denmark (Copenhagen, May 1978).

ON A WINNING COALITION OF THE CHARAKTERISTIC FUNCTION GAME AS A SOLUTION OF THE RESOURCE ALLOCATION PROBLEM

Jacek W. Mercik
International Research Institut of Management Science ,129090 Moscow,8,Schepkina Str.,USSR

Introduction

In up-to-date decision making problems modelling that has made use of the game theory one can notice that the problem of non-synonymous activities /strategies/ of an individual player has been left out. The ideas of this problem can be found for example in Morgenstern`s /1/ paper. The playerswho for reaching a definite goal join a coalition usually have /each of them/ more than one strategy ready to use. It is intuitively clear that the coalition in the result of its activities receives a pay-off which next /if possible/ should be divided between members of this coalition. We now possess many classical proposals on how to solve that kind of games /for example von Neumann--Morgenstern, Shapley, Nash, Maschler`s solutions/. These solutions aim to find a kind of compromise how to divide a coalition`s pay-off /to define player`s potential, influence, power and so on/ and straightfully from there to establish a composition of the coalition. The problem of coalition formation is still an open case. In Cassidy and Mangold`s /2/ model one can find that a fundamental property of the model is that it recognizes an interdependency of the coalition formation /the strategy level in an n-person conflict/ and pay-off determination /the tactical level within the n-person conflict/. We are going to deal with this direction and consider the problem of the coalition formation among players, each from among has a determinate set of strategies at his disposal. The strategies, which have been used in a framing of the coalition may increase or decrease a common coalition pay-off /1/. This is a second problem which has been separated by Contini /3/ who has defined two problems: /1/ maximization of the vector of expected pay-offs and /2/ maximization of the joint probability of achieving predeterminatedgoal values for all attributes. The proposals for the first problem solution one can find for example in Zeleny /4/. We will also make efforts to give a proposition / based upon arbitrary just feeling/ of the pay-off`s vector, however this is not our main goal and this proposition is questionable.

In conclusion, we consider the following problem: how to form a winning coalition if each of the game players has his own strategies which result of using may depend upon strategies chosen by other members of the coalition. The game like this, given in the form of cooperative, 0-1 /showing technological restrictions/ characteristic function game is in this same time a solution of a resource allocation problem, i.e. resources with specific properties: indispen-sable for all players /that makes them to take part in the game/ and which can be less than they need them /conflict/ and which besides can be used by players for their different strategies /strategy substitution/. The good example of this kind of resources may be water /or energy/ which can be used to fulfil some different strategies of one player in a sensible manner for other players /for example because of water pollution/.

It will also be very usefull if we describe clearly we think what mean by the maximization of the joint probability of achieving predeterminated goal value. Each game player tends to be a member of such a coalition which members have their own strategies possibly least discrepant one from others. It will, besides be better if they cooperated strictly /it makes increase of their pay-off/ but we think as minimum postulate we may use the mentioned above first one.

2. Description of strategies.

Each player "i" /i=1,2,...,N/ of the game has respectively his own set of strategies S_i. Each strategy $s^i_k \in S_i$ /i=1,2,...,N/ is in one of the three following binar relations with one of the strategies of other players /for $i \neq j$/ :

$s^i_k \sim s^j_l$ _ indifferent relation,

$s^i_k \uparrow s^j_l$ - strictly cooperative relation /players i and j cooperation quarantees higher pay-offs than separately/,

$s^i_k \downarrow s^j_l$ - antinomic relation /their common use of those strategies quarantees less pay-offs than seperately/.

We assume for these relations that for $R \in \{\sim, \uparrow, \downarrow\}$, $s^i_k R s^j_l \Rightarrow s^j_l R s^i_k$ for $i,j=1,\ldots,N$; $k=1,\ldots,|S_i|$; $l=1,\ldots,|S_j|$.

3. How to Form a Coalition.

Let $D \subseteq N$ be a winning coalition of the game. The player belonging to this coalition should not have his strategies /if possible/ in the binar relation $\nmid$ with strategies of other players from D. The player also agrees with the postulate, that the rest of game players will prefer the coalition without relations of $\nmid$ type /full information concerning the strategies is available/.

Let us introduce the following formulas:

$$r(i,j)=1/2 \text{ card}\left\{ (s_k^i,s_l^j):\ s_k^i \nmid s_l^j \ ,\ \text{for } 1 \leqslant k \leqslant |S_i| \ ,\ 1 \leqslant l \leqslant |S_j| \right\}. \qquad /1/$$

and for $D \subseteq N$

$$r(D) = \max_{i,j \in D} r(i,j) \ . \qquad /2/$$

We will call index $r(D)$ a rank of coalition D.

$$r(i) = \sum_{j \in D} r(i,j) \ , \quad \text{for } i \in D \ . \qquad /3/$$

$$R(D) = \sum_{i \in D} r(i) \overset{(3)}{=} \sum_{i \in D} \sum_{j \in D} r(i,j) \ . \qquad /4/$$

$R(D)$ - total number of relations of $\nmid$ type from the point of view of each player of coalition D.

$$Z(i,D) = \begin{cases} 0 & \text{if } R(D) = 0 \\ \dfrac{r(i)}{R(D)} & \text{if } R(D) \neq 0 \end{cases} \qquad /5/$$

$Z(i,D)$ - index of conformity of player i with coalition D.
Of cource

$$0 \leqslant Z(i,D) \leqslant 1$$

because of $\sum_{i \in D} Z(i,D) = 1$ and $r(i,j) \geqslant 0$ for all $i,j \in D$.

We say also, that player i is conformable to coalition D if $Z(i,D) = 0$. In another case the player i is conformable to coalition D in a $Z(i,D)$ degree.

Let further

$$\delta(i,j)=1/2 \text{ card}\left\{ (s_k^i,s_l^j):\ s_k^i \uparrow s_l^j \ ,\ \text{for } 1 \leqslant k \leqslant |S_i| \ ,\ 1 \leqslant l \leqslant |S_j| \right\}. \qquad /6/$$

and for $D \subseteq N$

$$\delta(D) = \sum_{i,j \in D} \delta(i,j) \qquad /7/$$

$$\delta(i) = \sum_{j \in D} \delta(i,j) \qquad /8/$$

$$A(i,D) = \begin{cases} 0 & \text{if } \delta(D) = 0 \\ \dfrac{\delta(i)}{\delta(D)} & \text{if } \delta(D) \neq 0 \end{cases} \qquad /9/$$

$A(i,D)$- index of superadditivity of player i in coalition D.

Of cource

$0 \leqslant A(i,D) \leqslant 1$

We say also,that player i is the most desireable for coalition D if $A(i,D) = 1$. In another case the player i is desireable for coalition D in a $A(i,D)$ degree.

The postulate of conformity does not quarantee the uniqneness of winning coalition D. In the case,when one using the postulate of conformity has chosen more than one winning coalition one may use the index of superadditivity to choose one coalition from a set of coalitions derived by the index of conformity.

Definition: The winning coalition,D,of the game $\Gamma = \langle N;v \rangle$ is the coalition which is winning by function v and which is optimal in a sense of the postulate of conformity.

From formula /5/ we receive that optimal coalition,D,is the coalition with /2/ $r(D) \longrightarrow 0$ and if $D_1, D_2 \subseteq N$, $D_1 \neq D_2$, where D_1, D_2 are winning in the game $\Gamma = \langle N;v \rangle$ and $r^*(D_1) = r^*(D_2)$, where $r^*(\cdot) = \min_{v(D)=1} r(D)$

we use the index of superadditivity and we choose this one for which $\delta(\cdot)$ is greater.

Taking into consideration both the postulates we receive a game $\Gamma_Z = \langle N,v,Z,A \rangle$. The solution of this game is a coalition,$D^* \subseteq N$, which is winning in a sense of the game $\Gamma = \langle N;v \rangle$, and which fulfills these two postulates mentioned above.

4. Proposal of a pay-off function for coalition D^*.

Let D^* be a solution of the game $\Gamma_Z = \langle N,v,Z,A\rangle$ and S^* - the set of strategies of players from D^*. Let $|D| = m, /2 \leq m \leq N/$. The set $S = S^* \times S^*$ consist of all pairs of strategies belonging to S^*. One can divide the set S into three disconnected subsets consisting of pairs of rategies according to relations $\sim, \uparrow, \downarrow$.

Let

$$S_0 = \{(s_i,s_j):\ s_i \in S^*,\ s_j \in S^* \setminus S_i,\ s_i \sim s_j \quad \text{for } i,j=1,\dots,m\}$$

where s_i, s_j are strategies of players i and j respectively.

$$S_+ = \{(s_i,s_j):\ s_i \in S^*,\ s_j \in S^* \setminus S_i,\ s_i \uparrow s_j \quad \text{for } i,j=1,\dots,m\}$$

$$S_- = \{(s_i,s_j):\ s_i \in S^*,\ s_j \in S^* \setminus S_i,\ s_i \downarrow s_j \quad \text{for } i,j=1,\dots,m\}$$

The players for whose /5/ $Z(\cdot,D^*)) > 0$ are "load" for coalition D^* because they have strategies which together with one of strategies of the rest of players of D^* may decrease the pay-off value for coalition D^*. So, we think that they would receive a part of "extra incomes" of coalition D^* according to their $Z(\cdot,D)$ as a price payed by D^* for excluding /if possible/ this kind of strategies by those players.

The "extra incomes" of coalition D^* are incomes which appear as a result of using strategies being in $\uparrow$ type binar relations by D^*.

Some part of these "extra incomes" would be divided between those players for whom /9/ $A(\cdot,D)) > 0$ according to /9/.

It also seems to be true, that in general it is impossible to divide absolutly the set S into subsets S_0, S_+, S_- in order to any element of pair (s_i,s_j) belonging to one of those subsets not to be in the same time an element of another pair belonging to the other subset S_0, S_+, $S_- \subset S$.

Then one can distinguish the following types of pay-off functions for coalition D^*:

$$V_B = \frac{1}{|S_1 \setminus S_-|\,|S_2 \setminus S_-| \dots |S_m \setminus S_-|} \sum_{(s_1,\dots,s_m)} f(s_1,\dots,s_m) \qquad /10/$$

where:

$f(s_1,\dots,s_m)$ - pay-off function if players from D^* use strategies $(s_1,\dots,s_m)$.

We named V_B an absolute pay-off where each /excluding if possible strategies from S_-/ vector $(s_1,\dots,s_m)$ has the same probability to be chosen. If one of $|S.\setminus S_-|$ is equal to zero then we put on its place to formula /10/ a value 1.

Definition: Strategies $(s_1,\dots,s_m)$ and $(s'_1,\dots,s'_m)$ of coalition D are equivalent each to the other if a number of $\sim,\uparrow,\downarrow$ type relations is the same in both vectors of strategies.

$$V_o = \min_{*} f(s_1,\dots,s_m \mid S_o,S_+,S_-) \qquad /11/$$

where

$f(s_1,\dots,s_m \mid S_o,S_+,S_-)$ - pay-off function when players from coalition D^* use strategies $s_1,\dots,s_m$ prefering strategies belonging first to S_o, then to S_+ and S_-. It means that if ther is no such $(s_1,\dots,s_m)$ where all pairs of strategies $\in S_o$ we look for $(s'_1,\dots,s'_m)$ where pairs of strategies belong to $S_o \cup S_+$ prior to S_o and so on.

$*$ - Means here and further the maximum over the set of equivalent strategies of D^*.

We name V_o an indifferent pay-off function of D^*.

$$V_+ = \max_{*} f(s_1,\dots,s_m \mid S_+,S_o,S_-) \qquad /12/$$

where $f(s_1,\dots,s_m \mid S_+,S_o,S_-)$ the same as in /11/ but with preferences generated by the order S_+,S_o,S_-.

We name V_+ a superadditive pay-off function of D^*.

We receive immeditely the following consequence:

$$V_o \leqslant V_B \leqslant V_+$$

Using /10/ - /12/ we can obtain a pay-off function for an individual player of game $\Gamma_Z = \langle N,v,Z,A\rangle$

$$x_i = (1/m)V_B + a\cdot Z(i,D^*)(V_+ - V_B) + (1-a)\,A(i,D^*)(V_+ - V_B) \quad \text{for } i\in D^* \qquad /13/$$

and $x_i \equiv 0$ for $i \notin D^*$.

where

a - coefficient determinating a partition of "extra incomes", $V_+ - V_B$, among players having $\downarrow$ type strategies and resigning them and players "making" these "extra incomes".

One can notice that

$$\sum_{i\in D} x_i = V_+$$

It is the fact of a silent assumption that coalition D^* is efficient what means that players will use only the strategies which can give a maximal coalition pay-off value. In another case we can measure their wishes to achieve the maximal coalition pay-off value and if neccessary we will reduce respectively the value of /13/ x_i for them.

5. Conclusions.

This type of decision making seems to be quite complicated,as it requires each player to compare the profits gained by his participation in a coalition with restrictions imposed on his own strategies by the commomh policy of this coalition. In other words,each game player always forms a matrix of binar relations between his strategies and strategies of possible partners in the coalition - but it seems to be the very natural way of decision making.

It also seems to be very natural that a decision maker usually prefers the situation where he is equal to others cooperetors /the postulate of conformity/.

And at last but not least the proposed model and procedure seems to be easy to programed and applied to practical decision making in conflict situations.

6. Literature.

/1/ Morgenstern,O.: Strategic allocation and integral games.In: Advances in game theory /Procc. of the Second All-Union Conference on Game Theory,Vilnius,USSR,1971/,Publ."Mintis",Vilnius,1973, pp.96-99.

/2/ Cassidy,R.G; Mangold,J.: Coalition Behaviour in n-Person Conflicts. J. Mathematical Sociology,vol 4/1975/,no.1,pp.61-82.

/3/ Contini,B.M.: A Decision Model under Uncertainty with Multiple pay-offs. In: Theory of Games.Techniques and Applications, A.Mensch ed.,N.Y.,1966.

/4/ Zeleny,M.: Games with Multiple Payoffs. Intern.J. Game Theory, vol.4,Issue 4,1975,pp.179-191.

A Package for Analytic Simulation of Econometric Models

Carlo Bianchi, Giorgio Calzolari and *Paolo Corsi*

Centro Scientifico IBM, Pisa

Abstract: Some analytic simulation techniques for the analysis of the reduced form and of the dynamic properties of econometric models are described. Comparisons are made with analytical methods available for linear models.

Keywords: Econometric models; structural form; reduced form; analytic simulation; stochastic simulation; impact multipliers; dynamic multipliers; forecast errors; asymptotic standard errors.

1. Introduction

The evaluation of an econometric model as a simultaneous equations system and the analysis of its dynamic properties are crucial steps in the model building process, particularly when the model is used for forecasting and for simulating alternative economic policies.
In contrast with linear econometric models, where analytic methods are always applicable, in nonlinear models one must generally resort to simulation techniques.

Purpose of this paper is to briefly describe some analytic simulation techniques (combination of numerical simulation and analytical methods, according to the definition by Howrey and Klein [10] and by Klein [12]) for the analysis of the reduced form and of the dynamic properties of the models. The proposed techniques, which integrate the program for stochastic simulation described in [2], extend, to nonlinear models, methods that are available in the literature for linear econometric models only. Also for linear models, however, these methods can be sometimes preferred, both for their greater simplicity in the input of data and for their computational performances. In particular, these techniques allow a fast and reliable computation of the following:

- Standard errors of the reduced form equations.
- Reduced form coefficients (impact multipliers, in particular) and

covariance matrix of their asymptotic distribution.

- Dynamic (interim) multipliers and related asymptotic covariance matrices.
- Asymptotic variances of the forecast errors.

An example of the application of the above mentioned techniques to a "test" model will be presented in section 2; in this section, dealing with the standard errors of the reduced form equations, the analytic simulation methodology will also be briefly described. The extension of the methodology to the computation of the covariance matrices of impact and dynamic multipliers and of forecast errors will be shortly discussed in sections 3, 4 and 5, together with a short comment on the computational performances; finally, in section 6 the main features of the package implemented by the authors at the IBM Scientific Center of Pisa will be presented.

2. Standard Errors of the Reduced Form Equations

Let

(2.1) $Ay_t + Bz_t = u_t \qquad t=1,2,\ldots,T$

be a linear econometric model in its structural form, where y_t, z_t and u_t are, respectively, the vectors of the endogenous and predetermined variables and of the structural stochastic disturbances at time t, A and B are matrices of structural coefficients (A nonsingular square matrix). Furthermore, let u_t be distributed as

(2.2) $u_t \sim N(0,\Sigma)$; $\quad \mathrm{cov}(u_t,u_s) = \delta_{ts}\Sigma \quad$ all t,s

δ_{ts} being the Kronecker delta; in other words the vectors u_t are supposed to be independent and identically distributed, with a multivariate normal distribution, zero mean and covariance matrix constant over time.

The estimated structural model is

(2.3) $\hat{A}y_t + \hat{B}z_t = \hat{u}_t$

where $\hat{u}_t$ are the estimated residuals and

(2.4) $\hat{\Sigma} = 1/T(\sum_{t=1}^{T}\hat{u}_t\hat{u}_t')$

is an estimate of the covariance matrix of the structural disturbances (or, that is the same, of the structural form equations).

The restricted reduced form (i.e. the reduced form derived from the structural, which takes into account all the restrictions on coefficients) is:

(2.5) $y_t = -A^{-1}Bz_t + v_t$

where

(2.6) $v_t = A^{-1}u_t$

is the vector of the reduced form disturbances at time t. It is clearly

(2.7) $v_t \sim N(0, A^{-1}\Sigma A'^{-1})$

so that an estimate of the reduced form covariance matrix (Ω) is immediately available as

(2.8) $\hat{\Omega} = \hat{A}^{-1}\hat{\Sigma}\hat{A}'^{-1}$

provided the estimated $\hat{A}$ is nonsingular.

If the model is nonlinear, equation (2.8) cannot be applied. In fact, in the nonlinear case, for the structural econometric model

(2.9) $f(y_t, z_t, a) = u_t$

(where a is a vector including all the structural coefficients, because a clear distinction between coefficients of y_t and coefficients of z_t, i.e. between the elements of A and B, is no more possible), an explicit analytic expression of the reduced form

(2.10) $y_t = g(z_t, a, u_t)$

is, in general, unknown. Nevertheless, the covariance matrix of the reduced form equations can be computed by an analytic simulation procedure. This procedure is based on a nonexplicit linearization of the model in the neighbourhood of the solution point corresponding to the period t under examination. It is clear from equations (2.5) and (2.6) that the elements of the matrix A^{-1} (such that $A^{-1}u_t = v_t$ are the reduced form disturbances) are the partial derivatives of the endogenous variables with respect to the structural disturbances at time t (elements of the vector u_t). These derivatives can be computed via numerical simulation, stored in a matrix $\hat{D}_t$ and the reduced form covariance matrix ($\hat{\Omega}$) at time t can be computed as:

(2.11) $\hat{\Omega}_t = \hat{D}_t \hat{\Sigma} \hat{D}_t'$

where $\hat{\Sigma}$ is estimated as in (2.4), given the additive hypothesis on the structural disturbances in (2.9). It must be pointed out that, while in case of linear models, being $D_t = A^{-1}$, D_t will be constant, for nonlinear models D_t (and consequently Ω_t) will be time-varying, so that, in equation (2.11), the subscript t has been introduced. The computation of the $\hat{D}_t$ matrix (partial derivatives of the endogenous variables, with respect to the structural disturbances, in the solution point at time t) required by the analytic simulation method, can be performed using finite increments on the structural disturbances. More exactly, first a deterministic control solution is computed, at time t, with all the $\hat{u}_t$ set to zero. Then a value ϵ is assigned to the disturbance of the first equation, all the other being still zero, and the model is solved again. The procedure is then repeated for all the structural stochastic equations and the differences between the disturbed solutions and the control solution, divided by the values adopted for ϵ, supply the numerical values of the partial derivatives. The way in which $\hat{D}_t$ is computed (by numerical simulation) and the use of equation (2.11) define the analytic simulation procedure as a combination of numerical simulation and analytical methods. The proposed procedure, when applied to linear models, is an alternative to the use of equation (2.8). The advantages in using analytic simulation techniques are related to the fact that, for medium or large-size models, the econometricians generally use the standard Gauss-Seidel iterative algorithm for the solution, expressing the model as a set of equations, each equation being normalized with respect to a different endogenous variable. In such a case it is much easier to compute the $\hat{D}_t$ $(=\hat{A}^{-1})$ matrix by simulation, rather than to invert $\hat{A}$; in fact, for the inversion of the $\hat{A}$ matrix, the model should be expressed according to (2.1), which involves the practical problem of the correct correspondence between variables and coefficients.

For nonlinear models, an alternative way of estimating the reduced form covariance matrix $(\hat{\Omega}_t)$ is based on the stochastic simulation of the model [4]. Without entering into the details of this methodology, it is important to remark that, by means of stochastic simulation, the accuracy of the estimates (asymptotically exact), increases with the number of replications; the analytic simulation procedure, on the contrary, is not exact, but involves (via linearization) a systematic approximation.

In order to check the size of this approximation, for some variables of the nonlinear Klein-Goldberger model (revised version by Klein [11], estimated by 2SLS with 4 principal components), the standard errors of the reduced form equations have been computed by means of the analytic simulation procedure and by means of the stochastic simulation approach after 50, 500, 5000 and 50000 replications. The results are displayed in table 1.

From table 1 one could get the strong impression that the stochastic simulation results converge to those of analytic simulation as the number of replications goes to infinity. This is of course impossible, due to the nonlinearity of the model, but clearly gives an idea of the great accuracy of the analytic simulation method, which requires one control solution and as many disturbed solutions as the number of stochastic equations (16, for this model).

In regard to the computational performances of the analytic simulation procedure in estimating the $\hat{\Omega}$ matrix, it must be remarked that, in case of linear models, the computation time required by the procedure is comparable with that required by the use of the (analytic) formula (2.8). In case of nonlinear models, the comparison

Table 1

Klein-Goldberger Model

Reduced Form Standard Errors at 1965

Cd = consumption of durables; X = gross national product; W = wages and salaries and supplements to wages and salaries; Pc = corporate profits including inventory valuation adjustment; p = implicit GNP deflator.

Variab. Name	Computed Value	Standard Errors: Stochastic Simulation, Number of Replications				Analytic Simulation
		50	500	5000	50000	
Cd	55.33	2.78	2.48	2.44	2.42	2.42
X	530.1	9.05	8.44	8.52	8.54	8.53
W	310.8	5.24	4.77	4.73	4.77	4.78
Pc	41.97	6.44	6.21	6.16	6.11	6.11
p	1.225	.040	.035	.036	.036	.036

between analytic and stochastic simulation is, in general, largely in favour of the analytic simulation procedure; in fact, this procedure supplies results whose accuracy is similar to that of the results obtained, in a much more expensive way, via stochastic simulation after a very large number of replications. For example, the results displayed in table 1 require, on a computer IBM/370 model 168, less than one second of CPU time for analytic simulation, 6 seconds for 500 replications of stochastic simulation and about 10 minutes for 50000 replications.

3. Impact Multipliers and Asymptotic Standard Errors

Following Dhrymes [6] and Goldberger [8], the reduced form coefficients are defined as the partial derivatives of the conditional expectation of each current endogenous variable with respect to each predetermined variable, with all other z_t's held constant; properly speaking, the impact multipliers are the subset of the reduced form coefficients corresponding to the current exogenous variables, but in this section the two terms will be indifferently used.

For linear models, the matrix of the reduced form coefficients (Π)

can be directly estimated by the matrix product $-A^{-1}B$, as follows from equation (2.5). For nonlinear models, as pointed out in the previous section, the reduced form is unknown; disregarding the effects of nonlinearities on the conditional expectation of the endogenous variables (effects which are always very small, according to the experience of the authors), the computation at time t, of the multiplier of the j-th exogenous variable with respect to the i-th endogenous variable ($\hat{\pi}_{ijt}$), can be performed, as in [12], using simulation techniques, by the ratio:

(3.1) $\hat{\pi}_{ijt} = (\hat{y}^d_{it} - \hat{y}^c_{it})/(z^d_{jt} - z^c_{jt})$

where $\hat{y}^c_{it}$ is the deterministic control solution corresponding to the control value z^c_{jt} and $\hat{y}^d_{it}$ is the disturbed solution corresponding to the value $z^d_{jt} = z^c_{jt} + \epsilon$, all the other predetermined variables being equal to their control values.

Attempts to derive the small-sample distribution of the impact multipliers have been performed using Monte Carlo methods (see, for example, [15]). These methods, however, have a major drawback in the possible non-existence of finite moments in the small-sample distribution of the structural and reduced form coefficients, even for linear models when estimated with simultaneous consistent methods [13], [14]. Truncation must be, therefore, performed on the distribution of the pseudo-random disturbances to be used in the Monte Carlo experiment [17,p.1004], thus involving some arbitrariness.

As suggested by Theil [18,p.377], resort to asymptotic distribution could be sometimes preferable; under quite general assumptions, it can be shown that the asymptotic covariance matrix of the reduced form coefficients is given by

(3.2) $\Psi_t = J_t \Delta J_t'$

where J_t is the matrix of the second order derivatives, properly arranged, of (2.10) with respect to z_j and a, $(\partial^2 g_i / \partial z_j \partial a_k)$, computed in the point $(z_{jt}, a, u_t = 0)$, and Δ is the asymptotic covariance matrix of all the structural coefficients.

The problem of computing the asymptotic covariance matrix of impact multipliers ($\hat{\Psi}$) was dealt with in 1961 by Goldberger, Nagar and Odeh [9] for linear models (in which case $\hat{\Psi}$ is not time varying); they have proposed the explicit formula

(3.3) $\hat{\Psi} = (\hat{A}^{-1} \otimes [\hat{\Pi}' \; I])\, \hat{\Delta}\, (\hat{A}^{-1} \otimes [\hat{\Pi}' \; I])'$

where $\otimes$ denotes Kronecker product; the above formula, in order to be applicable to nonlinear models, would require an explicit linearization of the model, thus making extremely laborious the process also for small models. Even in case of linear models the procedure is quite laborious; this is probably one of the reasons of the quite different and contradictory results displayed in the literature, for the same test model [3], [5], [7], [9], [16].

Analytic simulation overcomes most of the difficulties, allowing a fast and reliable computation, even for moderately complex models and with no difference between linear and nonlinear models; $(\partial^2 g_i / \partial z_j \, \partial a_k)$, in fact, can be simply computed in the point $(z_{jt}, \hat{a}, u_t = 0)$, using finite increments, as

(3.4) $\Delta(\Delta g_i / \Delta \hat{a}_k) / \Delta z_j$

thus requiring, for the complete computation of the $\hat{J}_t$ matrix, approximately as many solutions of the model at time t as the product of the number of exogenous variables with the number of estimated structural coefficients.

4. Dynamic Multipliers and Asymptotic Standard Errors

The dynamic properties of an econometric model are related to the presence of lagged endogenous variables in z_t (vector of the predetermined variables); in this case equation (2.1) could be more properly rewritten as:

(4.1) $Ay_t + Bx_t + Cy_{t-1} = u_t$

where B is the matrix of the structural coefficients of the exogenous variables x_t and C is the matrix of the structural coefficients of the lagged endogenous variables y_{t-1}.

In analogy with the impact multiplier, the k-lag dynamic (delay as in [8], or interim as in [16]) multiplier, could be defined as the partial derivative of the conditional expectation of an endogenous variable at time t with respect to an exogenous variable at time t-k.

The k-lag interim multipliers in linear models could be computed in the following way:

(4.2) $\hat{\Pi}_k = (-\hat{A}^{-1}\hat{C})^k(-\hat{A}^{-1}\hat{B})$;

for nonlinear models, the computation of the elements of $\hat{\Pi}_k$ is always performed using simulation techniques, by the simple ratio (3.1), where instead of z^d_{jt} and z^c_{jt}, the values of x_j at time t-k are used.

Analytic simulation can be profitably used for estimating the covariance matrix of the asymptotic distribution of dynamic multipliers. In fact, even if Schmidt [16] gave an analytic solution to the problem, the proposed method (revised by Brissimis and Gill [5], [7]) is applicable to linear models only, and has the practical drawbacks of requiring the use of large sparse matrices whose non-zero elements are hard to be filled automatically and of requiring a large computation time. For example, for the Klein-I model, the computation of the standard errors of interim multipliers up to lag 15 using the method proposed by Gill and Brissimis [7] requires about 6 minutes of CPU time on a computer IBM/370 model 168; an ad-hoc program which uses the analytic simulation approach performs the same computation in less than one second.

The analytic simulation procedure used for this purpose, is quite similar to the one discussed in the previous section; the only difference is that the partial numerical derivatives must be computed with respect to the values of x_j at time t-k.

5. Asymptotic Covariance Matrix of the Forecast Errors

Under the assumption of independence among structural disturbances in different periods, the analysis of the forecast error, or briefly of the forecast, can be performed decomposing the error into two independent components: a first component that depends only on errors on the estimated coefficients and a second component that depends only on the vector of the structural disturbances.

Following the mentioned decomposition, for linear models and in one-step simulation, Goldberger, Nagar and Odeh [9] have proposed, for the estimation of the covariance matrix of forecasts $\hat{\Phi}$, the following formula:

(5.1) $$\hat{\Phi} = F\hat{\Psi}F' + \hat{\Omega}$$

where F is a matrix containing, in a convenient order, the values of the predetermined variables in the forecast period, $\hat{\Psi}$ is the asymptotic covariance matrix of the reduced form coefficients defined in (3.3) and $\hat{\Omega}$ is the covariance matrix of the reduced form disturbances defined in (2.8).

The previous formula could be applied also to nonlinear models; in fact, using analytic simulation techniques, eqs. (3.2,3.4) and (2.11) will provide, respectively, an estimation of the matrices Ψ and $\hat{\Omega}$ to be used in (5.1). It is, however, much simpler and more precise, as suggested in [1], to compute directly the asymptotic covariance matrix of the component due to the errors on the structural coefficients (i.e. $F\hat{\Psi}F'$), using an analytic simulation approach, without the intermediate step of the computation of $\hat{\Psi}$, covariance matrix of the reduced form coefficients. This method is simply based on the numerical computation of first order derivatives of the endogenous variables with respect to the structural coefficients. In this way it is not necessary to compute the second order derivatives involved by the matrix $\hat{\Psi}$, thus ensuring a greater numerical accuracy and a considerable shortening of execution time (for example, 0.05 seconds, instead of 10 seconds of CPU time, for the linear Klein-I model).

6. Main Features of the Package

The package is written in FORTRAN IV and ASSEMBLER languages; its basic structure is similar to that described in [2].

For each model, two data sets must be prepared: one is a FORTRAN subroutine containing the model's equations in suitable form for the Gauss-Seidel solution algorithm; the other is a data set containing the time series of the endogenous and exogenous variables, the estimated coefficients, residuals and asymptotic covariance matrix of the structural coefficients.

The additional analytic simulation techniques have been introduced into the program in form of four separate subroutines (less than 1000 FORTRAN statements on the whole) which perform, respectively, the computations described in sections 2, 3, 4 and 5. 512K of main storage are generally sufficient for the simulation of medium-size models (for example, the already mentioned Klein-Goldberger model).

References

[1] Bianchi,C. and G.Calzolari, "The One-Period Forecast Errors in Nonlinear Econometric Models", International Economic Review, 21 (1980, forthcoming).

[2] Bianchi,C., G.Calzolari and P.Corsi, "A Program for Stochastic Simulation of Econometric Models", Econometrica, 46 (1978), 235-236.

[3] Bianchi,C., G.Calzolari and P.Corsi, "A Note on the Numerical Results by Goldberger, Nagar and Odeh", Econometrica, 47 (1979), 505-506.

[4] Bianchi,C., G.Calzolari and P.Corsi, "Some Results on the Stochastic Simulation of a Nonlinear Model of the Italian Economy", in Models and Decision Making in National Economies, ed. by J.M.L.Janssen, L.F.Pau and A.Straszak, Amsterdam: North Holland, (1979), 411-418.

[5] Brissimis,S.N. and L.Gill, "On the Asymptotic Distribution of Impact and Interim Multipliers", Econometrica, 46 (1978), 463-469.

[6] Dhrymes,P.J., Econometrics: Statistical Foundations and Applications, New York: Harper & Row, (1970).

[7] Gill,L. and S.N.Brissimis, "Polynomial Operators and the Asymptotic Distribution of Dynamic Multipliers", Journal of Econometrics, 7 (1978), 373-384.

[8] Goldberger,A.S., Econometric Theory, New York: John Wiley, (1964).

[9] Goldberger,A.S., A.L.Nagar and H.S.Odeh, "The Covariance Matrices of Reduced-Form Coefficients and of Forecasts for a Structural Econometric Model", Econometrica, 29 (1961), 556-573.

[10] Howrey,E.P. and L.R.Klein, "Dynamic Properties of Nonlinear Econometric Models", International Economic Review, 13 (1972), 599-618.

[11] Klein,L.R., "Estimation of Interdependent Systems in Macroeconometrics", Econometrica, 37 (1969), 171-192.

[12] Klein,L.R., "Dynamic Analysis of Economic Systems", Int. J. Math. Educ. Sci. Technol., 4 (1973), 341-359.

[13] McCarthy,M., "A Note on the Forecasting Properties of Two-Stage Least Squares Restricted Reduced Forms: The Finite Sample Case", International Economic Review, 13 (1972), 757-761.

[14] Sargan,J.D., "The Existence of the Moments of Estimated Reduced Form Coefficients", London School of Economics & Political Science, Discussion Paper, 46 (1976).

[15] Schink,G.R., "Small Sample Estimates of the Variance Covariance Matrix of Forecast Error for Large Econometric Models: the Stochastic Simulation Technique", Ph.D. Dissertation, University of Pennsylvania, (1971).

[16] Schmidt,P., "The Asymptotic Distribution of Dynamic Multipliers", Econometrica, 41 (1973), 161-164.

[17] Schmidt,P., "Some Small Sample Evidence on the Distribution of Dynamic Simulation Forecasts", Econometrica, 45 (1977),

[18] Theil,H., Principles of Econometrics, New York: John Wiley, (1971).

ON THE RECURSIVE ESTIMATION OF STOCHASTIC AND TIME-VARYING PARAMETERS IN ECONOMETRIC SYSTEMS

Kurt Brännäs and Anders Westlund
Department of Statistics
University of Umeå, Sweden

Abstract: Interdependent econometric models specified for the control of economic processes are often subject to structural changes, and characterized by stochastic and time-varying structural parameters. A two-stage procedure based on Kalman filtering is suggested for recursive estimation in structural forms. The first stage gives recursive estimates of the reduced form parameters, and the second stage deals with recursive structural form estimation. The estimation procedure does well as long as the a priori knowledge necessary for the Kalman filtering is true. As soon as some of this information is misspecified, the properties are aggravating. These robustness problems are illustrated in the case of mis-specified transition matrices.

1. Introduction

One basic device for economic planning, e.g. the control of economic processes, is the specification of econometric models. Such models are often specified as multirelational and simultaneously dependent, e.g.

$$(1)\quad y(t) = f\{y(t),y(t-1),\ldots,x(t),x(t-1),\ldots,e(t),e(t-1),\ldots;A(t),B(t),\ldots\},$$

where $y(t)$ is an (mx1) vector of endogenous variables, $x(t)$ is an (nx1) vector of fixed exogenous variables, $e(t)$ is an (mx1) vector of residuals; $A(t)$, $B(t)$ etc. are parameter matrices of corresponding order; $f(.)$ is a vector function, and t is the discrete time index.

One task in econometric modelling that has always occupied econometricians is the problem of how to efficiently estimate the unknown structural parameter matrices in (1). Unfortunately, there is no universally best estimation technique and the relative advantages and disadvantages depend on such factors as the size of the model, whether f(.) is linear or non-linear, if the model involves dynamics, if there are certain problems of satisfactorily specifying the model, whether or not the unknown structural parameters are time-invariant, etc. The ultimate goal of econometric modelling (whether it is pure economic analysis, prediction or control) is of great importance when selecting estimation technique.

Econometricians are seldom able to justify the assumption of time-invariant parameters in econometric models, and in (1) the parameters are allowed to be time-varying. Traditionally, however, in econometrics an assumption of time-invariant structure is put forward, and so far only some rather conditional results for estimating time-varying econometric models are known. Furthermore, a reasonable hypothesis is that the problem of structural instability is accentuated for econometric models used for the control of economic systems, i.e. econometric policy optimization is preferably carried out within the framework of time-varying systems.

An early work on the problem of estimating single equation econometric models with random parameters is Rubin (1950), which is followed e.g. by Quandt (1958), Hildreth and Houck (1968), Swamy (1970), and Cooley and Prescott (1973). Due to the degree of analytical complexity little is done for interdependent systems (see, however, e.g. Kelejian, 1974).

One important approach to detecting structural changes in econometric models is based on recursive estimation. Since Gauss gave the recursive formula for least squares estimation, several more general recursive formulae have been formulated. Recursive estimation algorithms for two-stage least squares are given in Pandya and Pagurek (1973) and Riddell (1975). The Kalman filtering technique, originally derived in Kalman (1960), has recently been suggested as a suitable recursive approach to estimation in structurally time-varying models. Sarris (1973), Athans (1974) and Otter (1978) are dealing with single equation models, and Schleicher (1975), Rustem et al. (1976) and McWhorter et al. (1976) consider alternative specifications of econometric systems (however, without treating the dependence between jointly dependent variables and disturbances explicitly).

The purpose of this paper is to suggest and to some extent evaluate the Kalman filtering technique as an approach to the recursive estimation of structural parameters in interdependent systems. In section 2 we introduce Kalman filtering for structural form estimation, and make some comments on the basic theory. In section 3 some robustness properties of structural form Kalman filtering are indicated, and in section 4 some conclusions are given.

2. Kalman filtering for structural form estimation

A characteristic feature in structural forms is the presence of several endogenous variables in each equation. We write a linear version of (1) as

$$
\begin{aligned}
(2)\qquad y(t) &= (I_m \otimes y'(t))\alpha(t) + (I_m \otimes x'(t))\beta(t) + e(t) = \\
&= (\tilde{y}(t),\ \tilde{x}(t))\ (\alpha'(t),\ \beta'(t))' + e(t) = \\
&= z(t)\ C(t) + e(t),
\end{aligned}
$$

where $\otimes$ denotes the Kronecker product and I_m the (mxm) identity matrix. We let the vector of unknown stochastic parameters C(t) vary as

(3) $C(t+1) = E(t)\ C(t) + w(t).$

The residuals e(t) and w(t) are assumed to be normal with zero means and covariances $E(e(t)e'(s))=R^s\delta_{ts}$, $E(w(t)w'(s))=Q^s\delta_{ts}$ and $E(e(t)w'(s))=0$, where δ_{ts} is the Kronecker delta.

The reduced form corresponding to (2)-(3) is given by

(4) $y(t) = \tilde{x}(t)\ \Pi(t) + u(t)$

(5) $\Pi(t+1) = D(t)\ \Pi(t) + v(t).$

Now, in (2) $\tilde{y}(t)$ is in general correlated with e(t). The removal of this dependency is the motivation for the instrumental variable methods in econometrics, since neglection of the dependency gives rise to inconsistent estimates. The approach taken here is parallel to two-stage least squares which amounts to replacing $\tilde{y}(t)$ by predictors $\hat{\tilde{y}}(t)$ obtained from the reduced form. Here, Kalman filtering is used in both stages.

Straightforward application of the Kalman filter (see e.g. Athans, 1974) to (4)-(5) gives

(6) $P(t+1/t) = D(t)\ P(t/t)\ D'(t) + Q$

(7) $P(t+1/t+1) = (I - K(t+1)\ \tilde{x}(t+1))\ P(t+1/t)$

(8) $\hat{\Pi}(t+1/t) = D(t)\ \hat{\Pi}(t/t)$

(9) $\hat{\Pi}(t+1/t+1) = \hat{\Pi}(t+1/t) + K(t+1)\ (y(t+1) - \tilde{x}(t+1)\ \hat{\Pi}(t+1/t)),$

where $K(t+1) = P(t+1/t)\ \tilde{x}'(t+1)\ (\tilde{x}(t+1)\ P(t+1/t)\ \tilde{x}'(t+1) + R)^{-1}$, $Q = E(v(t)v'(t))$, $R = E(u(t)u'(t))$, and P(.) is the covariance matrix of $\hat{\Pi}(.)$.

For each t we have $\hat{y}(t+1/t+1) = \tilde{x}(t+1)\ \hat{\Pi}(t+1/t+1)$, and then

(10) $\hat{z}(t+1/t+1) = (\hat{\tilde{y}}(t+1/t+1),\ \tilde{x}(t+1)\ ).$

Replacing z(t) in (2) by $\hat{z}(t/t)$, and using the Kalman filter in a second stage on (2)-(3) gives the structural form estimation equations

(11) $P^s(t+1/t) = E(t)\ P^s(t/t)\ E'(t) + Q^s$

(12) $P^s(t+1/t+1) = (I - K^s(t+1)\ \hat{z}(t+1/t+1)\ )\ P^s(t+1/t)$

(13) $\hat{C}(t+1/t) = E(t)\ \hat{C}(t/t)$

(14) $\hat{C}(t+1/t+1) = \hat{C}(t+1/t) + K^s(t+1)\ (y(t+1) - \hat{z}(t+1/t+1)\ \hat{C}(t+1/t)\ )$,

where $K^s(t+1) = P^s(t+1/t)\hat{z}'(t+1/t+1)\ (\hat{z}(t+1/t+1)P^s(t+1/t)\hat{z}'(t+1/t+1) + R^s)^{-1}$, and $P^s(.)$ is the covariance matrix of $\hat{C}(.)$.

We notice from the filtering equations (6)-(9) and (11)-(14) that they are structurally equivalent, which is an appealing fact for practical computation and use. Further, the Kalman filter is based on the assumption of fixed regressors. By assumption the exogenous variables $\tilde{x}(t)$ are fixed, whereas $z(t) = (\tilde{y}(t), \tilde{x}(t))$ is stochastic in the endogenous $\tilde{y}(t)$. From this it follows that equations (6)-(9) are exact and the estimator $\hat{\Pi}(t+1/t+1)$ is a best linear estimator, or a best estimator under normality assumptions, in the sense of minimum mean square error. For structural form estimation $\tilde{y}(t+1)$ is replaced by the predictor $\hat{\tilde{y}}(t+1/t+1)$, which is a function of $\hat{\Pi}(t+1/t+1)$ and therefore stochastic. In this case the fixed regressor assumption is clearly violated.

In Brännäs (1978) the randomness of $\hat{z}(t+1/t+1)$ is recognized and corresponding filtering equations are derived. The obtained estimator is structurally equivalent to (11)-(14), but with the gain matrix $K^s(t+1)$ now given by

(15) $$K^{s'}(t+1) = P^s(t+1/t)\hat{z}'(t+1/t+1)\ \{\hat{z}(t+1/t+1)P^s(t+1/t)\hat{z}'(t+1/t+1) + P^{C\tilde{z}}(t+1/t) + R^s\}^{-1}.$$

Compared to $K^s(t+1)$, (15) is seen to include a new covariance term $P^{C\tilde{z}}(t+1/t)$, which is a function of $\hat{C}(t+1/t)$ and the prediction error in $\hat{z}(t+1/t+1)$. Since $P^{C\tilde{z}}(t+1/t)$ is a positive semidefinite matrix the stochastic explanatory variables of the structural form imply a smaller gain matrix and consequently more conservative updating of estimates. It should further be noted that the estimator in Brännäs (1978) is based on orthogonal projections and is a best linear estimator. Consequently there may exist nonlinear estimators with lower mean square error.

The suggested estimator, which approximates the exact one, seems to be a suitable device for recursive estimation in time-varying econometric models. It can also be used as an approach to the detection of structural changes, and to the estimation of type and degree of structural change. In these cases the approximation is reasonable since it provides a less conservative updating of estimates.

In the filtering equations the transition matrix $E(t)$, the initial parameter vector $\hat{C}(0/0)$ and the covariance matrices $P^s(0/0)$, Q^s and R^s are assumed known. The corresponding reduced form values are obtained by transformation. The efficiency of the estimator apparently depends on the quality of the a priori knowledge of transition, parameter and covariance matrices. Robustness against these is hence very important. The robustness against a priori information is studied in Brännäs and Westlund (1979) and in Westlund (1979). Robustness against incorrectly assumed transitions is briefly studied in the next section by simulations.

3. Evaluation of robustness in structural form Kalman filtering

The present robustness study is based on simulations, on the small interdependent system;

$$\begin{aligned} y_1(t) &= a_{12}(t)y_2(t) + b_{10}(t) + b_{11}(t)x_1(t) + b_{12}(t)x_2(t) + e_1(t) \\ y_2(t) &= a_{21}(t)y_1(t) + b_{20}(t) + b_{23}(t)x_3(t) + b_{24}(t)x_4(t) + e_2(t) \ , \end{aligned} \tag{16}$$

where the parameters vary according to (3). The exogenous variables x(t) are real economic time series, and the initial values for the structural parameters are given in table 1.

a_{12}	a_{21}	b_{10}	b_{11}	b_{12}	b_{20}	b_{23}	b_{24}
.7	-2.5	149.5	-.8	-.7	374.	-1.5	1.

Table 1. Initial structural parameters (t=0)

The evaluation deals only with mis-specified transition matrices; for other aspects of robustness of the structural form estimator reference is given to Brännäs and Westlund (1979) and to Westlund (1979).

By (16) and (3) three sets of data (M^k, k=1,2,3) are generated. The three structures differ only in the applied transition matrices E(t), viz.

M^1: E(t) = I, for t=1,2,...,20,

M^2: E(t) = 1.02 I, for t=1,2,...,20,

M^3: E(t) = I, for t=1,2,...,9,11,12,...,20, and
E(10) = S I, where S is a diagonal matrix with diag(S) = (1,2,1,1; .5,1,1,1,1,1,1,.5).

The random vectors w(t) and e(t) are generated as normally distributed with zero mean and diagonal covariance matrices Q^s; diag(Q^s) = (0,.0001,.0016,0; 2.25,.0001,.0001, 0,0,16,0,0,.0004,.0001), and R^s; diag(R^s)=(100,100), respectively. For each data set, 30 replicates are generated.

The parameters are now estimated under different assumptions (J_k, k=1,2,3) on the tran-

ASSUMPTION, J_k	STRUCTURE M^k		
	k=1	k=2	k=3
J_k=1	E(t)=I	E(t)=I	E(t)=I
J_k=2	-	E(t)=1.02I	E(t)=1.02I
J_k=3	-	-	E(t)=I,t≠10 E(10)=S I

Table 2. Summary of the design

sition matrix E(t), see table 2. Corresponding alterations of transitions in the reduced form are obtained by transformations.

MEAN BIAS (k=1,2,3)						
	k=1	k=2		k=3		
Parameter	J_1=1	J_2=1	J_2=2	J_3=1	J_3=2	J_3=3
a_{12}	-.020	-.161	-.184	-.452	-.474	-.092
a_{21}	.055	.626	.117	-.436	-1.091	.057
b_{10}	-2.168	-32.260	-29.568	15.091	45.109	2.846
b_{11}	.031	.260	.367	.104	-.037	.013
b_{12}	.019	.186	.128	-.145	-.182	.021
b_{20}	-4.881	-76.809	-21.755	-11.766	38.482	-4.234
b_{23}	.000	.426	.138	-.016	-.238	-.011
b_{24}	.000	-.214	-.030	.230	.414	.002
MEAN SQUARE ERROR (k=1,2,3)						
	k=1	k=2		k=3		
Parameter	J_1=1	J_2=1	J_2=2	J_3=1	J_3=2	J_3=3
a_{12}	.001	.034	.054	.367	.439	.021
a_{21}	.016	.544	.042	.403	2.035	.015
b_{10}	31.207	1.4E+3	1.6E+3	442.210	2.9E+3	39.528
b_{11}	.003	.096	.224	.028	.021	.002
b_{12}	.001	.045	.026	.043	.060	.001
b_{20}	164.607	8.2E+3	3.6E+3	463.923	14.7E+3	166.700
b_{23}	.005	.239	.282	.008	1.306	.006
b_{24}	.001	.060	.004	.100	.267	.001

Table 3. Structural form estimation in M^k(k=1,2,3); mean biases and mean squared errors

The simulations are evaluated by measuring mean biases and the mean squared errors at each time t and over the entire estimation period.

Mean biases and mean squared errors for the structural form estimation in M^1 are given

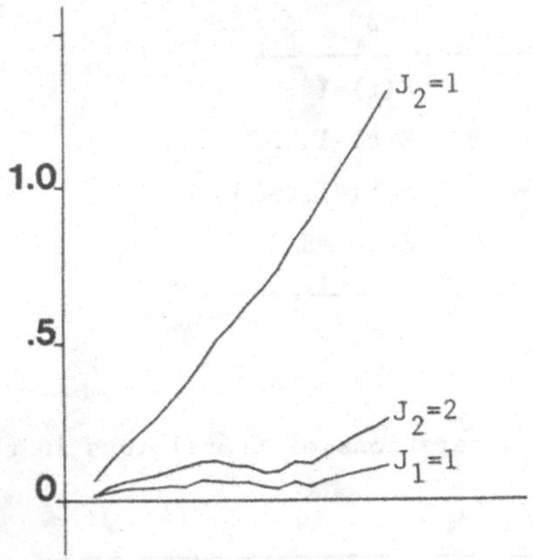

Figure 1a. Observed bias in M^1 and M^2; $a_{21}(t)$

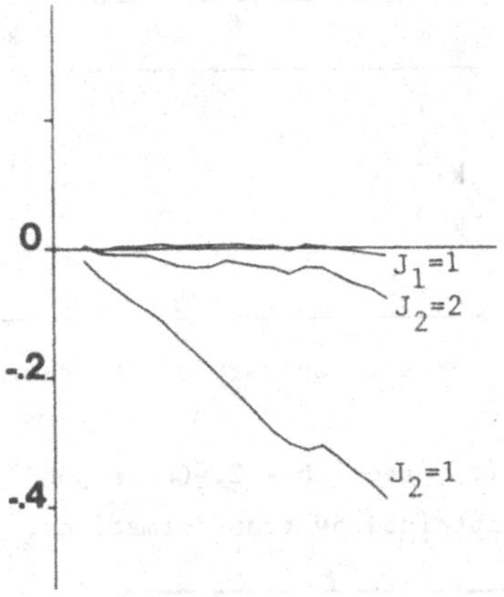

Figure 1b. Observed bias in M^1 and M^2; $b_{24}(t)$

in table 3. It is first noticed that the two-stage estimation approach performs very well for M^1. In figure 1 the evolution of mean biases is illustrated by $a_{21}(t)$ and $b_{24}(t)$.

In the M^2-case the structural parameters change with t according to a trend. When estimating M^2 we assume the transition matrices to be correctly specified (J_2=2), or

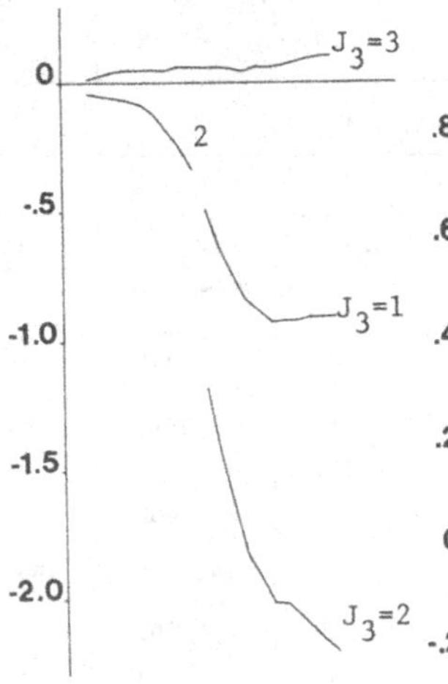

Figure 2a. Observed bias in M^3; $a_{21}(t)$

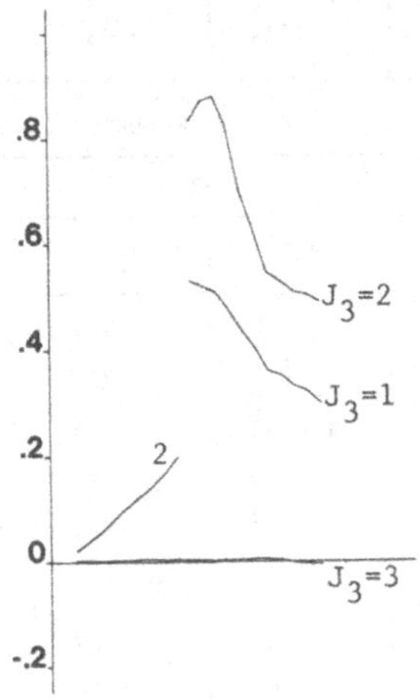

Figure 2b. Observed bias in M^3; $b_{24}(t)$

falsely specified to be an identity matrix (J_2=1). The results obtained for structural form estimation in M^2 are summarized in table 3. In the J_2=2 case, the mean bias and squared error results are satisfactory. Although the absolute biases are somewhat higher than for M^1, the relative sizes are almost similar; see figure 1. As to the misspecification J_2=1, there is a substantial sensitivity for most parameters, but not for all of them. In figure 1 it is seen that the bias in estimating $a_{21}(t)$ for J_2=1 is successively increasing, to 23 percent at t=10 and to 38 percent at t=20. A similar pattern is observed for the other parameters.

In generating structure M^3 we introduce a structural shift at t=10 for $a_{12}(t)$, $b_{10}(t)$ and $b_{24}(t)$. Table 3 indicates for J_3=3 very small increases in mean biases and mean squared errors compared to the correctly specified M^1. We notice a considerable sensitivity against transition mis-specifications. These conclusions are further supported by results illustrated in figure 2. For most parameters there is a tendency for fast adaption to more acceptable bias levels (see e.g. $b_{24}(t)$), but the increase of relative biases are in most cases of unacceptable sizes.

4. Conclusions

There is a demand in econometrics for models with stochastic and time-varying parameters. It is argued to be so especially when the model is directly used for policy purposes. The estimation in time-varying, or structurally changing, econometric models is preferably based on a recursive approach. The Kalman filter, developed in modern control theory, is one appealing technique for recursive estimation. A two-stage procedure based on Kalman filtering is suggested for recursive estimation in structural forms. The first stage gives recursive estimates of the reduced form parameters, and the second stage deals with recursive structural form estimation. As Kalman filtering in general, this approach is found to be computationally simple.

In structural forms the explanatory variables include endogenous variables. Then, even if the exogenous variables often are assumed fixed, there is stochasticity in $\hat{z}(t+1/t+1)$, as it is partially a function of the first stage Kalman estimator $\hat{\Pi}(t+1/t+1)$. As is illustrated in section 2 the two-stage procedure is an approximation of the exact filter. This approximation which reveals itself in the Kalman gain is assumed to be satisfactory.

In section 3 the robustness problems are illustrated in the case of mis-specified transition matrices. If we incorrectly use the identity matrix as transition matrices at the times of structural change, the quality of the recursive estimation is decreased. The evaluation of the two-stage procedure would probably be more positive with longer time series, but short series is a common problem in applied econometrics, and knowledge of small sample properties is important. The sensitivity to mis-specifications in transition matrices points to the necessity of searching for efficient estimation methods of this a priori information.

The recursive nature of the estimator as a by-product to estimation presents opportunity for structural change detection. It is important to find techniques for further improvement of this detection and for faster adaption to new structures in the estimation of structural parameters. One class of possible techniques is the limited memory filters.

References

Athans, M.: "The importance of Kalman filtering methods for economic systems". Ann. Econ.Soc.Meas. 3, 49-64 (1974).

Brännäs, K.: "On Kalman filtering in econometric systems". Statist.Res.Rep. 1978-8, University of Umeå (1978).

Brännäs, K. & Westlund, A.: "Robustness properties of a Kalman filter estimator in interdependent systems with time-varying parameters". Paper at the European Meeting of the Econometric Society, Athens (1979).

Cooley, T. & Prescott, E.C.: "Varying parameter regression: a theory and some applications". Ann.Econ.Soc.Meas. 2, 463-473 (1973).

Hildreth, C. & Houck, J.: "Some estimators for a linear model with random coefficients". J.Amer.Statist.Assoc. 63, 584-595 (1968).

Kalman, R.E.: "A new approach to linear filtering and prediction problems". J.Bas. Engng. 82, 34-45 (1960).

Kelejian, H.H.: "Random parameters in a simultaneous equation framework: identification and estimation". Econometrica 42, 517-527 (1974).

McWhorter, A., Spivey, W.A. & Wrobleski, W.J.: "A sensitivity analysis of varying parameter econometric models". Internat.Statist.Review 44, 265-282 (1976).

Otter, P.W.: "The discrete Kalman filter applied to linear regression models: statistical considerations and an application". Statistica Neerlandica 32, 41-56 (1978).

Pandya, R.N. & Pagurek, B.: "Two-stage least squares estimators and their recursive approximations". Proc. of the 3rd IFAC Conf., Hague (1973).

Quandt, R.E.: "The estimation of the parameters of a linear regression system obeying two separate regimes". J.Amer.Statist.Assoc. 53, 873-880 (1958).

Riddell, W.C.: "Recursive estimation algorithms for economic research". Ann.Econ.Soc. Meas. 4, 397-406 (1975).

Rubin, H.: "Note on random coefficients". In Statistical Inference in Dynamic Economic Models, Cowles Commission, ed. T.C. Koopmans, 419-421 (1950).

Rustem, B., Velupillai, K. & Westcott, J.H.: "Recursive parameter estimation using the Kalman filter: an application to analyse time-varying model parameters and structural change". Paper at the European Meeting of the Econometric Society, Helsinki (1976).

Sarris, A.H.: "Kalman filter models, a Bayesian approach to estimation of time-varying regression coefficients". Ann.Econ.Soc.Meas. 2, 501-523 (1973).

Schleicher, S.: "Prior information in forecasting with econometric models".Paper at the 3rd world congress of the Econometric Society, Toronto (1975).

Swamy, P.A.V.B.: "Efficient inference in a random coefficient regression model". Econometrica 38, 311-323 (1970).

Westlund, A.: "Robustness properties of Kalman filtering in two-stage recursive structural form estimation". Proc. of the 42nd ISI Conf., Manila (1979).

COMPUTING EQUILIBRIA IN AN INDUSTRY PRODUCING AN EXHAUSTIBLE RESOURCE.

G. M. Folie.,
Department of Economics, University of New South Wales,
Sydney, N.S.W. Australia.
and
A. M. Ulph.,
Department of Economics, University of Southampton,
Southampton, U.K.

1. Introduction.

The situation considered in this paper is of an industry depleting an exhaustible resource where some of the producers in the industry have formed together in a cartel, a situation that typifies many of the real world markets for exhaustible resources,such a copper, bauxite, and, most notably, crude oil.

From the practical viewpoint, an obvious motivation for such work is the interest in energy policy, particularly following OPEC's action in raising the world price of oil. The long term energy problem can be characterised by the transition from cheap, convenient but exhaustible forms of energy such as oil and gas to more expensive but abundant forms of energy. A key factor controlling this transition is the price path for oil, currently determined by OPEC. It is of some interest, therefore, to understand the behaviour of industries with a structure similar to the oil industry. Questions one could consider include - what are the likely price and production paths for such an industry; how stable is the cartel; what are the relative gains to the cartel and other firms in the industry? To answer such questions, one needs to compute equilibrium price and output paths for the industry under a range of assumptions.

From the theoretical point of view, the analysis of such questions was confined, until recently, to the familiar static models, i.e. for markets where the product is infinitely reproducible. However, the explicit analysis of the case of an exhaustible resource yields a number of results which cannot be derived from static models, so that the study of partially cartelised exhaustible resource markets is proving a rich field for economic analysis. To date the interesting results have been derived analytically for fairly special cases, and it is important to assess to what extent such results generalise to the more complex models that would characterise real world markets. Such generalisations, however, are a non-trivial matter, and results are most readily derived by simulating numerical examples. Again, an algorithm is required to compute the equilibria of such models.

2. The Structure of the Problem.

Consider an industry with an arbitrary number of firms producing an exhaustible resource. While different producers may face different costs of extraction, refining, or transportation, to the consumer the resource is homogenous, and demand is given by:

$$P = f(Q,t)$$

At some future date, t^*, a substitute resource will become available at price P^*, so that for $t \geq t^*$, $Q(t) = 0$ if $P(t) \geq P^*$.

There are two groups of firms, a cartel group and a competitive fringe group. For the purpose of this paper, no attempt is made to explain why firms belong to one group rather than another, membership is given exogenously. It would clearly be more satisfactory to explain the market structure endogenously, as a function of costs, demand and reserve conditions, and a few comments on this will be made at a later stage. For notation, a firm in the cartel group will be denoted by subscript c, a firm in the competitive fringe group by subscript f, with C and F being the index sets respectively for the two groups. Where it is immaterial which group a firm belongs to it will be denoted by subscript i, belonging to index set $I = C \cup F$.

At time t each firm i has used $X_i(t)$ of its total initial reserves $\bar{X}_i$ and faces convex extraction costs $K_i(Q_i, X_i, t)$, where Q_i is its current production rate. It is assumed that all reserves are economic and will be exhausted before the backstop technology becomes economic at price P^*.

There are four concepts of equilibrium that will be discussed in this paper. In a competitive equilibrium, all firms $i \in I$ take prices in all time periods as given and set their outputs over time so as to maximise the present value of their profits, subject to the constraint that aggregate production cannot exceed initial reserves. The paths of prices and output must clear the market in each time period. The competitive equilibrium serves as a benchmark for comparison with the imperfect market structures discussed below. In each of the three imperfect market structures the fringe continues to act competitively.

In a Nash-Cournot equilibrium the cartel takes the path of output set by the fringe firms as given and sets price so as to maximise the present value of the cartel's profits, subject to the constraint that the cartel's aggregate output does not exceed its initial reserves. An equilibrium consists of a path of prices set by the cartel and a path of outputs by the fringe such that the output set by fringe is that assumed by the cartel in determining its price, while the price set by the cartel is that assumed by the fringe in determining its output plan. The Nash-Cournot model requires the cartel to behave in a naive manner - ignoring the impact its price path has on the fringes' output.

As will be shown in the next section, the exhaustible nature of the resource means that in computing its optimal production path the fringe will add on to its marginal costs of production an appropriate user cost in each period. In the quasi-Stackelberg model, the cartel assumes that these user costs are given; in other words the cartel assumes that the fringe's supply function for each period is given. Using these period-by-period supply functions the cartel calculates a sequence of excess demand functions in the usual manner, and sets price to maximise the present value of its profits. What the cartel neglects is that as it changes its price plan, the finge's new output path will correspond to a different set of user costs. An equilibrium is a set of prices set by the cartel and a set of user costs set by the fringe which are mutually consistent in a manner analogous to the Nash-Cournot model.

Finally in a Stackelberg model, the cartel fully realises the response of the fringe to its price plans, and sets a path of prices to maximise the present value of the cartels resource, taking account of the fringe's supply response.

The algorithm to be described can be used to compute equilibria of the first three kinds, and we shall mention briefly the difficulties involved in computing Stackelberg equilibria.

3. The Algorithm for the Competitive Case.

This section will describe the algorithm for the competitive case, the extensions to the other models being described in the next section. We begin by formalising the problem facing a competitive firm. For simplicity it will be assumed that all producers $i \in I$ have the same discount rate δ.

A firm's problem can then be formalised as:

$$\max_{\{Q_i(t)\}} \int_0^{T_i} \{P(t)Q_i(t) - K_i(Q_i, X_i, t)\}e^{-\delta}dt$$

subject to

$$\dot{X}_i(t) = Q_i(t)$$
$$Q_i(t) \geqslant 0$$
$$X_i(0) = 0; \quad X_i(T_i) = \bar{X}_i$$

where T_i is the time at which firm i exhausts its reserves, to be determined.

Application of Pontryagin's pointwise maximum principle shows that the optimal production pattern for each firm must satisfy the conditions:

$$\left.\begin{aligned} & P(t) \leqslant MK_i + \lambda_i^*(t); \; Q_i(t) \geqslant 0 \\ & \lambda_i^*(t) = \lambda_i e^{\delta t} + \int_t^{T_i} \frac{\partial K_i}{\partial X_i} e^{-\delta(\tau-t)} d\tau \end{aligned}\right\} \qquad \ldots(1)$$

where MK_i is the marginal cost of production, $\frac{\partial K_i}{\partial Q_i}$, and $\lambda_i^*(t)$ is the user cost which consists of two parts. The first term is the exhaustion rent which takes account of the fact that the resource is finite, so that depletion now foregoes profits which could occur later; as Hotelling (1931) has made familiar, the present value of this exhaustion rent is constant over time. The second term takes account of the effect of depletion now in terms of increasing costs of extraction in subsequent periods. As Levhari and Liviatin (1977) note, if the resource is not fully depleted over time, the first term will be zero, while, obviously, if costs of production are independent of cumulative depletion the second term will be zero.

In (1) $\lambda_i^*(t)$ can be rewritten as:

$$\lambda_i^*(t) = \lambda_i e^{\delta t} + e^{\delta t}\left[\int_o^{T_i} \frac{\partial K_i}{\partial X_i} e^{-\delta\tau} d\tau - \int_o^t \frac{\partial K_i}{\partial X_i} e^{-\delta\tau} d\tau\right]$$

$$= (\lambda_i + Z_i)e^{\delta t} - e^{\delta t}\int_o^t \frac{\partial K_i}{\partial X_i} e^{-\delta\tau} d\tau$$

$$= R_i e^{\delta t} - e^{\delta t}\int_o^t \frac{\partial K_i}{\partial X_i} e^{-\delta\tau} d\tau \qquad \ldots(2)$$

where $Z_i = \int_o^{T_i} \frac{\partial K_i}{\partial X_i} e^{-\delta\tau} d\tau$ and $R_i = \lambda_i + Z_i$. In (2) the only unknown is $\underline{R} = (R_1, R_1 \ldots R_I)$. The problem then is to find the correct user costs R_i to assign to each producer. To do this, note that if the correct $\underline{R}$ were known then the dynamic resource allocation problem can be decomposed into a sequence of simple static competitive problems, with $\underline{R}$ knows the supply curve for each producer is known for every period, and by simply equating demand with the total industry supply curve, the price and output for each producer can be determined for each producer. To find the correct $\underline{R}$ we employ a simple tatonnement procedure. The algorithm can now be specified as follows:

(a) Assume initial values for R_i, $i \in I$. Set $D_i(0) = 0$ $i \in I$

(b) For each time period $t = 1,2\ldots$

 (i) Compute supply curves for each producer i

 (ii) Compute $P(t)$, $Q_i(t)$

 (iii) Compute $D_i(t) = D_i(t-1) + Q_i(t)$

(c) Continue (b) until a time period $\bar{T}$ such that

$$\sum_i Q_i(\bar{T}) = 0$$

(d) Compute Excess Demands $XD_i(R_1 \ldots R_I) = D_i(\bar{T}) - \bar{X}_i$

(e) Adjust R_i and return to (b)

The only remaining question is to determine whether such an iterative procedure is convergent. An intuitive proof is given here. There is a single market demand for this particular resource and the only intrinsic difference between each firm's output lies in the difference in the extraction costs. Hence each firm's output can be considered to be grossly substitutable for any other firm's output. The aggregate demand for each firm's stock of the resource is given by:

$$D_k = D_k(\underline{R})\ldots \; \forall k \varepsilon I$$

$$\frac{\partial D_k}{\partial R_\ell} \begin{cases} \geq 0 \; ; \; \ell \neq k \\ \qquad\qquad \forall k, \ell \; \varepsilon I \\ < 0 \; ; \; \ell = k \end{cases}$$

This follows from the property of the other firms' reserves being gross substitutes.

Aggregate excess demand for each firm's resource stock is simply

$$XD_k = D_k - \bar{X}_k \; \ldots \; \forall k \varepsilon I$$

and the Jacobian of XD_k is equal to the Jacobian of D_k. The problem is to find a set of $\underline{R}$, such that:

$$XD_k(\underline{R}) = 0.$$

Given the property of gross substitutability between the excess demand for different firms' reserves it is possible to invoke the well known theorem that the application of Walrasian tatonnement will ensure that the equilibrium values of $\underline{R}$ will be reached. Unfortunately, the step size must be very small to ensure convergence, and this usually requires an inordinately large number of iterations. Thus a variant of Newton's method for solving simultaneous non-linear equations was used here, since for guesses "reasonably" near to the solution, the procedure is quadratically convergent. This requires computing the Jacobian matrix numerically at each guess and then updating. The procedure is conceptually a more sophisticated tatonnement, since this approach implies that each firm recognises that an increase in its value of $\underline{R}_i$, will not only reduce the demand for its reserves, but will actually increase the demand for all other firms' reserves. By using this extra information, the equilibrium set of user costs will be found more rapidly.

4. Extensions of the Algorithm.

Computing competitive equilibria is of little interest given the partially cartelised nature of real world exhaustible resource markets, and the value of the algorithm outlined in section 3 is that it can also be used to compute Nash-Cournot and quasi-Stackelberg equilibria. The naive behavioural assumptions concerning the cartel means that the cartel believes it knows the fringe's supply function for each period. The cartel is then able to compute excess demand functions for each period and acts like a monopolist with respect to these excess demand functions. As is well-known (Hotelling, 1931), a monopolist exploiting an exhaustible resource equates marginal revenue in each period to marginal cost plus user cost, which has a form similar to that in (1). So the same decomposition argument that was applied to the competitive case applies here; once the correct user costs have been discovered, the dynamic problem can be considered as a sequence of one period Nash-Cournot or Stackelberg models.

Thus, all that is required to compute Nash-Cournot or quasi-Stackelberg models is to alter the procedure in step b (i) and (ii) for computing the equilibrium price and output for each period. The competitive model is replaced by a Nash-Cournot model or a Stackelberg model. It should be noted that the discontinuous nature of the marginal revenue function in the Stackelberg model requires careful handling at this stage. A more detailed account of these models can be found in Ulph and Folie (1978a), together with numerical results for a number of cases.

5. Stackelberg Equilibria.

The algorithm outlined in section 3 cannot be applied to Stackelberg models, because the decomposition argument no longer applies. The cartel knows that changing prices in one period will lead to the fringe changing its supply response in every period, so that one needs to solve for a complete path of prices. Analytically, Stackelberg models have been solved only for special cases (Gilbert, 1978, Ulph and Folie, 1979, Maskin and Newbery, 1978).

This is unfortunate, for the Stackelberg solution is more appealing than the Nash-Cournot model. Indeed, it has been shown that the naive behavioural assumptions of the Nash-Cournot model can lead to equilibria in which the profits of the cartel are lower than in a competitive equilibrium (Ulph and Folie, 1978a,b). This suggests that the quasi-Stackelberg model may be quite attractive, for computational experience suggests that it eliminates the undesirable features of the Nash-Cournot model, but is fairly easy to compute.

An interesting feature of Stackelberg equilibria is that there are problems of dynamic inconsistency (Ulph and Folie 1979, Maskin and Newbery, 1978). Essentially, the existence of the fringe constrains the cartel from setting full monopoly prices, but once the fringe's resources are actually exhausted the constraint is removed, and the cartel is free to announce monopoly prices immediately. Maskin and Newbery (1978) have formulated an interesting concept of rationally expected equilibria to deal with this problem, but analysis has been confined, so far, to fairly simple models.

6. Empirical Models.

The market to which the kind of analysis outlined in the earlier section of this paper has been mostly widely applied is that of crude oil, for obvious reasons. A survey of such studies has been carried out by Hammoudeh (1979). While the empirical models are able to handle general formulations of demand or cost conditions (incorporating lag structures and dynamic effects), they tend to bypass the difficulties of calculating true equilibrium solutions. The studies which are closest to the analysis found in this paper are those by Cremer and Weitzman (1976) and Pindyck (1978). Both studies claim to produce Stackelberg equilibria, although in the latter case Pindyck makes the fringe myopic, interested only in the current periods profits. This is more correctly interpreted as a Quasi-Stackelberg model with the fringe's rent constrained to be zero.

The Cremer and Weitzman study is a genuine Stackelberg model, although no details are provided about how the model is solved. We believe it was solved by imposing a finite time horizon and then searching across a finite dimensional price space, which could be a dangerous procedure if proper account is not taken of the value of reserves undepleted at the end of the time horizon.

By contrast, we will conclude this section by illustrating how the general algorithm outlined in section 3 can be applied to practical problems. We present some preliminary results from an attempt to study the oil industry. The demand for oil at the consuming centres was assumed to be characterised by the following dynamic demand relationship:

$$P(t) = 35 - 1.6667\, Q\, e^{-0.03t}$$

where Q is the total market demand at time t. The demand for oil falls to zero when the price reaches $35 per bl. The long run demand elasticity is -0.4 when the price is $10/bl and the demand is 15 bbl/yr, and the demand is considered to grow at 3% p.a.

On the supply side, it was assumed that OPEC consisted only of the Persian Gulf and North African nations. The fringe producers were the existing US and Western European producers (North Sea Oil) and the remainder of the world. Details of supply conditions and reserve levels are given in Table 1.

Table 1.

	Supply Functions.	Reserves(bbl).
OPEC	$1.15 + 0.001\ q_1^2$	500
US & W.Europe	$4.22 + 0.001\ q_2^2$	100
Rest of world	$2.21 + 0.001\ q_3^2$	180.

The required real rate of return on all assets was assumed to be 8%. Non-linear cost functions were used to reflect problems most nations would face in expanding output.

Comparing the perfectly competitive equilibrium with that of a Nash-Cournot equilibria, we obtain a doubling of the price of oil, following cartelisation, to a price of $7 per bl. The reason for the divergence between this price and the observed price of $13 per bl. is accounted for by the fact that our model captures long-run behaviour, and it is well known that the short-run demand elasticity for oil is very low. This is the point made by Pindyck's analysis, which showed that much of OPEC's current monopoly power arises from their ability to exploit the long lags in adjusting to the new price. Pindyck predicted a fall in the world oil price once major countries had eventually adjusted to the intitial price rise.

Some further analysis, again of a preliminary nature, was undertaken. Some commentators have claimed that OPEC itself is not the monopolist, but rather Saudi-Arabia, and that the rest of the OPEC countries are as much part of the fringe as US, Western Europe, and the rest of the world, (Erickson and Winokur, (1977)). We reran the model using Saudi-Arabia as the cartel, but found that the equilibrium price was not much different from the perfectly competitive price. This suggests that Saudi-Arabia by itself could not exert very significant monopoly power, so this analysis implies that OPEC is the monopolist rather than Saudi-Arabia. However, we stress the preliminary nature of these results; in particular it would be more plausible to examine dominant firm equilibria rather than Nash-Cournot. Table 2 illustrates the profits and exhaustion rents earned by the three groups under the three different market structures.

Table 2.

Profits and Rents for Different Market Structures.

		OPEC		US & W. Europe.	Rest of World.
Competitive Market.	Profits	1147		175	354
	Rents	2.17		1.71	1.90
OPEC as price setter.	Profits	1316		278	683
	Rents	0.95		2.76	3.29
		a	b		
Saudi Arabia as price setter.	Profits	440	784	181	385
	Rents	1.48	2.50	1.78	2.03

a Saudi Arabia
b Rest of OPEC

7. Conclusion.

This paper has presented an algorithm for calculating equilibria of three models of an industry depleting an exhaustible resource-competitive, Nash-Cournot and quasi-Stackelberg. Unfortunately the algorithm cannot be applied to a Stackelberg model, which is perhaps the most appealing model. However, for practical purposes the quasi-Stackelberg model might act as a sufficiently good approximation. More research on this problem is required.

References.

Cremer, J., and Weitzman, M.L., (1976), "OPEC and the Monopoly Price of World Oil", European Economic Review, Vol. 8, pp. 155-164.

Erickson, E., and Winokur, H., (1977),"Nations, Companies and Markets, International Oil and Multinational Corporations,"in Oil in the Seventies, ed. C.Watkins and M.Walker, The Fraser Institute, Vancouver, pp.170-201.

Gilbert, R.J., (1978), "Dominant Firm Pricing in a Market for an Exhaustible Resource", Bell Journal of Economics, Vol. 9, No. 2. pp.385-395.

Hammoudeh, S., (1979), "The Future Oil Price Behaviour of OPEC and Saudi Arabia: A Survey of Optimisation Models". Energy Economics, Vol. 1., No. 3.

Hotelling, H., (1931), "The Economics of Exhaustible Resources", Journal of Political Economy, 39, No. 2.

Levhari, D., and Liviatin, N., (1977), "Notes on Hotelling's Economics of Exhaustible Resources", Canadian Journal of Economics, Vol. 10, No. 2.

Maskin, E., and Newbery, D.C., (1978), "Rational Expectations with Market Power - the Paradox of the Disadvantageous Tariff on Oil", Discussion Paper 129, University of Warwick.

Pindyck, R.S., (1978), "Gains to Producers from Cartelization of Exhaustible Resources", Review of Economics and Statistics, Vol. 60.

Ulph, A.M., and Folie, G.M., (1978a), "Gains and Losses to Producers from Cartelization of an Exhaustible Resources", CRES Working Paper R/WP26, Australian National University, Canberra.

Ulph, A.M., and Folie, G.M., (1978b), "A Note on Nash-Cournot Equilibria", CRES Working Paper R/WP27, Australian National University, Canberra.

Ulph, A.M., and Folie, G.M., (1979), "Dominant Firm Models of Resource Depletion", Discussion Paper, Department of Economics, University of Southampton.

OPTIMIZATION OF A COUNTRY'S TRADE POLICIES

MIREK KARASEK
MIT, Ontario Government, Toronto
and
Mirek Karasek Associates,Toronto, Canada M4C 5L6

SUMMARY:

The "commodity-market" matrix framework together with a new "flip-side" algorithm of an integer programming technique are the two basic ingredients.

After transformation of all the basic data and information pertinent to the export operations under the given trade and economic conditions into the bi-parametric domain of the commodity-market matrix' cells, maximization of an ensuing non-linear problem yields the base solution (i.e. the base trade policy).

Comparison of the base solution with a solution based on different starting parameters then analyzes prospects for this alternative and/or new trade policy.

ASSUMPTIONS:

Assumption 1: We assume that the more important trading patterns (i.e. those that quantitatively provide a sizeable chunk of the total export figure) and their market-relative allocation[1)] are expected to remain (under ceteris paribus conditions) fairly constant[2)] in the years to come.

Reversely, the minor trades and/or spot markets do not hold much of importance in the overall picture anyway.

Assumption 2: Producers and/or exporters are trying to maximize their profit via the given prices (see e.g. [5] , p. 1860)

Assumption 3: Suppose that there are two basic tools of modelling the foreign trade operations:

a) within the framework of fixed technical (i.e. I-O) coefficients and

b) via the linear programming framework.

Then according to [3] , p. 53, the linear programming framework allows for greater variability to meet the overall intricacies of the "commodity-country" markets.

Assumption 4: Whatever level of aggregation in "commodity-country" matrix framework we work upon, the relative weights of all "commodity-market" cells must always sum up to unity. Thus, every aggregation level is assumed to be a closed system.

Assumption 5: Given the linear programming framework with a major objective and the set of activities modelled within the matrix format with rows and columns

---

1) in lit [3] on p. 67 we read: "... from the standpoint of a single exporter ..... there is a common assumption that the cost of reallocating the supply to each market ex post is prohibitive ..."

2) for confirmation see literature [9] , p. 59 when we read: "... patterns ...... are expected to remain ... constant for the next decade...."

constraints, then the system can be interpreted as a hierarchy along the lines of Saaty's "basic problem" in [11] , p. 65.

METHODOLOGY:

The following principles and features of the simulation procedure are adhered to.

The matrix ("commodity-market") format allows us to:

- obtain much better insight into most of critical elements of the country's export operations and such that is extremely useful for policy making purposes and a general overview of the situation. Since the aggregation level of commodities (and to certain extent even markets) is left entirely up to user, allowing for adding one or more of the most important commodities in great detail to even the most crudely aggregated matrix without any problem, the only serious constraints seem to be obtainability of data, time and technical limits of computing equipment.
- use the linear programming approach that was said in Assumption 3 to be beneficial to export studies.

In each "commodity-market" cell of the export simulation matrix we have two numerical parameters to reckon with:

- the one in lower-left (LL) corner, denoted by X_{ij} [3] (where i goes for commodity and j for market), is a multifactor parameter that incorporates most of the information listed in APPENDIX B.
- the parameter in upper-right (UR) corner, W_{ij} [3] , represents the relative weight this particular cell has within the total exports (as it was called for in Assumption 4).

Both LL and UR parameters constructed for each "commodity-market" cell thus define both, the somewhat overplayed constancy in trading patterns (as seen in Assumption 1), and the more recent changes in factors mentioned in item 1 of paragraph I as they occur at the producers and importers.

Then, we introduce the narrow intervals I_{ij}'s and K_{ij}'s built around X_{ij}'s and W_{ij}'s in each individual cell.

The introduction of these mini-intervals stems from Assumption 1 and as such:

- prohibits any ex post reallocation of production (and/or exports),
- prohibits drastic changes in exported (and produced) commodities assortments, and, more or less,
- suggests a constant production patterns (over a considerable time-period) as well,
- allows us to interpret the resulting $\{x_{ij}\}$ and $\{w_{ij}\}$ in the similar fashion we interpret the derivative as the slope of the curve and the indicator of functional increases or decreases.

---

3) For the sake of clarity: X_{ij} , W_{ij} will henceforth denote the original (or starting) numerical values of these parameters, whereas x_{ij} , w_{ij} represent their arbitrary values within the intervals I_{ij} , K_{ij} and $\{x_{ij}\}$, $\{w_{ij}\}$ are thus their resulting values.

In the next step, the "flip-side" algorithm (see APPENDIX A) finds an optimum solution, i.e. the sets $\{x_{ij}\}$ and $\{w_{ij}\}$ that provide a quantitative rendition of how individual commodities sold in individual markets rank in contributing to the overall pay-off of the investigated trade policy under the maximum pay-off criterion. To put it in another words, general idea is to let the optimization process establish a "BEST COMBINATION OF CELLS" pertaining to the given economic conditions (or trade factors).

The "BEST COMBINATION" then mapps such foreign trade policy that assures the maximum pay-off from the country's foreign trade operations under the given trade factors. The methodology thus starts from the, so called, base solution in which standard conditions, i.e. levels of the trade factors (mentioned in I/1) that are chosen to be those of the middle '70s, are simulated.

After the optimum solution is found, these cells' results $\{x_{ij}\}$ and $\{w_{ij}\}$ are converted via the intensity scale of importance (see e.g. [12] , pp. 253 - 256 or Table 5 of APPENDIX A) into the simple graphic expressions, such as $\uparrow$ ...(i,j)th cell's potential for increased pay-off, and $\downarrow$ ... (i,j)th cell's importance has decreased, and the policy assessments are enhanced by the Saaty's eigen-vectors of column importance weights.

In the second step, the change in the trade, marketing and economic conditions yields (probably) different starting parameters X_{ij} (for re-estimation of X_{ij}'s we use various methods, such as the one presented in APPENDIX C) and thus the different (i.e. alternative) solution. The crux of the simulation is then to compare the alternative solution(s) with the standard one and to record and analyze the most important differences in the individual cells' importance coefficients.

In the subsequent steps we may use the same technique, only this time we apply it to the detailed "commodity market" matrix in which one of the previously designated (on the pay-off scale) cells is disaggregated in order to analyze the individual exported commodities.

EXAMPLE OF THE SIMULATION TECHNIQUE

To illustrate the simulation algorithm a 7 x 4 "commodity-market" matrix format was chosen together with the decision to have 1974 Canadian export and import data-base (from [14]) acting for the current state of affair and a base for this experimental run.

Most of the tariff data and transportation costs data were taken from the literature [1] and even though they are admittedly outdated (maybe even outright off target) they have served their purpose, i.e. to test the susceptibility of the simulation technique to the assumed changes in the trade-economic factors, rather well.

The starting matrix with all its K_{ij}'s and I_{ij}'s and rows and columns constraints is presented in Table 1. Appropriate X_{ij}'s and W_{ij}'s lie in the centre of the cell's intervals.

Starting Matrix for Base Solution

Sector	U.S.A.		JAPAN		EEC		S. AMERICA		Constr.
0+1	100 82	3.30 2.70	57 52	1.90 1.50	73 59	3.50 2.90	88 72	1.70 1.30	9.40 289
2+4	94 76	11.70 9.50	56 46	4.70 3.90	73 59	7.30 6.00	84 68	0.44 0.36	22.0 278
3	98 80	17.30 14.20	50 40	1.00 0.80	85 69	0.44 0.36	91 75	0.001 0.000	17.10 294
5	81 67	2.60 2.20	37 31	0.11 0.00	88 72	0.60 0.40	225 185	0.22 0.18	3.20 393
6	85 69	14.90 12.10	59 49	0.33 0.27	68 56	3.70 3.10	80 66	1.10 0.90	18.20 266
7	81 67	27.70 22.70	70 58	0.11 0.09	90 74	1.70 1.30	81 67	1.30 1.10	28.00 294
8+9	29 23	1.80 1.40	122 100	0.11 0.09	64 52	0.44 0.36	103 85	0.11 0.09	2.20 289
Constr.	516	72.10	411	7.50	491	16.10	685	4.40	

TABLE 1.

Base Solution Matrix

Sector	Commodities	U.S.	Canadian Exports JAPAN	EEC	S.AMERICA
0+1	Food, Live Animals, Beverage, Tobacco, etc.	2	1/9	1/2	7
2+4	Crude Materials, excl. Fuels, Animal & Veg. Oil, Fat	1/5	6	6	1/8
3	Mineral Fuels (Petroleum Products)	6	1/2	1/8	1/5
5	Chemicals (Radioactive Materials)	1/7	3	2	7
6	Basic Manuf. (Paper, Textile)	6	1/9	1/3	1/9
7	Machines, Transport. Equipment (Auto)	2	2	7	1/8
8+9	Misc. Manuf. Goods (Clothing)	1/7	7	1/2	7
Importance-weights		0.32	0.21	0.20	0.27

TABLE 2.

Starting Matrix for an Alternative Export Policy

Sector	U.S.A.		JAPAN		EEC		S. AMERICA		Constr.
0+1	113 93	3.30 2.70	57 52	1.90 1.50	86 70	3.50 2.90	88 72	1.70 1.30	9.40 313
2+4	110 90	11.70 9.50	56 46	4.70 3.90	84 68	7.30 6.00	84 68	0.44 0.36	22.0 303
3	111 91	17.30 14.20	50 40	1.00 0.80	90 74	0.44 0.36	91 75	0.001 0.000	17.10 311
5	105 85	2.60 2.20	37 31	0.11 0.00	105 85	0.60 0.40	225 185	0.22 0.18	3.20 429
6	109 89	14.90 12.10	59 49	0.33 0.27	84 68	3.70 3.10	80 66	1.10 0.90	3.20 302
7	112 92	27.70 22.70	70 58	0.11 0.09	112 92	1.70 1.30	81 67	1.30 1.10	28.0 342
8+9	40 32	1.80 1.40	122 100	0.11 0.09	78 64	0.44 0.36	103 85	0.11 0.09	2.20 312
Constr.	636	72.10	411	7.50	580	16.10	685	4.40	

TABLE 3.

Alternative Export Policy Solution

Sector	Commodities	Canadian Exports U.S.	 JAPAN	 EEC	 S.AMERICA
0+1	Food, Live Animals, Beverage, Tobacco, etc.	1/9	2	6 ↑	1 ↓
2+4	Crude Materials, excl. Fuels, Animal & Veg. Oil, Fat	1/5	6	6	3
3	Mineral Fuels (Petroleum Products)	6	2	1/9	1/5
5	Chemicals (Radioactive Materials)	1/7	1/9	7 ↑	7
6	Basic Manuf. (Paper, Textile)	6	1/9	1/3	1/9
7	Machines, Transport. Equipment (Auto)	6 ↑	1/9	1/7 ↓	1/3
8+9	Misc. Manuf. (Clothing, Watch)	1/7	3 ↓	3	9
Importance-weights		0.37 ↑	0.15 ↓	0.26 ↑	0.22 ↓

TABLE 4.

In the first run, all tariffs were assumed to be holding all across the board and the final matrix (with the importance-scale entries) in TABLE 2. shows the vector of numerical assessment of the basic importance-weights for all four markets.

For the second run we made the assumption about the effective tariffs being removed in the U.S.[4)] and EEC markets. The appropriate starting matrix is seen in TABLE 3. and the final table is then presented in TABLE 4.

CONCLUSION AND DISCUSSION:

The comparison of TABLE 2. with TABLE 4. offers the following conclusions:

a) As far as the markets go the predictable consequence of the new trade conditions, i.e. an increased importance of exports to the U.S. and EEC, has been upheld.

b) Far more interesting, however, are the movements of the individual cells' importance assessments. Denoting by ↑ a substantial increase of a cell's importance intensity (same notation is used for importance weights column elements) and by ↓ a substantial decrease, there are six significant changes in TABLE 4. cells' assessments (as opposed to TABLE 2. entries): Increase of importance of sector 7 exports to the U.S. (if the optimum pay-off is to be realized) and sectors 0+1 and 5 to EEC. For the same reason the cuts in exports of sector 8+9 to Japan, sector 7 to EEC and 0+1 to South America (most of which are nonsensical anyway) should follow.

And this is, in a nutshell, what we can realistically expect from the new trade conditions. Now, whether all these anticipated changes are congruent with the country's blue-print of industrial strategy is altogether different matter. To include these criteria, we have to adjust the starting matrix intervals I_{ij}'s accordingly. One way would be to assess a prohibitive low X_{ij}'s to the "no-no" sectors (e.g. these we won't like to subsidize any more) via expression (B4).

REFERENCES:

[1] Balassa, B., "Tariff Protection in Industrial Nations and its Effects on the Exports of Processed Goods from Developing Countries", C. Journal of Economics, I, No. 3

[2] Baumann, H., "Structural Characteristics of Canada's Pattern of Trade", C. Journal of Economics, IX, No. 3 (August 1976)

[3] Chipman, J.S., "A Survey of the Theory of International Trade: Part 3, The Modern Theory", Econometrica, Vol. 34, No. 1

[4] Dubey, P., Shubik, M., "Trade and Prices in a Closed Economy with Exogenous Uncertainty, Different Levels of Information, Money and Compound Futures Markets", Econometrica, Vol. 45, No. 7

[5] Hatta, T., "A Recommendation for a Better Tariff Structure", Econometrica, Vol. 45, No. 8

[6] Hu, S.C., "Uncertainty, Domestic Demand, and Exports", C. Journal of Economics, VIII, No. 2

4) in [2] on p. 417 we read: "... there is no support for the hypothesis that

[7] Karasek, M., "Methods of Simulation of Economic Processes II", TESLA Res. Report, 1967, p. 9

[8] Karasek, M., "A simple Graphical Rendition of Problems in Exporting Canadian Manufactured Goods", ITAB Res. Report, MIT, 1979

[9] Postner, H.H., Factor Content of Canadian International Trade: An Input-Output Analysis, Econ. Council of Canada, 1975

[10] Wonnacott, R.J., "Industrial Strategy: A Canadian Substitute for Trade Liberalization", C. Journal of Econ., VIII, No. 4

[11] Saaty, T.L., "Modelling Unstructured Decision Problems: The Theory of Analytical Hierarchies", Proc. I. Internat. Conf. on Math. Modelling, University of Missouri-Rolla, 1977, pp. 59 - 76

[12] Saaty, T.L., Rogers, P., "Higher Education in the U.S. (1985-2000): Scenario Construction Using a Hierarchical Framework with Eigenvector Weighting, Socio-Econ. Plan. Sci., Vol. 10, pp. 251 - 263

[13] Williams, J.R., "Commodity Trade and the Factor Proportions Theorem", Can. Journal of Economics, X, No. 2

[14] OECD, Statistics of Foreign Trade

APPENDIX:

A) The Maximization Problem of Two Variables, the "Flip-Side" Algorithm and its Application to Policy Simulations

Formulated mathematically, the proglem of trade policies simulations is to find such solutions $\{x_{ij}\}$, $\{w_{ij}\}$, $x_{ij} \in I_{ij}$ and $w_{ij} \in K_{ij}$, for which the function

$$f(x,w) = \sum_i \sum_j w_{ij} x_{ij} \quad \text{reaches its maximum} \tag{A1}$$

subject to linear constraints:

$$\sum_j x_{ij} = a_i , \qquad i = 1 \ldots m , \tag{A2}$$

$$\sum_i x_{ij} = b_j , \qquad j = 1 \ldots n , \tag{A3}$$

$$\sum_i a_i = \sum_j b_j , \tag{A4}$$

$$\sum_j w_{ij} = g_i , \qquad i = 1 \ldots m , \tag{A5}$$

$$\sum_i w_{ij} = h_j , \qquad j = 1 \ldots n , \tag{A6}$$

$$\sum_i g_i = \sum_j h_j . \tag{A7}$$

4) cont'd the U.S. tariff per se prevents operations on a sufficient scale for Canadian plants to be competitive ... see also [2] , eq. (2e) on p. 416

Now, let M be the closed bounded subset of all the feasible solutions of the problem (A1) - (A7), with its sup M and inf M.

Consider the sequence $\{f_n\}$ in M, such that

$f_1 = f\left[x_1(w_o)\right]$.... is the solution of the LP (A1) - (A4) with costs $w_{iojo} \in K_{ij}$, i.e. the original weights W_{ij}.

$f_2 = f\left[w_1(x_1)\right]$.... is the solution of the problem (A1), (A5) - (A7) with "variables" $w_{ij} \in K_{ij}$ and new constant "costs" introduced into it are the results $\{x_{ij}\}$ from the computation of f_1,

$f_3 = f\left[x_2(w_1)\right]$.... is the solution of (A1) - (A4) with weights $\{w_{ij}\}$ resulting from f_2

$f_n = f\left[x_r(w_{r-1})\right]$ or $f\left[w_r(x_{r-1})\right]$ .... is the solution of the last "constants variables" flip, where $r+(r-1) = n$.

It becomes clear that for $\{f_n\} \in M$ holds

$$f_1 \leq f_2 \leq f_3 \leq \ldots f_{n-1} \leq f_n \tag{A8}$$

and, also that: 1) $\{f_n\}$ is somewhat dependent on the choice[5)] of initial values W_{iojo}, 2) the number of elements in (A8) is, according to the Heine-Borel Theorem, finite, and 3) (A8) converges either to a certain local or the absolute maximum.

The limit point (and thus the solution of (A1) - (A7) is found via the following theorem:

<u>THEOREM</u>: A necessary condition for function (A1) to attain its local or absolute minimum subject to (A2) - (A7) is the existence of at least two following elements of the sequence (A8), f_{k-1} and f_k, such that it holds $f_{k-1} = f_k$, $k = 1 \ldots .n$.

<u>Proof</u>: Suppose $f_{k-1} \neq f_k$

Then, from (A8) follows

$$f_{k-1} < f_k , \qquad \text{for } k = 1 \ldots .n. \tag{A9}$$

Let us consider some index number k_{max}, in which (A8) attains its limit point. From (A9) we have

$$f_{k_{max}} < f_{k_{max}+1} \tag{A10}$$

and the relation between the supposition and the contradiction in (A10) completes this scheme of proof.

---

5) it could be shown that to assure the steepest convergency of (A8) to a local (but still permissible) maximum or to the absolute maximum, the centres of intervals K_{ij}, we denote them W_{ij}, have to be chosen in f_1.

To transform the numerical optimization (A1) - (A10) into the terms of policy assessments (to be seen in TABLE 2. and 4.) we use the below conversion table. First, let us denote an individual (i,j)th cell's results such that we assign: $+$ to $x_{ij} \in (X_{ij}, x_{ijmax}\rangle$, $++$ to $x_{ij} \equiv x_{ijmax}$, $-$ to $x_{ij} \in \langle x_{ijmin}, X_{ij})$ and $--$ to $x_{ij} \equiv x_{ijmin}$. Then, we use the following conversion to arrive at the importance scale (see e.g. [12], p. 253).

Conversion Table

Definition	Intensity of Importance	Cells' Results of Optimization
Two activities contribute equally to the objective	1	$+^{-}$ or $-^{+}$
One activity is to be slightly favoured	2	$--^{++}$ or $++^{-}$
The judgment is to favour one activity over another	3	$-^{++}$
The judgment is to strongly favour one activity	5 6	$+^{+}$ $++^{+}$
Demonstrated Importance	7	$+^{++}$
Absolute Importance	9	$++^{++}$
If activity i has one of the above non-zero numbers assigned to it when compared with activity j, then j has the reciprocal value when compared with i	1/2 1/3 1/5 1/7 1/8 1/9	$++^{--}$ or $--^{+}$ $+^{--}$ $-^{-}$ $--^{-}$ $-^{--}$ $--^{--}$

TABLE 5.

PROPOSITION: Final solution matrices in TABLE 2. and 4. could be regarded as the workable approximations of the judgmental matrices of the pairwise comparison defined in [11], p. 61.

Proof: Follows easily from the linear constraints (A2) - (A7) and the role of X_{ij}'s and W_{ij}'s in TABLE 5. intensity of importance scale.

B) The All-Important Coefficient X_{ij}

Suppose we have a "one-exporter-one-importer" situation, where one commodity i is being traded (as seen in Figure 1)

Figure 1

and where EX_i ... is the exporter's share of total world exports of the i-th commodity, IM_i ... is the importer's share of total world imports of the i-th commodity, EX_{ij} ... is the share of total exporter's trades of i-th commodity going to the j-th importer, IM_{ij} ... is the share of total imports of i-th commodity by j-th exporter.

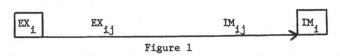

Under ideal conditions and factors equilibrium (see [13]), the "pure" hypothetical exported quantity

$$EX_{ij} = \left[IM_{ij} \times IM_i / EX_i \right] \qquad (B1)$$

Now, from [1] , eq. (3) - (4) on p. 585 and eq. (3) in [5] on p. 1861 we can construct a "real" (however, still hypothetical) quantity EX_{ij} , where to the original equilibrium conditions we add: the effective tariff z_{ij} , transportation cost d_{ij} , money exchange rate p_{ij} and eventual artificial barriers to grow for"unwanted" industries b_i (this could denote the productivity levels scale as well)

$$\widehat{EX}_{ij} = (1 + Z_{ij}) \cdot \left[IM_{ij} \times IM_i / EX_i \right] , \qquad (B2)$$

$$\text{where} \quad Z_{ij} = (d_{ij} + z_{ij}) \times p_{ij}/b_i \ . \qquad (B3)$$

As opposed to (B1) and (B2) , we have the actual $\overline{EX}_{ij}$ from trade statistics, and so the coefficient X_{ij} is now

$$X_{ij} = \overline{EX}_{ij} / \widehat{EX}_{ij} , \qquad (B4)$$

where $X_{ij} \in \langle 0, +\infty \rangle$ and all numerical values greater than 1 are desirable because they are better than even the ideal conditions prescribe for the particular (i,j)th cell.

C) Graphical Representation of Factors Involved in Equations (B2) - (B4)

An example of nomographic representation of main factors influencing the actual Canadian manufacturing exports to the U.S., i.e. $X_{man.,U.S.}$, is seen in Figure 2 (taken from [8]).

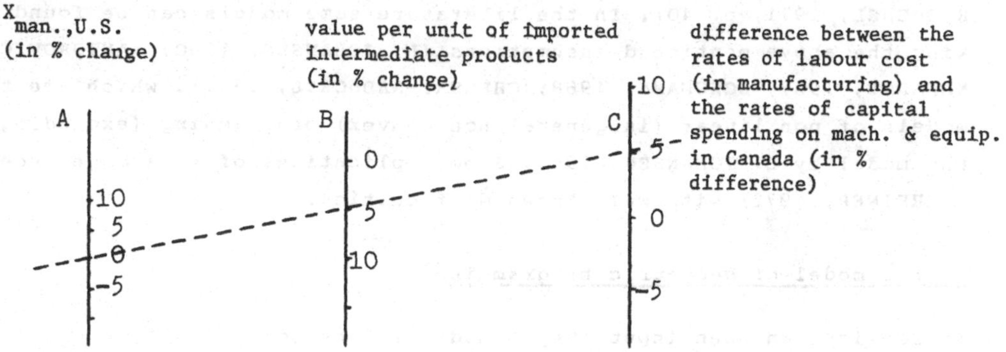

Figure 2

Since, by the definition of nomograms, a straight edge cutting the three lines (A, B, C) gives the related value of the three variables, from assumed changes in rates on lines B and C we can easily estimate the change in X_{ij} on line A.

AN OPEN INPUT-OUTPUT MODEL WITH CONTINUOUS SUBSTITUTION BETWEEN PRIMARY FACTORS AS A PROBLEM OF GEOMETRIC PROGRAMMING

M. Luptacik
Technical University Vienna
Argentinierstraße 8
A-1040 Vienna, Austria

1. Substitution in input-output models

The standard input-output models are characterized by constant input coefficients and constant labour and capital coefficients (Leontief production function). Under this assumption "is growth likely to be impeded by shortages of specific factors rather than by a general scarcity of resources ... . The empirical analysis of substitution possibilities is therefore critical to design of planning models and to the interpretation of their results" (CHENERY-RADUCHEL, 1971, p. 29). A.S. MANNE (1974, p. 451) writes in a survey on the multi-sector models: "... input-output is an unsatisfactory framework for dealing with the problems of choice and substitution. Without sufficient substitution, a planning model is likely to reveal spurious bottlenecks and to indicate erratic behaviour of the efficiency prices".

For these reasons we suppose furthermore the constant input coefficients, but a direct substitution possibility between labour and capital is added. Specially for the analysis of demand for primary factors is this substitution assumption crucial. "Direct substitution between capital and labor appears to be of greater significance for employment than indirect substitution under most of our assumptions" (CHENERY-RADUCHEL, 1971, p. 40). In the literature some models can be found making the above mentioned assumptions (L. JOHANSEN, 1960; KANTOROVIC-MAKAROV, 1967; SCHUMANN, 1968; CHENERY-RADUCHEL, 1971), which are the models of non linear (in general non-convex) programming (excluding the model by L. JOHANSEN, 1960. Some applications of this model see P. SCHREINER, 1972) with well known difficulties.

2. The model of geometric programming

We consider an open input-output model in the form

$$A.\underline{x} + \underline{y} = \underline{x} \qquad (1)$$

or

$$\underline{x} = (E-A)^{-1}.\underline{y} \qquad (2)$$

where $A = (a_{ij})$ is an nxn matrix of input coefficients, $\underline{x}$ is an n-dimensional gross production vector, $\underline{y}$ an n-dimensional final demand vector and E an nxn identity matrix.

We assume exogenously given final demand $\underline{y}$ and substitution posibilities for labour and capital inputs, according to a Cobb-Douglas function for each sector of production. The function is written as

$$x_j = \varepsilon_j L_j^{\alpha_j} K_j^{\beta_j} \qquad j = 1, 2, \ldots, n \tag{3}$$

where x_j, L_j and K_j indicate gross output, employment and total capital stock in sector j (Empirical estimation of such production functions see e.g. P. SCHREINER, 1972 for about thirty sectors of the Norwegian economy).

The balance constraints (1) can be written as

$$(E-A)\underline{x} \geqq \underline{y} \tag{4}$$

or

$$\sum_{j=1}^{n} (\delta_{ij} - a_{ij}) x_j - y_i \geqq 0 \qquad i = 1, 2, \ldots, n \tag{5}$$

where $\delta_{ij} = 0$ for $i \neq j$ and $\delta_{ij} = 1$ for $i = j$ (Kronecker delta). Now we substitute function (3) for x_j and after simple transformation we get

$$\sum_{j \neq i} d_{ij} L_j^{\alpha_j} K_j^{\beta_j} L_i^{-\alpha_i} K_i^{-\beta_i} + \frac{y_i}{(1-a_{ii})\varepsilon_i} L_i^{-\alpha_i} K_i^{-\beta_i} \leqq 1 \tag{6}$$

for $i = 1, 2, \ldots, n$

where

$$d_{ij} = \frac{a_{ij}\varepsilon_j}{(1-a_{ii})\varepsilon_i} \geqq 0 \qquad \text{for } i, j = 1, 2, \ldots, n$$

The form of constraints for capital stock depends on the transferability of capital between the particular sectors of the economy. Under the assumption of perfect transferability we have

$$K_1 + K_2 + \ldots + K_n \leqq \bar{K} \tag{7a}$$

where $\bar{K}$ indicates the total capital stock in the economy. In the opposite case (non-transferability of capital), we have constraints for capital stock in each sector of the economy

$$K_j \leqq \bar{K}_j \qquad j = 1, 2, \ldots, n \tag{8a}$$

where $\bar{K}_j$ indicate the disponable capital stock in sector j.

It would be possible to consider analogous constraints for the second primary factor labour. But there is a crucial difference. It is a goal of policy to decrease labour input (the working time) on one hand but to improve the conditions for labour use (the exploitation of capital) on the other. For this reason we consider the objective function as a minimization of labour input

$$\min L = L_1 + L_2 + \dots + L_n \tag{9}$$

where L indicates the total labour input in the economy. In other words, for exogeneously given final demand $\underline{y}$, the objective function (9) implies the maximization of labour productivity.

Now we have the following problem of geometric programming

$$\min L = L_1 + L_2 + \dots + L_n$$

under the constraints (6), the constraints

$$\frac{1}{\bar{K}}(K_1+K_2+ \dots +K_n) \leqq 1 \tag{7b}$$

or

$$\frac{1}{\bar{K}_j} \cdot K_j \leqq 1 \qquad \text{for } j = 1, 2, \dots, n \tag{8b}$$

and the constraints

$$L_j>0, \; K_j>0 \qquad \text{for } j = 1, 2, \dots, n \tag{10}$$

The economic interpretation of the conditions (7) or (8) is obviously. The conditions (5) imply that the gross production in each sector of the economy must be greater or equal to the deliveries of this sector to all other sectors of the economy and to the final demand. In other words - corresponding to the form (6) - the sum of the proportions of the deliveries from sector i - into all other sectors and to the final demand - to the net production of sector i must be equal or lower than one.

The solution of this model as a standard problem of geometric programming (which provides efficient algorithms with convergence to the global optimum compared with similar model in general of non-convex programming e.g. by SCHUMANN, 1968, pp. 138-164) gives us - for exogeneously given final demand - the optimal allocation of labour and capital (under the assumption of transferability of capital) to the particular sectors of the economy (or the optimal exploitation of capital in each sector of the economy in the case of non-transferability of capital). In other words, we get the optimal gross production for each sector in the case of neoclassical sectoral production function[1].

3. The dual model

A very useful instrument and important informations for economic analysis provides the duality theory of geometric programming[2]. It can be shown (LUPTACIK, 1977, pp. 65-71) that the dual variables are - from economic interpretation point of view - elasticity coefficients, as opposed to the interpretation of the dual variables as marginal coefficients in linear programming. The elasticity coefficients as important indicators for economic analysis are without dimension and therefore easily comparable. They give us the percentage change of the value of the primal function when the corresponding term in the primal problem changes by one percent and all other terms remain constant.

For reason of simplicity but without loss of generality we consider only two sectors in our model of geometric programming. The dual model has the following form:

$$\max v(\underline{\delta}) = \left(\frac{1}{\delta_{01}}\right)^{\delta_{01}} \cdot \left(\frac{1}{\delta_{02}}\right)^{\delta_{02}} \cdot \left(\frac{d_{12}}{\delta_{11}}\right)^{\delta_{11}} \cdot \left(\frac{c_1}{\delta_{12}}\right)^{\delta_{12}} \cdot$$

$$\cdot \left(\frac{d_{21}}{\delta_{22}}\right)^{\delta_{21}} \cdot \left(\frac{c_2}{\delta_{22}}\right)^{\delta_{22}} \cdot \left(\frac{\frac{1}{\bar{K}_1}}{\delta_{31}}\right)^{\delta_{31}} \cdot \left(\frac{\frac{1}{\bar{K}_2}}{\delta_{41}}\right)^{\delta_{41}} \cdot \lambda_1(\underline{\delta})^{\lambda_1(\underline{\delta})} \cdot$$

$$\cdot \lambda_2(\underline{\delta})^{\lambda_2(\underline{\delta})} \cdot \lambda_3(\underline{\delta})^{\lambda_3(\underline{\delta})} \cdot \lambda_4(\underline{\delta})^{\lambda_4(\underline{\delta})}$$

where $\lambda_m(\underline{\delta}) = \sum_{t=1}^{T_m} \delta_{mt}$ for $m = 1, \ldots, 4$ and $c_i = \dfrac{y_i}{(1-a_{ii})\,\varepsilon_i}$; $i = 1,2$

such that

---

1) "It would be useful if the algorithm (that means simplex algorithm, M.L.) could be extended to the case where the underlying sectoral production functions are neoclassical rather than of the fixed coefficients, activity analysis variety" (W.E. DIEWERT, 1974, p. 154). Another version of the model with CES sectoral production functions see M. LUPTACIK, 1977, pp. 53-61.

2) For duality theory of geometric programming see DUFFIN-PETERSON-ZENER, 1967, Chapters 3-4.

$$\delta_{01} \geq 0, \ \delta_{02} \geq 0, \ldots \delta_{41} \geq 0$$

$$\delta_{01} + \delta_{02} = 1 \qquad (11)$$

$$\left.\begin{array}{l} \delta_{01} - \alpha_1\delta_{11} - \alpha_1\delta_{12} + \alpha_1\delta_{21} = 0 \\ \delta_{02} + \alpha_2\delta_{11} - \alpha_2\delta_{21} - \alpha_2\delta_{22} = 0 \\ - \beta_1\delta_{11} - \beta_1\delta_{12} + \beta_1\delta_{21} + \delta_{31} = 0 \\ \beta_2\delta_{11} - \beta_2\delta_{21} - \beta_2\delta_{22} + \delta_{41} = 0 \end{array}\right\} \qquad (12)$$

Now we consider e.g. δ_{21}. If the deliveries of the first sector to the second remain constant, it holds (LUPTACIK, 1977, pp. 70 and 74):

$$\delta^o_{12} = \frac{\partial L^o}{L^o} \Big/ \frac{\partial y_1}{y_1}$$

what is exactly the definition of elasticity. According to this form δ^o_{12} give us the percentage change of total labour input (or employment) when the final demand of sector one changes by one percent (and the demand of the second sector remain constant).

In similar way can be interpreted the other dual variables δ_{mt} excluding the variables δ_{01} and δ_{02}. The latter give us the proportion of labour input (employment) of sector one and two to the total labour input (employment) in the economy. The orthogonality conditions (12) are the balance equations for the primary factors labour and capital used in sector one and two. Under the assumption of constant return to scale ($\alpha_j+\beta_j = 1$ for $j = 1,2$) the conditions (12) imply the equality of the value of final demand to the value of the primary factors labour and capital (see LUPTACIK, 1977, pp. 79-81). In other words the exhaustion theorem well known in economics (see theorem by Eisenberg and Van Moeseke in: SENGUPTA-FOX, 1971, pp. 93-94).

4. Surrogate constraints and the transferability of capital

Now we come to the above mentioned problem of transferability of capital between various sectors of the economy. For this reason we consider two problems of geometric programming: first minimize (9) under the constraints (6), (7b), (10) and second minimize (9) under the constraints (6), (8b) and (10). In generally

A. $\min g_o(\underline{x})$

such that

$$g_m(\underline{x}) \leqq 1, \qquad m = 1, 2, \ldots, M$$

and

$$g_{M+1}(\underline{x}) \leqq 1$$

$$g_{M+2}(\underline{x}) \leqq 1$$

$$\underline{x} > \underline{O}$$

where $g_m(\underline{x}) = \sum_{t=1}^{T_m} c_{mt} \prod_{j=1}^{N} x_j^{a_{mtj}} \qquad m = O, 1, \ldots, M$

and $g_{M+1}(\underline{x}) = c_{M+1} \prod_{j=1}^{N} x_j^{a_{M+1j}}$

$$g_{M+2}(\underline{x}) = c_{M+2} \prod_{j=1}^{N} x_j^{a_{M+2j}}$$

with $c_{mt} > O$ for $t = 1, 2, \ldots, T_m$, $m = 1, 2, \ldots, M+2$

B. min $g_o(\underline{x})$

such that

$$g_m(\underline{x}) \leqq 1 \qquad m = 1, 2, \ldots, M$$

and

$$\xi g_{M+1}(\underline{x}) + \eta g_{M+2}(\underline{x}) \leqq 1$$

$$\underline{x} > \underline{O}$$

where $\xi \geqq O$, $\eta \geqq O$ and $\xi + \eta = 1$ (the so-called surrogate constraint. See also BEIGHTLER - PHILLIPS, 1976, Chapter 6). Then can be proved the following theorem:

<u>Theorem 1</u>: $g_o(\underline{x}_A^o) = g_o(\underline{x}_B^o)$ then, if

$$\xi = k.\delta^o_{M+1,A}$$

$$\eta = k.\delta_{M+2,A}$$

for some positive k, where $\delta^o_{M+1,A}$ and $\delta^o_{M+2,A}$ are the optimal values for the dual variables of the problem A.

<u>Proof</u>: From the Kuhn-Tucker conditions for both problems and by using the duality theory of geometric programming we get:

$$\frac{\partial g_o(\underline{x})}{\partial \underline{x}} + \sum_{m=1}^{M} \mu_m \cdot \frac{\partial g_m(\underline{x})}{\partial \underline{x}} + \delta^o_{M+1,A} \cdot g_o(\underline{x}^o_A) \frac{\partial g_{M+1}(\underline{x})}{\partial \underline{x}} + + \delta^o_{M+2,A} \cdot g_o(\underline{x}^o_A) \cdot \frac{\partial g_{M+2}(\underline{x})}{\partial \underline{x}} = 0 \tag{13}$$

$$\frac{\partial g_o(\underline{x})}{\partial \underline{x}} + \sum_{m=1}^{M} \mu_m \cdot \frac{\partial g_m(\underline{x})}{\partial \underline{x}} + \xi\lambda^o_B \cdot g_o(\underline{x}^o_B) \cdot \frac{\partial g_{M+1}(\underline{x})}{\partial \underline{x}} + + \eta\lambda^o_B \cdot g_o(\underline{x}^o_B) \cdot \frac{\partial g_{M+2}(\underline{x})}{\partial \underline{x}} = 0 \tag{14}$$

If now $\xi = \frac{1}{\lambda^o_B} \cdot \delta^o_{M+1,A}$

and $\eta = \frac{1}{\lambda^o_B} \cdot \delta_{M+2,A}$

it follows from conditions (13) and (14) that $g_o(\underline{x}^o_A) = g_o(\underline{x}^o_B)$[1].

The theorem 1 give us the conditions under which the solution of the model with transferability of capital between various sectors is identical with the solution of the model under non-transferability of capital. If the surrogate multipliers - or the "evaluations" of transferable capital used in particular sector of the economy - correspond to the dual variables or elasticity coefficients for nontransferable capital in particular sectors, the solutions of both models are identical.

5. The nonlinear employment multiplier

A very important question in the framework of input output analysis is, how the changes in the exogenously given final demand influence the demand for primary factors or for the employment. Under the assumption of Leontief production function the answer is well known.

However consider this question in the framework of our model with direct substitution between labour and capital. Suppose, the final demand will be increased and we have a new vector of final demand $\underline{y}^{new}$. That means we have new coefficients $c_1^{new}, c_2^{new} \ldots c_n^{new}$ (under the assumption that all other parameters remain constant). How will be changed now the value of objective function or employment in the economy?

According to the duality theory of geometric programming we can write:

---

1) A little other formulation of this theorem see also BEIGHTLER - PHILLIPS, 1976, Chapter 6, theorem 3, p. 216 and corollary 2 p. 218.

$$\frac{L^{new}}{L^{old}} \geqq \left(\frac{c_1^{new}}{c_1^{old}}\right)^{\delta_{12}^{old}} \left(\frac{c_2^{new}}{c_2^{old}}\right)^{\delta_{22}^{old}} \cdots \qquad (15)$$

or

$$L^{new} \geqq L^{old} \cdot \left(\frac{c_1^{new}}{c_1^{old}}\right)^{\delta_{12}^{old}} \left(\frac{c_2^{new}}{c_2^{old}}\right)^{\delta_{22}^{old}} \cdots \qquad (16)$$

The right side of the form (15) give us a lower bound on the relative increase of employment due to the increase of the final demand. In other words a lower bound on the nonlinear employment multiplier.

Using the form (16) we get a lower bound for new levél of employment corresponding to the new vector of final demand.

References:

BEIGHTLER CH.S. - PHILLIPS D.T., 1976, "Applied Geometric Programming", John Wiley & Sons, Inc. New York-London-Sydney-Toronto.

CHENERY H.B. (ed.), 1971, "Studies in Development Planning", Harvard University Press, Cambridge, Massachusetts.

CHENERY H.B. - RADUCHEL, 1971, "Substitution in Planning Models" in H.B. CHENERY (ed.), 1971.

DIEWERT W.E., 1974, "Applications of Duality Theory" in INTRILIGATOR M.D. - KENDRICK D.A. (eds.), 1974.

DUFFIN R.J. - PETERSON E.L. - ZENER C., 1967, "Geometric Programming - Theory and Applications, John Wiley & Sons, Inc. New York-London-Toronto.

INTRILIGATOR M.D. - KENDRICK D.A. (eds.), 1974, "Frontiers of Quantitative Economics", Vol. II, North-Holland Publishing Company, Amsterdam-Oxford.

JOHANSEN L., 1960, "A Multi-sectoral Study of Economic Growth", North-Holland Publishing Company, Amsterdam.

KANTOROVIC L.V. - MAKAROV V.L., 1967, "Optimálne modely perspektivneho plånovania" in: Pouzitie matematiky v ekonomike, III. diel, Bratislava.

LUPTACIK M., 1977, "Geometrische Programmierung und ökonomische Analyse" (Geometric Programming and Economic Analysis), Mathematical Systems in Economics, Verlag Anton Hain, Meisenheim am Glan.

MANNE A.S., 1974, "Multi-sector Models for a Development Planning: A Survey" in: M.D. INTRILIGATOR - KENDRICK D.A. (eds.), 1974, S. 449-481.

SCHREINER P., 1972, "The Role of Input-Output in the Perspective Analysis of the Norwegian economy" in BRÓDY A. - CARTER A.P. (eds.), "Input-Output Techniques", Proceedings of the Fifth International Conference on Input-Output Techniques, Geneva, January, 1971, North-Holland, Amsterdam-London.

SCHUMANN J., 1968, "Input-Output Analyse", Springer-Verlag, Berlin, Heidelberg, New York.

SENGUPTA, J.K. - FOX K.A., 1971, "Optimization Techniques in Quantitative Economic Models", Studies in Mathematical and Managerial Economics, Vol. 10, North-Holland Publishing Company, Amsterdam-London.

AN EQUILIBRIUM MODEL FOR AN OPEN ECONOMY WITH INSTITUTIONAL CONSTRAINTS ON FACTOR PRICES

Lars Mathiesen and Terje Hansen

The Norwegian School of Economics and Business Administration

N-5000 Bergen, Norway.

Abstract.

Our purposes with this paper are twofold. First we want to show that the complementarity format is convenient for economic equilibrium modelling. Next we extend the traditional general equilibrium model to incorporate institutional constraints on prices like for example a minimum wage law. Such constraints will normally cause unemployment, which in most developed economies triggers off some compensation to the unemployed. We make these transfers explicit within the model by introducing a public sector with taxes on commodities, tariffs on imports, an unemployment compensation program and a public expenditure plan.

A worked numerical example illustrates how the model may be used to study the consequences of a certain increase in the nominal wage.

1. Introduction.

Economic planning in an open economy has been approached through a variety of models. For multi-sectoral modelling the input-output technique and linear programming are often used. Excellent reviews are given by Manne (1974) and Blitzer et.al.(1975). A great advantage of linear programming is that it provides a means for efficient, systematic exploration of the economy's choice set as delimited by technological and other constraints.

Public policies consistent with such a model will largely be limited to resource allocation. For countries with developed markets there is, however, a whole set of public policies related to the price side. These policies are concerned with taxes, subsidies, tariffs and the exchange rate. As such these policies relate to the dual of the resource planning model. A model to accommodate for both types of policies should thus contain behavioral relationships which constrain response of individual economic agents to policy changes as well as the physical constraints on material balance and resource availability.

Using the complementarity format we present a multisectoral model of an economy engaged in international trade. One central feature of our model is that demand and supply are endogenous which implies that prices depend directly on demand and supply. In this way the model allows for greater substitution between factors of production and between commodities. The price system will thereby reflect a larger set of its alleged determinants.

The format of complementarity treats quantities and prices in a symmetric manner. This allows institutional constraints on endogenous prices as well as quantitites and even ties between these two sets of variables. Consequently, such phenomena as minimum wage laws may be incorporated in the analysis. Without such constraints, models cannot explain the simultaneous existence of excess supply of a factor of production and yet a positive market price. Our model will in this respect resemble fixed price models; i.e., Barro and Grossman (1971) and Drèze (1975).

The presence of such constraints may cause unemployment. We assume that unemployed workers receive a compensation such that their disposable income is the same as those who are employed at the minimum wage. For the Scandinavian countries this is a reasonable description of the labor market. This assumption simplifies the characterization of market demand in the model.

Making transfers to the unemployed explicit in the model necessitates the inclusion of a transfer scheme. Hence we introduce a public sector with taxes on goods and factors of production, tariffs on imports, an unemployment compensation program and a program for expenditures. Like in other analyses, notably Shoven and Whalley (1973) and Shoven (1974), tax rates etc. are exogenous to the model.

In the next section we briefly review the concepts of equilibrium and complementarity. In section 3 we present the basic model and make economic interpretations. In section 4 we extend the model to include institutional constraints on prices and a public sector. Finally, in section 5, we illustrate the applicability of the model by a numerical example.

2. General equilibrium and the complementarity format.

We will in this section review the notion of general equilibrium and the complementarity format. These two concepts constitute the building blocks of our model and are essential to the subsequent discussion.

Consider an economy with production described by the technology matrix B, where $b_{kj} > 0$ ($b_{kj} < 0$) represents an input (output) of the k'th commodity in the j'th process. Further let ω represent initial holdings, let π denote prices and let

$\xi(\pi)$ denote market demand. Because of the generality of the theory of economic equilibrium there exists numerous definitions of a general equilibrium though their content very much coincide. We shall use the following definition of a general competitive equilibrium:

Definition: A price vector π^* and a vector of activity levels y^* constitute a general competitive equilibrium if:

i) Excess supply is non-negative in all markets; i.e. $-By^* + \omega \geq \xi(\pi^*)$.

ii) No activity earns a positive profit; i.e., $B'\pi^* \geq 0$.

iii) Prices and activity levels are non-negative, i.e., $\pi^* \geq 0, y^* \geq 0$.

iv) An activity that earns a deficit is not run, and operated processes run at a balance; i.e., $\pi^{*\prime}By^* = 0$

v) A commodity in excess supply has zero price and a positive price implies market clearance; i.e., $(-By^* + \omega - \xi(\pi^*))'\pi^* = 0$

The complementarity format is as follows: Find z such that

$F(z) \geq 0,\ z \geq 0$ and $z'F(z) = 0$.

When the mapping F of R^n into itself is an affine transformation, say $F(z) = q + Mz$ the corresponding complementarity problem is said to be linear , otherwise it is nonlinear. The complementarity format has long been known to economists through the equilibrium conditions of a general equilibrium and the Kuhn Tucker conditions of a local extremum of a mathematical programming problem. The former is seen by the association.

$$z = \begin{pmatrix} \pi \\ y \end{pmatrix},\quad F\begin{pmatrix} \pi \\ y \end{pmatrix} = \begin{pmatrix} -\xi(\pi) - By + \omega \\ B'\pi \end{pmatrix}.$$

From this it is also seen that our definition of a general equilibrium will be a linear complementarity problem when market demand is linear in prices, i.e., $\xi(\pi) = d + D\pi$.

The complementarity format has gained considerable attention recently. Much effort is put into exploring the conditions for existence of solutions and devicing algorithms for computing such solutions. For an up to date review, see Balinsky and Cottle (1978). The similarity between complementarity and fixed point theory, as for example applied to the computation of general equilibrium, is also generally acknowledged, see Karamardian (1977) and Balinsky and Cottle (1978).

We find the complementarity format convenient for direct modelling of economic equilibrium problems, examples are Hansen and Manne (1977) and Mathiesen (1978). The traditional approach, however, is the indirect one where this format is derived as the first order conditions for an extremum of a specified optimization problem.

3. A multisectoral planning model for an open economy.

The model presented below is a static general equilibrium model for an open economy. Goods in foreign trade are assumed to be exported and imported at exogenously stipulated prices, while prices for resources, goods and services traded domestically are endogenously determined. We shall use the following notation:

P	will denote internationally stipulated prices expressed in a foreign currency, with subscripts x and m referring to export and import prices respectively.
π	will denote endogenously determined prices of goods and services.
ω	will denote endogenously determined prices of resources.
γ	will denote the endogenously determined price of the local currency in terms of the foreign currency.
A,B	will denote technology-matrices. $a_{ij} > 0$ (<0) indicate that the i'th good is an input to (output of) the j'th process. $a_{ij} = 0$($b_{kj} = 0$) indicate that the i'th good (k'th resource) is not involved in the j'th process. $b_{kj} > 0$ indicate that the k'th resource is an input to the j'th process.
$D(\pi,\omega,\gamma)$	will denote final demand for goods and services.
$S(\pi,\omega,\gamma)$	will denote supply of resources.
y	will denote activity levels of domestic production.
y_x, y_m	will denote activity levels of export and import respectively.

An equilibrium can be characterized by activity levels y, y_x and y_m and prices π,ω and γ, such that the conditions (1)-(7) are satisfied.

(1) $$-D(\pi,\omega,\gamma) - Ay - y_x + y_m \geq 0,$$
$$S(\pi,\omega,\gamma) - By \geq 0,$$

i.e., imports plus domestic output is at least as great as exports, domestic input and final demand in markets for all goods and services, and for resources total demand must not exceed supply. In sum, excess demand is non-positive in all markets.

(2) $$P_x'y_x - P_m'y_m \geq E,$$

which constrains the balance of payments (in foreign currency) to be no worse than E. The parameter E will depend on this specific economy's potential for obtaining credit ($E < 0$) or its obligations to repay loans ($E > 0$) in the planning period.

(3) $$A'\pi + B'\omega \geq 0,$$

i.e., no process earns a positive profit.

$$P_x\gamma \leq \pi \leq P_m\gamma,$$

i.e., domestic prices for goods and services are bounded by internationally stipulated prices expressed in the local currency. Observe that this is equivalent to

(4) $$\pi - P_x\gamma \geq 0,$$
$$-\pi + P_m\gamma \geq 0.$$

We must further have

(5) $\pi \geq 0, \omega \geq 0, \gamma \geq 0, y \geq 0, y_x \geq 0, y_m \geq 0,$

i.e., non-negative prices and activity levels.

$$\begin{aligned} &\{- D(\pi,\omega,\gamma) - Ay - y_x + y_m\}'\pi = 0, \\ (6) \quad &\{\ S(\pi,\omega,\gamma) - By\}'\omega = 0, \\ &\{\ P_x'y_x - P_m'y_m - E\}\gamma = 0, \end{aligned}$$

i.e., a commodity or resource in excess supply has a zero price and a positive price implies a zero excess supply.

$$\begin{aligned} &\{A'\pi + B'\omega\}'y = 0, \\ (7) \quad &\{\pi - P_x\gamma\}'y_x = 0, \\ &\{-\pi + P_m\gamma\}'y_m = 0, \end{aligned}$$

i.e., an activity that earns a deficit is not used and an activity that is operated runs at a balance.

From what is said in the preceeding section there is nothing remarkable with this model. We have just extended the interpretation of activities and markets of an equilibrium model of a closed economy to take account of characteristics of an open economy. Before we incorporate additional features like institutional constraints on endogenous prices we will make some observations.

Summing up the conditions in (6) and (7) we get $\omega'S(\cdot) = \pi'D(\cdot) + \gamma E$, which is Walras Law and in national accounting terms equivalent to:

Factor income = Domestic demand + Value of net exports in local currency.

Assume that $D(\cdot)$ and $S(\cdot)$ according to economic theory are homogenous of degree zero in π, ω and γ. Then if $(\pi^*, \omega^*, \gamma^*)$ are equilibrium prices, so are $(\lambda\pi^*, \lambda\omega^*, \lambda\gamma^*)$ for $\lambda > 0$. Hence only relative prices are determined by the model (1)-(7).

Equilibrium prices for this model could be calculated using the fixed point algorithms of Eaves (1972) and Scarf and Hansen (1973). The problem with these algorithms, however, is that if the number of markets is large, then they may not be computationally feasible. With the loss of some generality, however, large problems may be solved using linear complementarity in the following way. We guess initially on equilibrium values of π, ω and γ and replace $D(\cdot)$ and $S(\cdot)$ by their first order Taylor expansions. (1)-(7) then correspond to a linear complementarity problem. The solution to this problem $\bar{\pi}$, $\bar{\omega}$, $\bar{\gamma}$, $\bar{y}$, $\bar{y}_x$ and $\bar{y}_m$ will then in a sense represent an approximation of a competitive equilibrium. The goodness of the approximation will obviously depend on the price sensitivity of the demand and supply functions and the technology as well as the "goodness" of the initial guess.

Let

$$(8)\qquad \begin{aligned} D(\pi,\omega,\gamma) &= d + D_1\pi + D_2\omega + D_3\gamma, \\ S(\pi,\omega,\gamma) &= s + S_1\pi + S_2\omega + S_3\gamma. \end{aligned}$$

Substituting (8) into (1) we get a linear complementary problem. Two observations should be made. First the linearization has in some sense normalized the prices which now are absolute. Hence demand is no longer homogenous of degree zero. This may be theoretically unattractive, but does not reduce the practical applicability of this format.

The other observation concerns the mathematical structure of the linear complementarity problem. If the matrix

$$M = \begin{pmatrix} -D_1 & -D_2 & -D_3 \\ S_1 & S_2 & S_3 \\ 0 & 0 & 0 \end{pmatrix}$$

is positive semidefinite then Lemkes almost complementary algorithm will compute the solution if one exists, or else show that none exists.

4. Constraints on factor prices and a public sector.

In the analysis so far we have assumed that there are no constraints on resource prices. The presence of such constraints implies, for example, that the market prices and the shadow prices of the factors of production will not necessarily coincide. Unless such constraints are introduced, models cannot explain the simultaneous existence of excess supply of a factor of production and yet a positive market price. Hansen and Manne (1977) demonstrated that constraints on market prices easily could be incorporated in an equilibrium model through linear complementarity. Though the focus of their paper was somewhat different from ours, the following discussion is closely related to their work.

In order to illustrate how bounds on factor prices may be incorporated in our model, we shall consider the case with lower bounds on wages. We shall in that connection make the following assumption:

<u>Assumption:</u> Unemployed labor receives unemployment compensation such that disposable income is the same for those that are unemployed and those that are employed at the minimum wage.

We make this assumption in order to simplify the model since we then do not have to distinguish in the demand functions between those who are employed and those who receive unemployment compensation if there is unemployment and the minimum wage applies.

The minimum wage requirements are reflected by the constraints $\omega \geq \bar{\omega}$. For factors of production other than labor the corresponding component of $\bar{\omega}$ is obviously 0. If necessary the model can accomodate more complicated minimum wage requirements, for example one that ties wages to a cost of living index. In the subsequent discussion we shall have to distinguish between shadow prices and market prices for factors of production. We shall let λ denote shadow prices and let ω denote market prices. Let v denote the non-negative wedges between the market prices and the shadow prices, i.e., $\omega = v + \lambda$, and substitute $v + \lambda$ for ω throughout the model. The complementarity constraint associated with $v + \lambda \geq \bar{\omega}$ will then be $(v + \lambda - \bar{\omega})'v = 0$. Thus if a component of v is strictly positive, i.e., there is a positive wedge between the market and the shadow price, then the corresponding minimum wage constraint is effective. On the other hand, if the minimum wage constraint is not effective, then the wedge is zero.

Our market demand functions presuppose that unemployed factors of production are paid the market wage. The necessary transfer of income behind this assumption could be left outside the model and there would be no need for a public sector in the model. We will, however, make these transfers explicit. Hence we introduce a public sector with its fiscal means, taxes and tariffs, and its ends, an unemployment compensation program and a public consumption plan.

Let t be the rate of a value added tax and let r be the tax that applies to factors of production. Let further c be tariff rates that apply to imports. Obviously a more complicated tax system could be incorporated in the model. Let

$\hat{\pi} = (1+t)\pi$ denote the price paid for the end use of goods,

$\hat{\omega} = (1-r)\omega$ denote the after tax remuneration to the factors of production,

$\hat{P}_m = (1+c)P_m$ denote the prices paid in foreign currency for imported goods,

$\bar{\omega}$ denote the exogenously stipulated minimum market price for factors of production,

$D(\hat{\pi},\hat{\omega},\gamma)$ denote private demand for goods and services.

$\bar{G} + g\alpha - g\beta$ denote the public expenditure plan where $\bar{G}$ and g are vectors and α and β (≥ 0) are variables representing an increase and a reduction of the marginal part of the plan.

With these symbols, the income part of the public budget can be described as:

$$T = t\cdot D(\hat{\pi}\ \hat{\omega}\ \gamma)'\pi + r\cdot S(\hat{\pi},\omega,\hat{\gamma})'\omega + c\cdot\gamma\cdot P_m' y_m$$

and the expenditure part:

$$G + Y = (\bar{G} + g\alpha - g\beta)'\ \binom{\pi}{\omega} + \bar{\omega}'u.$$

The vector u denotes the excess supply of resources. $\bar{\omega}'u$ thus represents the cost of the unemployment compensation program. The public budget-constraint is given by

$$T - G - Y = F,$$

where $F > 0$ denotes the surplus income in an overbalanced budget, and $F < 0$ is interpreted conversely. The complementarity condition

$$\alpha \cdot \beta = 0$$

completes the model.

The public budget constraint is non-linear because of products of endogenous variables. We shall therefore have to linearize the constraint by taking its first order Taylor expansion. The model will structurally be similar to the one presented in the preceeding section and the same comments apply. Because of the added features of this section, i.e., the wage constraints, the public budget etc., it is less likely that the resulting coefficients matrix is positive semidefinit. Fortunately Lemke's algorithm seems to be rather robust and may be able to compute a solution. This, however, can no longer be guaranteed.

5. A numerical example.

In order to illustrate the applicability of the model we shall consider an economy with 3 resources (capital, skilled and unskilled labor). Technology is described by the coefficients in Table 1.

TABLE 1 Technological coefficients.

Text	Activity 1	2	3	4	5	6	7
Good 1	-1	-1	-1				
Good 2				-1	-1		
Good 3						-1	-1
Capital	.7	.2	.1	.6	.2	1.0	1.4
Skilled labor	.2	.2	.1	.2	.3	.4	.1
Unskilled labor	.3	.6	.7	.1	.2	.2	.6

Private investment demand and public demand for goods and resources are described by the coefficients in table 2.

TABLE 2 Private investment demand and public demand.

Text	Private investment (Fixed)	Public demand Fixed	Public demand Variable
Good 1	.5	3.7	$.3(\alpha-\beta)$
Good 2	.5	2.7	$.3(\alpha-\beta)$
Good 3	1.	.3	$.2(\alpha-\beta)$
Capital	2.	2.9	$.1(\alpha-\beta)$
Skilled labor	1.	.9	$.1(\alpha-\beta)$
Unskilled labor	2.	.6	$.4(\alpha-\beta)$

Tax rates t on goods and r on resources are .2 and .4 respectively. Private non-investment demand for goods in terms of producer prices are as follows:

$$D(\pi,\omega,\gamma) = \begin{pmatrix} 5.3 - 4.8\pi_1 + 1.2\pi_2 + .6\pi_3 + .18\omega_2 + .24\omega_3 \\ 2.8 + 1.2\pi_1 - 3.6\pi_2 + .6\pi_3 + .30\omega_2 + .30\omega_3 \\ 8.6 + .6\pi_1 + .6\pi_2 - 4.8\pi_3 + .12\omega_2 + .06\omega_3 \end{pmatrix}$$

Supply of capital, skilled and unskilled labor are given by:

$$S(\pi,\omega,\gamma) = \begin{pmatrix} 10.2 & & & & \\ 5 & -1.2\pi_1 & -1.2\pi_2 & -.6\pi_3 & +1.8\omega_2 \\ 11.6 & -1.2\pi_1 & -1.2\pi_2 & -.6\pi_3 & +1.8\omega_3 \end{pmatrix}.$$

Finally we have trade balance (E = 0), a balanced public budget (F = 0) and import and export prices are given by:

$$P_m = \begin{pmatrix} 1.2 \\ 1.0 \\ 2.2 \end{pmatrix}, \qquad P_x = \begin{pmatrix} 0.8 \\ 0.9 \\ 2.0 \end{pmatrix}$$

In the base case there are no constraints on factor prices. In this case unskilled labor receives a compensation of 1 whereas skilled labor is paid 2. We shall consider two alternatives to the base case.

Case 1: A minimum wage of 1.1 is introduced. Unemployed labor receives an unemployment compensation of 1.1.

Case 2: A minimum wage of 1.2 is introduced. Unemployed labor receives an unemployment compensation of 1.2.

The numerical example thus illustrates how the model may be used to study the consequences of a 10% alternatively 20% increase in the nominal minimum wage. The model solution for the 3 cases are given on the next page. There are several interesting aspects of the solution. International trade increases considerably. There is a drastic change in the structure of production. The sectors that produce goods 1 and 2 contract considerably, whereas the sector producing good 3 (the exported good) expands significantly.

As expected unemployment of unskilled labor increases and public demand is reduced in order to finance the unemployment compensation program. The numerical example thus illustrates the kind of questions the model may answer.

Let us finally point out that the error due to the linearization of the public budget constraint was insignificant. The Taylor expansion of this constraint was taken around the solution for the base case.

SOLUTIONS

		Base case	Case 1	Case 2
Domestic activities in use	3	8	7.56	7.07
	5	2	1.33	.64
	6	4	4.23	4.48
Import good 2		4	4.61	5.20
Export good 3		2	2.30	2.60
π_1		1	1.06	1.13
π_2		1	1	1
π_3		2	2	2
ω_1		1	1.01	1.01
ω_2		2	1.93	1.84
ω_3		1	1.1	1.2
γ		1	1	1
$\alpha - \beta$		1	.51	-.13
Unemployment of unskilled labor		0	.69	1.50

REFERENCES:

M.L. Balinski and R.W. Cottle (eds.) (1978) : Complementarity and fixed point problems, Mathematical Programming Study 7, North Holland Publishing Co., Amsterdam.

R.J. Barro and H.I. Grossman (1971): A general disequilibrium model of income and employment, American Economic Review, pp. 82-93.

C.R. Blitzer, P.B. Clark and L. Taylor (eds.) (1975): Economy-wide models and development planning, Oxford University Press, London.

J. Drèze (1975): Existence of an exchange equilibrium under price rigidities, International Economic Review, pp. 301-320.

B.C. Eaves (1973): Homotopies for computation of fixed points, Mathematical Programming, 3, pp. 1-22.

T. Hansen and A.S. Manne (1977): Equilibrium and linear complementarity: An economy with institutional constraints on prices, in G. Schwødiauer (ed.), Equilibrium and Disequilibrium in Economic Theory, pp. 227-237, D Reidel Publ. Co., Dordrecht.

S. Karamardian (ed.)(1977): Fixed points, algorithms and applications, Academic Press Inc., New York.

A.S. Manne (1974): Multi-sector models for development planning: A survey, in M.D. Intriligator and D.A. Kendrick, Frontiers of quantitative economics, vol. II., North Holland Publ. Co., Amsterdam.

L. Mathiesen (1977): Marginal cost pricing in a linear programming model: A case with constraints on dual variables, The Scandinavian Journal of Economics, No. 4.

H. Scarf, with the colloboration of T. Hansen (1973): The computation of economic Equilibria, Yale University Press, New Haven.

J.B. Shoven (1974): A proof of the existence of a general equilibrium with ad valorem commodity taxes, Journal of Economic Theory, 8, pp. 1-25.

J.B. Shoven and J. Whalley (1973): General equilibrium with taxes: A computational procedure and an existence proof, Review of Economic Studies, pp. 475-489.

CONTROLLABILITY AND OBSERVABILITY OF DYNAMIC ECONOMIC SYSTEMS

Reinhard Neck
University of Economics
Vienna,Austria

I. INTRODUCTION

In recent years mathematical systems and control theory has shown a strong interest in the qualitative properties of dynamic systems. There have been major theoretical advances in understanding the structural properties of dynamic models, such as stability (which is not treated here), controllability, and observability. Loosely speaking, controllability (reachability) is the ability and effectiveness of a control (instrument) to influence and modify the behavior of the dynamic system, whereas observability (reconstructability) is the ability to uncover unobservable systems data from a set of observed data. Although these concepts have been developed by control theorists and engineers, their applicability is much more widespread. For economics, in particular, it is important for modelling the economy to know the qualitative differences between the behavior of different models. The properties of controllability and observability are, however, also indirectly important as the possibility of finding a stabilization policy (by means of optimization theory, for example) depends upon a model having these properties.

The present paper has a communicative aim. Its purpose is twofold:In the first part we give a short review of the concepts of controllability and observability and of some necessary and sufficient conditions which dynamic systems must fulfill in order to have these properties. In the following section we show an application of these concepts to an economic problem, namely to the dynamic theory of economic policy.

II.THE THEORY OF CONTROLLABILITY AND OBSERVABILITY OF DYNAMIC SYSTEMS

1. The State Space Model.

There are several possibilities of constructing systems of dynamic equations for describing dynamic systems. Two important ones are:

a) The input-output-model:The system is given by relations between inputs (controls) and outputs. For instance, in continous time such a system can be described by a first-order vector differential equation:

$$\underline{\dot{x}}(t) \equiv \frac{d\underline{x}(t)}{dt} = \underline{f}(x(t), \underline{u}(t), t), \tag{1}$$

where $\underline{x}(t) \in R^n$ (output), $\underline{u}(t) \in R^r$ (control), $\underline{f}$ (....) is a vector-valued function. In discrete time, we have a difference equation:

$$\underline{x}(t+1) = \underline{f}(\underline{x}(t), \underline{u}(t), t). \tag{2}$$

b) The state-space model:Here we introduce intermediate variables, so-called state variables:

$$\dot{\underline{x}}(t) = \underline{f}(x(t), \underline{u}(t), t),$$
$$\underline{y}(t) = \underline{g}(\underline{x}(t), \underline{u}(t), t), \quad (3)$$

where now $\underline{x}(t) \in R^n$ is the state and $\underline{y}(t) \in R^m$ is the output. (3) defines trajectories in the state space. In the special case of a linear system, (3) is specialized to:

$$\dot{\underline{x}}(t) = \underline{A}(t)\,\underline{x}(t) + \underline{B}(t)\,\underline{u}(t),$$
$$\underline{y}(t) = \underline{C}(t)\,\underline{x}(t) + \underline{D}(t)\,\underline{u}(t), \quad (4)$$

where $\underline{A},\underline{B},\underline{C}$, and $\underline{D}$ are matrices of appropriate dimensions. The system may be time-varying or time-invariant (constant); in the last case, $\underline{A}(t) = \underline{A},\dots,\underline{D}(t) = \underline{D}$ for all t. In discrete time, a linear system can be written as:

$$x(t+1) = \underline{A}\,\underline{x}(t) + \underline{B}\,\underline{u}(t),$$
$$\underline{y}(t) = \underline{C}\,\underline{x}(t) + \underline{D}\,\underline{u}(t). \quad (5)$$

Sometimes the output equation can be simplified to

$$\underline{y}(t) = \underline{C}\,\underline{x}(t) \quad (6)$$

The input-output- and the state-space-representations are equivalent in the sense that under suitable conditions one can be deduced from the other and vice versa. In particular, a state-space representation is equivalent to a dynamic system given by higher order differential (or difference) equations. In the case of linear systems a change of variables always permits us to discuss linear differential (difference) equations of all orders uniformly as first-order differential (difference) equations of vector variables of various dimensions. Regard, for instance, the following linear econometric model given in structural form:

$$\underline{y}(t) = \underline{A}_o\,\underline{y}(t) + \underline{A}_1\,\underline{y}(t-1) + \dots + \underline{A}_k\,\underline{y}(t-k) +$$
$$+ \underline{G}_o\,\underline{u}(t) + \underline{G}_1\,\underline{u}(t-1) + \dots + \underline{G}_j\,\underline{u}(t-j) + \underline{f}(t), \quad (7)$$

where $\underline{y}(t)$ is a vector of endogenous economic variables, $\underline{u}(t)$ is a control (instrument) vector and $\underline{f}(t)$ is exogenous. If $(\underline{I} - \underline{A}_o)$ is nonsingular, we can derive the reduced form:

$$\underline{y}(t) = \underline{A}_1'\,\underline{y}(t-1) + \dots + \underline{A}_k'\,\underline{y}(t-k) + \underline{C}_o\,\underline{u}(t) +$$
$$+ \underline{C}_1\,\underline{u}(t-1) + \dots + \underline{C}_j\,\underline{u}(t-j) + \underline{d}(t) \quad (8)$$

with $\underline{A}_i' = (\underline{I} - \underline{A}_o)^{-1}\underline{A}_i,\ i = 1,\dots,k,$

$\underline{C}_i = (\underline{I} - \underline{A}_o)^{-1}\underline{G}_i,\ i = 1,\dots,j,$

$\underline{d}(t) = (\underline{I} - \underline{A}_o)^{-1}\underline{f}(t)$. Then the state space model can be derived as follows (cf. Aoki 1976): $\underline{x}^1(t) = \underline{y}(t) - \underline{C}_o\,\underline{u}(t) - \underline{d}(t)$

$$\underline{x}^{l+1}(t) = \underline{x}^l(t) - \underline{A}_l\,\underline{y}(t-1) - \underline{C}_l\,\underline{u}(t-1),\ l = 1,2,\dots,k-1.$$

The state $\underline{x}(t)$ is a (km)-vector (^{T} denotes the transpose): $\underline{x}(t) = [\underline{x}^{1T}(t),\dots,\underline{x}^{kT}(t)]^T$. The state space representation is

$$\underline{x}(t+1) = \underline{A}\,\underline{x}(t) + \underline{B}\,\underline{u}(t) + \underline{E}\,\underline{d}(t),$$
$$\underline{y}(t) = \underline{C}\,\underline{x}(t) + \underline{D}\,\underline{u}(t) + \underline{d}(t) \quad (9)$$

where $\underline{A} = \begin{bmatrix} \underline{A}_1' & \underline{I} & \underline{O} \dots \underline{O} \\ \underline{A}_2' & \underline{O} & \underline{I} \dots \underline{O} \\ \dots & \dots & \dots \\ \underline{A}_{k-1}' & \underline{O} & \underline{O} \dots \underline{I} \\ \underline{A}_k' & \underline{O} & \underline{O} \dots \underline{O} \end{bmatrix}$, $\quad \underline{B} = \begin{bmatrix} \underline{H}^1 \\ \underline{H}^2 \\ \vdots \\ \underline{H}^{k-1} \\ \underline{H}^k \end{bmatrix}$,

$$\underline{E} = \begin{bmatrix} \underline{A}_1{}' \\ \vdots \\ \underline{A}_k{}' \end{bmatrix}, \qquad \underline{H}^i = \underline{A}_i{}'\,\underline{C}_o + \underline{C}_i,\ i = 1,\ldots, k,$$
$$\underline{C} = [\,\underline{I}\ \underline{O}\ \ldots\ \underline{O}\,],\ \underline{D} = \underline{C}_o .$$

In general there are several state-space representations for a model in the reduced form; the one given above results in the state space with minimal dimension, which has important consequences in terms of realization theory (for the realization problem see Silverman 1971, Myoken 1976). For an extensive discussion of the relationships between state space form and structural, final, and reduced forms, see Preston and Wall (1973). The advantage of the close relationships between these forms for linear systems is that structural properties which can be easily derived using the state space form can be interpreted in terms of the instrument multipliers of the reduced or final form.

2. Concepts and Criteria for Controllability.

We start with some definitions. A state $\underline{x}_1$ of a deterministic dynamic system (1) is controllable if all initial conditions $\underline{x}_o$ at any previous time t_o can be transferred to $\underline{x}_1$ in a finite time interval by some control function $\underline{u}(t,\underline{x}_o)$. If all states $\underline{x}_1$ are controllable, the system is (completely) (state) controllable. If controllability is restricted to depend on t_o, the state is controllable at time t_o. If the state can be transferred from $\underline{x}_o$ to $\underline{x}_1$ as quickly as desired independent of t_o, the state is totally controllable. If all states are totally controllable, the system is totally (state) controllable. Another concept is perfect state controllability: A system (1) is perfectly state controllable if there exists a control $\underline{u}(t)$, $t_o \leq t \leq t_1$, such that $\underline{x}(t) = \bar{\underline{x}}(t)$ for all $t \in [t_1', t_1]$, $t_1' \geq t_o$, where $\bar{\underline{x}}(t)$ is a given state trajectory. One can also characterize $\bar{\underline{x}}(t)$ not in terms of functional values but by the requirement that it lies in some subspace of the state space. Analogous definitions for output controllability can be given by replacing the state $\underline{x}_1$ by the output $\underline{y}_1$ in the above definitions. For example, system (4) is output controllable if the output vector $\underline{y}$ can reach the target $\underline{y}(t_1) = \underline{y}_1$ at some time $t_1 \geq t_o$ starting from an arbitrary initial condition $\underline{x}(t_o) = \underline{x}_o$, $\underline{y}(t_o) = \underline{C}\,\underline{x}_o + \underline{D}\,\underline{u}(t)$, by manipulating the control vector $\underline{u}(t)$, $t_o \leq t \leq t_1$. A similar definition applies for perfect output controllability (functional reproducibility). From the solution of linear dynamic systems it follows that to determine complete state controllability at time t_o for those systems it is necessary and sufficient to investigate whether the zero state instead of all initial states can be transferred to all final states. The same is true of output controllability of linear systems.

In this section we will state some criteria for complete and total controllability of a linear system without proof (for proofs see, among others, Wiberg 1971, Willems and Mitter 1971, Aoki 1976, Murata 1977). We will concentrate upon linear systems since for nonlinear systems very few useful results are obtainable so far. Global criteria have to be replaced by local ones in the case of nonlinear systems. Usually nonlinear systems therefore are treated by linearising the nonlinear equation.

a) Time-invariant systems: The system

$$\dot{\underline{x}}(t) = \underline{A}\,\underline{x}(t) + \underline{B}\,\underline{u}(t) \qquad (1o)$$

is totally state controllable if and only if the (n x nr)-matrix $\underline{P} \equiv [\underline{B}\ \underline{AB} \ldots \underline{A}^{n-1}\ \underline{B}]$ has rank $\rho(\underline{P}) = n$. $\underline{P}$ is called the state controllability matrix . Since $\rho(\underline{P}) = \rho(\underline{PP}^T)$ for $\rho(\underline{P}) = n$ it is sufficient that det $(\underline{PP}^T) \neq 0$. To check state controllability we need not, however, calculate $\underline{P}$, but only a matrix with a smaller number of columns: Define the (n x jr) -matrix $P_j \equiv [\underline{B}\ \underline{AB}\ \ldots\ \underline{A}^{j-1}\ \underline{B}]$. If j is the least integer such that $\rho(\underline{P}_j) = \rho(\underline{P}_{j+1})$, then $\rho(\underline{P}_k) = \rho(\underline{P}_j)$ for all integers k>j; j is called the controllability index of $(\underline{A}, \underline{B})$. It should be noted that completely controllable stationary linear systems can be tranferred to any desired state as quickly as possible. We have not imposed any restrictions on $\underline{u}(t)$; if the magnitude of $\underline{u}(t)$ is bounded, the set of states to which the system can be transferred by t_1 is called reachable set at t_1. Conditions can be derived also for this case.

The criterion given for the continuous-time system (1o) is also true (replacing totally by completely) for the discrete-time system

$$\underline{x}(t+1) = \underline{A}\,\underline{x}(t) + \underline{B}\,\underline{u}(t): \qquad (11)$$

Such a system is completely state controllable, i.e. $\underline{x}$ can reach a preassigned target vector $\bar{\underline{x}}(t)$ at some time $t = t_1 \in R^{1+}$ starting from an arbitrary initial state by manipulating the control $\underline{u}(t)$, $1 \leq t \leq t_1$, if and only if the $(n \times t_1 r)$- controllability matrix $\underline{P}_{t_1} \equiv [\underline{B}\ \underline{AB} \ldots \underline{A}^{t_1-1}\underline{B}]$ has rank $\rho(\underline{P}_{t_1}) = n$.

Another necessary and sufficient condition is given by the coupling criterion: Consider system (1o). If $\underline{A}$ has distinct eigenvalues then the system is totally controllable if and only if $\hat{\underline{B}} = \underline{N}^{-1}\underline{B}$ has at least one non-zero element in each row, where $\underline{N}$ is the modal matrix (nonsingular) with eigenvectors of $\underline{A}$ as its column vectors. From this follows a sufficient condition for the system to be state controllable, namely that there exists for the columns of $\underline{B}$ some combination that is linearly dependent on all n right eigenvectors of $\underline{A}$ (eigenvector condition). If $\underline{A}$ has not distinct eigenvalues the Jordan form must be calculated and a corresponding criterion in terms of Jordan blocks is available. Analogous criteria hold for complete controllability of discrete time systems.

The coupling and eigenvector criteria can be interpreted that connectedness between the input and all elements of the state vector is equivalent to total controllability. Furthermore, from the eigenvector condition follows that if any $\underline{u}_i^T$ is not orthogonal to a column of the matrix $\underline{B}$, where $\underline{u}_i^T$ is the i-th left eigenvector of $\underline{A}$ (the i-th row of $\underline{N}^{-1}$), then the coupling criterion is satisfied and we need only one control (a scalar input) in order to have the system (1o) with continuous time be state controllable. This can be extended to the calculation of minimal sets of instruments for controllability (cf. Preston 1974, Theorem 3). For discrete-time systems with a scalar input this is also true, but it takes at least n steps to transfer $\underline{x}$ to an arbitrary desired state.

Still another equivalence criterion for state controllability says that the system (1o) is state controllable if and only if the symmetric matrix $\underline{P}(t_o,t) \equiv \int_{t_o}^{t} e^{-\underline{A}T}\underline{BB}^T e^{-\underline{A}^T T}dT$ is nonsingular for some time $t \geq t_o$.

Consider now the corresponding criteria for output controllability. The system

$$\dot{\underline{x}}(t) = \underline{A}\,\underline{x}(t) + \underline{B}\,\underline{u}(t), \qquad\qquad (12)$$
$$\underline{y}(t) = \underline{C}\,\underline{x}(t)$$

is output controllable if and only if the (n x nr)-matrix $\underline{Q} \equiv [\underline{CB}\ \underline{CAB}\ \dots\ \underline{CA}^{n-1}\underline{B}] = \underline{CP}$ called output controllability matrix has rank $\rho(\underline{Q}) = m$. The corresponding discrete-time system

$$\underline{x}(t+1) = \underline{A}\,\underline{x}(t) + \underline{B}\,\underline{u}(t), \qquad\qquad (13)$$
$$\underline{y}(t) = \underline{C}\,\underline{x}(t)$$

is output controllable if and only if the ($m \times t_1$ r)-matrix $\underline{Q}_{t_1} \equiv [\underline{CB}\ \underline{CAB}\ \dots \underline{CA}^{t_1-1}\underline{B}] = \underline{CP}_{t_1}$ has rank $\rho(\underline{Q}_{t_1}) = m$.

An alternative condition says the system (12) is output controllable if and only if the symmetric matrix $\underline{Q}(t_o,t) \equiv \int_{t_o}^{t} \underline{C}e^{\underline{A}(t-T)}\underline{BB}^T e^{\underline{A}^T(t-T)}\underline{C}^T dT$ is nonsingular for some $t \geq t_o$.

We can also give a criterion for perfect output controllability: The system (12) is perfectly output controllable if and only if $\rho(\underline{M}) = mn$, where

$$\underset{(mn \times (2n-1)r)}{\underline{M}} \equiv \begin{bmatrix} \underline{CB} & \underline{CAB} & \dots & \underline{CA}^{2n-2}\underline{B} \\ \underline{O} & \underline{CB} & \dots & \underline{CA}^{2n-3}\underline{B} \\ \vdots & \vdots & \ddots & \vdots \\ \underline{O} & \underline{O} & \dots \underline{CB} \dots & \underline{CA}^{n-1}\underline{B} \end{bmatrix}.$$

A necessary condition for this to hold is $m \leq (2-\frac{1}{n})r$. Analogous conditions can be stated for perfect state controllability and for discrete-time systems. It is interesting to note that for m = 1 the condition about $\underline{M}$ reduces to the one for output controllability.

There are some relationships between state controllability and output controllability: Consider system (12); assume $m \leq n$ and $\rho(\underline{P}) = n$, i.e. the system is state controllable. Then the system is output controllable, $\rho(\underline{CP}) = m$, if and only if $\underline{C}$ has rank $\rho(\underline{C}) = m$. The relationship can also be expressed in terms of the controllability index: If $m \leq \rho(\underline{P}_j) = \rho(\underline{P}_{j+1}) \leq n$ holds for some $j \leq n$, then the system (12) is output controllable if and only if $\rho(\underline{C}) = m$. A similar relation holds for the coupling criterion: If $\rho(\underline{C}) = m \leq n$ and $\underline{A}$ has all distinct eigenvalues, a sufficient condition for the system (12) to be output controllable is that there exist for the columns of $\underline{B}$ some combination that is linearly dependent on all n right eigenvectors of $\underline{A}$. For the case in which not all eigenvalues of $\underline{A}$ are distinct a more complicated condition is required.

b) Time-varying systems: The system

$$\dot{\underline{x}}(t) = \underline{A}(t)\,\underline{x}(t) + \underline{B}(t)\,\underline{u}(t), \qquad\qquad (14)$$

where $\underline{A}(t)$ and $\underline{B}(t)$ are piecewise differentiable n - 2 and n - 1 times, respectively, is totally state controllable if and only if $\underline{Q}(t)$ has rank $\rho(\underline{Q}(t)) = n$ for times everywhere dense in $[t_o,t_1]$, i.e. there are only isolated points in t with $\rho(\underline{Q}) < n$. Here

$\underline{Q}(t) \equiv [\underline{Q}_1\ \underline{Q}_2\ \dots\ \underline{Q}_n]$ with $\underline{Q}_1 \equiv \underline{B}(t)$ and $\underline{Q}_{k+1} \equiv -\underline{A}(t)\underline{Q}_k + \dot{\underline{Q}}_k$ for $k = 1,2,\dots,n-1$. For systems where $\underline{A}(t)$ and $\underline{B}(t)$ are given by analytic functions complete controllability implies total controllability, i.e. $\rho(\underline{Q}(t)) = n$ is equivalent to complete controllability; nonanalytic systems with $\rho(\underline{Q}(t)) < n$ might be completely controllable but not totally controllable.

3. Observability.

We define a state $\underline{x}(t_1)$ at time t_1 of a system to be observable if knowledge of the input $\underline{u}(t)$ and output $\underline{y}(t)$ over a finite time segment $t_o < t \leq t_1$ completely and uniquely determines $\underline{x}(t_1)$. If all states $\underline{x}(t)$ are observable, the system is (completely) observable. If observability depends on t_o, the state is observable at t_o. If the state can be determined for t in any arbitrarily small time segment independent of t_o, it is totally observable. Observability when $\underline{u}(t) = \underline{O}$ is called zero-input observability.

To determine complete observability for linear systems it is necessary and sufficient to see whether the initial state $\underline{x}(t_o)$ of the zero-input system can be completely determined from $\underline{y}(t)$, because knowledge fo $\underline{x}(t_o)$ and $\underline{u}(t)$ permits $\underline{x}(t)$ to be calculated from the solution of the system.

Again we state some criteria for observability.

a) Time-invariant systems: The system

$$\begin{aligned} \dot{\underline{x}}(t) &= \underline{A}\ \underline{x}(t) + \underline{B}\ \underline{u}(t), \\ \underline{y}(t) &= \underline{C}\ \underline{x}(t) + \underline{D}\ \underline{u}(t) \end{aligned} \tag{15}$$

is totally observable if and only if the (rn x n)-matrix $\underline{R}$ has rank $\rho(\underline{R}) = n$, where

$$\underline{R} \equiv \begin{bmatrix} \underline{C} \\ \underline{CA} \\ \vdots \\ \underline{CA}^{n-1} \end{bmatrix}$$ is the observability matrix.

An analogous criterion holds for discrete-time systems for complete observability. It should be noted again that $\underline{u}(t)$ is not restricted; with restrictions on $\underline{u}(t)$ we have the concept of recoverable state at t_1.

An alternative necessary and sufficient condition for the system (15) to be observable is that the symmetric matrix $R(t_o,t) \equiv \int_{t_o}^{t} e^{\underline{A}^T T}\ \underline{C}^T \underline{C} e^{\underline{A}T}\, dT$ is nonsingular for some time $t \geq t_o$.

b) Time-varying systems: A system

$$\begin{aligned} \dot{\underline{x}}(t) &= \underline{A}(t)\ \underline{x}(t) + \underline{B}(t)\ \underline{u}(t) \\ \underline{y}(t) &= \underline{C}(t)\ \underline{x}(t) + \underline{D}(t)\ \underline{u}(t) \end{aligned} \tag{16}$$

with $\underline{A},\underline{B}$ piecewise differentiable n-2,n-1 times, respectively, is totally observable if and only if $\rho(\underline{R}(t)) = n$ for times everywhere dense in $[t_o,t_1]$, where $\underline{R}^T(t) = [\underline{R}_1^T\ \underline{R}_2^T\ \dots\ \underline{R}_n^T]$ with $\underline{R}_1 = \underline{C}(t)$ and $\underline{R}_{k+1} = \underline{R}_k\underline{A}(t) + \dot{\underline{R}}_k$ for $k = 1,2,\dots,n-1$. A similar condition holds for the discrete-time case.

The symmetry between the criteria for controllability and observability reveals that there is a relationship of duality between these concepts. Consider two systems:

System 1: $$\dot{\underline{x}}(t) = \underline{A}(t)\ \underline{x}(t) + \underline{B}(t)\ \underline{u}(t), \quad \underline{y}(t) = \underline{C}(t)\ \underline{x}(t) + \underline{D}(t)\ \underline{u}(t). \tag{17}$$

System 2: $$\dot{\underline{w}}(t) = -\underline{A}^T(t)\ \underline{w}(t) + \underline{C}^T(t)\ \underline{v}(t), \quad \underline{z}(t) = \underline{B}^T(t)\ \underline{w}(t) + \underline{D}^T(t)\ \underline{v}(t). \tag{18}$$

Then it can be shown that system 1 is totally controllable (observable) if and only if system 2 is totally observable (controllable).

III. SOME ECONOMIC APPLICATIONS OF CONTROLLABILITY AND OBSERVABILITY

1. Controllability Concepts in the Theory of Economic Policy.

The static theory of economic policy which originated with Tinbergen (1952) considers linear systems of the form

$$\underline{A}^\times \underline{x}^\times + \underline{B}^\times \underline{u}^\times + \underline{C}^\times \underline{z}^\times = \underline{O}\ , \tag{19}$$

where $\underline{x}^\times \in R^n$ is a vector of target variables, $\underline{u}^\times \in R^r$ a vector of instrument variables, and $\underline{z}^\times \in R^l$ a vector of exogenous non-controllable variables. In the case of the fixed targets specification it is assumed that the targets have prescribed fixed values $\underline{x}^\times = \bar{\underline{x}}$. Tinbergen asks essentially two questions: a) Existence: When has the policy problem defined by (19) and $\bar{\underline{x}}$ a solution, i.e. when is (19) compatible with $\bar{\underline{x}}$? It turns out that it is equivalent to ask when has (19) a solution for every possible vector $\underline{s} = -(\underline{A}^\times \underline{x}^\times + \underline{C}^\times \underline{z}^\times)$, and this is the case if and only if $\rho(\underline{B}^\times) = n \leq r$. The last inequality is the static counting rule demanding there being at least as many linearly independent instruments as there are linearly independent targets. b) Uniqueness: (19) has a unique solution for every $\underline{s}$ if and only if $\rho(\underline{B}^\times) = n = r$. Then $\underline{u}^\times = \underline{B}^{\times -1}\ \underline{s} = -\underline{B}^{\times -1}(\underline{A}^\times \underline{x}^\times + \underline{C}^\times \underline{z}^\times)$.

These problems have to be distinguished from the one of design, where the existence conditions are assumed to be fulfilled and it is asked which values $\underline{u}$ has to take, especially if some target function has to be optimized.

Preston (1974) has provided a dynamic generalization of this problem. In a dynamic theory adjustment paths and adjustment times are explicitly introduced, i.e. we have a situation of disequilibrium. This is important especially for a theory of economic policy, since disequilibrium may be one cause of justification of policy measures within the framework of a market economy. We assume that the system (19) is disturbed from a position of policy equilibrium achieving its desired target $(\bar{\underline{x}},\bar{\underline{u}})$. Considering only linear deterministic time-invariant systems, we model the disturbance by a first-order process in continuous time:

$$\dot{\underline{x}}(t) = \underline{A}\ \underline{x}(t) + \underline{B}\ \underline{u}(t) + \underline{L}, \quad \underline{x}(t_o) = \underline{x}_o \neq \underline{O}, \tag{20}$$

where $\underline{x}$ and $\underline{u}$ can be interpreted as deviations from the equilibrium values : $\underline{x}(t) = \underline{x}^\times(t) - \bar{\underline{x}}$, $\underline{u}(t) = \underline{u}^\times(t) - \bar{\underline{u}}$, $\underline{L}$ = const. such that

$$\dot{\underline{x}}^\times(t) = \underline{G}\left[\underline{A}^\times\underline{x}^\times(t) + \underline{B}^\times\underline{u}^\times(t) + \underline{C}^\times\underline{z}^\times(t)\right],$$

$$\underline{x}^\times(t_o) \neq \bar{\underline{x}},\ \underline{A} = \underline{GA}^\times,\ \underline{B} = \underline{GB}^\times,\ \underline{L} = \underline{GC}^\times\ \underline{z}^\times + \underline{A}\ \bar{\underline{x}} + \underline{B}\ \bar{\underline{u}}.$$

One possible dynamic formulation of an existence problem would be that of "target point objective": Does there exist a policy vector $\underline{u}(t)$ that transfers $\underline{x}(t_o) \neq \underline{O}$ to $\underline{x}(t_1) = \bar{\underline{x}}$ in arbitrary adjustment time t_1 ? A solution to this existence problem is given by a straightforward application of the criterion for state controllability: The desired $\underline{u}(t)$ exists if and only if $\underline{P} = \left[\underline{B} \;\; \underline{AB} \ldots \underline{A}^{n-1}\underline{B}\right]$ has rank $\rho(\underline{P}) = n$. A generalization to higher-order adjustment processes would be possible by using state-space models; in this case the concept of output controllability would become relevant. It is easily seen that static controllability, i.e. the existence of a static policy, implies dynamic one in the sense defined above, but not vice versa. Furthermore, the theory of the minimal set of instruments for dynamic existence shows an interesting asymmetry between static and dynamic controllability: From the eigenvector condition follows that in many cases dynamic controllability could be achieved with much fewer independent instruments; in the extreme case static controllability requires $r=n$ instruments but dynamic controllability only $r=1$.

The idea of "target point objective" has also been criticized because it does only imply that the system is able to move through the target point after an arbitrary period of time but not that it stays there. However, the policy maker is also interested in keeping targets on a desired path once achieved (Nyberg and Viotti 1978, Aoki 1975). This leads to the problem of "target path objective" : Does there exist, for arbitrary initial conditions and arbitrary but known exogenous influences, a policy vector $\underline{u}(t)$ or a sequence of such vectors that transfers the system $\underline{x}(t_o) = \underline{O}$ to $\underline{x}(t) = \bar{\underline{x}}$ for all $t > t_o$? Now the answer is given by the criterion of perfect state controllability: The desired $\underline{u}(t)$ exists if and only if the matrix

$$\underline{R} = \begin{bmatrix} \underline{B} & \underline{AB} & \ldots & \underline{A}^{2n-2}\underline{B} \\ \underline{O} & \underline{B} & \ldots & \underline{A}^{2n-3}\underline{B} \\ \ldots & \ldots & \ldots & \ldots \\ \underline{O} & \underline{O} & \ldots & \underline{A}^{n-1}\underline{B} \end{bmatrix}$$

has rank $\rho(\underline{R}) = n^2$. A necessary condition for this is $n^2 \leq (2n-1)r$ or $n\left(\frac{n}{2n-1}\right) \leq r$, from which follows $n \leq r$. For $n = 1$ the condition reduces to that of state controllability; if $r = 1$, $n = 1$ is necessary for perfect controllability. Thus, perfect controllability seems to be a more proper dynamic generalization of Tinbergen's theory of policy than state controllability.

Uebe (1976) and Preston and Sieper (1977) have developed conditions of "target path controllability" for discrete-time systems. Uebe considers a system

$$\underline{x}(t+1) = \underline{A}\,\underline{x}(t) + \underline{B}\,\underline{u}(t) \qquad (21)$$

with $\underline{x}(t_o) = \bar{\underline{x}}(t_o)$; the target path is given by

$$\underline{x}(t+i) = \bar{\underline{x}}(t+i),\; i = o,1,\ldots,t_1-1.$$ Then the desired $\underline{u}(t)$ exists if and only if the rank $\rho(\underline{R}) = n.t_1$, where in this case

$$\underline{R} = \begin{bmatrix} \underline{B} & \underline{AB} & \ldots & \underline{A}^{t_o+t_1-2}\underline{B} \\ \underline{O} & \underline{B} & \ldots & \underline{A}^{t_o+t_1-3}\underline{B} \\ \ldots & \ldots & \ldots & \ldots \\ \underline{O} & \underline{O} & \ldots & \underline{A}^{t_o-1}\underline{B} \end{bmatrix}.$$

For $t_1 = 1$ and $t_o = n$ this again implies state controllability. The necessary condition here is

$$n\, t_1 \leq (t_o + t_1 - 1)r \text{ or } t_1 \left[\frac{n}{r} - 1\right] + 1 \leq t_o;$$

this means that t_o has to be advanced sufficiently into the future.

That the results for discrete-time systems show that the achievement of a target-path objective considerably dependes upon the degree to which policy action anticipates this policy objective has been stressed by Preston and Sieper (1977). They use a general state-space model (5) and solve their policy problem by using the criterion for perfect output controllability. Their approach is especially well suited for economists, since the elements of the perfect output controllability matrix $\underline{R}$ in their problem:

$$\underline{R} = \begin{bmatrix} \underline{D} & \underline{CB} & \underline{CAB} & .. & \underline{CA}^{t_o+t_1-2}\underline{B} \\ \underline{O} & \underline{D} & \underline{CB} & ... & \underline{CA}^{t_o+t_1-3}\underline{B} \\ \ldots & \ldots & \ldots & \ldots & \ldots \\ \underline{O} & \ldots & \ldots & \ldots & \underline{CA}^{t_o-1}\underline{B} \end{bmatrix}$$

are the multipliers of the final form of the linear econometric model from which the state-space model (5) has been derived. For instance $\underline{CA}^{i-1}\underline{B}$ is the dynamic matrix multiplier of the instruments at lag i, i.e. the total effect on the targets, i periods later, of a unit change in the instrument now. Furthermore, in the case of target path controllability for all t_o and t_1 the conditions reduce to the static Tinbergen ones. If this is not fulfilled, however, policy must anticipate its objective ($t_o > 0$, i.e. there must be some interval called policy lead between the policy origin 0 and the target path origin t_o) if it is to achieve its target path objective; but if it does so, it will be successful and will therefore not have to resort to optimization techniques. Policy anticipation is a compensation for instrument deficiencies, although there is an upper bound on the policy lead beyond which further anticipation cannot introduce further independent instruments.

In this approach the theory of economic policy is concerned mainly with the trade-off between four parameters: the number of instruments r, the number of targets n, the policy lead t_o and the target path interval t_1, expressed by the condition that the number of "time-indexed instruments" (instruments at each time interval) $r(t_o + t_1)$ must be greater than or equal to the number of "time-indexed targets" $n\, t_1$. In addition, there is a trade-off between anticipation in this case and the welfare loss due to an optimization formulation of the problem without anticipation but not reaching its targets exactly. A more general theory of economic policy would have to take into account this, too, in formulating its problem.

2. Economic Interpretation of Observability.

There is no economic interpretation of observability available far which is as close to the mathematical concept as the one described above for controllability. Aoki (1976) gives some examples in which the formal properties of observability are used to determine whether an equilibrium of a market is unique and whether in a monetary disequilibrium model markets for some goods remain cleared over some time period.

An interpretation which seems to give more insights would start from the idea that observability refers to variables that are not directly available to the model builder or not accessible to direct measurement. In economics, expectations of variables and utility would be examples of variables of this kind. If we can assume a fixed linear relation of such variables with observed variables the question of whether the conditions of observability are fulfilled becomes relevant. This is especially true when the concept is used in a stochastic context:In this case observability becomes a condition for the behavior of the estimation error of some parameter or state vector as the size of observation data grows, that is a condition of consistent estimation in the sense of probability convergence (Aoki 1967, 1976). This becomes especially relevant for the economic problem of estimation in the presence of errors in variables (observational error). Even the duality between observability and controllability could be interpreted as reflecting the strong relationship between forecasting and estimating on the one hand and controlling or policy on the other one, for instance in business cycle theory.

Thus the concepts of controllability and observability seem to have great actual and even greater potential importance for the formulation and analysis of economic models.

References

M.Aoki (1967), Optimization of Stochastic Systems. New York.

M.Aoki (1975), On a Generalization of Tinbergen's Condition in the Theory of Policy to Dynamic Models. Review of Economic Studies 42, 293-296.

M.Aoki (1976), Optimal Control and System Theory in Dynamic Economic Analysis. New York et al.

Y.Murata (1977), Mathematics for Stability and Optimization of Economic Systems. New York et al.

H. Myoken (1976), A Dynamical Existence Problem of Macroeconomic Policy Model. International Journal of Systems Science 7, 1227-1237.

L.Nyberg, S.Viotti (1978), Controllability and the Theory of Economic Policy: A critical view. Journal of Public Economics 9, 73-81.

A.J.Preston (1974), A Dynamic Generalization of Tinbergen's Theory of Policy. Review of Economic Studies 41, 65-74.

A.J.Preston, K.D.Wall (1973), Some Aspects of the Use of State Space Models in Econometrics. IEE Conference Publication 1o1, 226-239.

A.J.Preston, E.Sieper (1977), Policy Objectives and Instrument Requirements for a Dynamic Theory of Policy. In:J.D.Pitchford, S.J.Turnovsky (eds.), Applications of Control Theory to Economic Analysis, Amsterdam et al., Essay 9, 215-253.

L.M.Silverman (1971), Realization of Linear Dynamical Systems. IEEE Transactions on Automatic Control AC-16, 554-567.

J. Tinbergen (1952), On the Theory of Economic Policy. Amsterdam et al.

G.Uebe (1977), A Note on Aoki's Perfect Controllability of a Linear Macro-economic Model. Review of Economic Studies 44, 191-192.

D.M. Wiberg (1971), State Space and Linear Systems. New York.
J.C. Willems, S.K. Mitter (1971), Controllability, Observability, Pole Allocation, and State Reconstruction. IEEE Transactions on Automatic Control AC-16, 582-595.

THE DEVELOPMENT OF ECONOMIC SYSTEM IN CASE OF DIFFERENTIAL OPTIMIZATION (FOR ONE-SECTOR DYNAMIC MODEL)

V. ZHIYANOV
Institute for Systems Studies
Moscow USSR

In present report the one-sector dynamic model is considered. The capital stocks are divided into generations. Such models are called "putty-clay models" or "models with embodied technological change.

Variable quantities of the model are governed by criterium of differentiel optimization (d.o.). Accorrding to this criterium the policy of substitution of non-effective capital stocks is optimal provided it ensures maximal rate of national income increase.

The principle of D.o. mathematically furmulated in this report. The model is described by a system of delay differentional equations. The delay itself being an internal variable of the model. In some cases of interest explicit solutions can be found. These give an insight into technological change influence upon the dynamics of economical parameters.

Helpful assistance of A. Khovansky in the preparation of this report is gratefully acknowledged.

First of all the author gives the abridged description of the model considered in more details in lecture of professor L. Kantorovich in this book.

In an economic system manufacturing a single product (the one-sector model), two main productive factors are distinguished - (i) capital stocks differentiated by the time of their creation and measured in product units, and (ii) labour, measured in labour units. Denote by $T(t)$ the total labour in the system at time t. This function is assumed to be given.

The efficiency of production is characterised by production function $\mathcal{U}(x,y,\tau)$ which is the net product created in a time unit by labour y using capital x. $\mathcal{U}$ is assumed to be a convex positive homogeneous function of first order.

Investment into capital growth and replacement is difined in terms of its intensity: $\varkappa(t)\,dt$ is investment during the time interval $[t, t+dt]$. The function $\varkappa(t)$ is given in the model, but it may be made dependent, on the national income at time t or on the other parameters. In the version of the model considered below, $\varkappa(t)$ is equal to a constant share of the national income:

$$\varkappa(t) = \gamma P(t)$$

Quantity $\varphi(t)\,dt$ is number of employees engaged by new funds, which is to be found. The funds replasment policy given by the function $m(t)$. Where $m(t)$ is by definition the creation time of funds substituted at time t.

The national income is obtained according to

$$P(t) = \int_{m(t)}^{t} \mathcal{U}(\varkappa(\tau), \varphi(\tau), \tau)\, d\tau \tag{1}$$

The labour balance equation is

$$\varphi(t) = T'(t) + \varphi(m(t))\, m'(t) \tag{2}$$

The capital stocks equation is

$$\varkappa(t) = \gamma \int_{m(t)}^{t} \mathcal{U}(\varkappa(\tau), \varphi(\tau), \tau)\, d\tau \tag{3}$$

Beyond balance equations the model variables must satisfy the equation of d.o. We mathematically formulate the criterium of d.o. and derive the corresponding equation for our model.

Principle of differential optimization

Consider a manifold A of vector functions $\gamma(t)$ of variable t,

$$\gamma(t) = \{ \gamma_0(t), \gamma_1(t), \ldots, \gamma_n(t) \}$$

Vector-function $\gamma(t)$ from A is called a trajectory; each trajectory describes one of the possibilities for the evolution of the system. We assume that the trajectories $\gamma(t)$ are smooth functions with finite number of discontinuities. In the discontinuity point t_0 we consider two vectors

$$\gamma^+(t_0) = \lim_{t \to t_0^+} \gamma(t), \quad \gamma^-(t) = \lim_{t \to t^-} \gamma(t)$$

The component $\gamma_0(t)$ of the trajectory plays a special role. It is assumed to be continuous with $\gamma_0(t) < t$. The interval $[\gamma_0(t), t]$ will be called "the influence interval" of the trajectory $\gamma(t)$ at the moment t.

Let $F_\gamma(t)$ be the optimizated functional whose value depends upon the trajectory γ and time t. The following conditions are assumed to be satisfied:

i) if the trajectory $\tilde{\gamma}$ tallies with γ over its influence interval at the moment t, then $F_{\tilde{\gamma}(t)}(t) = F_{\gamma(t)}(t)$. This condition means that the functional F is independent of the future development of the system and is completely defined by its pre-history;

ii) for any γ, there exists the right-hand derivative of F (t) as the function of time.

Definition 1. The trajectory $\gamma(t)$ is called to be differentially optimal in respect to the functional F, if for any other trajectory $\tilde{\gamma}(t)$ such that $\gamma(t) = \tilde{\gamma}(t)$ for $\gamma_0(t) \le t \le t_0$ the following condition is satisfied:

$$F'^{+}_{\gamma}(t_0) \geqslant F'^{+}_{\tilde{\gamma}}(t_0)$$

Definition 2. The trajectory $\gamma(t)$ is called to be differentially optimal on the interval $[a.\ b]$ in respect to F if it is differentially optimal everywhere on this interval. We will say that the differentially optimal trajectories satisfy the criterium of differential optimisation. The differentially optimal trajectory can be said to be moving towards the maximal growth of the functional F.

Let us derive the equation for differential optimization within the framework of economic model considered by us. In this model the trajectories are formed by the pairs of functions $\gamma(t)=\{m(t),\varphi(t)\}$ that satisfy equation (2) (in the version where the function $\varkappa(t)$ is an exogeneous variable), and combinations $\gamma(t)=\{m(t),\varphi(t),\varkappa(t)\}$ governed by (2), (3) (for the version where the function $\varkappa(t)$ is equal to the constant part of national income. Such trajectories give the balanced (in labour resources and funds) development of the economy. For trajectories considered the influence interval is the time interval $[m(t), t]$. The function $\varphi(t)$ is assumed to be discontinuous piece wize smooth, while the functions $m(t)$ and $\varkappa(t)$ are assumed to be continuous piece wize smooth functions. The functional is obtained according to (1).

First we consider the version in which the function $\varkappa(t)$ is defined as exogeneous. We compute the right-hand derivative for the trajectory $P'_\gamma(t)$ at arbitrary point t_0:

$$P'_\gamma(t_0)=\mathcal{U}[\varkappa(t_0),\varphi(t_0),t_0]-\mathcal{U}[\varkappa(m(t_0)),\varphi(m(t_0)),m(t_0)]\,m'(t_0)$$

If the trajectories coincide until the moment t_0, for them the values of $\varkappa(m(t_0))$, $\varphi(m(t_0))$ and $m(t_0)$ are equal. For the functions $\varkappa$ and φ it follows from $m(t)<t$, for m -- from its continuity. With the help of (2) the functional $P'^{+}_\gamma(t_0)$ can be considered as the function of the single variable $m'^{+}(t_0)$:

$$P'_\gamma(t_0)=\mathcal{U}\left[\varkappa(t_0),T'(t_0)+\varphi(m(t_0))\,m'^{+}(t_0)\right]-\mathcal{U}\left[\varkappa(m(t_0)),\varphi(m(t_0)),m(t_0)\right]m'^{+}(t_0) \tag{4}$$

From equation (4) we see that the function P'^{+}_γ is convex in m'^{+} Therefore the maximum of P'^{+}_γ is reached at the point where the derivative is zero. Hence we obtain the following equation:

$$0=\frac{dP'^{+}_\gamma}{dm'^{+}}=\frac{\partial\mathcal{U}[\varkappa(t),\varphi^{+}(t),t]}{\partial\varphi^{+}}\varphi(m(t))-\mathcal{U}[\varkappa(m(t)),\varphi(m(t)),m(t)]$$

or

$$\frac{\partial\mathcal{U}[\varkappa(t),\varphi^{+}(t),t]}{\partial\varphi^{+}}=\frac{\mathcal{U}[\varkappa(m(t)),\varphi(m(t)),m(t)]}{\varphi(m(t))} \tag{5}$$

In the following we'll be interested in continuous solutions only. For such solutions $\varphi^{+}(t) = \varphi(t)$ and the right-handed sign can be omitted. For the production function of Cobb-Douglas type

$$\mathcal{U}(\varkappa(t), \varphi(t), t) = f(t)\, \varkappa^{\alpha}(t)\, \varphi^{\beta}(t), \quad \alpha + \beta = 1$$

(here the exogeneously defined function $f(t)$ describes the progress in technology embodied in funds of period t) the equation of differential optimisation is

$$\beta \Pi(t) = \Pi(m(t))$$

where

$$\Pi(t) = f(t) \cdot \left[\frac{\varkappa(t)}{\varphi(t)}\right]^{\alpha}$$

has clear economic interpretation: the productivity of labour attained with funds accumulated by the moment t.

In the version of the model with exogeneously defined function $\varkappa(t)$ the principle of differential optimization also leads to equation (5). Indeed, in the case considered the quantity $P_{\gamma}'^{+}$ on the trajectory

$$\gamma(t) = \{\varkappa(t), \varphi(t), m(t)\}$$

is defined by the formula

$$P_{\gamma}'^{+} = \mathcal{U}[\varkappa^{+}(t), \varphi^{+}(t), t] - \mathcal{U}[\varkappa(m(t)), \varphi(m(t)), m(t)]\, m'^{+}(t)$$

Function $\varkappa$ is continuously dependent upon time, as seen from (3). Consequently, the number $\varkappa^{+}(t_0) = \varkappa(t_0)$ is the same for all trajectories, which do coincide for $t < t_0$, and cannot be varied. We are led to the same extremal problem and to the old equation of differential optimization.

<u>Linearisation of model equations</u>

In this section the version is treated where $\varkappa(t)$ is defined as exogeneous. In this case the equations of the model are:

$$T(t) = \int_{m(t)}^{t} \varphi(\tau)\, d\tau \qquad (6)$$

$$\beta \Pi(t) = \Pi(m(t)) \qquad (7)$$

where

$$\Pi(t) = f(t)\, \varkappa^{\alpha}(t)\, \varphi^{-\alpha}(t)$$

Let the functions $\varphi(t)$ and $m(t)$ be the solutions of the model equation system (6), (7) for given $\varkappa(t)$ and $f(t)$. We want to know what happens to these solutions as $\varkappa(t)$ and $f(t)$ are varied a little. To be more precise, let $f(t)$ be substituted by $f(t)+\delta f(t)$ starting from t_0, while for $\delta f(t)=0$ (introduction of new technology). Let from t_0 on the function $\varkappa(t)$ be substituted with $\varkappa(t)+\delta\varkappa(t)$, $\delta\varkappa(t)=0$ for $t \leq t_0$ (introduction of additional capital investments). The variations are assumed to be small, i.e.

$$\frac{\delta f(t)}{f(t)} \ll 1 \quad , \quad \frac{\delta \varkappa(t)}{\varkappa(t)} \ll 1$$

We are interested in the corresponding variations of $\delta\varphi(t)$ and $\delta m(t)$ (in the lowest order). In order to find the small variations $\delta\varphi(t)$ and $\delta m(t)$ we linearise the equations (6), (7) near the solution $\varphi(t)$, $m(t)$.

Perturbed equation (6) is

$$\int_{m+\delta m}^{t} [\varphi(\tau)+\delta\varphi(\tau)]\, d\tau = T(t)$$

or

$$\int_{m}^{t} \varphi(\tau)d\tau - \int_{m}^{m+\delta m} \varphi(\tau)d\tau + \int_{m}^{t} \delta\varphi(\tau)d\tau - \int_{m}^{m+\delta m} \delta\varphi(\tau)d\tau = T(t)$$

As the functions $m(t)$, $\varphi(t)$ satisfy the eq. system (6)-(7), we have

$$\int_{m}^{t} \varphi(\tau)\, d\tau = T(t)$$

We extract the dominant terms:

$$\int_{m}^{m+\delta m} \varphi(\tau)d\tau = \varphi(m(t))\,\delta m(t) + \text{(second order term)}$$

$$\int_{m}^{m+\delta m} \delta\varphi(\tau)d\tau = \text{(second order term)}$$

Linearised equation is obtained as we neglect the second order terms in (6). It has the form

$$\int_{m}^{t} \delta\varphi(\tau)d\tau = \varphi(m(t))\,\delta m(t)$$

Recalling that $\delta\varphi(t) = 0$ for $t \leq t_0$ and confining ourselves to the interval $t_0 \leq t \leq t_0 + A$, in which $m(t) < t_0$, we obtain

$$\int_{t_0}^{t} \delta\varphi(\tau)\,d\tau = \varphi(m(t))\,\delta m(t)$$

It is convenient for us to introduce the new inknown function $j(t)$, equal to $\int_{t_0}^{t}\delta\varphi(\tau)d\tau$. Both $\delta\varphi$ and δm are easily expressed in terms of $j(t)$:

$$\delta\varphi = j'(t) \quad \text{and} \quad \delta m = \frac{j'(t)}{\varphi(m(t))}$$

Now we linearise equation (7). Perturbed equation (7) is

$$\Pi(m + \delta m) = \beta f(f + \delta f)(\varkappa + \delta\varkappa)^{\alpha}(\varphi + \delta\varphi)^{-\alpha}$$

From Taylor's expansion,

$$\Pi(m) + \Pi'(m)\,\delta m \quad + \text{(second order terms)} =$$

$$= \beta\Pi(t) + \delta f\,\varkappa^{\alpha}\varphi^{-\alpha} + f\alpha\varkappa^{-\alpha-1}\varphi^{-\alpha}\,\delta\varkappa + \text{(second order terms)} - \alpha f\varkappa^{\alpha}\varphi^{-\alpha-1}\delta\varphi .$$

Neglecting the second order terms and making use of

$$\Pi(m(t)) = \beta\Pi(t)$$

we obtain the linearised equation (7) :

$$\Pi'(m)\,\delta m = \beta\Pi(t)\left[\frac{\delta f}{f} + \alpha\frac{\delta\varkappa}{\varkappa} - \alpha\frac{\delta\varphi}{\varphi}\right]$$

Let us transform this equation. :

$$\frac{1}{\alpha\beta}\,\frac{\Pi'(m)}{\Pi(t)}\,\frac{\varphi(t)}{\varphi(m)}\varphi(m)\delta m = \varphi(t)\left[\frac{\delta f}{\alpha f} + \frac{\delta\varkappa}{\varkappa}\right] - \delta\varphi$$

Recalling that $\varphi(m)\,\delta m = j$ and $\delta\varphi = j'$, we obtain the equation for j :

$$j' + \frac{1}{\alpha\beta}\,\frac{\Pi'(m)}{\Pi(t)}\,\frac{\varphi(t)}{\varphi(m)}\,j = \varphi(t)\left(\frac{\delta f}{\alpha f} + \frac{\delta \varkappa}{\varkappa}\right)$$

We summarise our computations: let $\varphi(t)$, $m(t)$ satisfy equations (6), (7) for given $f(t)$, $\varkappa(t)$ and $\Pi(t) = f(t)\,\varkappa^{\alpha}(t)\,\varphi^{-\alpha}(t)$ Let $f(t)$ and $\varkappa(t)$ undergo small variations $\delta f(t)$ and $\delta\varkappa(t)$ so that $\frac{\delta f}{f} \ll 1$, $\frac{\delta\varkappa}{\varkappa} \ll 1$ and $\delta f = \delta\varkappa = 0$ for $t \leq t_0$. Then in order to find the small variations $\delta\varphi(t)$ and $\delta m(t)$ one must do the following:

i) solve a first order equation in respect to the function $j(t)$

$$j'(t) + \frac{1}{\alpha\beta}\,\frac{\Pi'(m)}{\Pi(t)}\,\frac{\varphi(t)}{\varphi(m)}\,j = \varphi(t)\left(\frac{1}{\alpha}\,\frac{\delta f}{f} + \frac{\delta\varkappa}{\varkappa}\right)$$

with initial conditions $j(t_0) = 0$

ii) compute $\delta\varphi$ and δm with the formulas

$$\delta m = j'/\varphi(m)\,, \qquad \delta\varphi = j'$$

Now let us study the solution upon the interval $t_0 \leq t + t_0 + \tau$ which is a small fraction of the characteristic size $m(t_0)$. We recall that $j(t_0) = 0$. Therefore over the small interval j is relatively small and can be neglected. We obtain the approximate equation

$$j' \approx \varphi(t)\left(\frac{1}{\alpha}\,\frac{\delta f(t)}{f(t)} + \frac{\delta\varkappa(t)}{\varkappa(t)}\right)$$

Recalling that $j' = \delta\varphi$ we obtain:

$$\frac{\delta\varphi}{\varphi} \approx \frac{1}{\alpha}\,\frac{\delta f}{f} + \frac{\delta\varkappa}{\varkappa} \qquad (8)$$

Equality (8) has an economic interpretation. As

$$\Pi(t) = f(t)\,\varkappa^{\alpha}(t)\,\varphi^{-\alpha}(t)$$

we have

$$\delta \Pi(t) = \Pi(t)\left[\frac{\delta f}{f} + \alpha\frac{\delta \varkappa}{\varkappa} - \alpha\frac{\delta \varphi}{\varphi}\right]$$

Formula (8) shows that

$$\frac{\delta f}{f} + \alpha\frac{\delta \varkappa}{\varkappa} - \alpha\frac{\delta \varphi}{\varphi} \approx 0$$

Consequently for perturbed solution the function $\Pi(t) + \delta\Pi(t)$ approximately coincides with the old function $\Pi(t)$, i.e., the additional capital investments $\delta \varkappa(t)$ and additional technology progress effect $\delta f(t)$ in the conditions of differentially optimal development result mostly not in the labour productivity increase on the newly introduced funds $\Pi(t) \approx \Pi(t) + \delta\Pi(t)$, but rather in realignment of greater amount of labour resources from the old to up-to-date funds.

More precisely, we formulate the behaviour of $\varphi(t)$ and $m(t)$ upon the perturbed trajectory: old funds should be closed down to such an extent as to compensate with the released labour resources the labour productivity increase on the new funds (that has taken place due to additional capital investments and introduction of new technology) and bring the productivity on new funds to the old level (i.e. the labour productivity level with existing funds of the unperturbed system).

MODELLING AND COMPUTATION OF WATER QUALITY PROBLEMS IN RIVER NETWORKS

H. Baumert, P. Braun, E. Glos
Institute of Water Management
119 Berlin/GDR

W.-D. Müller, G. Stoyan
Central Institute of Mathematics
and Mechanics
108 Berlin/GDR

Abstract:
A computer program, allowing the computation of stationary and transient water quality in river networks, and considering different possibly non-linearly interacting water quality components has been developed. The graph of the network is part of the input data. The model of the hydrophysical and ecological problem under consideration is a coupled system of generalized one-dimensional convection-diffusion equations along with boundary conditions at the nodes of the graph. To solve the system numerically, a maximum norm stable and monotone difference scheme is used, allowing any Peclét numbers, in particular vanishing diffusion and/or convection. The computational power and flexibility of the program is described.

Introduction

Since the industrialization of agriculture, the development of industries and the flat building are fastly going on in GDR, and on the other hand the water resources are limited, the water situation becomes more and more difficult. Otherwise, due to growing pollution the investments in water quality management become more and more expensive. Therefore and along with the quite complex charakter of water quality problems, the decision maker cannot longer employ traditional thumb rules but ought to make use of specific tools such as packages of scientific computer programs.

In the following the program GRAPH is described. It allows simulations of effects due to different decisions in water quality problems in any river system. GRAPH is a generalized program in the sense of /1/. A quite detailed users manual exists /14/.

The Hydrophysical and Ecological Processes and Variables

In the program the following water quality components may be taken into account as state variables of the ecological problem, assuming that they are sufficiently homogeneously distributed within the cross sections of the network reaches (density problem!):

- soluted minerals or organic substances, e.g. salt, nutrients
- suspended matter
- algae, bacteria, fungi, zooplankton etc.
- heat (temperature)
- toxic or hygienically harmful substances, e.g. concerogenes
- radioactivity

The following transport and mixing processes are considered as the needed hydrophysical basis of the ecological problem:
- convective transport
- longitudinal mixing
- dilution
- homogeneous mixing at the internal nodes of the graph

The following external influences may be considered as forcing functions or input quantities:
- matter import into the system from
  - point sources (tributaties, waste water outlets etc.)
  - line sources (e.g. diffusive mineral instrusions from the fields)
  - inflow-boundary conditions
- sun radiation in the photosynthetic active range
- temperatur (which may be considered as an internal state variable, too), e.g. connected with cooling water problems.

The program is able to answer e.g. questions of the following types:
- What is the effect of an artificial river aeration installation on water quality under different seasonal conditions? (quasi-stationary long-term problems)
- How does an amount of a highly toxic or radioactive substance propagate through the river network? (fully non-stationary short-term problems)

The Basic Eqations

Considering the processes mentioned above, mass balances for an infinitely small disk element of the river give the following P.D.E. system (k= 1 ... n)

$$A \frac{\partial c_k}{\partial t} + Q \frac{\partial c_k}{\partial x} - \frac{\partial}{\partial x} (A \cdot D \cdot \frac{\partial c_k}{\partial x}) + q_w \cdot c_k = A \cdot f_k(c_1 \ldots x,t) \qquad (1)$$

(I) (II) (III) (IV) (V)

where

c_k - value of the k-th water quality component, e.g. a concentration or an excess temperature

A - cross sectional area

Q - flow rate in the river reach
D - longitudinal mixing coefficient (monotone function of Q)
q_w - rate of lateral water inlow

The terms (I–IV) describe the following processes:

(I) - temporal change of the water quality component at a fixed point
(II) - convective transport
(III) - longitudinal mixing (dispersion, cf. /10/)
(IV) - dilution by freshwater addition from tributaries or from ground water

The term (V) is a free and easy programmable real procedure in GRAPH. It may e.g. have the following structure:

$$\underset{(V)}{A\cdot f_k(\,..\,)} = \underset{(VI)}{P_k\cdot\delta(x-x_o)} + \underset{(VII)}{L_k(x,\ \ldots)} + \underset{(VIII)}{\sum_i(\nu_{ik}-\nu_{ki})} + \underset{(IX)}{\Omega_k} \qquad (2)$$

The terms (VI–IX) describe the following processes:

(VI) - matter import by a point source at $x=x_o$
(VII) - matter import by a spatially distributed line source
(VIII)- internal ecological interactions in the water body (matrix of trophic interactions)
(IX) - surface interactions with sediment and atmosphere

ν_{ik} is the rate of the generalized "reaction" $c_i \longrightarrow c_k$.

The structure of the river network is described by an abstract graph and the related incidence matrix which is part of the related incidence matrix which is part of the input file of the program.

Along with the boundary conditions

a) inflow node: $c_k(x,t) = c_k^b(t)$

b) outflow node: $\partial c_k/\partial x = 0$ (3)

c) internal node: $\sum_{p\ni\mathcal{N}} \lim_{x\to x_p}(Q\cdot c_k - A\cdot D\,\frac{\partial c_k}{\partial x})_p = 0$

and the initial conditions

$$c_k(x,0) = c_k^o\ (x) \qquad (4)$$

we have now a full description of many classes of water quality problems. ($\mathcal{N}$ - set of all reaches directly connected with the internal node under consideration)

Fig.1 Abstract graph representation of a river network
•/o : internal/external node

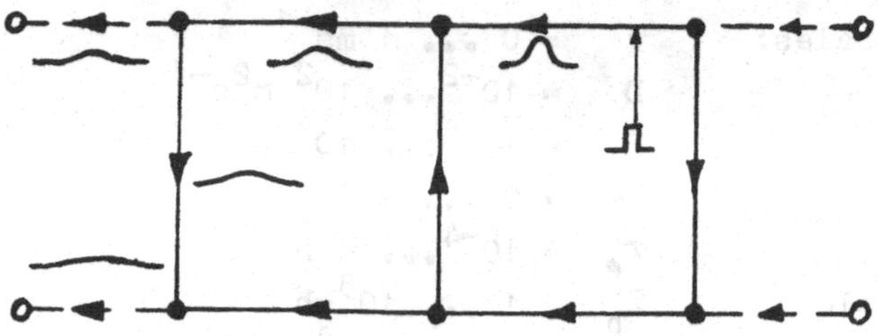

Time Scales of Water Quality Problems

As shown in the paper /6/, the time scales of the processes and variables are important for the choice of simplified versions of the general model (1). In /6/ the solution of linear non-stationary convection-diffusion problems is described by the amplitude damping

$$\alpha(H)=\exp\left\{-\frac{P}{2}\left[(1+H^2)^{1/4}\cos(\Psi/2)-\frac{1}{2}\left(1+\frac{\tau_*}{\tau_e}\right)^{-1/2}-\frac{1}{2}\left(1+\frac{\tau_*}{\tau_e}\right)^{1/2}\right]\right\} \quad (5)$$

and by the phase shift in relation to the pure convection problem

$$\varphi(H)=\frac{P}{2}\left[(1+H^2)^{1/4}\sin(\Psi/2)-\frac{H}{2}\left(1+\frac{\tau_*}{\tau_e}\right)^{1/2}\right] \quad (6)$$

where

P – generalized Peclét number, $P = Pe + 2L/(D \cdot \tau_e)^{1/2}$

Pe – Peclét number, $Pe = vL/2D$

H – special similarity number of linear convection-diffusion problems, $H = \frac{\tau_*}{\tau_b}/(1+\frac{\tau_*}{\tau_e})$

τ_b – time scale of the dominant first bondary condition

τ_e – time scale of the dominant ecologic process, in the sense of an optimum linear substitute of the term (VIII) in (2)

τ_* – dynamic time scale of the river, $\tau_* = 4D/v^2$

L – length of the river reach under consideration

Ψ – auxiliary symbol, Ψ = arc tan H

v – flow velocity, v = Q/A

To avoid stiffness problems we make use of the hierarchy of time scales in ecological processes (Tichonov theorem, cf. /27/) and treat only meso-scale phenomenae in the time domain.

A rough estimation, considering real world data and the whole application range of the simulation program, reveals the following parameter ranges:

Hydraulic scales:	v	$= 0 \ldots 1 \ ms^{-1}$
	D	$= 10^{-2} \ldots 10^{2} \ m^2s^{-1}$
	L	$= 10 \ldots 10^{2} \ m$
	Pe	$= 0 \ldots 10^{4}$
	τ_*	$= 10^{-4} \ldots \quad h$
boundary scale:	τ_b	$= 1 \ldots 10^{3} \ h$
ecological scale:	τ_e	$= 10 \ldots 10^{2} \ h$

Following that we find the ranges of H and P to be

$$H = 0 \ldots \frac{\tau_e}{\tau_b} \quad , \quad P = 10^2 \ldots 10^4$$

The resulting damping and phase shift values are shown in the table below:

P \ H	10^{-7}	10^{-1}
10^2	1.0 0.0	0.94 0.18
10^4	1.0 0.0	0.002 17.8

upper value: a
lower value: φ, in degrees

While problems with small P and H could be desdribed by the simple transport equation (no damping and no phase shift occurs), problems with larger P and H have to be described by the full parabolic equation (1). For stationary problems we have $\tau_b \rightarrow \infty$ $(H \rightarrow 0)$ and equation (1) turns over to an elliptic one. If both velocity and dispersion vanish, it degenerates to an ordinary differential equation. All these features of the model should be included in the numerical algorithm, too.

Numerical Methods

The numerical solution of the initial-boundary value problem (1-4) is quite well described in the literature only in two cases: if velocity is small and dispersion comparably large, or if dispersion can be neglected at all. Many simulation programs (see /1/) switch from one numerical scheme to another one, depending on the parameters. A great part of literature is devoted to overcome these limitations, using both finite elements and finite differences.

The difficulty in developing the desired numerical scheme covering the whole parameter range mentioned above, comes from the physically justified demand that such a method should be stable, conservative, preserving positiveness and should allow calculations with small numerical diffusion. For finite elements, it has been shown /13/ that mass lumping in transient problems is not advisable.

For finite differences, the usual central difference approximation of the first derivative leads to instability if $|q|>1$, where $q = vh/2D$ is the so-called cell Reynolds number (h = step size in x-direction). To overcome this, upwind differencing and several combinations of central and one-sided differences have been recommended /7,18,22, 23/ which are equivalent to taking a greater dispersion coefficient.

The exact difference approximation to the spatial differential operator in the case of constant coefficients has been used in a number of papers /3,8,17,20/. This approximation can also be rewritten as a combination of the central and the one-sided difference /3/. All the mentioned schemes have the draw-back to sacrifice accuracy for a gain in stability - this is also true for the exact difference approximation as used in /3,8,17,20/, if the equation is non-stationary and has a non-zero source term, as has been mentioned in /3/ and /20/ and becomes clear from /24/. At the same time a quite remarkable numerical diffusion is introduced into these schemes growing up to $|v| \cdot h/2$.

In this water quality research project, the difference scheme described in /25/ has been used. The scheme has been developed from that of /26/. In the linear case it possesses the following properties:

- the scheme is maximum norm stable on any non-equidistant spatial and time grid
- preserves positiveness
- represents a conservative approximation of the differential equation
- accepts first and second kind boundary conditions and
- allows for any ratio of diffusion to convection, including degenerated cases.

Particularly, if the diffusion vanishes, the scheme approaches a maximum norm stable scheme of gas dynamics. For special choice of the ratio of time and spatial steps the scheme is of second order of accuracy in space and time. This same choice, for vanishing diffusion, turns the scheme over to the method of characteristics. Due to this property it is possible to calculate steep concentration fronts with small numerical diffusion. The scheme also allows for degeneration of the coefficient multiplying the time derivative and admits simultaneously vanishing of diffusion and convection - which is of practical interest, too.

Fig.2 Computed & observed dispersion patterns

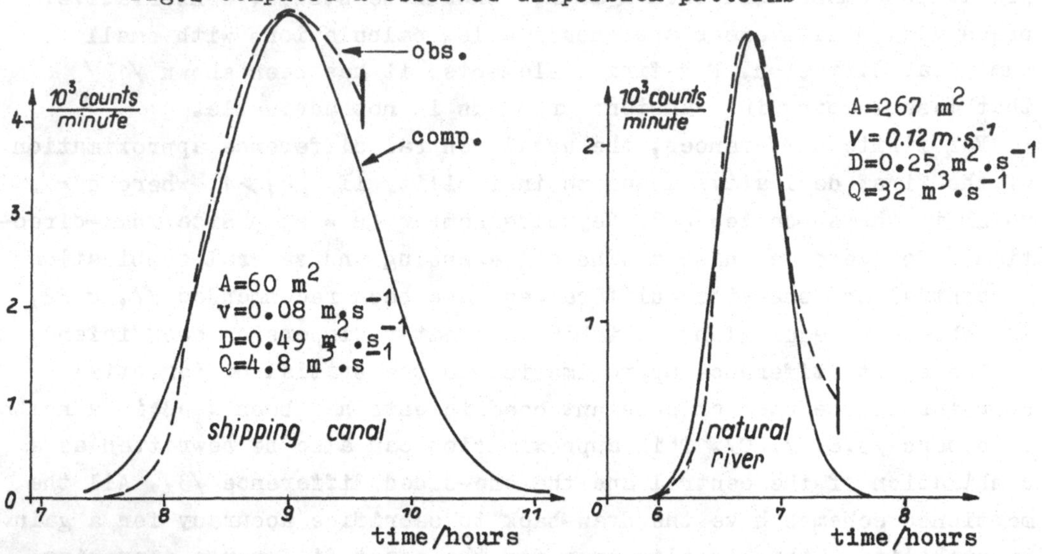

The difference scheme is of the two-level, three-point type. Therefore, at the inner points of the lines of the graph, it is equivalent to a tridiagonal system of linear equations. At the nodes of the graph more than three unknowns are connected. The approximation is selected to guarant conservativity. In the stationary case, if dispersion and sources are zero, this approximation turns out to be identical to the "classical" mixing rule.

Making use of the algorithm of /12/, the whole set of linear equations with a large sparse matrix can be reduced to a small set with dense matrix, the dimension being equal to the number of nodes. This linear system is solved by a standard algorithm /11/, where the concentrations in all nodes are known. Hence, the whole problem is splitted in to a series of first kind boundary value problems for the lines of the graph. These problems then are treated by the standard shortened Gauss elimination.

For nonlinear sources (2) the equations (1) are solved using a simple iteration technique.

GRAPH has a number of additional features, e.g. it allows the solution of stationary problems in a first step and the subsequent solution of the full transient problem as a second step, further, after the data input (the structure of the graph and the governing right hand sides of the differential equations), a part of the graph can be selected on which the problem has to be solved actually. The water balance will not be destroyed. For more details, including the realized instationary imbedding technique, storage and CPU time requirements, see /14,15/.

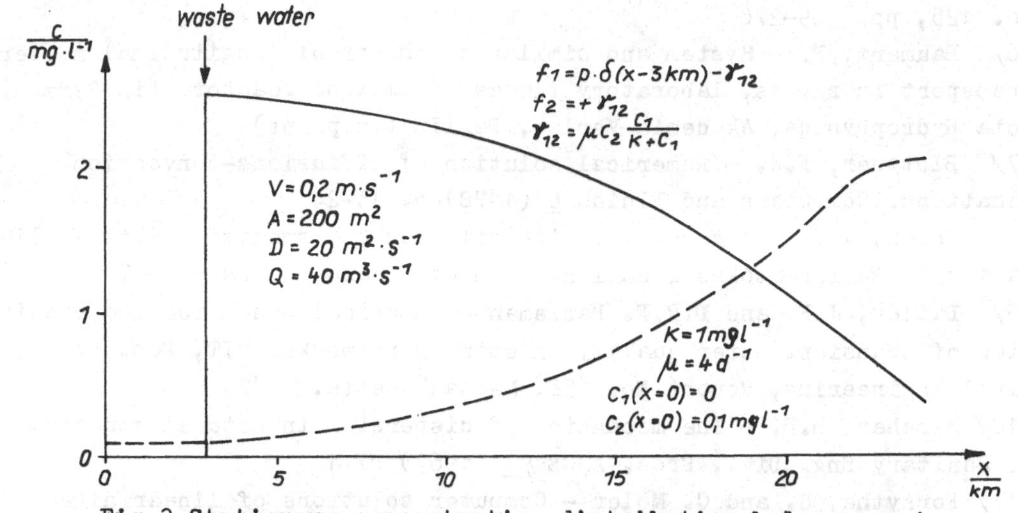

Fig.3 Stationary concentration distribution below a waste outlet, considering a nonlinear kinetic

Due to the described numerical and computational properties and the several possibilities contained in the program, a broad range of practical problems as mentioned at the beginning of the paper has been solved by the authors and other users: short term and long term forecasting of temperature and concentration distributions, simulation of accident situations, and ecological studies. The CPU time was always in the range of minutes (BESM-6).

The program GRAPH is, in diferent versions, available by licence.

References:

/1/ Abbott, M.B., Preissmann,A. and R.Clark - Logistics and benefits of the European Hydrologic System. Proc. of the IIASA/IBM symposium on "Logistics and benefits of using mathematical models in hydrologic and water resources systems", Pisa, Italy, 24-26 october 1978

/2/ Abbott, M.B. - Scientific and commercial aspects of applied mathematical modelling. Proc. of the Int.Conf. on "Applied mathematical modelling", Madrid 1978

/3/ Barrett, K.E. - The numerical solution of singular-perturbation boundary-value problems. Quart. J.Mechn.Appl.Math. 27(1974)pp. 57-68

/4/ Baumert, H. - Investigations on the consideration of mixing processes in water quality models of river ecosystems (in German). Dr.rer.nat. Thesis, Technical University Dresden, GDR, 1978

/5/ Baumert, H. and A. Becker - Water quality modelling in surface water networks with special regard to quality breakdowns. Proc. of the IIASA/IAHS symposium on "Modelling the water quality of the hydrological cycle", Baden, Austria, september 1978, IAHS-AISH Publ.

No. 125, pp. 269-276

/6/ Baumert, H. - System and similarity theory of longitudinal matter transport in rivers, laboratory flumes and mixing reactors (in German). Acta Hydrophysica, Akademie-Verlag, Berlin (in print)

/7/ Blottner, F.G. - Numerical solution of diffusion-convection equations. Computers and Fluids 6 (1978)pp. 15-24

/8/ Chien, J.C. - A general finite-difference formulation with application to Navier-Stokes equations. Computers and Fluids 5(1977)pp.15-31

/9/ Dailey, J.E. and D.R.F. Harleman - Numerical model for the prediction of transient water quality in estuary networks. MIT, Dep. of Civil Engineering, Report No. 158, Massachusetts, 1972

/10/ Fischer, H.B. - The mechanics of dispersion in natural streams. J. Sanitary Eng. Div., Proc. ASCE 93 (1967) HY 1

/11/ Forsythe, G. and C. Moler - Computer solutions of linear algebraic systems. Prentice-Hall 1967

/12/ Frjasinov, I.W. - Algorithm for the solution of finite difference problems on graphs (in Russian). J. vycisl.matem.i matem.fis. 10(1970) No. 2, pp. 474-477

/13/ Gresho, P.M., Lee, R.L. and R.L. Sani - Advection-dominated flows, with emphasis on the consequences of mass lumping. Finite elements in fluids, Vol. 3, pp. 335-350, Wiley-Interscience 1978

/14/ Müller, W.-D. and G. Stoyan - Users manual of the BESM-ALGOL program GRAPH-1 (in German). R/D report, Academy of Sciences of GDR, Central Institute of Mathematics and Mechanics, Berlin 1979

/15/ Müller, W.-D. - On a program for the computation of matter transport in river networks. In: Nonlinear Analysis - Theory and Applications (Proc.Int.Summer School, Berlin 1979, Editor: R.Kluge) Abh.Akad. Wiss. DDR (in print)

/16/ Nihoul, J.C.J. and J. Smitz - Mathematical model of an industrial river. Proc.of the IFIP conference on "Biosystems simulation in water resources and waste problems" (Ed.:Vansteenkiste), pp. 333-342, North-Holland 1976

/17/ Iljin, A.M. - A difference scheme for differential equations with small parameters at the higher derivatives (in Russian). Mat.sametki 6 (1969) No. 2, pp. 237-248

/18/ Raithby, G.D. and K.E. Torrance - Upwind-weighted differencing schemes and their application to elliptic problems involving fluid flows. Computers and Fluids 2 (1974) pp. 191-206

/19/ Rutherford, J.Ch. and M.J.O'Sullivan - Simulation of water quality in Tarawera River (Hawaii). J.Envir.Eng.Div., Proc.ASCE 1972, EE2, pp. 369-389

/20/ Roscoe, D.F. - The solution of the three-dimensional Navier-Stokes equations using a new finite difference approach. Int.J.Num. Meth.Engng. 10 (1976) No. 6, pp. 1299-1308

/21/ Samarski, A.A. - The theory of difference schemes (in Russian). Nauka, Noscow 1971

/22/ Samarski, A.A. - On monotone difference schemes for elliptic and parabolic equations in the case of non-selfadjoint elliptic operators (in Russian). J. vyčisl.matem.i matem.fis. 5 (1965) No. 3, pp. 548-551

/23/ Spalding, D.B. - A novel finite difference formulation for differential expressions involving both first and second derivatives. Int. J.Num.Meth.Engng. 4 (1972) pp. 551-559

/24/ Stoyan, G. - Monotone difference schemes for diffusion-convection problems. ZAMM 59 (1979) No. 8, pp. 361-372

/25/ Stoyan, G. - On a maximum norm stable, monotone and conservative difference approximation of the one-dimensional diffusion-convection equation. Proc.of the conference on "Simulation of coupled transport exchange and conversion processes in ground water", Editor:L.Luckner, Dresden, GDR, november 1979

/26/ Thomann, R.V. - Effect of longitudinal dispersion on dynamic water quality response of streams and rivers. Water Resources Research 9 (1973) No. 2, pp. 355-366

/27/ Tichonov, A.N. - Mat.Sb. 22 (1948) pp. 13 ff.

/28/ Yemerenko, Y.V. - Modelling the water quality in river basins. Proc. of the Soviet-American sympos. on "Use of math. models to optimize water quality management", Kharkov and Rostov-on-Don, USSR, december 1975, EPA-600/9-78-024, pp. 55-91

AN APPLICATION OF OPTIMAL CONTROL THEORY TO THE ESTIMATION OF THE DEMAND FOR ENERGY IN CANADIAN MANUFACTURING INDUSTRIES*

Michael Denny, Melvyn Fuss and Leonard Waverman

Institute for Policy Analysis
University of Toronto
Toronto, Canada, M5S 1A1

1. Introduction

This paper utilizes optimal control theory to model a manufacturing firm's demand for factors of production, with special emphasis on energy inputs. Unlike previous empirical estimation of the derived demand for energy which builds on either static models or steady state solutions to dynamic models ([1], [3]), the model presented in this paper explicitly incorporates the dynamic adjustment path between steady-states as a consistent, integral part of the modelling and estimation processes.

The point of departure is the internal cost of adjustment model of capital accumulation [6]. In our version of this model, the capital stock is the state variable while energy, labour, non-energy materials and investment are control variables. The objective of the firm is to minimize the present discounted costs of producing a flow of output, where included in costs are convex internal costs of adjusting the capital stock along a path between two steady-state equilibria. Pontryagin's maximum principle is used to solve for the optimal value of the controls conditional on the state variable, and the optimal transition path for the state variable. The great advantage of the optimal control derivation of this model is that the adjustment process is an endogenous part of the optimization procedure and not an ad hoc addendum to a static solution.

---

* This paper is drawn from a much larger report [2] in which an extensive analysis of the empirical results is presented. Financial support provided by the Ontario Ministry of Industry and Tourism is gratefully acknowledged. However, any opinions expressed in this paper are those of the authors alone.

The theoretical model is transformed into an econometric model by recognizing that the optimization of the control variables conditional on the state variable yields a normalized restricted cost function [5]. A quadratic approximation to an arbitrary restricted cost function yields an estimable system of factor demand functions.

The system of dynamic demand equations is fitted to data drawn from nineteen 2-digit Canadian manufacturing industries for the years 1962-75. The general characteristics of the estimated dynamic structure is discussed. Of particular interest is the speed of adjustment between steady states. In no case is the adjustment complete in one year. This result demonstrates the importance of a dynamic approach to energy demand modelling. Response to large shocks may take considerable time to fully influence behaviour.

2. A Theoretical Model with Internal Costs of Adjustment

A firm is assumed to have an implicitly defined production function,

$$F(v,x,\dot{x},Q,t) = 0$$

that describes the combinations of variable inputs, $v = \{v_j\}$, $j=1,\ldots m$, and quasi-fixed inputs, $x = \{x_i\}$, $i=1,\ldots n$, that can be chosen to produce optimally output Q at time t. If the quasi-fixed input levels are changed, $\dot{x} \neq 0$, then output falls for any given level of the variable and quasi-fixed inputs. This is the internal cost of adjustment model analyzed extensively by Treadway [6].

The firm is assumed to minimize the present value of the future stream of costs:

$$L(0) = \int_0^\infty e^{-rt}\left(\sum_{j=1}^{m} \hat{w}_j v_j + \sum_{i=1}^{n} \hat{q}_i z_i\right) \qquad (1)$$

where r is the firm's discount rate, $\hat{w}_j$ and $\hat{q}_i$ are the prices of the variable and quasi-fixed inputs and $z_i = \dot{x}_i + \delta x_i$ is the net addition to the stock of the i-th quasi-fixed factor, where δ_i is the depreciation rate of the i-th stock. The minimization is accomplished by choosing the time paths of the control variables $v(t)$, $\dot{x}(t)$ and the state variable $x(t)$ so as to minimize $L(0)$, given any initial $x(0)$ and $v(t), x(t) > 0$.

Given strict quasi-concavity of the production function in the variable factors, the first order conditions for these control variables, $\frac{\partial L(0)}{\partial v(t)} = 0$, can be solved for the optimal $v(t)$ as functions of $w(t)$, $x(t)$, $\dot{x}(t)$, $Q(t)$ and t; say $\bar{v}(t) = h(w(t), x(t), \dot{x}(t), Q(t), t)$. The vector $w(t) = (w_2(t) \ldots w_j(t) \ldots w_m(t)) = (\hat{w}_2(t)/\hat{w}_1(t), \ldots \hat{w}_j(t)/\hat{w}_1(t), \ldots \hat{w}_m(t)/\hat{w}_1(t))$ is a vector of normalized factor prices.

To link the control problem to our empirical specification, define the function

$$G(t) = \sum_{j=1}^{m} w_j \bar{v}_j(t) = G(w(t), x(t), \dot{x}(t), Q(t), t)) . \tag{2}$$

This is the normalized restricted cost function whose value is the minimum variable cost of producing Q at time t conditional on the level of the quasi-fixed factors or state variables $x(t)$ and the control variable $\dot{x}(t)$. This cost function can be shown to be: (a) increasing and concave in w, (b) increasing and convex in $\dot{x}$, (c) decreasing and convex in x and (d) $\partial G/\partial w_j = \bar{v}_j$, the conditional cost minimizing input level.

Substituting the cost function $G(t)$ into (1) and integrating by parts, we obtain

$$\bar{L}(0) - \sum_{i=1}^{n} q_i x_i = \int_0^\infty \hat{w}_1 e^{-rt}[G(w,x,\dot{x},Q,t) + \sum_{i=1}^{n} u_i x_i]dt \tag{3}$$

where $u_i = q_i(r + \delta_i)$ is the normalized user cost of the i-th quasi-fixed factor. Minimizing (3) w.r.t. $x(t)$, $\dot{x}(t)$ is the same as minimizing (1) w.r.t. $x(t)$, $\dot{x}(t)$ and $v(t)$ since (3) incorporates the optimal $v(t)$ conditional on $x(t)$, $\dot{x}(t)$.

A solution may be obtained using Pontryagin's maximum principle. Assuming static expectations with respect to factor prices ($\hat{w}_1$, w constant) we can construct the Hamiltonian,

$$H(x,\dot{x},\lambda,t) = e^{-rt}[G(w,x,\dot{x},Q,t) + \sum_i u_i x_i] + \lambda\dot{x} \quad .$$

After eliminating $\dot{\lambda}$ from the necessary conditions, we have

$$-G_x - rG_{\dot{x}} - u - B\ddot{x} - C\dot{x} = 0 \tag{4}$$

where $u = (u_1,\ldots u_n)$, $B = [-G_{\dot{x}\dot{x}}]$, and $C = [-G_{x\dot{x}}]$. A steady state solution x^* is given by

$$-G_x(w,x^*) - rG_{\dot{x}}(w,x^*) - u = 0 \tag{5}$$

and is unique if $|A^* + rC^*| \neq 0$ where $A = [G_{xx}]$ and $*$ indicates evaluation at $x = x^*$ and $\dot{x} = 0$. Rewrite (5) as $-G_x(w,x^*) = u + rG_{\dot{x}}(w,x^*)$. The left hand side is the marginal benefit to the firm of changing capital while the RHS is the marginal cost (user cost plus the marginal adjustment cost of a change in the flow of capital services at $\dot{x} = 0$).

Treadway [6] links this model to the flexible accelerator literature by showing that the demand for the quasi-fixed factor can be generated from (4) and (5) as an approximate solution (in the neighbourhood of $x^*(t)$) to the multivariate linear differential equation system

$$\dot{x} = M^*(x^* - x) \tag{6}$$

where, assuming C^* is symmetric, M^* satisfies the condition

$$B^*M^{*2} + rB^*M^* - (A + rC^*) = 0 \quad . \tag{7}$$

In our empirical work we will approximate $G(t)$ by a quadratic function which implies C^* is symmetric. This condition is sufficient for Le Chatelier's principle as applied to short-run and long-run demand functions to be satisfied.

For a single quasi-fixed factor (6) and (7) become,

$$\dot{x}_1 = \beta^*(x_1^* - x_1) \tag{8}$$

$$-G^*_{\dot{x}_1\dot{x}_1}\beta^{*2} - rG^*_{\dot{x}_1\dot{x}_1}\beta^* + (G^*_{x_1x_1} + rG^*_{x_1\dot{x}_1}) = 0 \tag{9}$$

Solving (9) for the stable root yields

$$\beta^* = -\tfrac{1}{2}[r - ((r^2 + 4(G_{x_1x_1} + rG_{x_1\dot{x}_1}))/G_{\dot{x}_1\dot{x}_1})^{\frac{1}{2}}]$$

3. An Econometric Model of the Dynamic Demand for Energy with Capital as a Quasi-Fixed Factor

Gross output (Q) is produced using aggregate energy (E), labour (L), capital (K) and materials (M). The production function is

$$F(E,L,M,K,\dot{K},Q,t) = 0 \tag{10}$$

The duality between cost and production functions [4] implies that the technology can be represented by a normalized restricted average cost function, $\frac{G}{Q}(p_E, p_M, K, \dot{K}, Q, t) = \frac{L}{Q} + p_E\frac{E}{Q} + p_M\frac{M}{Q}$, where p_E and p_M are prices normalized by the price of labour.

The quadratic approximation to the average cost function used in this study is

$$\begin{aligned}
\frac{G}{Q} = {} & \sum_{i=1}^{4} \alpha_{oi}\, D_i + \alpha_{ot}\cdot t + (\sum_1^4 \alpha_{Ei}\, D_i)p_E + (\sum_1^4 \alpha_{Mi}\, D_i)p_M \\
& + (\sum_1^4 \alpha_{Qi}\, D_i)Q + (\Sigma\alpha_{Ki}\, D_i)(\frac{K_{-1}}{Q}) + \alpha_{\dot{K}}(\frac{\Delta K}{Q}) \\
& + \tfrac{1}{2}[\gamma_{EE}p_E^2 + \gamma_{MM}p_M^2 + \gamma_{QQ}Q^2 + \gamma_{KK}(\frac{K_{-1}}{Q})^2 + \gamma_{\dot{K}\dot{K}}(\frac{\Delta K}{Q})^2] \\
& + \gamma_{EM}p_Ep_M + \gamma_{EQ}p_EQ + \gamma_{MQ}p_MQ + \gamma_{EK}p_E(\frac{K_{-1}}{Q}) + \gamma_{MK}p_M(\frac{K_{-1}}{Q}) \\
& + \gamma_{QK}Q(\frac{K_{-1}}{Q}) + \gamma_{E\dot{K}}p_E(\frac{\Delta K}{Q}) + \gamma_{M\dot{K}}p_M(\frac{\Delta K}{Q}) + \gamma_{Q\dot{K}}(\frac{\Delta K}{Q}) + \gamma_{K\dot{K}}K_{-1}(\frac{\Delta K}{Q}) \\
& + \alpha_{Et}\cdot p_E\cdot t + \alpha_{Mt}\cdot p_M\cdot t + \alpha_{Kt}(\frac{K_{-1}}{Q}) + \alpha_{t\dot{K}}\cdot t(\frac{\Delta K}{Q})
\end{aligned} \tag{11}$$

where the dummy variable D_i equals one if the observation is in region i and is zero otherwise. The data used in this study are drawn from four regions of Canada: Quebec, Ontario, the Prairies, and British Columbia. The dummy variables were introduced to allow for the heterogeneous nature of regional data in Canada. The specification (11) allows some of the parameters of the quadratic function to vary across regions, and controls for the differences in aggregate 2-digit manufacturing production levels in the different regions.

The number of parameters can be reduced by assuming that marginal adjustment costs are zero when there is no adjustment. This implies

$$\alpha_{\dot{K}} = \gamma_{E\dot{K}} = \gamma_{M\dot{K}} = \gamma_{Q\dot{K}} = \gamma_{K\dot{K}} = \alpha_{t\dot{K}} = 0 \quad .$$

Differentiating the average cost function with respect to the prices of energy and materials yields the input-output short-run demand equations

$$\frac{E}{Q} = \sum_i \alpha_{E_i} D_i + \gamma_{EE}p_E + \gamma_{EM}p_M + \gamma_{EQ}Q + \gamma_{EK}\left(\frac{K_{-1}}{Q}\right) + \alpha_{Et}\cdot t \tag{12}$$

$$\frac{M}{Q} = \sum_i \alpha_{M_i} D_i + \gamma_{EM}p_E + \gamma_{MM}p_M + \gamma_{MQ}Q + \gamma_{MK}\left(\frac{K_{-1}}{Q}\right) + \alpha_{Mt}\cdot t \tag{13}$$

Using our previous results, the optimal path of the quasi-fixed factor K is characterized by

$$\left(\frac{K - K_{-1}}{Q}\right) = -\tfrac{1}{2}\left[r - \left(r^2 + (4\gamma_{KK}/\gamma_{\dot{K}\dot{K}})\right)^{\frac{1}{2}}\right]$$
$$\cdot\left[\gamma_{KK}^{-1}\left\{-\sum_i \alpha_{Ki} D_i - \gamma_{EK}p_E - \gamma_{MK}p_M - \alpha_{Kt}\cdot t - u_K\right\} - \left(\frac{K_{-1}}{Q}\right)\right] \tag{14}$$

Rather than estimating the cost function (11), we form the short-run demand function

$$\frac{L}{Q} = \frac{G}{Q} - p_E\cdot\frac{E}{Q} - p_M\cdot\frac{M}{Q} = \sum_i \alpha_{oi} D_i + \alpha_{ot}\cdot t - \tfrac{1}{2}\gamma_{EE}p_E^2 - \tfrac{1}{2}\gamma_{MM}p_M^2 - \gamma_{EM}p_Ep_M + \alpha_Q Q$$
$$+ \alpha_K\left(\frac{K_{-1}}{Q}\right) + \alpha_{Kt}\left(\frac{K_{-1}}{Q}\right)\cdot t + \tfrac{1}{2}\gamma_{QQ}Q^2 + \tfrac{1}{2}\gamma_{KK}\left(\frac{K_{-1}}{Q}\right)^2$$

$$+ \tfrac{1}{2}\gamma_{\dot{K}\dot{K}}(\frac{\Delta K}{Q})^2 + \gamma_{QK}Q(\frac{K_{-1}}{Q}) \qquad (15)$$

Four equations are estimated: the short-run labour demand function (15), the short-run energy (12) and materials (13) demand equations and the capital formation equation (14).

The aggregate model described above is completed by appending an energy submodel that describes the choice of specific energy sources to minimize unit energy costs. Given this submodel choice, the second stage aggregate model chooses the cost minimizing quantities of total energy, labour and materials conditional on the beginning of period capital stock. Combining the two models, the normalized restricted cost function takes the form,

$$G(t) = G(\hat{p}_E(p_{E1}, \ldots p_{En}, K_{-1}, Q, t)/\hat{p}_L, K_{-1}, \Delta K, p_M, Q, t) \quad . \qquad (16)$$

The function $\hat{p}_E(\cdot)$ is a restricted price aggregator function which is approximated by a version of the translog unit cost function

$$\ln \hat{p}_E = \sum_{i=1}^{6} \sum_{k=1}^{4} \beta_i^k D_k \ln \hat{p}_{Ei} + \sum_{ij}\sum \beta_{ij} \ln \hat{p}_{Ei} \cdot \ln \hat{p}_{Ej} + \sum_i \beta_{iK} \cdot \ln \hat{p}_{Ei} \ln(\frac{K_{-1}}{Q})$$

$$+ \sum_i \beta_{it} \ln \hat{p}_{Ei} \cdot t \qquad (17)$$

There are six energy sources (i) and four geographic regions (k) in our models. The estimated submodel consists of the price aggregator function (17) and the demand functions (18) for the energy sources (obtained from Shephard's Lemma) in terms of the cost shares, M_{Ei} ,

$$M_{Ei} = \sum_{k=1}^{4} \beta_i^k D_k + \sum_j \beta_{ij} \ln \hat{p}_{Ei} + \beta_{iK} \ln (\frac{K_{-1}}{Q}) + \beta_{it} \cdot t \; , \quad i=1,\ldots 6 \qquad (18)$$

The energy submodel is estimated first and the estimated $\hat{p}_E$, from (17), is used as an instrumental variable for the aggregate energy price in the estimation of the aggregate model. For a detailed analysis of this two-stage modelling of energy demand see Fuss [3].

4. Empirical Results

The energy demand model, with costs of adjustment, was estimated for nineteen industries at the 2-digit SIC level for four Canadian regions over the years 1962-75. Energy consumption is divided into six commodity types: coal, LPG, fuel oil, natural gas, electricity and motor gasoline. The demand for each type is determined by the minimization of total energy costs conditional on the output level and stock of capital. Only a general overview of the extensive detailed results can be given here. For additional details see [2].

There are significant possibilities for energy "conservation" through rising energy prices. The mean response across all industries is a 7.2% reduction in energy demand for a 10% increase in energy prices after substitution among energy sources. Substitution possibilities exist between natural gas and fuel oil, and to a lesser extent between these fuels and electricity. The long-run response of total energy consumption to energy price increases is very similar to the short-run response. The largest differences exist in some of the industries, e.g. Primary Metals and Chemicals, which are particularly heavy energy users. These industries have some potential for larger long-run energy reductions. Since energy costs are a very small proportion of total costs, large increases in energy prices raise manufacturers' costs by only a small percent.

In contrast with either the disaggregated energy sources or aggregate energy, the long-run responsiveness of labour and materials to price changes is substantially larger than in the short-run. Their price elasticities often increase by one-half between the short and long-run. However, the long-run elasticities are less than one indicating that total expenditures will rise with price increases.

How quickly do industries adjust their capital stock to new situations? Our estimates suggest that they adapt slowly. The adjustment coefficient, β^* , equals the proportion of adjustment to long-run equilibrium that occurs in one year. The average value is about 0.4 but four industries have values lower than 0.2. At the average value about three-quarters of the adjustment occurs within three years. Remember that industries need never actually attain long-run equilibrium at current prices. In a world subject to exogenous shocks relatively slowly adjustment to new information may be sensible.

A protracted debate has occurred about the relationship between energy consumption and capital usage. Engineers have suggested that energy and capital are complements, i.e., more capital intensive production implies more energy consumption. Our results are mixed. In eleven of the nineteen industries, capital and energy are substitutes, not complements. Any simple general conclusion about capital-energy substitutability is likely to be wrong.

In Canada, possible losses of employment due to higher energy prices concern policy makers. In all industries we find that capital and labour are substitutes which suggests that policies or shocks that raise the price of capital will increase employment. A more complex relationship exists between the demand for labour and the price of energy. In the short-run fifteen industries show a positive change in employment when energy prices rise. However, in the long-run only eleven show a positive response. Further inspection indicates a wide variety of responses. The demand for labour rises in the short-run in some industries and falls in the long-run and vice-versa. In many cases the short-run response is larger in absolute terms than in the long-run.

Our results indicate the importance of constructing energy demand models that incorporate theoretically consistent adjustment paths for industries. Moreover, the models estimated here do not contradict earlier contributions about the importance of conservation through higher prices. Industries can adjust the quantities of particular fuels that they consume and can alter the total energy to labour, capital and materials ratios used in producing any output level. Further research is needed to investigate the validity of and reasons for the relatively long adjustment lags evident in our results.

REFERENCES

[1] Berndt, E.R. and D.O. Wood, "Technology, Prices, and the Derived Demand for Energy", Review of Economics and Statistics, August, 1975, pp. 259-68.

[2] Denny, M., M. Fuss and L. Waverman, Energy and the Cost Structure of Canadian Manufacturing Industries, Institute for Policy Analysis Technical Paper No. 12, University of Toronto, Toronto, Canada, August 1979.

[3] Fuss, M.A., "The Demand for Energy in Canadian Manufacturing: An Example of the Estimation of Production Structures with Many Inputs", Journal of Econometrics, January 1977, pp. 89-116.

[4] Fuss, M. and D. McFadden, Production Economics: A Dual Approach to Theory and Applications, North Holland Publishing Company, Amsterdam, 1978.

[5] Lau, L.J., "A Characterization of the Normalized Restricted Profit Function", Journal of Economic Theory, February, 1976, pp. 131-163.

[6] Treadway, A.B., "The Globally Optimal Flexible Accelerator", Journal of Economic Theory, Vol. 7, 1974, pp. 17-39.

OPERATIONAL MULTIPLE GOAL MODELS FOR LARGE ECONOMIC ENVIRONMENTAL MODELS

J.A. Hartog*, P. Nijkamp**, and J. Spronk*

* Erasmus University, Rotterdam
** Free University, Amsterdam

1. Introduction

Already a few decades ago, a number of scholars (Tinbergen [1952, 1965], van Eijk and Sandee [1957], Theil [1964], have brought macro economic policy making into a mathematical programming framework. More recently, it has been emphasized that this problem can be handled in a much more flexible way by means of multiple criteria decision methods (Spivey and Tamura [1970], Despontin and Vincke [1977] and Wallenius et al. [1978]).

In this paper we give an illustration of the use of a new multiple criteria decision method applied to an existing input-output model. The multiple criteria decision method used, Interactive Multiple Goal Programming (cf. Nijkamp and Spronk [1978a, b, c]), is described in section 2. In the third section we describe the structure of the input-output model, which has been used to demonstrate this method. For this purpose we defined six different and mutually conflicting goal variables, which are described in section 4. The decision-maker - using a terminal display - repeatedly proposes combinations of the goal variables for which the consequences are calculated by a set of computer programs (section 5). An example of a session with a decision-maker solving his decision problem in the indicated fashion, is presented in the sixth section.

2. Interactive Multiple Goal Programming

Recently, interactive methods have become rather popular in decision analyses. They are based on a mutual and successive interplay between a decision-maker and an analyst. These methods do neither require an explicit representation or specification of the decision-maker's preference function nor an explicit quantitative representation of trade-offs among conflicting objectives. Obviously, the solution of a decision problem requires that the decision-maker provides information about his priorities regarding alternative feasible states, but in normal interactive procedures only a set of achievement levels for the various objectives have to be specified in a stepwise manner. The task of the analyst is to provide all relevant information especially concerning permissible values of the criteria and about reasonable compromise solutions.

Interactive Multiple Goal Programming (IMGP) was developed to combine some of the advantages of multiple goal programming (as devised and further developed by Charnes and Cooper) with some of the advantages of interactive procedures. Because of its use of aspiration levels and preemptive priorities, multiple goal programming is in close agreement with decision-making in practice. Although it is one of the stronger methods available, an important drawback should be mentioned: multiple goal programming requires a considerable amount of a priori information on the decision-maker's preferences. That is why we are proposing an interactive variant of multiple goal programming (IMGP).

In IMGP the decision-maker has to provide information about his preferences on basis of a solution and a potency matrix presented to him. A solution is a vector of optimum values for the respective goal variables. The potency matrix consists of two vectors, representing the ideal and the pessimistic solution, respectively. The ideal solution shows for each of the goal variables separately the maximum value, given the solution concerned. The pessimistic solution lists for each of the goal variables separately the worst value obtained during the successive maximizations needed to obtain the ideal solution. The decision-maker only has to indicate whether a solution is satisfactory or not, and if not, which of the minimum goal values should be improved, and by what amount. Then a new solution is presented to him together with a new potency matrix. He then has to indicate whether the shifts in the solution are outweighed by the shifts in the potency matrix. If not, a new solution is calculated and so on. IMGP may be characterized as a systematic procedure (guided by the decision-maker) of imposing constraints on the set of feasible actions. A flow chart of the procedure is given in Figure 1.

We conclude this section by mentioning some key properties and possibilities of IMGP. In IMGP the goal variables are assumed to be known and concave in the instrumental variables. The preference function of the decision-maker is not assumed to be known. However, it is assumed to be concave, both in the goal variables and in the instrumental variables. Given these assumptions, both optimizing and satisficing behaviour can be incorporated.

The decision-maker has to give only information on his local preferences. However, all available a priori information can be incorporated within the procedure. The decision-maker has the opportunity to reconsider this a priori information during the interactive process. In order to include more of such learning effects, it is wise to repeat the procedure several times.

As shown in Nijkamp and Spronk [1979b], IMGP converges within a finite number of interactions to a final solution, which does exist and is feasible. Apart from an ε-neighbourhood, this solution is optimal. Whether this solution is unique or not, depends on the decision-maker's preferences (for instance, if the decision-maker is a satisficer having formulated targets which are attainable within the feasible region, a unique final solution does not exist in general).

Figure 1. A simpliflied flow chart of interactive multiple goal programming.

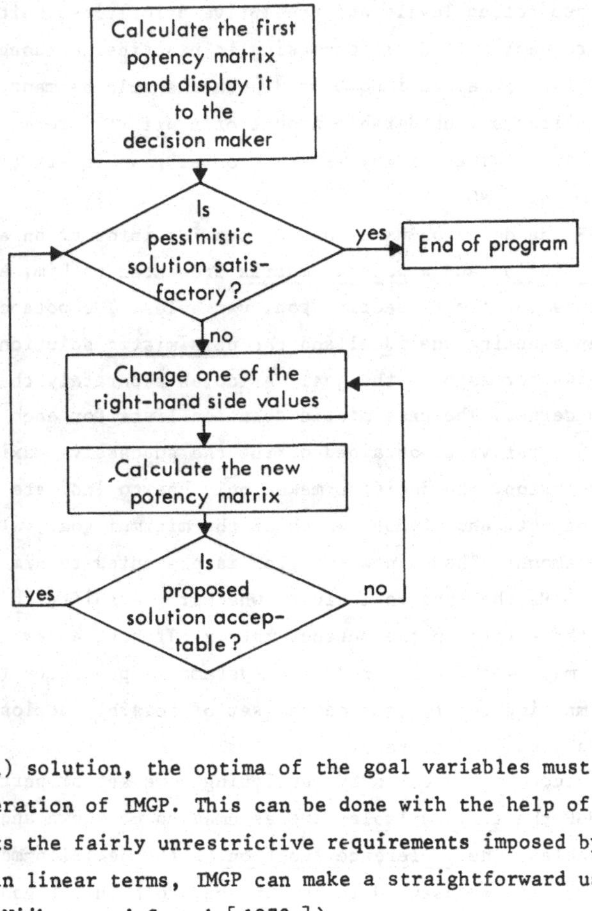

Given a new (proposal) solution, the optima of the goal variables must be (re)-calculated during each iteration of IMGP. This can be done with the help of any optimization method which meets the fairly unrestrictive requirements imposed by IMGP. If the problem is stated in linear terms, IMGP can make a straightforward use of goal programming routines.(see Nijkamp and Spronk [1979c]).

3. Structure of the model

The restrictions that define the boundaries of the feasibility region of the goal variables are dynamic Leontief-type inequalities:

(3.1) $$x_t \geq (A + D)x_t + K(w_{t+1} - w_t) + v_t$$

$$x_t \leq w_t$$

All variables are expressed in constant prices. The vector w is the vector of production capacity in every year, and x is the vector of actual production. The vector v is defined as the sum of final consumption and export surplus.

The matrix A of technical coefficients is derived from the input-output tables published by the Statistical Office of the European Communities. The matrix of capital coefficients K was computed using the vintage model method, from which the depreciation coefficients D result as a byproduct. A more detailed description of the construction of the matrices A, K and D, can be found in van Driel et al. [1979, sections I.2. and I.5.].

The full model contains the 17-sectors of the NACE-CLIO classification of the Statistical Office of the European Communities, to which five pollution sectors have been added. The pollution problem was treated by means of the emission-approach. In this approach the nuisance, i.e. the unabated pollution, can be evaluated at its abatement costs. Five columns are added to the matrix of technical coefficients. The elements in the upper part of each of these columns are the technical coefficients that represent the relative expenses on conventional goods needed to abate one unit of the pollution concerned. The abatement sectors themselves pollute too. These abatement costs form the lower part of the columns in exactly the same fashion as is done for the conventional sectors. Five rows were added to the technology matrix to represent the amounts paid per unit of activity to each of the abatement sectors. The data were taken from a study of the Central Planning Bureau of the Netherlands [1975].

To start the experiments with IMGP we used an aggregated version, which consists of three conventional sectors and one pollution sector. The main components of these aggregates are:

sector 1: building and commerce
sector 2: chemical products, metal products and means of transport
sector 3: agriculture, foods, textiles and the services sectors (exclusive of commerce)
sector 4: all abatement sectors

A further discussion on the choice of these aggregates can be found in van Driel et al. [1979, section III.2.]. The numerical data of this aggregated version of the model were computed in such a fashion that in each sector the export surplus equals zero. The model describes the industrial region lying within a radius of 300 km around Rotterdam, consisting of the Netherlands, Belgium, Nordrhein-Westfalen and France Nord. As a consequence of the extreme extent of the aggregation, the assumption of no export surplus is not far beyond the truth.

The simulations cover a period of ten years, together with the relations that define the goal variables, we end up with a model consisting of 160 relations in 130 structural variables. The computations involved in manipulating this model are not too expensive. One iteration consisting of solving 6 of these LP-problems takes some 30 seconds of central processing unit time. On the other hand, the system is not that aggregated that its behaviour becomes obvious. Experience indicates that the outcome of each iteration shows unexpected traits. The interrelations even in so small a system cause the prediction of its behaviour to be a hazardous task. Nevertheless, experiments with the full model will remain necessary. Because of its greater scope the lessons to be learned from the larger model will be much richer. Furthermore, the description of reality by means of 22 sectors is optimum in the sense that an equilibrium is attained between the advantages and disadvantages of more detail (Ibid., section IV.1.), while the four sectors of the aggregated model are statistical constructs, combining essentially dissimilar sectors into one aggregate.

4. Selection of goal variables

In our experiment we chose six goal variables, thus not exceeding Miller's magical number seven (cf. Miller [1956]). Our choice was to some extent arbitrary, because at this stage of the experiments we could not consult 'real-life' decision-makers. At the same time, we wondered whether these decision-makers, while using our procedure, would propose changes in the set of goal variables (see section 6). It should be stressed that, when our procedure is being used in less experimental situations, the decision-maker must have the opportunity to formulate this set at the start of the procedure and to change it whenever he likes. In this experiment we have chosen the following goal variables.

(1) Wages - Defined as the sum of all wages over all ten years of the planning period. Because the model is formulated in real terms, this goal variable can be considered to be a proxy for employment. This goal variable was indexed in terms of the wages of the year just before the planning period. If the annual wages would not change during the planning period, this goal variable would have a value of 1000. This goal variable is to be maximized.

(2) Consumption - Defined as the sum of the consumption of products from sectors 1, 2 and 3 (see section 3) over all ten years of the planning horizon. This goal variable was indexed in terms of the consumption in the year before the planning horizon. This goal variable is to be maximized too.

(3) Minimum Growth of Consumption - This goal variable was defined as the minimum over the planning period of the annual rise of the sum of consumption of products from sectors 1, 2 and 3. The goal variable was indexed in the same fashion as goal variable two. Also this goal variable is to be maximized. Implicitly, we restricted the value of this goal variable to be non-negative. If the decision-maker would not like such a restriction, it can easily be removed.

(4) The goal variable "Nuisance" - Nuisance is defined as the amount of unabated pollution. Amounts of pollution are defined by means of the production costs

of the abatement industries (see section 3) over all ten years of the planning horizon. This goal variable was indexed in terms of the nuisance in the year before the planning horizon. This goal variable is to be minimized.

(5) Maximum Growth of Capacity - This goal variable, which is to be minimized in order to eliminate too large jumps in the series, has been included as a means to 'stabilize' the growth path of the economy. This goal variable was indexed by reference to the total production capacity in the year before the planning horizon.

(6) Production of the Anti-Pollution Industry - This goal variable was not indexed. It is measured in millions of 1965 Eurodollars production worth (like originally all variables in the input-output model). Although this goal variable is not the object of economic endeavour, we included it deliberately to obtain information about the learning aspects of our procedure.

In section six, discussing the results of the experiment, we shall show that the experiences with the interactive procedure include learning effects concerning the relevance of the goal variables.

5. A brief description of the computer programs

The computer programs for IMGP have been designed in a way, such that the decision-maker - sitting at a computer terminal - is in conversational contact with the computer system (in the case of our experiments the IBM 370/158 of the University of Technology in Delft, the Netherlands). Structured programming was used, having the advantage that parts of the program can be tested (and changed) independently of other parts. The programs were solved by means of calls to the IBM's MPSX/370-package, imbedded in PL/I computer programs. These modules were coordinated by means of command procedures.

We give a sketch of the system of programs in figure 2. Given a new problem, the following programs have to be carried out once. The data have to be transformed into the required MPSX input format by means of the matrix generator. Then a PL/I computer program using MPSX, calculates the first potency matrices. The outcomes of the linear programs, which have to be solved in order to calculate this potency matrix, are stored in the dataset PROBFILE A. The potency matrix itself is stored in a dataset which can be displayed to the decision-maker.

After these initial operations, the decision-maker can choose between two command procedures, 'START' and 'RESUME', which are essentially the same, except for one thing. START copies the data of the linear programs underlying the first potency matrix (stored in PROBFILE A) to the dataset PROBFILE B and displays the first potency matrix to the decision-maker. RESUME does not include such a copy command, thus leaving the dataset PROBFILE B as it was after the last iteration of the preceding session. Accordingly, it displays the accompanying potency matrix to the decision-maker. Clearly, START is used when a new decision-maker starts tackling the problem, or when a decision-maker wants to restart the whole interactive procedure from the beginning. RESUME is used when a decision-maker wants to continue the session after a break.

Figure 2. The system of computer programs used for the implementation of IMGP.

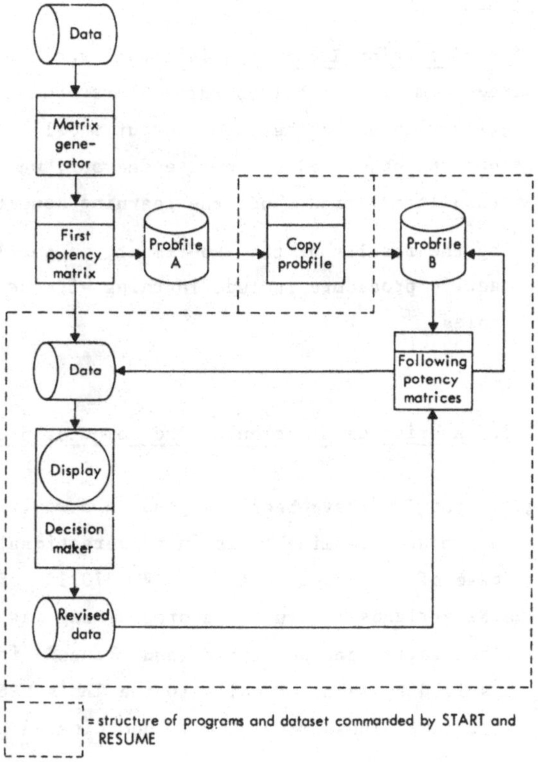

Thus both START and RESUME display a potency matrix together with a sequence of questions, which have to be answered by the decision-maker. The first question is whether the presented solution is satisfactory or not. If the decision-maker states it is satisfactory, he can subsequently ask for a detailed (hardcopy) description of the results. If not, he has to indicate which goal variable should be changed in value and to what amount. These data and the data of PROBFILE B are then used in a PL/I program (again using the MPSX-package) which calculates the new potency matrix. The dataset PROBFILE B is changed. It now contains the data of the linear programs underlying the last calculated potency matrix. The potency matrix itself is stored in a dataset, which again is displayed to the decision-maker. The procedure terminates when the decision-maker states that the presented solution is satisfactory.

6. Some results

In this section we describe a session with a decision-maker using IMGP as described in section two, by means of the computer programs described in section five. We assumed his problem to be given by the model described in the third section and the goal variables specified in the fourth section. The session described was the third of the decision-maker in question. In our description of this session we shall also point to the learning effects obtained from the earlier sessions. The starting solutions are the following:

iteration 1	optimal (ideal) values	accepted (pessimistic) value
(1) wages (≃ employment)	3292	793
(2) consumption	2810	1000
(3) min. growth consumption	20	0
(4) nuisance	79	2751
(5) max. growth capacity	0	290
(6) anti-pollution production	28189	0

Inspecting the pessimistic values, which are lower bounds of the values the decision-maker has to accept, we see that - in this worst case - he has to accept a considerable reduction of employment, while the consumption does not necessarily grow, there may be a tremendous amount of nuisance, and there may be years in which the capacity triples. Also, in this unfavorable case, the anti-pollution industry does not produce anything. (Note - as the decision-maker did in one of the earlier experiments - that the production of this industry is not a proper goal variable because its value can be raised by switching to heavily polluting sectors). The first goal variable to be changed was chosen to be the wages, being a proxy for employment. The proposed value for this goal variable was 1500, corresponding with an average yearly growth of about 8 percent. It was estimated by the decision-maker that the existing unemployment could be removed in this way, while also a good deal of the housewifes could get a job. The consequences of this desire are shown below:

iteration 2	optimal (ideal) values	accepted (pessimistic) values
(1) wages (≃ employment)	3293	1500
(2) consumption	2810	1000
(3) min. growth consumption	20	0
(4) nuisance	79	2751
(5) max. growth capacity	12	290
(6) anti-pollution production	28189	0

It can be seen, that this alteration only influences the 'optimal' value of the capacity variable. Thus, where in the former solution it was conceivable that the capacity did not grow, there is now at least one year in which the capacity grows with 12 percent (of the capacity in year 0). Accepting this consequence, the decision-maker next wants to limit the nuisance to at most a value of 500. This implies the following characteristics:

iteration 3	optimal (ideal) values	accepted (pessimistic) values
(1) wages (≃ employment)	3091	1500
(2) consumption	2643	1000
(3) min. growth consumption	19	0
(4) nuisance	79	500
(5) max. growth capacity	12	255
(6) anti-pollution production	28189	9704

These results show that it now becomes necessary that the anti-pollution industry starts producing. Furthermore, consumption and wages can not increase as much as in the earlier iterations. Next, the decision-maker wants to limit the maximum capacity growth to a value of 30, in view of the estimated capital market conditions and to avoid too large instabilities within the economic system.
This leads to the following results:

iteration 4	optimal (ideal) values	accepted (pessimistic) values
(1) wages (≃ employment)	2024	1500
(2) consumption	2274	1000
(3) min. growth consumption	19	0
(4) nuisance	79	500
(5) max. growth capacity	12	30
(6) anti-pollution production	19977	9704

Our decision-maker continued the procedure until a solution was found which appeared satisfactory to him. The complete results of this experiment are given in Hartog et al. [1979]. The final solution is shown below.

iteration 14	optimal (ideal) values	accepted (pessimistic) values
(1) wages (≃ employment)	1600	1600
(2) consumption	1890	1890
(3) min. growth consumption	10	10
(4) nuisance	251	252
(5) max. growth capacity	15	15
(6) anti-pollution production	13556	13545

Obviously, it is not necessary for a decision-maker to continue the interactive procedure so far as the present decision-maker did, i.e. to proceed until a unique (or nearly unique) final solution occurs. One may as well stop at an earlier iteration, being left with a number of 'scenarios' all satisfying the minimum conditions specified by the decision-maker. The choice out of these scenarios can be made e.g. by a committee or otherwise.

An examination of the detailed results associated with the final result attained by this decision-maker showed that nearly all instrumental variables within the model behaved according to smooth growth paths. However, because no goal variable had been included to take care of a balanced distribution of activity over the industrial sectors, some undesired effects occured in this respect. In fact, this was one of the learning effects, which resulted in discussions and proposals for new goal variables. Other learning effects have led a.o. to the proposal to delete the sixth goal variable as being irrelevant. Furthermore, the nuisance goal variable was proposed to be changed in a per year maximum nuisance level.

REFERENCES

1. Central Planning Bureau of the Netherlands, Economische gevolgen van bestrijding van milieuverontreining, Monograph 20, The Hague, 1975.
2. Despontin, M. and P. Vincke, Multiple Criteria Economic Policy, in Advances in Operations Research, North-Holland, 1977.
3. Driel, G.J. van, J.A. Hartog and C. van Ravenzwaaij, Limits to the Welfare State, Martinus Nijhoff, Boston, 1979.
4. Eijck, C.J. van and J. Sandee, Quantitative Determination of an Optimum Economic Policy, Econometrica, 1959, pp. 1-13.
5. Hartog, J.A., P. Nijkamp, and J. Spronk, Operational Multiple Goal Models for Large Economic Environmental Systems, Report 7917/A, Centre for Research in Business Economics, Erasmus University, Rotterdam, 1979.
6. Iserman, H., The Relevance of Duality in Multiple Objective Linear Programming, in Starr, N.K. and M. Zeleny (eds), Multiple Criteria Decision Making, TIMS Studies in Management Sciences, Vol. 6, North-Holland, 1977.
7. Miller, G., The Magical Number Seven Plus or Minus Two: Some Limits on our Capacity for Processing Information, Psychological Review, Vol. 63, pp. 81-97.
8. Nijkamp, P. and J. Spronk, Analysis of Production and Location Decisions by Means of Multi-Criteria Analysis, Engineering and Process Economics, Vol. 4, (1979a).
9. Nijkamp, P. and J. Spronk, Interactive Multiple Goal Programming, Evaluation and Some Results. Centre for Research in Business Economics, Report 7916/A, Erasmus University, Rotterdam, (1979b).
10. Nijkamp, P. and J. Spronk, Goal Programming for Decision-Making. Ricerca Operative, Autumn 1979 (1979c).
11. Spivey, W.A. and H. Tamura, Goal Programming in Econometrics, Naval Research Logistics Quarterly, 1970, p. 183 ff.
12. Theil, H. Optimal Decision Rules for Government and Industry, Rand McNally, Chicago, 1964.

13. Tinbergen, J., On the Theory of Economic Policy, North-Holland, 1952.

14. Tinbergen, J., Economic Policy, Principles and Design, North-Holland, 1965.

15. Wallenius, H., J. Wallenius, and P. Vartia, An Experimental Investigation of an Interactive Approach to Solving Macro-economic Policy Problems, in OR'78 (ed. by K.B. Haley), North-Holland, 1978.

RESOURCE DISTRIBUTION COMBINATORIAL MODELS IN AIR POLLUTION PROBLEMS

L. Kruś, M. Libura, L. Słomiński

Polish Academy of Sciences
Systems Research Institute
Newelska 6, 01-447 Warszawa, POLAND

INTRODUCTION

The first and foremost purpose for the development of the models described in this paper was to provide an aid for planning and decision making in air pollution abatement problems. The paper deals with an optimal resource distribution for purification of air pollutants. As the resources one may consider purifying devices, new technologies, new fuels and other investment policies which are possible within limited funds.

It is assumed that an area is given in which the set of computational points (receptors) is determined (for these point an air pollution state is calculated). In this area some subareas are selected which are under particular protection (e.g. residential districts, national parks, ect.). A statistical description of climatological conditions is given for whole the area.

The computerized model of air pollutant dispersion is used to calculate statistics characterizing a pollution state in the receptors. The state depends on climatological conditions and location and parameters of emission sources.

The problem of resource allocation can be described as follows: A set of plants which pollute the air in the considered area is given. Sanitary norms for the air quality in the receptors are known. One attempts to assigne for a suitable subset of plants a limited number of devices (technologies) in such a way that the mentioned norm are guaranted in all receptors. We propose two objective functions in order to state the above problem as an optimization one. The first criterion, so called additive objective, is aimed at minimization of total costs given in a discounted form. The second criterion is of a bottleneck type and is oriented on minimization of a greatest frequency of exceeding of a sanitary level of pollution. When this objective is used a limit on the fund is introduced in to the model as a constraint.

Two approximative algorithms for solution of these combinatorial models are briefly outlined. The first heuristics involves a considerable reduction of the cardinality of the set of receptors. Successive change in the content of this set leads to an acceptable solution obtained at the moment when further selection of the receptors is impossible.

The second heuristics applies to the problem with the addtive objective function. Proposed algorithm runs in two main phases. At the first phase a feasible solution is constructed. Second phase is devoted to decreasing the cost of the solution obtained in the first phase.

1. PHYSICAL DESCRIPTION OF THE PROBLEM

We consider given area which is polluted by a set of air pollution sources (emitors)(see Fig. 1).

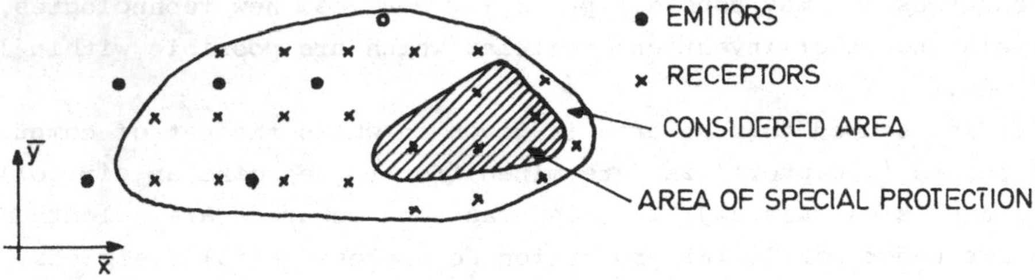

FIG. 1

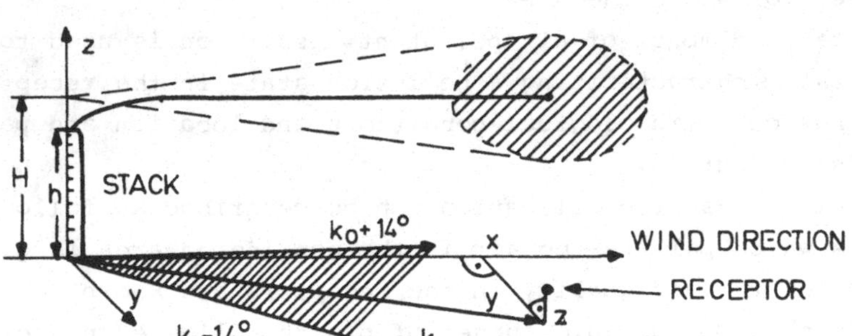

FIG. 2.

In the area a subarea of special protection with respect to air pollution state is separated. The areas are defined by sets of receptors computational points in which the air polution state is being determined. A subset of emitors is given for which pollution emission can be reduced by installation of purifying devices, by using higher quality fuels or by changes of technologies. Meteorological conditions are assumed to be uniform in all the area. They are described by speed and direction of wind and a given number of dispersion conditions - classes of atmospheric stabilities.

In the paper, the following notation is used:

I - the set of indices of the receptors in the area,

$\hat{I}$ - the set of indices of the receptors in the area under special protection,

i - the index of a particular receptor ($i \in I$ or $i \in \hat{I}$),

E - the set of indices of the emitors,

$\bar{E}$ - the set of indices of the emitors in which purifying devices can be installed,

e - the index of a particular emitor,

W - the set of indices of the meteorological conditions,

w - the index of particular mateorological conditions,

$P(w)$ - the probability of the occurrence of the meteorological conditions w,

$B_w(i)$ - the background pollution concentration in the point i under the conditions w.

In the considered resource distribution combinatorial problems, the computerized model of air pollution dispersion [1] is used. The model calculates given statistics of air pollution state in the considered areas as a function of locations and parameters of emitors, and of meteorological conditions.

The following data are needed in the calculations:

1. Statistics of meteorological conditions, namely the $P(w)$ vector of probabilities of particular wind derections and speed ranges and of atmospheric stabilities.
2. Parameters characterizing the particular emitors:
   - location (coordinates),
   - geometrical parameters of the stack (height and diameter),
   - average emission,
   - velocity of the emitted gas,
   - time of measurement of the average emission,
   - intensity of the emitted gas flow,
   - stack exit temperature,
   - ambient temperature,
   - average specific heat of the gas,
   - area of the spacial emitor (equal to zero for point emitors).
3. Cordinates of receptors.
4. Pollution concentration levels, the frequencies of exceeding of

which are calculated.

The Gaussian distribution [2-4] was used to describe the pollution despersion of a single emitor. In the Fig. 2 the stack of the heigh h is located in an origin of axes x,y,z. The axis x corresponds to the wind direction. The direction connecting the emitor and the receptor is denoted by k_o. The receptor has the coordinates x,y,z. The pollution concentration at the receptor is assumed to be zero for wind directions outside the angle $k_o \pm 14^o$. Otherwise the pollution concentration is defined by the following basic relation:

$$S(x,y,z,H) = \frac{Q}{2\pi \vartheta \sigma_y(x) \sigma_z(x)} \exp\left(-\frac{y^2}{2\sigma_y^2(x)}\right) \cdot \qquad /1/$$

$$\cdot \left[\exp\left(-\frac{(z-H)^2}{2\sigma_z^2(x)}\right) + \exp\left(-\frac{(z+H)^2}{2\sigma_z^2(x)}\right)\right] \cdot \Lambda$$

where:

Q is the intensity of the pollutant emission;

$\sigma_y(x)$, $\sigma_z(x)$ are the standard deviations in the direction y and z respectively; they are functions of the distance x and of parameters dependent on meteorogical conditions;

H is the effective stack height (taking into account a raising of the emited gas). It is a function of the stack geometrical parameters and of meteorological conditions;

ϑ is the speed of wind on the effective height;

Λ is the coefficient of the natural desintegration of the pollutant. It is a function of the distance, wind speed H, meteorological condition and features of the particular pollutant.

In the model, the effect of mixing layer is taken into account by appropriate description of $\sigma_z(x)$ and of H. The area sources are replaced by virtual (with respect to wind)point emitors. The formula (1) allows to calculate the contribution $p_{we}(i) = S(x,y,z,H)$ of the particular emitor e in the pollution concentration at the point i under the meteorological conditions w. Combining the contributions for all the emitors and meteorological conditions, the statistics of pollution state et each receptor are derived as follows:

Concentration of the pollutant

$$p_w(i) = \sum_{e \in \bar{E}} \left(p_{we}(i) + b_w(i)\right)$$

maximal concentration

$$p_{max}(i) = \max_{w} \sum_{e \in \bar{E}} \left(p_{we}(i) + b_w(i) \right) \qquad /2/$$

average concentration (per year)

$$\bar{p}(i) = \sum_{w} P(w)\, p_w(i) \qquad /3/$$

frequency of exceeding of the given concentration level L(i)

$$f_L(i) = \sum_{w \in \bar{W}} P(w) \qquad /4/$$

where

$$\bar{W} = \left\{ w: \sum_{e \in E} \left(p_{we}(i) + b_w(i) \right) > L(i) \right\}$$

and L(i)is the concentration level (given by sanitary norms).

2. COMBINATORIAL PROGRAMMING MODEL

We are going to describe a combinatorial (pure 0-1 integer programming) model which is aimed at improving the efficiency of decisions for air pollution decrease. Two different objective functions are used. The first criterion involves the minimization of the total cost considered as a doscounted cost of the investments and operational expenditures. The second criterion minimizes the greatest excess of the standart frequency, established by regulations. An optimal solution is searched in the space given by the set of constraints.

Before giving the details of the models let us introduce some further notations:

F(i) is an admissible frequency of exceeding of the pollution concentration level L(i);

M(i) is the maximal admissible level of the pollution concentration in the i-th receptor (a sanitary norm stated by the authority);

U(e) is the set of indices of the purifying devices (technologies, ect.) applicable to the e-th emitor (note that $U(e_1) \cap U(e_2)$ can not be empty for $e_1 \neq e_2$);

u is a particular index, $u \in U(e)$;

g_u is the number of devices of type u;

d_{eu} is the cost (discounted) of choosing u-th type of device for the e-th emitor;

D is the total found;

r_{eu} is a coefficient ($0 \leqslant r_{eu} < 1$) which reflects an effectiveness of the u-th device in the e-th emitor;

x_{eu} is a binary variable:

$$x_{eu} = \begin{cases} 1 \text{ if e-th emitor is assigned u-th device,} \\ 0 \text{ othervise;} \end{cases}$$

$z_w(i)$ is a binary variable:

$$z_w(i) = \begin{cases} 1 \text{ if } p_w(i) > L(i) \\ 0 \text{ othervise.} \end{cases} \qquad /5/$$

Now we are able to write down the formal mathematical models.

A. Constraints:

$$\sum_{u \in U(e)} x_{eu} \leqslant 1, \quad e \in \bar{E} \qquad /6/$$

- each emitor can be assigned no more than one purifying device choosen from the appropriate set U(e);

$$\sum_{u \notin U(e)} x_{eu} = 0, \quad e \in \bar{E} \qquad /7/$$

- the devices which don't belong to the set U(e) can not be used for the e-th emitor;

$$\sum_{e \in \bar{E}} x_{eu} \leqslant g_u, \quad u \in \bigcup_{e \in \bar{E}} U(e) \qquad /8/$$

- the number of assigned devices of the given type u can not be greater than the total number of devices of this type;

$$\sum_{w \in W} P(w)\, z_w(i) \leqslant F(i), \quad i \in (I \cup \hat{I}), \qquad /9/$$

- the probability of exceeding of the given standard level L(i) by the pollution in point i under all possible meteorological conditions should not be greater than the standard frequency F(i);

$$p_w(i) \leqslant M(i), \quad i \in (I \cup \hat{I}),\ w \in W, \qquad /10/$$

- the pollution concentration at each point i under any meteorological conditions should not exceed the established maximal level M(i).

The following two sets of the constraints express in an equivalent linear form the inequality of the type "if... then ..." given by (5):

$$\sum_{e \in \bar{E}} \sum_{u \in U(e)} p_{we}(i)\, r_{eu}\, x_{eu} + \sum_{e \in \bar{E}} \sum_{u \in U(e)} p_{we}(i)(1 - x_{eu}) +$$

$$+ b_w(i) - L(i) - C\, z_w(i) \leqslant 0, \quad i \in (I \cup \hat{I}), \qquad /11/$$

$$- \sum_{e \in \bar{E}} \sum_{u \in U(e)} p_{we}(i)\, r_{eu} x_{eu} - \sum_{e \in \bar{E}} \sum_{u \in U(e)} p_{we}(i)(1 - x_{eu}) +$$

$$- b_w(i) + (L(i) + c)\, z_w(i) \leqslant 0, \quad i \in (I \cup \hat{I}) \qquad /12/$$

where C and c are constants defined as follows:

C > maximum value o the pollution concentration that may happen in any receptor i, under any conditionow;

0 < c < an accuracy with which one can claim that the pollution concentration level in any point i exceeds L(i).

One can check that the constraints (11) and (12) are equivalent to(5) if

$$x_{eu},\ z_w(i) = 0 \text{ or } 1, \quad e \in \bar{E}, \quad u \in U(e), \quad w \in W. \qquad /13/$$

B. Objective Functions

The additive objective function:

$$\sum_{e \in \bar{E}} \sum_{u \in U(e)} d_{eu} x_{eu} \longrightarrow \text{minimum}. \qquad /14/$$

The bottleneck objective function:

$$\max_{i \in (I \cup \hat{I})} \Big(\sum_{w \in W} P(w)\, z_w(i) \longrightarrow \text{minimum}, \qquad /15/$$

Now, two problems of our particular interest are stated in the following way:

a/ objective function (14),
constraints: (6) - (13);

b/ objective function (15),
constraints: (6) - (13), supplemented by

$$\sum_{e \in \bar{E}} \sum_{u \in U(e)} d_{eu} x_{eu} \leqslant D, \qquad /16/$$

(the total expenditures on purification should not exceed the given found D).

3.HEURISTIC ALGORITHMS

The problems a/ and b/ of the preceeding paragraph are difficult combinatorial problems with thousands of constraints and variables. It is unlikely to solve the problems by an exact algorithm (say of branch and bound type) in a reasonable amount of time. For this reason we propose two different heuristics, which purpose is to reduce the dimension of the problem to a menagable size.

The first heuristics is based on replacing of the original set of receptors $(I \cup \hat{I})$ by much smaller set $(I \cup \hat{I})_s$, where s is the current phase of the algorithm. This heuristics is applicable to both problems.

The second heuristics consists of two main phases. At the first phase a feasible assignment of the set of purifying devices to the set of emitors is searched. Then we make an attempt to improve the obtained solution by decreasing its cost.

A. Heuristics Based on Modification of $(I \cup \hat{I})$

This heuristics works for both problems. The algorithm runs in a finite number of phases. A simplified flow-diagram is shown in Fig. 3.

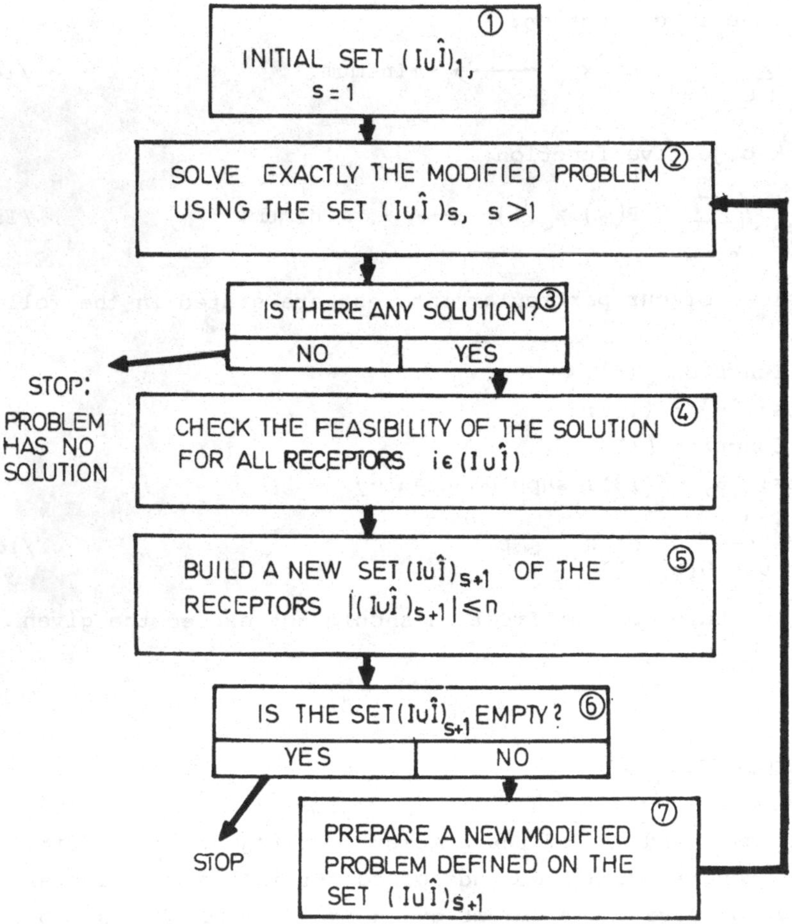

FIG.3.

The most important element of the approach is the rule according to which receptors are chosen for the current problem to be solved. The following rule has been adopted: for given number u choose no more than n receptors with the greatest positive value of:

$$\sum_{w \in W} P(w) \left[p_w(i) - L(i) \right]$$

The above receptors constitute the reduced set $(I \cup \hat{I})_s$. Current (modified) problem is the initial problem defined on the set $(I \cup \hat{I})_s$. In order to avoid the possibility of cycling, an auxiliary constraint is introduced. Namely, emitors which have got an assignment at an previous phase, save the same assignment or can get only a device of a higher efficiency (of a higher cost) at the present or further phase.

B. Two Phase Heuristics

This heuristics is applicable for additive objective function only. The solution process consists of two main phases: construction (block ⑤) of a feasible aolution (blocks ①, ②, ③) and then an improvement of this solution. A feasible solution is constructed by assigning puryfying devices to emitors using the measure $\mu(e)$ of effectiveness (see Fig. 4). In each step only one device is assigned to only one emitor. The meaning of $\mu(e)$ is the following: we invest in the emitor for which the ratio of decrease of the average pollution concentration (represented by $\bar{\Theta} - \bar{\Theta}(e)$) to the cost of this investment is maximal.

In the improvement phase an attempt is made at decreasing the cost by replacing in each emitor the installed device by a cheaper one. This is done by a simple enumeration of all possibilities. In this heuristics it is assumed that the sets U(e) have enough purifying devices to satisfy the sanitary norms. In contrast to the previously described model in this heuristic the constraints (8) and (10) are neglected.

REMARKS

In the Systems Research Institute of Polish Academy of Sciences a project on modelling of economic development of Silesia region, with respect to environment problems is carried on. This is done in strict cooperation with the Institute of Environmental Engineering in Zabrze. The project (see [5]) consists of three main directions: macro-modelling of regional development, modelling of energy subsystem, and problems of air pollution dispersion and resource distribution for improving the air quality state. The last problems have stimulated the elaboration of combinatorial models that have been described in this paper.

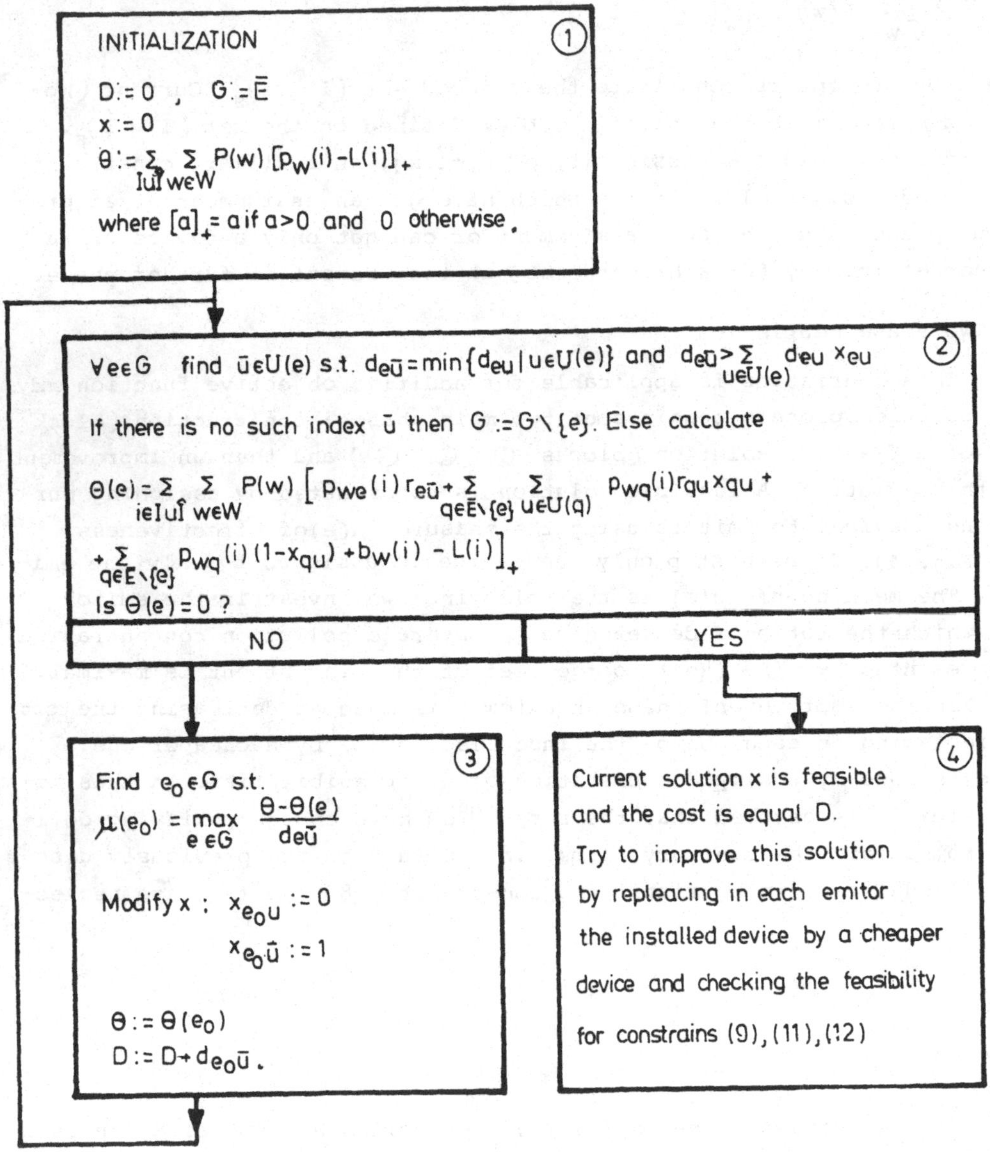

FIG.4

REFERENCES

[1] Sądelski M., Kruś., Makowski M., Markowiak L., Olinger W., Włodarski W., FORTRAN program (on Odra 1300 computer) for calculating dispersion of gas and dust pollutants in atmosphere (in Polish). In: Prace i Studia Instytutu Podstaw Inżynierii Środowiska PAN. Wrocław: Ossolineum 1976.

[2] Pasquill F., Atmospheric Diffusion. New York, Nostrand, 1962.

[3] Koogler I.B. et al., Multivariable Model for Atmospheric Dispersion Prediction. J. Air Poll. Control Ass. 4 (1967), 211-214.

[4] Manier G., Berechnung der Häufigkeitsverteilungen der Schadgaskonzentrationen in der Umgebung einer Einzelquelle. Staub 7 (1970), 298-303.

[5] Kruś L., Normative modelling of regional development: an example of a modelling project. In Models for Regional Planning and Policy Making. Proc. Joint IBM/IISA Conference, Vienna Sept. 1977. Eds. A. Straszak, B.V. Wagle.

THE ENERGY ECONOMICS OF THE UNITED KINGDOM, THE FEDERAL REPUBLIC OF GERMANY, AND BELGIUM

A Comparison by Means of a Linear Optimisation Model

K. Leimkühler and G. Egberts

Programmgruppe Systemforschung und Technologische Entwicklung
Kernforschungsanlage Jülich
5170 Jülich, Bundesrepublik Deutschland

1. Introduction

At present, one of the main objectives of highly industrialised countries facing problems generated by the dependence on imported oil is to be as flexible as possible concerning a future technology mix, both to provide new sources of supply and to safeguard the environment. In the long run, technological research and development will play the central part in achieving this objective. Energy systems analysis of the kind reported here provides a qualitative basis for planning such research and development.

2. Institutional Framework

The twenty member countries of the International Energy Agency (IEA) in Paris decided to set up a cooperative project to establish a common strategy for energy research, development, and demonstration (R,D&D). The results of the project should indicate which energy technologies are likely to be essential or promising, and which are likely to have a maximum impact on the national energy systems.

3. Organisation

Systems Analysis Teams have been established at Brookhaven National Laboratory (BNL), Brookhaven, New York (USA), and at the Nuclear Research Centre (KFA), Jülich, Federal Republic of Germany. These host laboratories are responsible for management and provision of computer facilities and support staff. Countries participating in the project have assigned national delegates to one of the two centres. They are to keep close contact with their national energy ministries or other organisations responsible for energy, research and planning, and to prepare the data which are necessary to perform systems analysis.

4. Energy Modelling for R&D Planning

Five major stages can be distinguished in the process of developing an R&D strategy, involving energy modelling both as a computational tool and as an analytical basis.

1) Energy model application, i.e., computation of energy scenarios including sensitivity cases.

2) Scenario evaluation, i.e., identification of preferred and less-preferred cases.

3) Technology evaluation, i.e., finding technology priorities based on the preferred scenarios and insights from analysis done outside the model.

4) Establishing a technology development program comprising objectives, a time scale and targets for market penetration by new technologies.

5) Recommendation of an R&D strategy based on the previous steps. This could logically fall into two classes: general and specific strategic action statements about technology groups as well as individual technologies.

The work reported here comprises parts of points 1)-3), i.e., application of a linear optimisation model to be described below, scenario evaluation, and technology evaluation.

The linear optimisation model (called MARKAL, an acronym for "MARKet ALlocation") which has been used by both laboratory teams was developed from models already in use at both locations. There are minor differences between the two versions due to the specific computer installations.

The amount of detail describing the model will be limited in the following, but a comprehensive documentation will be prepared towards the end of this year.

5. Characteristics of the Model

MARKAL is a flexible multi-time-period model of a generalised energy system. It is designed for the evaluation of the possible impacts of new energy technologies on national or regional systems. It can be applied under a variety of assumptions or restrictions.

MARKAL performs an integral optimisation over a specified number of time steps. The step size (period length) and number of steps is at the discretion of the user. Currently nine time steps with five-year intervals are used, covering the period from 1980 to 2020. Thus, instead of optimising an energy system at a certain time, the development of such a system is optimised over a time-span of forty years. Restrictions and constraints may be valid for one five-year interval only ("static" constraints), or for the total planning period ("dynamic" constraints.) The dynamic constraints are responsible for the logical connections between time intervals. Figure 1 shows the matrix structure of the multi-time-period model; here, hatched areas indicate non-zero matrix coefficients; the block-diagonal lower part contains static constraints, and the right-hand side contains the projections for the development of end-use energy demand. Although driven by the demand for end-use energy, the MARKAL model simulates the flow of energy in various forms from the sources of supply (primary energy carriers) through transformation systems to the demand devices which satisfy the end-use demands. Alternative supplies, processes and devices compete for the end-use markets. Figure 2 shows the energy flow in the MARKAL model, as well as the inputs and outputs which need to be specified.

6. Technology Data

Another most important factor in the application of the MARKAL model is the data base of technologies. Two categories of technologies can be distinguished:

a. Demand Device Technologies

These are technologies that provide a useful service, such as oil burners, heat pumps or electric motors. The input data consists of

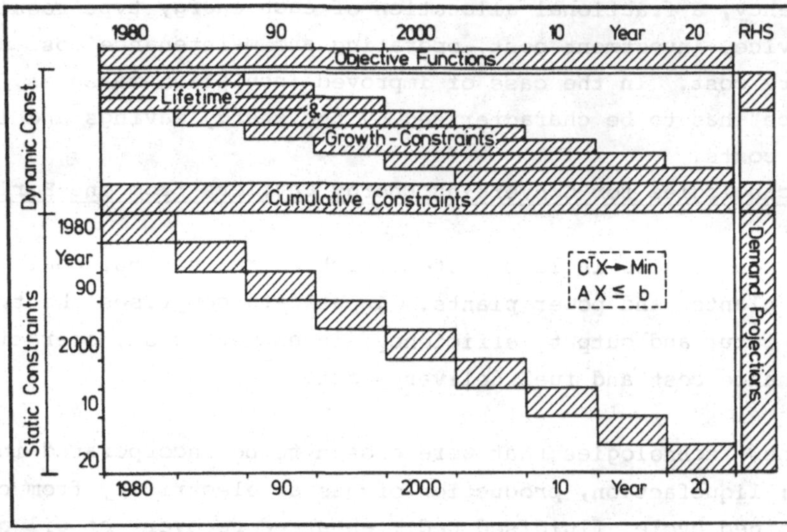

FIGURE 1: Matrix Structure of a MARKAL Model

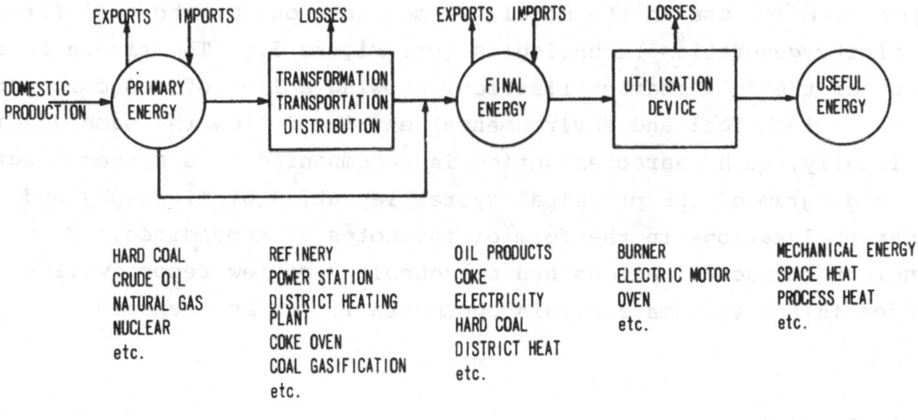

FIGURE 2: Energy Flow in a MARKAL Model

efficiency, a fractional allocation of each energy type required by the device, investment cost, operating and maintenance cost and fuel delivery cost. In the case of improved insulation measures, this "service" has to be characterised by the energy savings and installation costs.

b. Technologies for Converting Energy Carriers from one Form to Another

These technologies include systems such as refineries, coal gasification plants, and power plants. Input data comprises the type of energy input and output, efficiency, investment cost, operating and maintenance cost and fuel delivery cost.

The main new technologies that were chosen to be incorporated into the model are: liquefaction, production of gas or electricity from coal (MHD, combined cycle, fluidised bed); enhanced recovery of oil and gas; shale oil and tar sands; renewable energy sources (wind, wave, OTEC, biomass, etc.); nuclear energy (LWR, HTR, LMFBR, etc.); new and improved transportation systems and various conservation technologies. All countries' runs could draw on any or all of 34 kinds of technologies unless excluded by a policy constraint or by a separate analysis indicating that the technology excluded would never be a viable option.

To illustrate how a new energy technology of the conversion category is specified for use in the model, we may take one of the coal-fired electricity generation technologies (cf. Figure 3.) The figure is an excerpt of the full characterisation, showing a list of economic parameters; the technical and environmental entries follow the same scheme. Additionally, each characterisation is accompanied by a system description, a diagram of the principal system layout, a bibliography and further explanations in the form of footnotes or appendices. These technology characterisations and a technology review report will be included in the systems analysis documentation to be prepared.

7. Scenarios

At present, sixteen scenarios have been computed for several countries. The scenarios are characterised by a few main indicators, as for example:

- -different objective functions
- -constraints on the net import of oil
- -forced implementation of new technologies

	ECONOMIC PARAMETERS (1975 VALUES)	UNIT	REF. YEAR: 1995		PERCENT CHANGE BY 2020	COMMENTS OR REFERENCE TO FOOTNOTE
			MEAN VALUE	$\pm$ %		
305	TOTAL CONSTRUCTION COST	$10^6\$$	498.6	$\pm$ 25		(3)
306	CONSTRUCTION TIME	YEARS	5			
307	INTEREST DURING CONSTRUCTION	$10^6\$$	64.7			
308	TOTAL INVESTMENT COST	$10^6\$$	563.3	$\pm$ 25		
309	SPECIFIC INVESTMENT COST (PER CU+)	$\$$/kW	563.3	$\pm$ 25		
311	LONG TERM DISCOUNT RATE	%	5			
312	ECONOMIC LIFETIME	YEARS	20			
313	CAPITAL RECOVERY FACTOR (CRF)	%	8			
314	ANNUALIZED CAPITAL COST (PER CU)	$\$$/kW	45.1			
316	ANNUAL FIXED CHARGES (AFC)	%	3			
	AFC (COST/CU)	$\$$/kW	16.9			
317	ANNUAL FIXED OPERATING & MAINTENANCE COST (PER CU)					INCLUDED IN ITEM 321
320	FUEL COST (PER OUTPUT UNIT)	$\$$/GJ	5.7			(4)
321	VARIABLE OPERATING & MAINTENANCE COST (PER OUTPUT UNIT)	$\$$/GJ	1.7			0.8 $\$$/GJ FOR LOW SULFUR COAL (5)
325	TOTAL PRODUCTION COST (PER OUTPUT UNIT)	$\$$/GJ	10.02			0.036 $\$$/kWh

+CU = UNIT OF NET INSTALLED CAPACITY

FIGURE 3: Standard Table of Characteristics of Technological Processes

-limitations of fossil fuel use
-limitations of nuclear capacity.

In detail we have:

a. <u>Constraints on the net import of oil</u> (PS-, SP-Scenarios)
Two variables and their linear combinations have served as objective functions for the MARKAL model so far:

PRICE indicator: P = total discounted cost of the energy system for the whole time horizon,

SECURITY OF SUPPLY indicator: S = total net oil import over the whole time horizon.

The so-called "PS-scenarios" were obtained by minimising P with no constraint on S. One of these scenarios is taken as the reference scenario (PS-1.) The remaining scenarios are called "SP-scenarios" and were obtained by minimising P under the constraint that S is not allowed to exceed a given upper limit S^+:

$$S < S^+.$$

Enforced reduction of net oil import below the PS-1 value will gradually increase the total discounted energy system cost as oil is displaced by more expensive technologies, until a limit point is reached below which imports cannot be reduced without a shortfall in some energy demand (scenario SP-1.)

The corresponding values of system cost and net oil import constitute the PRICE-SECURITY trade-off curve of Figure 4. In order to obtain scenarios that were comparable between countries with different trade-off curves, scenario runs were taken at identical values of the marginal cost of S, i.e., the dual activity of the constraint $S < S^+$.

These scenarios were obtained by minimising the objective function

$$P + \lambda S,$$

where λ is the slope of the trade-off curve. This procedure leads to a meaningful aggregation of results for a group of countries. Suppose that two countries had different marginal costs for the security indicator S; the total cost for the group could be reduced without changing the total net oil import for the group, just by increasing the imports of the country with the highest value of λ and reducing the imports of the one with the lowest value of λ. (In our scenarios λ is measured in \$/GJ.)

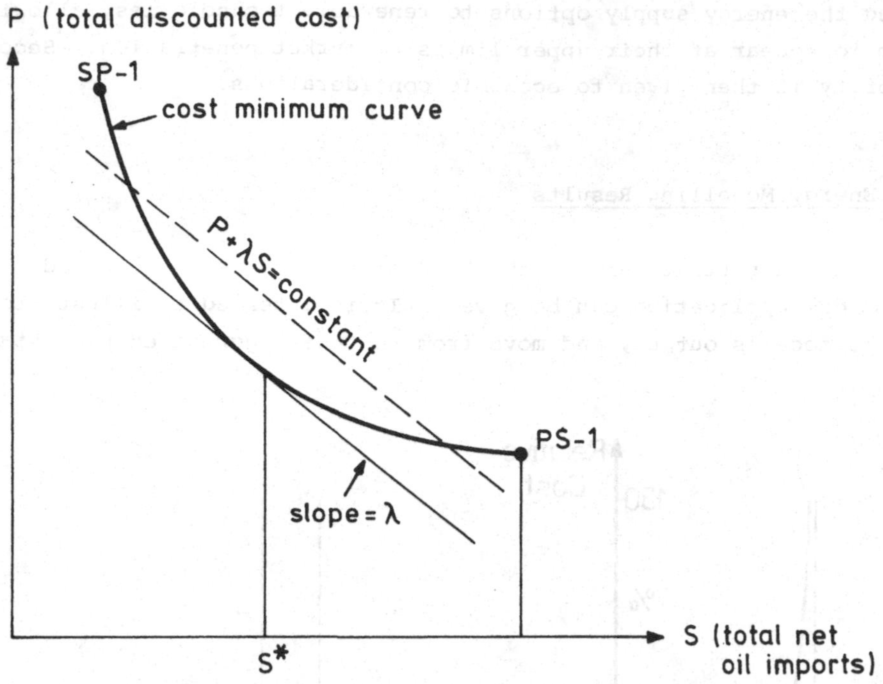

FIGURE 4: Cost-Security Trade-Off Concept

b. <u>Unaccelarated scenarios</u> (PS-1, SP-1/)
In these scenarios the date of availability and the constraints on the implementation of each new technology are kept the same as in the reference scenario PS-1.

c. <u>Accelerated scenarios</u> (PS-4, SP-4/)
Accelerated scenarios are those in which new technologies are given the chance to enter the system earlier and with higher implementation levels than in the unaccelerated scenarios.

d. <u>Sensitivity case scenarios</u> (suffix OIL A, OIL C, LIM NUC, LIM FOS)
The sensitivity case scenarios differ from the base case by either a different assumption for the oil price schedule or by a limitation on the total amount of nuclear or fossil energy produced.

e. <u>Renewable scenario</u> (RP-4)
Another variant, the RP-scenario, is created by giving first priority

among the energy supply options to renewable technologies, allowing them to appear at their upper limits of market penetration. Second priority is then given to economic considerations.

8. Energy Modelling Results

Here only a partial veiw of the kind of results being obtained from the model application can be given. It is intended to illustrate some of the model's output, and move from the most aggregated presentation

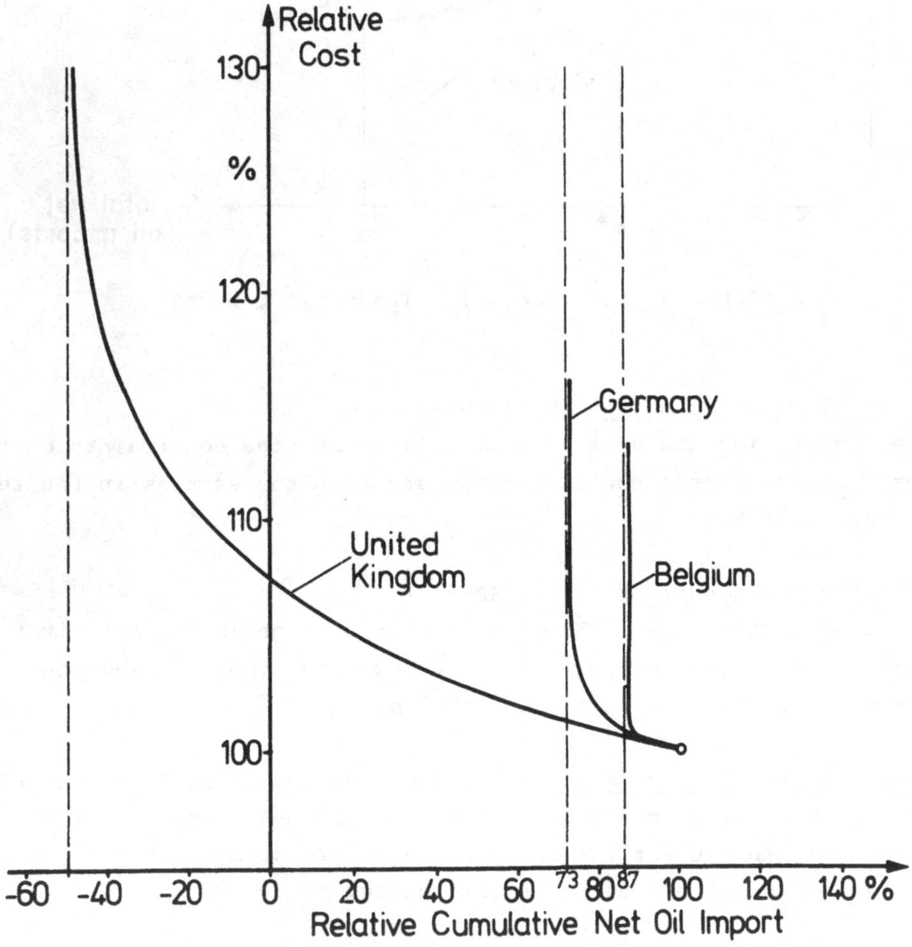

FIGURE 5: Trade-Off Curves for Belgium, Germany, and the United Kingdom

of results to a level at which the "success" of technologies from an R&D point of view may be seen. All results here are preliminary and will be restricted to Germany, the United Kingdom, and Belgium. Analogous results exist for Austria, Switzerland, Denmark, Spain, Italy, and the countries for which similar runs have been performed in Brookhaven.

Figure 5 shows the trade-off curves for the United Kingdom, Germany, and Belgium, with all curves normalised to the reference scenario PS-1 (see above.) To give an impression of the absolute differences, values for cumulative oil imports and system cost for both the reference scenario and a point of highly restricted oil consumption are as follows:

1. for Germany:

1.1.	PS-1:	Net Cumulative Oil Import:	ca. 255.000 PJ
		System Costs :	ca. 1.388.000 $10^6 \$_{75}$
1.2.	SP-1:	Net Cumulative Oil Import:	ca. 192.000 PJ
		System Costs :	ca. 1.460.000 $10^6 \$_{75}$

2. for the United Kingdom:

2.1.	PS-1:	Net Cumulative Oil Import:	ca. 50.000 PJ
		System Costs :	ca. 306.000 $10^6 \$_{75}$
2.2.	SP-1:	Net Cumulative Oil Import:	ca. -25.000 PJ
		System Costs :	ca. 391.000 $10^6 \$_{75}$

3. for Belgium:

3.1.	PS-1:	Net Cumulative Oil Import:	ca. 43.850 PJ
		System Costs :	ca. 156.000 $10^6 \$_{75}$
3.2.	SP-1:	Net Cumulative Oil Import:	ca. 35.700 PJ
		System Costs :	ca. 159.500 $10^6 \$_{75}$

This comparison gives a fair indication of the overall difference in the energy situation of all three countries: Germany - and Belgium even more so - is to a large and undesirable extent dependent on imported oil and will stay so in the time period considered; the percentage of imported oil, measured by total primary energy supply, may decrease to 30%-50% in 2020, depending on the rate of substitution by other technologies, mainly nuclear.

The United Kingdom, on the other hand, is for part of the time independent of imported oil, due to domestic production of North Sea oil. This situation, however, will not last forever, as in the last decade Great Britain will have to face the same problems as Germany and Belgium do now. But with a source like North Sea oil, it seems easier to face the situation of decreasing domestic oil resources some ten years ahead, as there is still time to make the right decisions and opt for the most convenient technology mix.

Now, the trade-off concept may not be restricted to unaccelerated scenarios; the trade-off curves of the accelerated scenarios have one thing in common for all countries: they lie considerably lower than the ones for the reference scenarios, and start with less imported oil, due to a higher number of degrees of freedom in choosing advanced technologies and due to their faster growth rates allowed.

As for Germany and Belgium, the trade-off quite clearly demonstrates that decisions have to be made more urgently.

What conclusions may now be drawn from the runs and scenarios evaluated so far? A first attempt would be to look for the impact various technologies within a certain technology group may have, as is done in Figure 6 for Germany, where technologies are classified after their share of total primary energy (TPE) in 1990, 2000, and 2010 respectively; but the development of ranking criteria for new technologies is just in progress, and we hope to have results for each of the member countries in the near future.

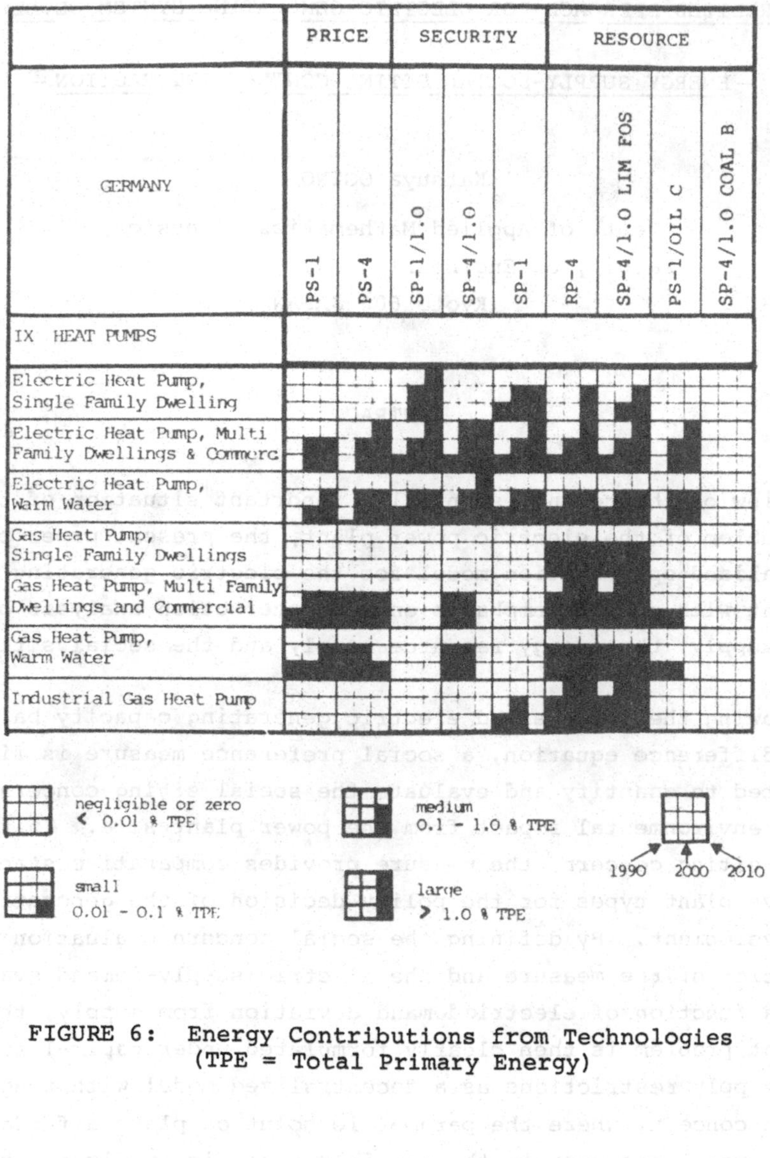

FIGURE 6: Energy Contributions from Technologies (TPE = Total Primary Energy)

DECENTRALIZED APPROACH FOR ELECTRIC GENERATING SYSTEM DEVELOPMENT

– ENERGY SUPPLY-SOCIAL SITING CONCERN INTERACTION –

Katsuya OGINO

Dept. of Applied Mathematics & Physics
Faculty of Engineering, Kyoto University
Kyoto 606 JAPAN

ABSTRACT

In view of the recent essentially important situation of the siting problem of the electric power plant, the present paper presents a decentralized optimization model for the electric generating system development with special emphasis on the interaction analysis of the electric supply, the energy resource supply and the social siting concern.

Following the interperiod electric generating capacity balance of a linear difference equation, a social preference measure is firstly investigated to quantify and evaluate the social siting concern mainly about the environmental impact from the power plant site. In the sense of social siting concern, the measure provides comparative standard to alternative plant types for the policy decision of the generating system development. By defining the social concern evaluation index as a function of the measure and the electric supply-demand evaluation index as a function of electricdemand deviation from supply, the development problem is then clearly formulated under capital and energy resource supply restrictions as a decentralized model with resource allocation concept, where the parametric solution plays a fundamental role. The model presupposes the possible electric supply shortage and evaluates the effects of the social concern on the electric supply-demand relation and on the energy resource allocation. The distinctive feature of the model is in permitting the regional participation in the generating system development through the social preference measure, pursuing the effective resource allocation policy. A simulation study is illustrated for the investigation of the model.

INTRODUCTION

In order to attain the stable supply of the electric power energy, the integrated policy analyses of the recent energy, resource and environment problems are fundamental. Especially, in view of the recent tight status of the primary energy resource supply and the prolonged trend of the siting problem of electric power plant facilities, the possible electric supply shortage would be apprehended, and the investigation on the social and regional acceptance of the siting problem and the effective allocation of the limited primary energy resource are urgently expected.

Regarding the electric generating system development, several linear programming models [1, 2] have been proposed so far. However, as far as the author's knowledge is concerned, any model investigating the electric supply shortage, the effective resource allocation and the social siting concern(mainly about the environmental problems around the electric power plant site) has not been reported.

For the effective development of the electric generating system, the author has made trials to develop the multiobjective programming models [4 - 6] and the decentralized optimization model [3], introducing the social siting concern into the decision phase of the development policy. The present paper represents another decentralized optimization model which mainly analyzes the effects of the social siting concern on the electric supply-demand relation and on the energy resource allocation, presupposing the possibility of the electric supply shortage. Assuming that the overall region under investigation consists of several subregions, the interperiod electric generating capacity balance is firstly presented in a form of the linear difference equation. The social siting concern evaluation index as a function of the social preference measure and the electric supply-demand evaluation index are secondly investigated. The social preference measure gives an order in regional preference to each alternative electric power plant type and provides the basis for comparative standard in selecting alternative power plant type mix in the generating system development. A decentralized optimization model with allocation concept of the available energy resource is then developed under the overall capital(cost) restriction, where the parametric solution plays a fundamental role. The distinctive feature of the model is in grasping the overall problem of the generating system development as a cooperation between the overall decision unit and the subregional decision units such that the subregions participate in the overall system development through the regional social preference measure, while the overall region pursues the stable electric supply over the overall region through the effective allocation of the limited energy resource. A simulation study is illustrated to investigate the present model.

INTERPERIOD GENERATING CAPACITY BALANCE

Assume that the overall region under investigation consists of r (sub)regions, and consider an overall planning period [0, T] which is broken down into n periods of equal time interval dt as

$$T = n \cdot dt. \tag{1}$$

Then, the electric generating capacity in a period k and in a region j is equal to the effective amount of the capacity available from the previous period plus the expanded amount of capacity in the period k. Thus, the <u>interperiod generating capacity balance</u> in the region j (j = 1, 2,..., r) can be represented by a linear difference equation

$$c_k^j = (I - A_k^j)c_{k-1}^j + e_k^j, \qquad (k = 1, 2,..., n), \qquad (2)_1$$

$$c_0^j : \text{given}, \qquad I : \text{mxm-unit matrix}, \qquad (2)_2$$

where

m : number of power plant types under investigation such as nuclear type, fossile type and so on,

c_k^j : electric generating capacity by each power plant type at the end of the period k and in the region j ; col.$(c_{1k}^j, c_{2k}^j,..., c_{mk}^j)$,

e_k^j : amount of the generating capacity by each power plant type, expanded in the period k and in the region j ; col.$(e_{1k}^j, e_{2k}^j, ..., e_{mk}^j)$,

A_k^j : diagonal mxm-matrix with the diagonal element a_{ik}^j of the attrition rate of each power plant type i in the period k and in the region j.

The new type i of power plant with new technology under development now can be included in the above investigation of the capacity balance by adding the following restrictions to $(2)_1$ and $(2)_2$;

$$c_{i0}^j = 0, \qquad (2)_3$$

$$e_{ik}^j = a_{ik}^j = 0, \qquad (k < v_i), \qquad (2)_4$$

where

v_i : period in which the commercial operation of the new type i power plant can be commenced.

SOCIAL CONCERN AND ELECTRIC SUPPLY EVALUATIONS

In view of the recent essentially important situation of the siting problem, the possible electric supply shortage would be apprehended due to the social siting concern besides the supply problem of the limited primary energy resource. Thus, the evaluation of the social siting concern and its effect on the electric supply-demand relation becomes one of the indispensable factors to be investigated for the smooth development of the electric generating system.

The social siting concern in region j mainly about the environmental problems from the electric power plants can be investigated in the most general form by the social concern evaluation index

$$J_2^j = J_2^j(prf_1^j, c_1^j, e_1^j; \dots; prf_n^j, c_n^j, e_n^j), \quad (j = 1, 2, \dots, r), \quad (3)$$

where

prf_k^j : comparative measure of the social(regional) preference from the point of view of the social siting concern to each power plant type in the period k and in the region j ; col.$(prf_{1k}^j, prf_{2k}^j, \dots, prf_{mk}^j)$.

The social preference, measured by the social poll and/or the questionnaire, gives an order in the social siting concern to alternative power plant types and provides the basis for the comparative standard in selecting alternative power plant type mix in the system development [3 - 6].

One of the concrete forms of the social concern evaluation index is given as

$$J_2^j = \sum_{k=1}^{n} prf_k^{j\prime} e_k^j, \quad (4)$$

corresponding to the expanded amount of the generating capacity, where the symbol "′" denotes the transpose of a vector.

Define the electric demand deviation from the electric supply in the period k over the overall region as

$$dp_k = pd_k - \sum_{j=1}^{r} \sum_{i=1}^{m} c_{ik}^j, \quad (5)$$

$$de_k = ed_k - \sum_{j=1}^{r} cpf_k^{j\prime} c_k^j \cdot dt, \quad (6)$$

where

dp_k : overall shortage amount of the generating capacity in the period k,

de_k : overall shortage amount of the electric power energy in the period k,

pd_k : overall peak power demand in the period k,

ed_k : overall energy demand in the period k,

cpf_k^j : capacity factor of each power plant type in the period k and in the region j ; col.$(cpf_{1k}^j, cpf_{2k}^j, \ldots, cpf_{mk}^j)$.

Then, the electric supply over the overall region can be evaluated by the supply-demand evaluation index in the most general form

$$J_1 = J_1(\| dp_1 \|, \| de_1 \|; \ldots; \| dp_n \|, \| de_n \|), \tag{7}$$

where the symbol $\| \cdot \|$ denotes a norm suitably chosen.

One of the concrete forms of the supply-demand evaluation index is given as

$$J_1 = \sum_{k=1}^{n} (dp_k/pd_k + de_k/ed_k), \tag{8}$$

which is to be adopted together with the electric demand restrictions

$$\sum_{j=1}^{r} \sum_{i=1}^{m} c_{ik}^j \leq pd_k, \qquad (k = 1, 2, \ldots, n), \tag{9}$$

$$\sum_{j=1}^{r} cpf_k^{j\prime} c_k^j \cdot dt \leq ed_k, \qquad (k = 1, 2, \ldots, n). \tag{10}$$

It is to be noted that the evaluation index (8) and the electric demand restrictions in (9) and (10), quite different from the conventional ones, indicate the premise of the possible electric supply shortage.

DECENTRALIZED OPTIMIZATION MODEL

Taking account of the limited supply of the primary energy resource and the possible electric supply shortage, the present section proposes a decentralized optimization model for the smooth development of the electric generating system, where the introduction of the social (regional) siting concern into the decision phase is fundamental for the regional understanding to the siting problem. Compared with the siting problem, the electric supply and the energy resource supply are to be investigated at the upper level. Thus, for the integrated policy analyses of the elecric supply, the energy resource allocation

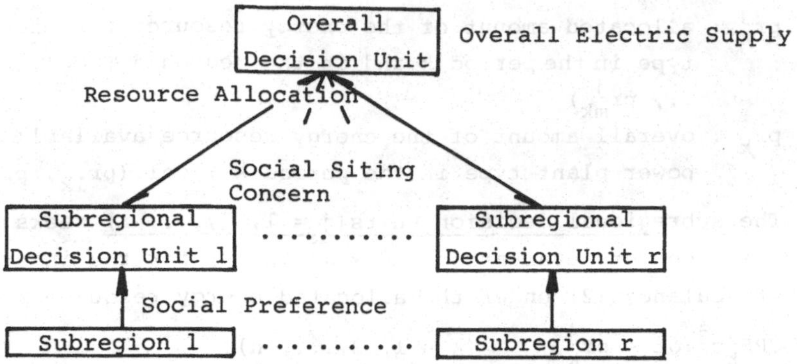

Fig. 1 Decentralized Scheme of The Model

and the siting problems, the problem of the electric generating system development is to be investigated in a decentralized and cooperative framework (Fig. 1) between the overall region and the subregions such that the overall decision unit pursues the overall stable electric supply through the effective allocation of the available energy resource to the subregions, while the subregions participate in the generating system development to seek the preferable power plant mix from the point of view of the regional siting concern.

The problem is now formulated as a decentralized optimization model as follows:

The overall decision unit seeks to minimize the electric supply-demand evaluation index J_1 in (7) under the restrictions on the available capital and the energy resource

$$\sum_{k=1}^{n}\sum_{j=1}^{r} (cp_k^{j\prime} e_k^j + om_k^{j\prime} CPF_k^j c_k^j \cdot dt) \leq cr, \tag{11}$$

$$\sum_{j=1}^{r} pr_k^j \leq pr_k, \qquad (k = 1, 2, \ldots, n), \tag{12}$$

where

cp_k^j : present worth of the capital cost per unit capacity of each power plant type in the period k and in the region j ; col. $(cp_{1k}^j, cp_{2k}^j, \ldots, cp_{mk}^j)$,

om_k^j : present worth of the o & m cost per unit power output of each power plant type in the period k and in the region j ; col. $(om_{1k}^j, om_{2k}^j, \ldots, om_{mk}^j)$,

CPF_k^j : mxm-diagonal matrix with the diagonal element cpf_k^j,

cr : overall capital available for the generating system development,

pr_k^j : allocated amount of the energy resource for each power plant type in the period k and in the region j ; col.$(pr_{1k}^j, pr_{2k}^j, \ldots, pr_{mk}^j)$

pr_k : overall amount of the energy resource available for each power plant type in the period k ; col.$(pr_{1k}, pr_{2k}, \ldots, pr_{mk})$.

The subregional decision units$(j = 1, 2, \ldots, r)$ seeks to maximize the social concern evaluation index J_2^j in (3) for the interperiod capacity balance (2) under the allocated energy resource restriction

$$CPF_k^j c_k^j . dt \leq pr_k^j, \qquad (k = 1, 2, \ldots, n). \tag{13}$$

The model reinvestigates the effective allocation policy of the limited energy resource and evaluates the effects of the social siting concern on the electric supply-demand relation and further on the inter-fuel(type) competition in the generating system development. It is to be noted that the optimal solution of the subregional decision units is a parametric one, which implements the overall decision on the energy resource allocation to the subsystems. Thus, the parametric solution plays a fundamental role in the present model.

To investigate the characteristics of the present model, a simulation study is illustrated in the following, where two types of power plant are investigated(i.e. $m = 2$) and the social concern evaluation index (4) and the supply-demand evaluation index (8) together with (9) and (10) are adopted. Other data assumed are $T = 15$; $n = 3$; $r = 2$; $c_0^j = 0$, $(j = 1, 2)$; $a_{ik}^j = 0.39 - 0.44$; $cpf_{ik}^j = 0.5 - 0.6$; $cr = 10^{13}$; and prf_k^j, pd_k and ed_k, $cp_k^j(=cp_k)$ and $om_k^j(=om_k)$, pr_k are given by Fig.2-Fig. 5 respectively. Fig. 2 indicates that the power plant type I vs. type II in the social preference is at the ratio of 0.8 to 0.2 in the subregion 1 throughout the overall planning planning period. Fig. 6

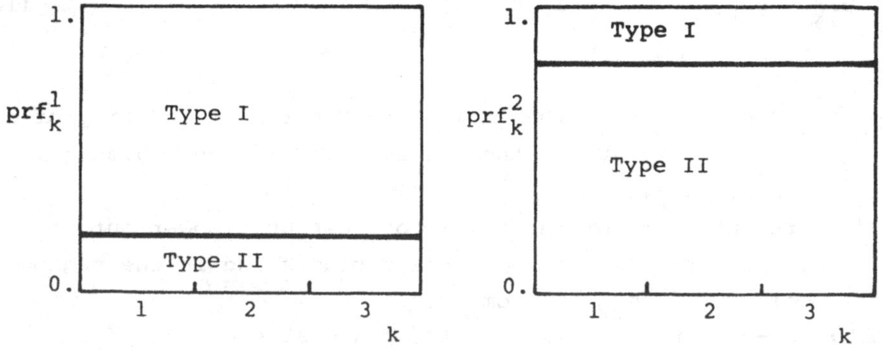

Fig. 2 Social Preference Measures

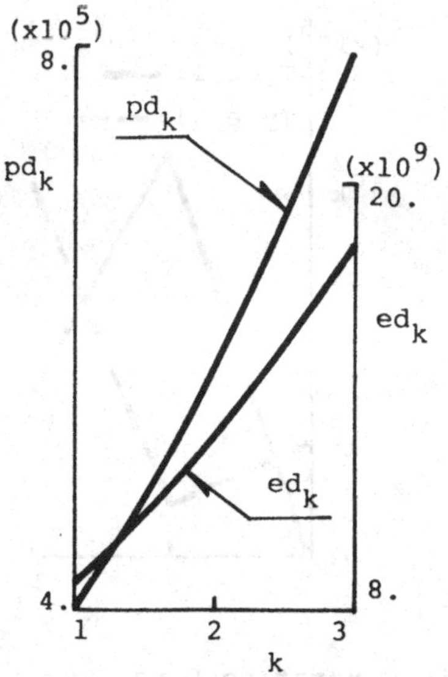

Fig. 3 Peak Power Demand and Energy Demand

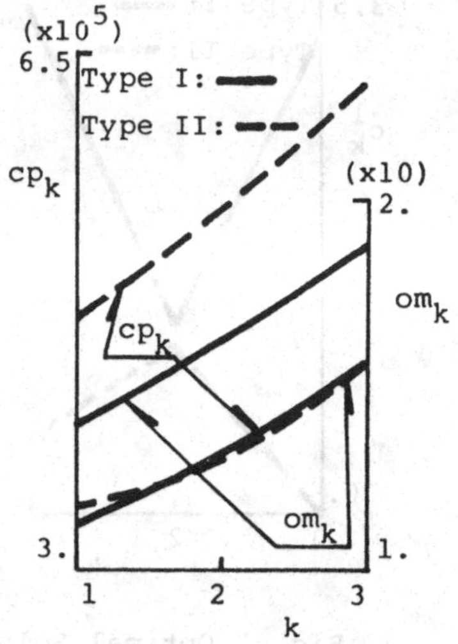

Fig. 4 Capital cost and O & M Cost

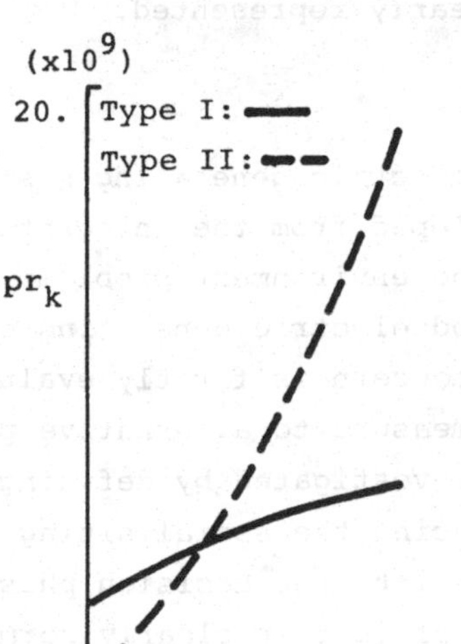

Fig. 5 Available Energy Resource

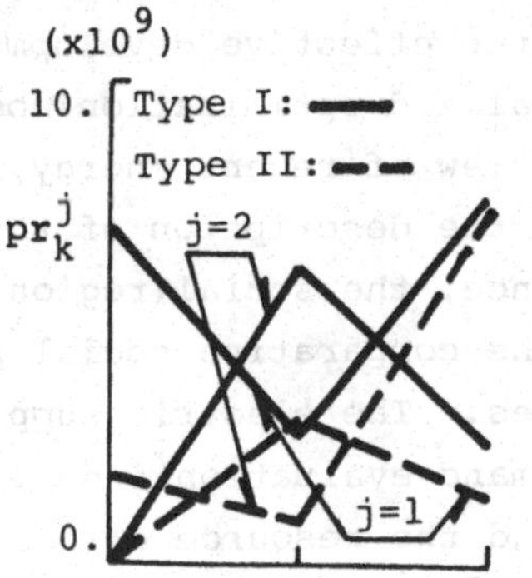

Fig. 6 Optimal Allocation of Energy Resource

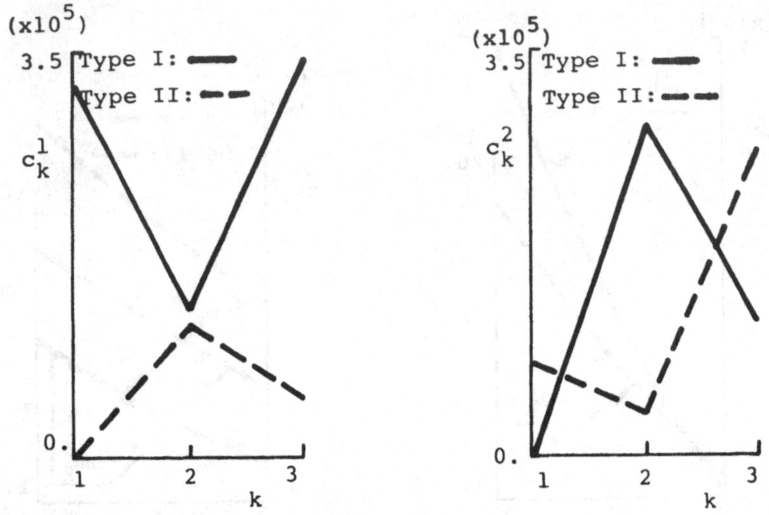

Fig. 7 Optimal Solutions in Generating Capacity

represents the optimal solution in the energy resource allocation. The transition of the corresponding installed electric generating capacity is given by Fig. 7, where the effect of the social preference on the generating capacity expansion is clearly represented.

CONCLUSIONS

For the effective development of the electric generating system, a decentralized optimization model is developed from the integrated point of view of recent energy, resource and environment problems. Following the description of the interperiod electric generating capacity balance, the social(regional) siting concern is firstly evaluated through the comparative social preference measure to alternative power plant types. The electric supply is also investigated by defining the supply-demand evaluation index. By introducing the social siting concern and the resource allocation concept into the decision phase of the generating system development, the model is then clearly formulated to analyze the effect of the social siting concern on the electric supply and to pursue the effective resource allocation policy in a cooperative framework of the overall region and the subregions, where the parametric solution plays a fundamental role. A simulation study is illustrated to investigate the characteristics of the present decentralized optimization model.

REFERENCES

[1] D. L. Farrar and F. Woodruff, Jr. : A model for the determination of optimal electric generating system expansion patterns, NTIS, Springfield, Va., 1973.

[2] F. Begrali and M. A. Laughton : Model building with particular reference to power system planning; the improved Z-substitutes method, in Energy Modelling, IPC Business Press Ltd., 1974.

[3] K. Ogino : Optimal expansion of generating capacity in national electric power energy system, Lecture notes in Control Information Sciences, Vol. 6 : Optimization Techniques, pp.489-499, Springer-Verlag, 1978.

[4] K. Ogino : Cost-environment concern interactions in capacity expansion policies of large-scale electric power energy generating system, Special project research on detection and control of environmental polltion, Vol. 4, pp.38-41, 1979.

[5] K. Ogino : Multiobjective programming models introducing social siting concern to electric generating capacity expansion policy analysis(in Japanese), Transactions of the Society of Instrument and Control Engineers, Vol. 15, No. 6, pp.839-844, 1979.

[6] K. Ogino : Electric supply-social siting concern tradeoff in optimizing electric energy system development, to appear in the Proceedings of the IV. Symposium uber Operations Research, Saarbrucken, 1979.

ON A STOCHASTIC MODEL OF RESERVOIR SYSTEM SIZING*

János Pintér
International Research Institute
of Management Sciences
129090 Moscow, Schepkina ul. 8.

1. Introduction

The probabilistic character of future events often plays significant role in water resources management. A well-known example of this is the problem of reservoir system design and operation, when - in the evaluation process of some decision (e.g. on reservoir-configuration or operational policy) - future random inflows and water demands have to be considered. Though various deterministic approaches have been applied to these problems with computational advantages (see e.g. [1], [2]), their results usually provide less detailed information on the system performance, than those obtained by stochastic models.

A powerful method of stochastic analysis is the Monte Carlo technique when suitably long simulated series of events provide base for evaluating the planned system and/or its operational policy. As it is stated in [3], p. 820., "Any complex set of constraints or operating rules can be included without too much difficulty, provided that they can be clearly defined in a deterministic mathematical form and can be clearly identified with either the reservoir storage or a decision to be taken in any given month". The stochastic simulation combined with some mathematical programming optimization method gives a possible way of selecting an improving sequence of decisions.

Recently, a large number of papers discussed various theoretical and practical aspects of stochastic reservoir system design and operation, see e.g. [3] - [18]. In a significant part of these works first formulate stochastic (chance-constrained) programming models, which - applying some structural simplifications e.g. separated stochastic constraints for the involved random variables, and/or special assumptions

---

* This model was elaborated in the framework of a joint research of the North Hungarian District Water Authority (NHDWA), Miskolc and the Computing Center for Universities, Budapest. The author expresses his sincere thanks to Mr. Ferenc Domonyik of the NHDWA for his active cooperation in the development and numerical evaluation of this model.

e.g. independent, identically distributed inflows - can be deduced to deterministic equivalents. A more general approach is applied in the works of Prékopa and his collaborators. This is based on a general result in the theory of stochastic programming, from which follows that certain properly constructed stochastic constraints with a special class of random variables can be fitted into convex programming problems (see [5]). Applying this result, in many cases joint stochastic contraints can be considered also for inhomogenous, interdependently structured random variables (e.g. water inflows and demands of a region), still yielding convex models) [9], [16-18]. As it is well-known, these problems - according to the global optimality of any local optimum - can be solved by effective computational procedures (see e.g. [20], [21]).

The aim of this paper is to give a summarized picture of a water resources management problem in Northern Hungary, with its mathematical model formulation and numerical solution. In consequence of the economic development, water resources presently available will hardly meet the rapidly increasing water demands. For this reason a system of reservoirs is planned by the North Hungarian District Water Authority (NHDWA) [22]. The reservoirs to be built have their water intake from rather small mountain streams with abrupt flowrate changes. The planned system has multiple purposes such as flood control, municipal, industrial and agricultural water supply, providing also local (e.g. fishing, recreational) possibilities. We remark that further possible water use objectives e.g. hydropower generation or water quality improvement by low-flow augmentation are not studied in this paper. (The former possibility was considered as of not primary significance in the region, while the other aspect was investigated in a separate analysis of regional water quality management by jointly applied waste water treatment and low-flow-augmentation [13]).

The objective of the presented model is to find optimal reservoir capacities, i.e. to select a system of reservoirs which jointly meet future water demands of a fixed time-horizon with minimal summed costs. As the examined problem is heavily dependent on random variables, the model is formulated as a stochastic programming problem, having constraints which can not be transformed to deterministic equivalents, and moreover, even their convexity properties are not settled. Therefore the performance of any "candidate" reservoir system is evaluated on the base of simulated series of random water yields and operational policies. On the other hand, considering possible non-convexities of the problem, the applied optimization method is based on a heuristic screening procedure and a special random search algorithm.

The model is presented with the primary aim to outline general features of the applied methodology. Therefore numerical details are only shortly referred here. An illustrative system of interdependently operated reservoirs is displayed on Fig. 1.

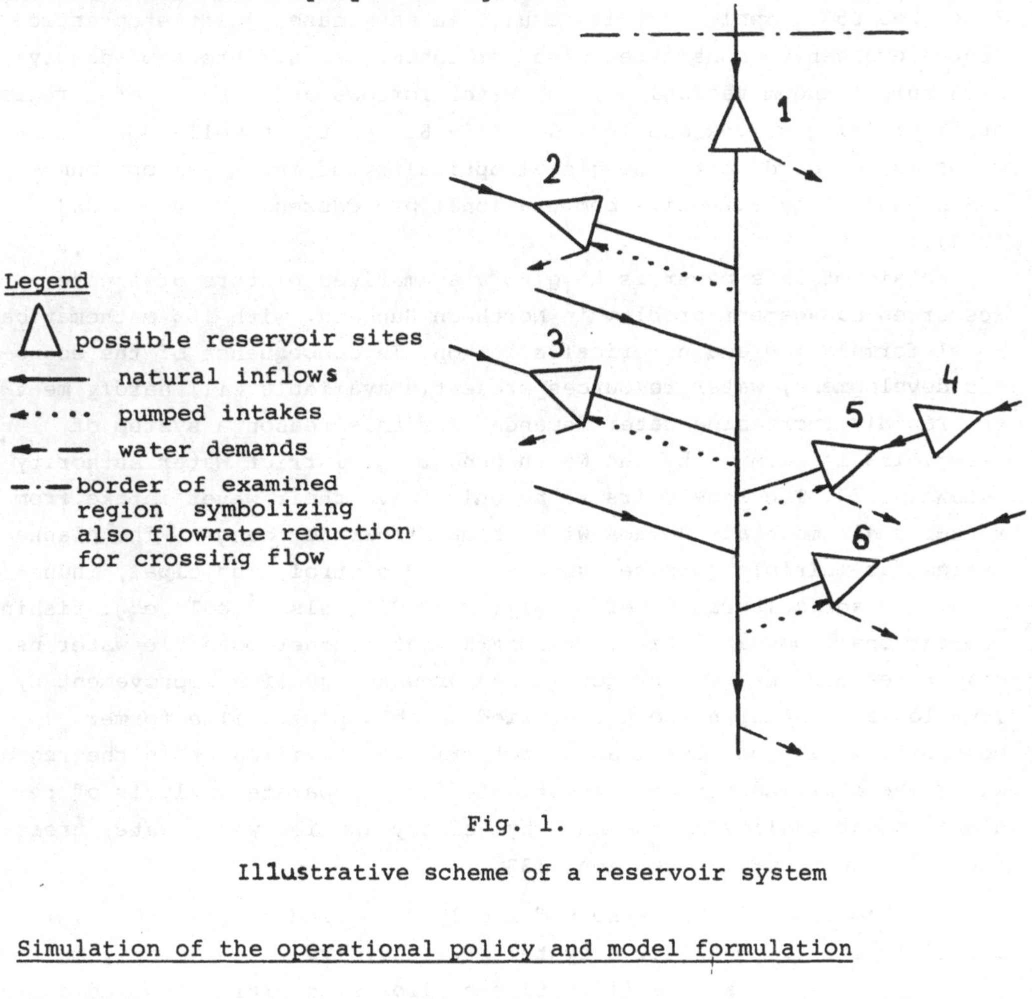

Fig. 1.

Illustrative scheme of a reservoir system

2. Simulation of the operational policy and model formulation

The simulation model is constructed with the purpose to provide information on the long-run characteristics of any studied system of reservoirs. This information, expressed by the value of the objective function and constraints of some mathematical programming problem to be formulated, will be used to select iteratively an improving sequence of decisions (reservoir-configurations).

The simulated events can be subdivided into three parts, namely:

streamflow generation, water intake and water release. In the following first we give a brief description of these processes, then formulate the mathematical model.

2.1. Streamflow model

The water intakes of the studied system are supplied mainly by small mountain streams, with abruptly varying quarterly flow rates, which can be considered as serially independent and having no cross-correlations. Based on (for some streams rather scarce) time series of flows and regional hydrologic analogies, first the estimations of quarterly mean, variance, minimal and maximal possible values of flows were calculated. Then, according to a statistical analysis of streamflow data, the quarterly streamflows were modelled having one of the following type distribution functions: three-parameter (translated) gamma distribution, two parameter (translated) exponential distribution and normal distribution. We note that the generated quarterly flows were finally adjusted, taking into account also the minimal and maximal possible flows, thus truncating the above distributions.
From a given quarterly flow monthly streamflows were generated as the sum of a main component (a fixed proportion of the quarterly flow), and a random noise term generated from a truncated normal distribution. The mean and variance parameters of these components could be estimated from the data provided by NHDWA. These preliminary monthly flows were then normalized, in order to give in sum the initial quarterly streamflow.

We note that though - in consequence of scarcity of hydrological data in the studied region - the outlined streamflow model has many heuristic features, the generated synthetic flows were considered by NHDWA experts as fairly acceptable.

2.2. Water intake process

Having generated for a given period the random inflows, these were transformed considering also external water demands.

The water intake of the reservoirs is modelled in two steps. First their natural (i.e. gravitational) water intake, then the possible filling up by pumping stations is considered. (Based on a previous analysis of the studied case, we concluded that it is worthwhile to retain all possible water quantities, except, of course, in flood situations).

The natural water intake process for each reservoir can be described in the following form. Given the initial entering flows, we have at the reservoirs recursive relations, which express the respective inflow as the sum of flows on their first order antecedent streams.If any of these streams arrives from a reservoir then - according to the policy of retaining water in the system at the uppermost storage possiblities - the maximum of the overflow of this reservoir and the permissible minimal streamflow is considered.

After realization of the natural inflow process, pumped filling up at some reservoirs may take place. Supposing that we have water surplus after all gravitational inflows, we try to fill up some reservoirs by pumping. The pumped intake of the reservoirs is modelled according to the order of increasing unit pumping costs.

We remark that in reality pumping takes place, of course, not from the very last reservoir, but from the nearest water surpluses to the reservoirs in increasing order of costs. Our presented description, however, models quite well the studied case, where pumping occurs only from a main stream to reservoirs on its tributaries. In more general cases, special transportation problems could be formulated for optimal pumped distribution of water surpluses, considering also the respective water demands.

2.3. Water release process

The selection of optimal operation rules for any fixed system of reservoirs, in principle, should be based on an imbedded, open-loop dynamic type stochastic programming model (see for general description [19], applications are given e.g. in [9], [17]). When solving this problem, the probabilistic structure of subsequent water yields, requirements and loss functions (from unsatisfied demands) should be taken into account. Based on this information, sequential operational decisions could be settled. As this type of modelling technique in this case is not applicable (according to computational difficulties in solving hierarchically structured stochastic programming problems, and data uncertainties), the water release process for any period is specified, considering only the water demands of this period. For the decision, preference order of water demands, target releases and loss functions, corresponding to deviations from targets are necessary (for construction of such functions see [7]). We emphasize that the preferential ordering of water demands is a function of the earlier demand-release process, therefore it may change from period to period.

The demands in the given period are then met, according to their actual ordering. This can be realized on the base of the following principle (for an algorithmic description see [10]):

Consider the reservoirs in the sequence of their increasing numbers (note that all antecedents of any reservoir have smaller serial numbers). Let us suppose that all water demands are also ordered uniformly at all reservoirs (i.e. identically numbered demands are identically important), according to their priorities.

Let us take the most important unsatisfied demand at the actually studied reservoir. This can be met if and only if all demands of higher priority can be satisfied at all reservoirs - from their potential water supply -, to which water can be transferred from this reservoir. (The potential water supply at any reservoir is the sum of its supply plus all water quantities which can be transferred to this reservoir. Here pumped water transports are only considered if they are economical, considering the benefits from supplied water.)

We remark that this principle can be "streamlined" by some practical considerations, e.g. first satisfying all demands on an acceptable minimal level, and then optimizing the distribution of remaining water supplies. These or other aspects, of course, have significant effects on the system performance characteristics.

In the studied case the demands were considered for each reservoir in the sequence of minimal necessary amount of streamflow , municipal, industrial and agricultural water needs, local water uses. We note that flood control was separately considered by reserving some capacity for abrupt short-term floods, usual in the studied region. We also remark that before turning to a new simulational cycle, seepage and evaporation losses were also considered for the remaining amounts of water to be stored until the next period.

For every cycle of the above outlined processes (streamflow generation, water intake and release) the occasional water shortages by their types were classified and registered. Thus, based on a suitable number (see later) of simulation runs for a given system of reservoirs, estimation for its reliability could be calculated.

2.4. Mathematical model formulation

Let us introduce the following notations :

$i \in I$	indices of possible reservoir sites
$j \in J$	operational periods of the system (regarded as months)
$k \in K$	water demand types (e.g. needs of municipalities, industries or agriculture)
z_j^i	water content of reservoir i at the end of period j (z_o^i is given as a function of reservoir capacity)
d_{jk}^i, D_j^i	water demand of type k and set of these demands against reservoir i in period j
r_{jk}^i	actual water release for demand k from reservoir i in period j
C^i	capacity of reservoir i
C_{min}^i, C_{max}^i	minimal and maximal possible capacities of reservoir i
C_f^i	capacities of reservoirs, kept exclusively for short-term flood control purposes
C_r^i	minimal retained amount of water in reservoir i (for local uses such as fishing, recreation etc.)
$B^i(C^i+C_f^i)$, $P^i(C^i)$	building and pumping station costs of reservoir i
$R^i(C_r^i)$	overall benefits from retained water at reservoir i

Here C^i, C_f^i, C_r^i represent decision variables to be optimized, z_j^i, d_{jk}^i, r_{jk}^i are random variables, C_{min}^i, C_{max}^i are fixed parameters of the system, while $B^i(C^i+C_f^i)$, $P^i(C^i)$ and $R^i(C_r^i)$ are monotonically increasing functions of their variables. We remark that the flood control volumes C_f^i were separately calculated on the base of short-term flood characteristics of the involved streams and thus had only indirect effects on the optimization procedure in C^i and C_r^i.

With the above notations, the following underlying deterministic model is considered:

minimal water contents of reservoirs provided for local use

(2.1.) $$z_j^i \geq C_r^i \qquad i \in I \quad j \in J$$

satisfaction of all water demands:

(2.2.) $$r_{jk}^i \geq d_{jk}^i \qquad k \in D_j^i \quad i \in I \quad j \in J$$

capacity and local water use constraints

(2.3.) $$C_{min}^i \leq C^i \leq C_{max}^i \qquad i \in I$$

$$0 \leq C_r^i \leq C^i$$

Considered investment (reservoir and pumping station construction) costs minus benefits of local water uses:

$$\sum_{i \in I} [B^i(C^i + C_f^i) + P^i(C^i) - R^i(C_r^i)] \to \min \tag{2.4.}$$

As in the above relations beside deterministic parameters random variables are also involved, we have to formulate their stochastic extensions, e.g. prescribing minimum permissible probabilities for the fulfillment of the constraints. We emphasize that satisfaction of different constraints may be of different importance (e.g. flood damages, municipal, industrial and agricultural water shortages usually mean decreasingly dangerous situations in this order), therefore separate stochastic constraints have to be formulated, even if the same random variables play role in each constraint. (In the studied case, the prescribed reliability bounds on system performance, according to the above requirements, were respectively 99, 95, 90 and 80 per cent.) Moreover, expected values of stochastic loss functions are also added to the objective function to be minimized.

Summing up the mentioned principles, the following stochastic extension of the underlying deterministic problem can be given. The objective function is

$$\min\{ \sum_{i \in I} [B^i(C^i+C_f^i)+ P^i(C^i)- R^i(C_r^i)] + \tag{2.5.}$$

$$+ \sum_{j \in J} \sum_{i \in I} E[\sum_{k \in D_j^i} L_{jk}^i(S_{jk}^i) + L_j^i(t_j^i)]\}$$

Where $S_{jk}^i = \max(d_{jk}^i - r_{jk}^i, 0)$ represents the possible shortage in meeting demand k at reservoir i in period j, $t_j^i = \max(C_r^i - z_j^i, 0)$ is the possible shortage of the retained amount of water, and L_{jk}^i, L_j^i are monotonously increasing loss functions of S_{jk}^i and t_j^i. As it was noted earlier, the functions B^i, P^i and R^i are also monotonously increasing functions (consisting of convex and concave pieces in our case).

The constraints (satisfaction of water demands and local water requirements) are symbolized in the form

(2.6.) $P(z_j^i \geq C_r^i \qquad j \in J,\ i \in I) \geq p_o$

(2.7.) $P(r_{jk}^i \geq d_{jk}^i \qquad j \in J,\ i \in I) \geq p_k \quad k \in K$

while the deterministic constraints

(2.3.) $C_{min}^i \leq C^i \leq C_{max}^i \qquad i \in I$

$0 \leq C_r^i \leq C^i$

remain unchanged.

3. Solution procedure

First we connect the probabilistic constraints - as penalty terms - to the objective function, e.g. in the form:

$$(3.1.)\quad \min\{ \sum_{i\in I} [B^i(C^i + C_f^i) + P^i(C^i) - R^i(C_r^i)] + \sum_{j\in J} \sum_{i\in I} E[\sum_{k\in D_j^i} L_{jk}^i(S_{jk}^i) + L_j^i(t_j^i)] +$$

$$+ \sum_{k\in K} C_k[\min(P(r_{jk}^i \geq d_{jk}^i \quad j\in J\ i\in I) - p_k, 0)]^2 +$$

$$+ C_o[\min(P\ (z_j^i \geq C_r^i \quad j\in J\ i\in I) - p_o, 0)]^2\}$$

where $C_k > 0$, $C_o > 0$ are given penalty multipliers, in principle tending to infinity, in a sequence of unconstrained problems. (For general methodology of sequential unconstrained minimization, see [23].) We remark that considering the usually occurring computational problems - and resulted substantial costs of computer runs - of the sequential solution methods, we used only a fixed set of penalty parameters. These multipliers were estimated from interpolation of the economic loss values at certain degrees of unsatisfied requirements. In consequence of this approach, slight deviations from the feasible domain were permitted in the course of the optimization. This attitude is supported also by the inherent inaccuracies of the stochastic function evaluations, to be analysed later.

Summing up the aforesaid, we have to solve following type stochastic problem

$$(3.2.)\quad \min\ f(x, y)$$
$$x_{min} \leq x \leq x_{max} \qquad x \in R^n \qquad y \in R^q$$

where x and y represent respectively the decision and random variables of the problem (both being finite dimensional vectors), and f is a stochastic function the values of which for any x of the n-dimensional interval $[x_{min}, x_{max}]$ can be calculated only approximately. Note that - according to its possible nonconvexities - f may have many local optima.

As - even if f might be in principle differentiable - in this or similar cases its gradients can not be easily computed, we assumed only continuity of f and used only its values through the minimization process. We remark that though there are existing methods to calculate global optima of arbitrary continuous functions on compact sets (see the review of some global search procedures e.g. in [24] and [25]), except for some simpler special problems, relatively few computational experience is reported so far for real-world applications. Therefore the following, partly heuristic procedure was applied for approximating the global solution of our problem:

First a preliminary screening of the possible solutions was accomplished by a "streamlined" random sampling method, which was intensified in the neighbourhood of some reservoir configurations proposed by NHDWA. In connection with this we remark the well-known fact that for arbitrarily small p and α ($0<p, \alpha<1$) there exist an integer $m_o(p,\alpha)$ such that a pure (simultaneous) random sample consisting of $m \geq m_o$ uniformly distributed points on the compact set of feasible decisions contains at least one decision from the best $100 \cdot p$ per cent of the possible decisions, with probability $1-\alpha$. (For example, $m_o(0.05, 0.01)=90$.) It is also worth to note that though the difference between the optimal solution and its approximation (based on m trials) is not known, this can be estimated from the empirical distribution function of the sample objective function values (see e.g. [26], where a sequential polynomial approximation of the value distribution function is constructed).

The above outlined screening procedure resulted some "promising" starting points for further examination. The improvement of these solutions could have been based on any locally convergent mathematical programming method. Considering the random noise effects in the evaluations of the objective function and the mere continuity assumption on it, a special random search type algorithm was used. This algorithm converges (under some regularity conditions) to the set of local minimizers of a continous function, and was tested on a number of standard deterministic and some noise-corrupted test functions of nonlinear programming [27].

The algorithm is an infinite iterative procedure, i.e. its steps are of the form

$$x_{k+1} = x_k + \alpha_k d_k \qquad k = 1,2,3,\ldots \tag{3.3.}$$
$$f(x_{k+1}) \leq f(x_k)$$

Here the selection of the directional vectors d_k (of unit length) is based on the comparison of some trial search steps, which are orthonormally transformed from independent uniformly distributed random vectors on the n-dimensional unit hypersphere. From these search vectors a stochastic estimation of the antigradient, projected into the subspace generated by the vectors d_k, is also computed, which - for suitably smooth objectives - tends to the locally most efficient direction. Selecting the locally most favourable direction among the above trial steps, the step length α_k is obtained from a one-dimensional extrapolation and (quadratic) interpolation procedure which is aimed at the approximate calculation of that local minimizer of f which is nearest to x_k on the line $x_k + \alpha d_k$. This approximate solution is denoted by x_{k+1} and the algorithm proceeds further from this point, until some termination criteria are not met. The obtained solution approximates a local minimizer of f.

Accomplishing the outlined procedure from different starting points, the best local solution was accepted as an estimation of the global optimum.

Some remarks are in order about the desired exactness of function evaluations in the course of the optimization process. It is heuristically plausible and in fact can be proved (see the sufficient criteria for convergence of iterative stochastic algorithms in [28]) that convergence to the local minimizers generally can be hoped only on the condition that random noise effects on the iteration procedure tend to zero. As in our case the function values can be estimated from a series of theoretically independent simulated events (experiments), this means that in order to decrease noise effects the number of these events should be increased. On the other hand, we know from Bernstein's classical inequality (see e.g. [29]) that the estimation of the probability $p = P(A) > 0$ of event A from the relative frequency r_m based on m independent trials (with outcomes A and non-A) needs

$$m \geq \frac{2p(1-p)\left(1+\frac{\varepsilon}{2p(1-p)}\right)^2 \ln \frac{2}{\delta}}{\varepsilon^2} \tag{3.4.}$$

to assure $P(|r_m-p| \geq \varepsilon) \leq \delta$ ($\delta > 0$, $0 < \varepsilon < p(1-p)$ arbitrary numbers). (We note that (3.4.) is not a necessary, but the best known general sufficient condition for the desired accuracy.) Therefore the exactness of simulations was gradually increased in the course of the random sampling and through the local optimization procedures.

The results of the calculations showed that in our case some reservoir configurations, consisting of two-three larger reservoirs, are more advantageous than systems with a larger number of smaller capacity reservoirs (proposed originally by NHDWA). This result can be explained by economies of scale, significant seepage losses, pumping costs and smaller local water use benefits at some sites. We finally note that some different configurations yielded quite close locally optimal values: therefore the system to be constructed will be selected considering also other regional economic and social aspects which were not investigated in our model.

References

1. Opricovic, S. - Djordjevic, B.: Optimal Long Term Control of a Multipurpose Reservoir with Indirect Users, Water Resources Research, Vol. 12. No 6., 1286-1290, Dec. 1976.
2. Gouevsky, I.V.: On optimum control of multi-reservoir systems, IIASA RR-74-25., Laxenburg, Dec. 1974.
3. Askew, A.M. - Yeh, W.W. - Hall, W.A.: Use of Monte Carlo Techniques in the Design and Operation of a Multipurpose Reservoir System, Water Resources Research, Vol. 7., No 4. 819-826., Aug. 1971.
4. Revelle, C. - Kirby, W.: Linear Decision Rule in Reservoir Management and Design 2. Performance Optimization, Water Resources Research, Vol. 6. No 4. 1033-1044., Aug. 1970.
5. Prékopa, A.: Stochastic programming models for inventory control and water storage problems, 229-245., in Coll. Math. Soc. J. Bolyai 7. Inventory Control and Water Storage, Győr, 1971. (Ed. A. Prékopa), North Holland Publ. Co., Amsterdam, 1973.
6. Eisel, L.M.: Chance Constrained Reservoir Model, Water Resources Research, Vol. 8. No 2., 339-347. April, 1972.
7. Loucks, D.P. - Jacoby, H.D.: Flow Regulation for Water Quality Management, 362-432. in: Dorfman, R. - Jacoby, H.D. - Thomas, H.A. (Eds.): Models for Managing Regional Water Quality, Harvard Univ. Press, Cambridge, Mass., 1972.
8. Becker, L. - Yeh, W.W.-G.: Optimal Timing, Sequencing and Sizing of Multiple Reservoir Surface Water Supply Facilities, Water Resources Research, Vol. 10., No 1. 57-62., Febr. 1974.
9. Prékopa, A. - Szántai, T.: On Multi-Stage Stochastic Programming (with Application to Optimal Control of Water Supply), 733-755., in Coll. Math. Soc. J. Bolyai, 12. Progress in Operations Research, Eger, 1974. (Ed. A. Prékopa), North Holland Publ. Co. Amsterdam, 1976.

10-11. Rosanov, Yu. A.: Some system approaches to water resources problems.

I. Operation under water shortage, IIASA-RR-74-17 Laxenburg, Oct. 1974.

II. Statistical equilibrium of processes in dam storage, IIASA-RR-75-4 Laxenburg, Febr. 1975.

12. Kaczmarek, Z.: Storage systems dependent on multivariate stochastic processes, IIASA-RR-75-20, Laxenburg, July, 1975.
13. Pintér, J.: A Stochastic Programming Model Applied to Water Resources Management, Technical Report No 11., Computing Center for Universities, Budapest, Aug. 1975.

14. Anis, A.A. - Lloyd, E.H.: Stochastic reservoir theory: An outline of the state of art as understood by applied probabilists, IIASA RR-75-30., Laxenburg, Sept. 1975.

15. Loucks, D.P. - Dorfman, P.J.: An Evaluation of Some Linear Decision Rules in Chance-Constrained Models for Reservoir Planning and Operation, Water Resources Research, Vol. 11., No 6. 777-782, Dec. 1975.

(16-19.) in:

Prékopa, A. (Ed.): Studies in Applied Stochastic Programming, Comp. and Aut. Inst. of the Hung. Acad. of Sci. 80/1978.

16. Prékopa, A. - Rapcsák, T., - Zsuffa, I.: A new method for serially linked reservoir system design using stochastic programming, 75-97.

17. Prékopa, A. - Szántai T.: On optimal regulation of a storage level with application to the water level regulation of a lake, 119-154.

18. Prékopa, A. - Szántai T.: Flood control reservoir system design using stochastic programming, 155-177.

19. Prékopa, A.: Dynamic type stochastic programming models, 179-209.

20. Simmons, D.A.: Nonlinear Programming for Operations Research, Prentice-Hall Inc., Englewood Cliffs, N.J. 1975.

21. Karmanov, V.G.: Mathematical Programming (in Russian) Nauka, 1975.

22. North Hungarian District Water Authority: A study on the Bódva-valley reservoir system (in Hungarian), Manuscript, Miskolc, 1977.

23. Fiacco, A.V. - Mc Cormick, G.P.: Nonlinear Programming: Sequential Unconstrained Minimization Techniques, John Wiley and Sons, Inc. New York, 1968.

24. Mc Cormick, G.P.: Attempts to Calculate Global Solution of Problems that may have Local Minima, in: F.A. Lootsma (Ed.): Numerical Methods for Nonlinear Optimization, Academic Press, London, 1972.

(25-26.) in:

Dixon, L.C.W. - Szegő, G.P.: Towards Global Optimization, North Holland Publ. Co., Amsterdam, 1975.

25. Dixon, L.C.W. - Gomulka, J. - Szegő, G.P.: Towards a Global Optimization technique, 29-54.

26. Archetti, F.: A sampling technique for global optimization, 158-165.

27. Pintér, J.: On the Convergence and Computational Efficiency of Random Search Optimization, paper presented at III. Symp. on Operations Research, Mannheim 1978. published in: Methods of Operations Research, 33 (1979), 347-362.

28. Poljak, B.T.: Convergence and rate of convergence of iterative stochastic algorithms I. General case (in Russian), Avtomatika u Telemekhanika, 12 (1976) 83-94.

29. Rényi, A.: Calculus of probability (in Hungarian), Tankönyvkiadó, Budapest, 1968.

AN LP ENERGY SUPPLY MODEL FOR WORLD REGIONS

Leo Schrattenholzer
International Institute for Applied Systems Analysis
Laxenburg, Austria

1. INTRODUCTION

One focus of the work of the Energy Systems Program at the International Institute for Applied Systems Analysis (IIASA) is a global energy model. This global energy model consists of a linked set of models describing different aspects of the energy system. The set of models is applied to each of seven (exhaustive) world regions. The balance between these regions is represented by a global trade model for primary energy.

One part of this model system is a Dynamic Linear Programming (DLP) model which finds the cost-optimal supply strategy for meeting a given set of secondary energy demand vectors (over the next 50 years). The most important constraints are the availability of energy conversion technologies and the amount of recoverable primary resources.

This paper will concentrate on the mathematical description of the DLP model. The results shown here have the character of examples. A comprehensive description of the model applications will be part of a book by the Energy Systems Program. Its publication can be expected in 1980.

2. STANDARD FORM OF A DLP MODEL

Dynamic Linear Programming (DLP) models can be formulated in many ways. A.Propoi [1] proposes to use a formulation similar to optimal control theory. According to Propoi, a DLP model is described by five groups of functions. Choosing among several alternatives within one group the one appropriate for the model described later, these groups are (in matrix/vector notation):

I. State Equations:

$$x(t+1) = \sum_{i=1}^{\nu} A(t-n_i)\ x(t-n_i) + \sum_{j=1}^{\mu} B(t-m_j)\ u(t-m_j); \quad t=0,\ldots,T-1 \qquad (1)$$

x Vector of state variables

u Vector of control variables

A,B Matrices (constants)

II. Constraints

$$G(t)x(t)+D(t)u(t) \lesssim (=) f(t) \tag{2}$$

G,D Matrices (constants)

f Vector (constants)

III. Boundary Conditions

$$x(0) = x^o \tag{3}$$

IV. Planning Period

T is fixed (4)

V. Performance Index

$$J(U)=(a(T),x(T))+\sum_{t=o}^{T-1}[(a(t),x(t))+(b(t),u(t))] \tag{5}$$

a,b Vectors (constants)

Remarks:

- More than one constraint can be imposed on variables. Thus, (2) can be repeated with different values for G, D, and f.
- Unless otherwise stated, x and u are assumed to be non-negative.

3. A DLP ENERGY SUPPLY MODEL

Since its beginning in 1975, the Energy Systems Program at IIASA has been using LP models [2,3]. The one now used is called MESSAGE (Model for Energy Supply Systems Alternatives and Their General Environmental Impact). This model is formulated in different versions, the most general of which is described in [4]. The following describes a version that was part of a family of models that was used to formulate scenarios of the global energy system in the next 50 years [5].

3.1. Capacities of Technologies (State Equations)

For modeling purposes the planning period was divided into steps (time periods) of equal length (5 years). Together with the lifetime of energy conversion plants of 30 years (= 6 periods), the following state equations for the capacities of technologies were derived:

$$c(t+1) = c(t)+5z(t)-5z(t-6) \tag{6}$$

where: c is the vector of installed capacities (LP variables

z is the vector of annual additions to capacity (LP variables)

t is the index of time period.

Thus formulated, the model requires the capacities of the technologies at the beginning of the planning period ($t \leq 0$) as well as the historical construction rates [$z(t-6)$ for $t-6 \leq 0$] as boundary conditions.

3.2. Resources (State Equations)

This group of equation keeps the balance of materials relevant to the system modeled. These include primary energy resources (coal, oil, etc.) and man-made materials (e.g. plutonium).

$$s(t+1) = s(t)-5r(t) \tag{7}$$

where: s is the vector of reserves (stocks) of primary energy carriers or man-made fuels (LP variables)

r is the vector of annual consumption of primary energy carriers (LP variables).

As described later in greater detail, each kind of resource is divided into several categories according to different (extraction) costs. The above balance holds for each of these categories.

The assumption of non-negativity of variables usually works as a binding constraint here. Together with the specification of s(0), the total amount of resource category at the beginning of the planning period, these equations limit the total consumption of any resource category over the planning period. The non-negativity constraint is removed for the r vector in the case of man-made materials so as to allow for <u>production</u> as well as for consumption of these materials.

Renewable energy sources (solar, hydro, etc.) are treated differently. No balances are calculated but limits on availability are imposed (see the description of bounds below).

3.3. Demand/Supply Balance (Constraints)

Energy demand is exogenous to the model. In present applications this demand is defined in terms of secondary energy, divided into the following sectors: electricity, liquid, solid, and gaseous fuels, heat, and soft (local) solar. The capability of including technologies that require one of the secondary energy carriers as input is modeled separately (e.g. electric power plant using liquid fuels).

$$Dx(t) \geq d(t) + Hx(t) \tag{8}$$

where: x is the vector of supply activities (LP variables)

D is the matrix describing supply/demand paths (constants)

H is the matrix with the coefficients for the use of secondary energy by technologies

d is the vector of secondary energy demand (exogenous inputs).

For electric energy, quantities demanded and supply activities are divided into different load regions. Although different variables are attributed to different load regions, these variables are linked to <u>one</u> capacity variable as described below.

3.4. Resource Consumption (Constraints)

This group of constraints links primary and secondary energy.

$$Gr(t) \geq Q_1x(t) + Q_2z(t) - Q_3z(t-6) \qquad (9)$$

Where: G is the binary matrix which aggregates resource categories
Q_1, Q_2, Q_3 are matrices of parameters describing specific consumption of resources by conversion technologies (constants).

The G matrix aggregates different cost categories for given kinds of resources (as these categories can be used for the same purpose) thus representing the nonlinear relationship between cost and total availability of a resource. Optionally, one of these categories can be defined as an import category. In (9), this distinction between indigenous and imported primary energy is not relevant. Later (see below) it becomes important.

Q_2 describes fuel requirements connected with power plant construction (inventory requirement). Q_3 describes inventory recovery from phased-out plants. Q_2 and Q_3 apply to nuclear power plants only.

3.5. Resource Extraction (Constraints)

The focus of the energy problem is shifted more and more from the physical availability of a resource to the limitations on its economic availability. The latter is characterized by constraints on annual production (for indigenous resources) or by upper bounds on imports (described later). The constraints on annual production of indigenous resources is expressed in limits by kind of resource:

$$G_1r(t) \leq p(t) \qquad (10)$$

where: G_1 is the matrix for the aggregation of indigenous resource categories (constants)
p is the vector of production limits for any resource kind (exogenous inputs).

The difference between matrices G and G_1 is that G aggregates indigenous and import categories whereas G_1 aggregates only the indigenous categories.

3.6. Capacity Utilization (Constraints)

The idea of the following constraints is simple: no production can increase beyond installed capacity. However, because of the disaggregation of total demand for electricity into load regions, the final form of the corresponding constraints is derived in the following way:

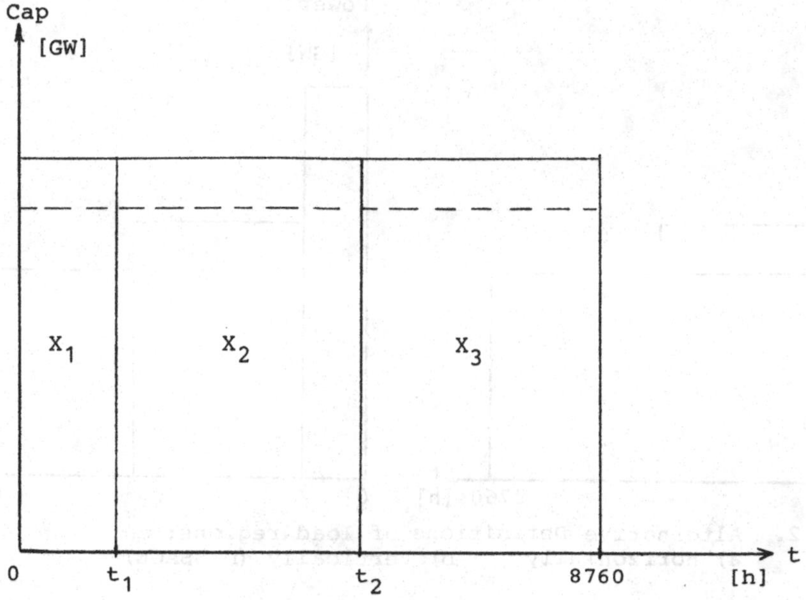

Figure 1. Illustration of Capacity Utilization

According to Figure 1, the output of any technology in any one load region cannot exceed the total capacity, reduced by some margin allowing for a certain fraction of time in which a facility is not available. Such a "safety factor" multiplied by the duration of a load region $(t_{i+1}-t_i)$ is used to calculate the information required for:

$$\begin{array}{l} B_1 x(t) \leq c(t) \\ \vdots \qquad \vdots \\ B_n x(t) \leq c(t) \end{array} \tag{11}$$

where: B_1 are the matrices defining load regions and availability of technologies in the load regions 1,...,n (constants).

Although the model (in this particular case, the matrix generator) is capable of dividing any demand sector into "any" number of load regions, this feature is presently used only for the demand for electric energy demand, where total demand is divided into three load regions.

It should be noted that a load regions are often defined in a different way. Here, a load region is defined by a segment of time; in contrast, it can be defined by segments of capacity in the area under the load curve. A graphic comparison of these two definitions is shown in Figure 2. (The load curve is ordered and approximated by a step function.)

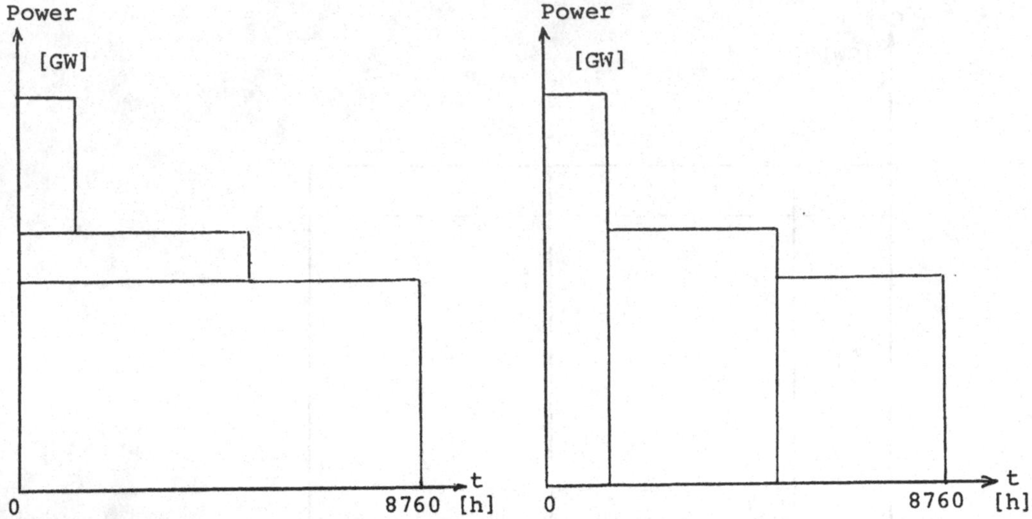

Figure 2. Alternative Definitions of load regions:
a) Horizontally b) Vertically (MESSAGE)

3.7. Build-Up Constraints (Constraints)

The fact that no technology can expand arbitrarily fast, is expressed by the following constraints on the annual build-up rates:

$$z(t) \leq \gamma z(t-1)+g \qquad (12)$$

where: γ is a (diagonal) matrix of growth parameters (constants)
g is a vector of start-up values allowing z to reach positive values after having been zero before (constants).

Constraints of this kind are optionally. In present applications, they are used almost exclusively for future technologies.

The second effect of this constraint is that it prevents overly large oscillations of the z-variables.

There is a second group of constraints on build-up rates:

$$\sum_{i \varepsilon I_1} z_i(t) \leq GUB(t) \qquad (13)$$

where: GUB(t) is a vector of absolute upper limits (exogenous parameters).

Presently, these constraints are used to limit the total construction of nuclear capacity in any one period.

3.8. Bounds (Constraints)

Bounds are constraints on single variables. As they are treated separately in many LP software systems, they are summarized here under a common heading. Furthermore, since only single variables are involved, a verbal description of the bounds seems sufficient.

Such bounds can be imposed on total output of a technology (x-variables) and on the z-variables (independently of the build-up constraints). On the resource side, total amounts of resource categories are limited [s(0)], as is annual availability of imports.

3.9. Environment (Constraints)

These equations formally belong to the group of constraints, but they are used primarily for accounting of pollutants. There is no formal difficulty in letting environmental damage enter the effective constraints or the objective function; the reason for dropping these possibilities lies in the data problem, i.e., there seems to be no sufficient agreement on the quantification of environmental impact that would allow for such an inclusion.

Tow kinds of impacts are considered: total emission of a number of pollutants and ambient concentration of some of these. The equations for emissions are:

$$e(t) = Ex(t) \tag{14}$$

where: e is the vector of emissions of pollutants (LP variables)

E is the matrix of specific emissions.

The equations for the ambient concentrations are:

$$b(t) = \sum_{\tau=1}^{t} \lambda(t-\tau)e(\tau) \tag{15}$$

where: b is the vector of concentration of pollutants (LP variables)

λ is a (diagonal) matrix of coefficients expressing the resting time of pollutants in the environment (input parameters).

3.10. Objective Function (Performance Index)

The objective function (to be minimized) is total discounted costs of energy supply. Total costs consist of capital costs (i.e., construction costs of new capacities), current costs (operating and maintenance), and fuel costs:

$$\sum_t \{\beta_1(t)(a_1(t), x(t)) + \beta_2(t)(d_2(t), z(t)) + \beta_3(t)(a_s(t), r(t))\} \rightarrow \min \tag{16}$$

where: β_i are discount factors (scalars)

a_i are vectors of cost coefficients (input parameters).

This discount factors are based on an annual discount rate.

4. SIZE OF A SAMPLE PROBLEM

For actual calculation, some of the equalities described were eliminated by inserting them into other equalities or constraints thus reducing the size of the LP matrix. Hence, the size of the following sample problem is smaller than one based on the model equations as described above. The sample problem was set up with:

6 demand sectors (one with 3 load regions, others with 1)
6 resources, 3 categories each
18 technologies
8 pollutants (for 3 of them, concentrations were calculated)
11 time periods.

This problem (and all other model runs) was solved by using standard LP packages (i.e., not using special techniques that make use of the dynamic structure of the model). The size of the resulting (single) matrix was

645 rows
737 columns

for the part with the actual constraints. The part of the accounting of the environmental impact adds

121 rows
88 columns.

This problem was solved in 90 CPU seconds of a CYBER 74 using the APEX system.

5. MODEL APPLICATION

Since any detailed description of concrete results would go beyond the framework of this paper, only aspects of present applications are being discussed, followed by an example of actual output. The effect that different parameters driving the model have on the model result differs with the size of the feasible region. In our applications, the feasible region is so small that its location in the state space is more important than the point of the optimal solution. Anyway, in this type of application of a DLP model a single optimal solution is always less important than the difference between solutions depending on different sets of input parameters (sensitivity analysis). As the location of the feasible region itself is the primary concern here, the parameters defining the constraints are more important than the cost coefficients. The most important constraints are primary energy reserves and their production, energy demand, and buildup rates of technologies.

The model output is primarily viewed as a consistent picture of the energy supply defined by the input assumptions. Important pieces of

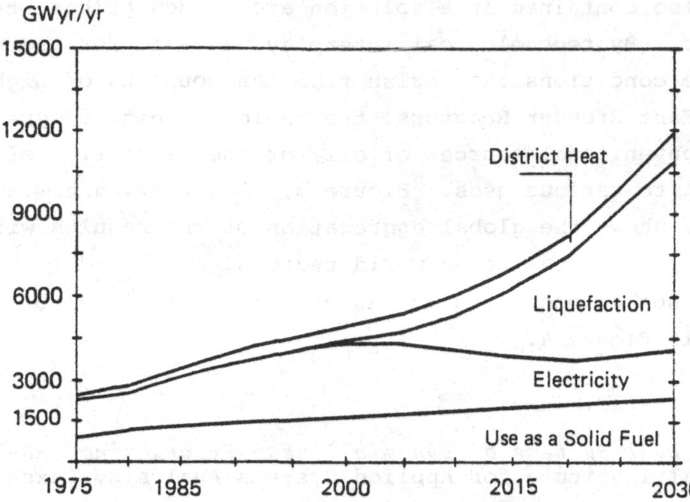

Figure 3. Global use of coal.

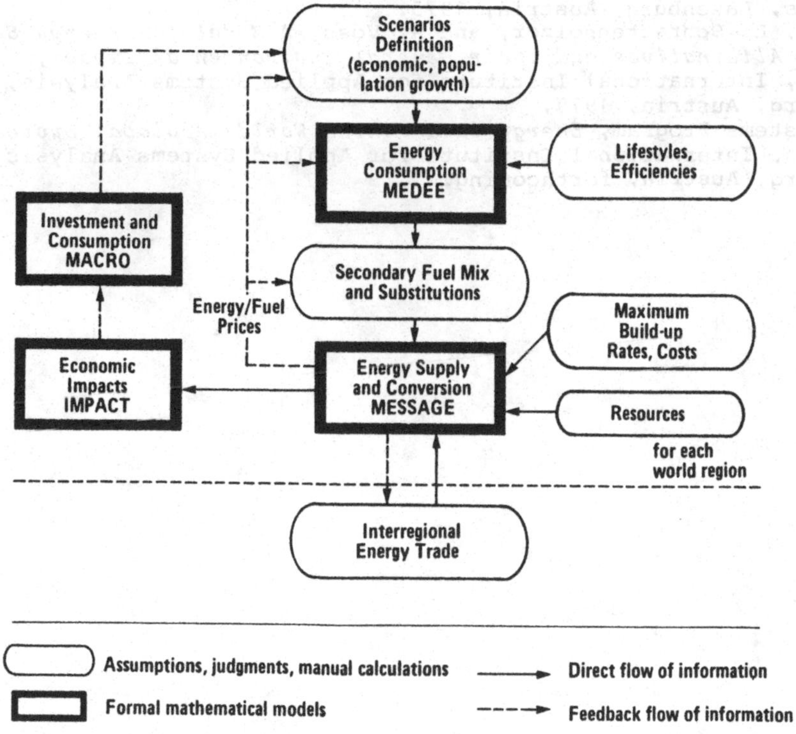

Figure 4. IIASA's Set of Energy Models: A Simplified Representation

information also contained in a solution are shadow prices, technological interplay, etc. By technological interplay we mean, for example, the results of the conditions that arise from the coupling of Light Water Reactors and Fast Breeder Reactors, the choice of either synthetic liquid fuels or nonconventional sources of oil, or the allocation of a limited amount of coal to various uses. Figure 3, illustrating how the results are processed, shows the global aggregation of the results with respect to the use of coal in the seven world regions.

The full set of models as it was used for IIASA's global analysis is described in Figure 4.

BIBLIOGRAPHY

Propoi, A.I., *Dual Systems of Dynamic Linear Programming*, RR-77-9, International Institute for Applied Systems Analysis, Laxenburg, Austria, 1977.

Häfele, W., and A.S. Manne, *Strategies for a Transition from Fossil to Nuclear Fuels*, RR-74-7, International Institute for Applied Systems Analysis, Laxenburg, Austria.

Suzuki, A., *An Extension of the Häfele-Manne Model for Assessing Strategies for a Transition from Fossil Fuel to Nuclear and Solar Alternatives*, RR-75-47, International Institute for Applied Systems Analysis, Laxenburg, Austria, 1975.

Agnew, M., L. Schrattenholzer, and A. Voss, *A Model for Energy Supply Systems Alternatives and Their General Environmental Impact*, WP-79-6, International Institute for Applied Systems Analysis, Laxenburg, Austria, 1979.

Energy Systems Program, *Energy in a Finite World--A Global Systems Analysis*, International Institute for Applied Systems Analysis, Laxenburg, Austria, forthcoming.

AN APPLICATION OF NONLINEAR PROGRAMMING TECHNIQUES TO THE ENERGY-ECONOMIC OPTIMIZATION OF BUILDING DESIGN

F. Archetti - C. Vercellis

Istituto di Matematica - Università di Milano - Milano - Italy

INTRODUCTION: A closer analysis of the process of heating (cooling) of buildings (which, e.g. in Italy, accounts for nearly 25% of the overall energy consumption) has been clearly becoming imperative, given the existing situation and the predictable trends in the cost and availability of energy.

Some mathematical models and the related computer codes [6], [7] have been developed in the last years, which can be used to simulate, in different weather conditions, the energy performance of different building designs and of different controls of the heating plant.

These simulation programs can be exploited also in a search for optimum performance designs but the numerical feasibility of this approach is severely limited by the large amount of computer time needed to scan the feasible domain accurately enough. This is particularly the case when more variables are introduced in order to enhance the sensitivity of the model to weather data.

In order to overcome these limitations of the simulation approach, the use of optimization techniques has been recently advocated by various authors [1] , [4] , [5] .

In this paper the authors, relying on a mathematical model of the heating process, set up an objective function whose optimization yields a sequence of designs of decreasing "cost" (a precise definition of cost is worked out in sect. 1) which converges to that design yielding the optimal economic balance between some "energy sensitive" parameters of the building (thickness of insulation layers, size of the windows, etc).

- This work has been performed in the framework of the "Progetto Finalizzato Energetica" of the Consiglio Nazionale delle Ricerche.

and related energy saving computed over the assumed number of years. The optimization process can be easily framed in different scenarios changing a set of parameters related to the assumed cost of money and rate of increase in the cost of energy.
A technique for constrained sensitivity analysis has been widely applied during the computations, reported in sect. 2, which yields valuable information about the influence in the value of the objective function of feasible perturbations in a neighborhood of the optimal design.

Sect. 1 Formulation of the optimization problem.

As we have stated in the introduction the main aim of this model is to help the decision maker to strike the right balance between the cost of additional insulation and other architectural parameters and the related energy saving computed over the assumed lifetime -N years- of the building.
Thus the "cost function" of our problem must account both for the operating cost (heating cost) CH, and that part of the cost of the building CM due to insulation materials, glass and concrete.
As far as CH is concerned we assume an yearly energy consumption E, constant along the years, and a constant yearly increase, by a rate ε, of the unit energy cost in the first year C_0. After the formula given in [9] it turns out

$$CH = EC_0 \frac{1-\exp((N+1)\varepsilon)}{1-\exp(\varepsilon)} .$$

As far as CM is concerned this too has to be computed over N years by the following formula

$$CM = C_R \frac{N\alpha(1+\alpha)^N}{(1+\alpha)^N-1}$$

where α is the cost of money and C_R the amount of money actually spent in glass, concrete and insulation.
Thus, in the cost function C = CH + CM, both C_R and E depend on the parameters of the building which are assumed as the control variables in the optimization, as described later on in this section.
The complex part in the computation of the "cost function" C is the

computation of E whose value depends on the design of the building, the weather conditions of the site and the control of the heating plant. The consumption E is computed as the yearly integral $E = \int E(t)dt$ where E(t) is the energy consumption at time t and the integration is performed with a step $t = 3 \sim 4$ hrs. for a synthetized year of $50 \sim 100$ days.

A basic index of the energy performance of a building is its thermal load Q(t): in our model Q(t) is computed as the quantity of heat actually given by the heating plant at time t, in order to keep the inner air temperature between 19° C and 21° C.

The energy consumption E(t) depends on Q(t) via the relation $E(t) = Q(t) \cdot \rho(Q(t))$, where $\rho(Q(t))$ is a nonlinear function of Q(t) whose values are derived experimentally and subsequently tabulated, expressing the performance of the heating plant for different values of the heating load.

For the actual computation of Q(t) we have been using the computer code NBSLD (National Bureau of Standards Load Determination) with some minor improvements in some numerical routines (the solution of the linear system, the computation of the response factors, etc.) and some more relevant modifications aimed at an effective plugging of NBSLD into an optimization software.

In the code NBSLD the building is assumed to be a modular structure: each module, later on termed "room" is bounded by NS surfaces S_i.

From the set of technological and architectural parameters which contribute to the thermal load of the "room" we are considering in the optimization, we have singled out 2 basic sets of control variables: x_1^i, i=1,2,...,NS, is the thickness of the insulation layer of the wall S_i; x_2^i, i=1,2,...,NS, the thickness of the concrete layer of S_i.

Moreover to allow for a stronger sensitivity of the optimization model to weather data the control variables x_3^i, i=1,2,...,NS, have been introduced as the ratio of window to wall surface on S_i, and x_4 as the ratio of lenght to width of the "room".

Upper and lower bounds are provided for each variable and an overall thickness bound is provided for any wall:

$$L_j^i \leq x_j^i \leq U_j^i \; ; \qquad \sum_{j=1}^{2} x_j^i \leq b_i .$$

The values of CH and CM depend on the variables $\{x_j^i\}$ respectively throughout E and C_R so that we can set the constrained optimization problem:

$$\min_{\{x_j^i,\ j=1,2,3,i=1,\dots,NS;\ x_4\}} \qquad C = CH + CM$$

subject to the above constraints.

Of course in an actual optimization a subset of control variables is fixed a priori to characterize the particular "room" we are considering (windows are allowed only in some "walls", etc.).

The heating load Q(t) of the "room" is computed solving the system $B \cdot X(t) = C(t)$ where $X_i(t)$, i=1,...,NS is the inside temperature at time t of S_i, $X_{NS+1}(t)$ is the inner temperature of the "room" and $X_{NS+2}(t) = Q(t)$.

The equations $\sum_{j=1}^{NS+2} b_{ij} \cdot x_j(t) = C_i(t) \qquad i=1,\dots,NS+1$

express the thermal equilibrium condition of S_i and of the air in the "room"; the last equation depends on the control of the heating plant and states its equilibrium condition.

The matrix B and the vector C(t) depend on the geometry of the "room" and the different heat contributions: those given by convection and radiation are expressed by explicit empirical relations. Also in C(t) is expressed the heat gain due to domestic equipment, lights, occupants and the heat loss due to air leakage.

The contribution due to heat conduction is derived from the values of temperature and heat flow which are obtained solving numerically for any layer the first order differential system equivalent to the heat conduction equation. The solution of this differential system is accomplished in the NBSLD by transform techniques: even if some numerical problems are posed also by this approach nevertheless it results in a significantly shorter computer time than that required by the finite difference methods.

Moreover a crucial advantage of tranform methods versus finite differences in the framework of this optimization model is that analytical derivatives of Q(t) with respect to the control variables of the optimization can be computed applying the implicit function theorem [1].

Sect. 2 Computational results.

The computations reported in this paper have been performed for a "room" characterized by the following general specifications.
Windows are allowed only in two walls (S_1 and S_3 oriented respectively towards North and South). The walls whose thickness is controlled in the optimization are S_1, S_3, S_2 (West) and ceiling S_5. In these walls the thickness of the concrete layer is assumed to be the same.
The wall S_4 and the floor S_6 adjoin other "rooms" of the building and do not contribute, in our model, control variables to the optimization.
Thus the following eight control variables are considered in the optimization while the other variables of the set $\{x_j^i\}$ are fixed according to the above specifications:

$w_1 = x_1^1$, $w_2 = x_1^2$, $w_3 = x_1^3$, $w_4 = x_1^5$, $w_5 = x_2^1 = x_2^2 = x_2^3$, $w_6 = x_3^1$,
$w_7 = x_3^3$, $w_8 = x_4$.

The weather data have been synthetized out of a six months period, a typical heating season in Northern Italy.
For the constrained optimization of C = CH + CM the authors have been using the program OPRQP of the OPTIMA package developed at the Numerical Optimization Centre of the Hatfield Polytechnic [2], [3].
The sensitivity analysis of the results has been performed using the program OPSEC of the same OPTIMA package [8].

Here we briefly recall the basic ideas for performing the sensitivity analysis of constrained optimization problems.
Let x* be a solution of the following problem:

$$\min f(x),\ x \in R^n, \text{ subject to the constraints } h_i(x)=0 \quad i=1,2,\dots t$$
and $g_i(x) \leq 0$, $i=t+1,t+2,\dots p$.

We assume that m constraints ($m \leq t$) are active at x* and let A be the their Jacobian matrix.
The further assumption of the linear independence of the gradients of the active constraints implies that A has rank m.
A perturbation δx is said to be feasible in x* if x* + δx satisfies the constraints active in x*, at least to a first order approximation.

Thus the feasible perturbations in x* are given by $A\delta x=0$ and define a (n-m) dimensional space.

If we restrict our attention to feasible perturbations it can be shown [8] that the sensitivity analysis of the constrained optimization problem is reduced to the unconstrained sensitivity in R^{n-m} of the problem whose Hessian matrix is obtained projecting the Hessian matrix of f(x) into the subspace of feasible perturbations.

In what follows we report the computational results of 3 cases: for each case several optimization runs have been tried from different starting points, which converged to the same optimal design.

For all cases the variables w_1, w_2, w_3 and w_7 have been set by the optimization to their upper bound, and w_6 to its lower bound.

N has been set to 20 in all cases.

1) $\varepsilon = 0.10$, $\alpha = 0.12$.

The variable w_5 assumes its minimum feasible value. The analysis performed by OPSEC results in a greater sensitivity to a perturbation in w_4 than in w_8: the same loss in the objective function is given, in a neighborhood of the optimal design, by δw_4 and by $\delta w_8 \simeq 2\delta w_4$.

2) $\varepsilon = 0.18$, $\alpha = 0.20$.

The variable w_4 is set to its maximum feasible value. The analysis performed by OPSEC results in a greater sensitivity to w_5 than to w_8. The same loss in the objective function is given, in a neighborhood of the optimal design by δw_5 and by $\delta w_8 \simeq 15\delta w_5$.

3) $\varepsilon = 0.14$, $\alpha = 0.16$.

The variable w_5 again assumes its minimum feasible value. The optimal values of w_4 and w_8 are larger than in case 1, and the sensitivity analysis yields the same results as in case 1.

Concluding remarks.

The activity reported in this paper is only a first step in the application of optimization techniques to the identification of energy optimal building designs.

The assumptions about the technological, architectural and economic parameters are, at the best, only a rough approximation to the complexities of a satisfactory model of the problem.
However the authors think that the numerical feasibility of the optimization approach has been clearly validated by the results obtained and that the model outlined in this paper could be developed into a general tool for the energy-economic analysis of the heating of a building.

REFERENCES

[1] - ARCHETTI, F., BALLABIO, D., VERCELLIS, C.: "Cost-benefit analysis of insulation in buildings via nonlinear programming" - to appear in "Numerical Optimization of Dynamic Systems" - L. Dixon and G. Szegö eds., North Holland.

[2] - BIGGS, M.C.: "Constrained minimization using recursive equality quadratic programming", in "Numerical Methods for Nonlinear Optimisation", ed. by Lootsma, Academic Press, 1971.

[3] - BIGGS, M.C.: "A numerical comparison between two approaches to the nonlinear programming problem", Technical Report N. 77, Numerical Optimisation Centre, Hatfield Polytechnic, 1976.

[4] - JUROVICS, S.A.: "Optimization applied to the design of Energy Efficient Building", I.B.M. Journal of Research and Development, Vol. 22, N. 4, 1978.

[5] - JUROVICS, S.A.: "Solar Radiation Data, Natural Lighting and Building Energy Minimization", I.B.M. Scientific Center, Los Angeles, 1979.

[6] - KUSUDA, T.: "NBSLD, the computer program for heating and cooling loads in buildings", NBS Building Science Series 69, 1976.

[7] - KUSUDA, T.: "Thermal Response Factors for multi-layer stuctures of various heat conduction systems", Paper N. 2108, ASHRAE Semi-annual Meeting, 1969.

[8] - MCKEOWN, J.J.: "Methods for sensitivity analysis", Technical Report N. 94, Numerical Optimisation Centre, Hatfield Polytechnic, 1978.

[9] - SILVESTRINI: "Il clima come elemento di progetto", Liguori, Napoli, 1978.

OPTIMIZATION OF THE SIGNAL-TO-NOISE RATIO IN THE OPTICAL DATA PROCESSING

R.Homescu

Department of Economic Cybernetics,
Academy of Economic Studies,
Calea Dorobanti 15-17, Bucharest, Romania

One of the major problems in the field of optical data processing is how to get an optimal signal-to-noise ratio. This paper will take into account the optical data processing based on holography.
Essentially, noise in holography is caused either by the coherence of light, or by the random variations of the transmittances of the processing system components, the recording medium included[1]. We shall further deal with the noise generated by the recording medium (i.e. the emulsion of the holographic plate) and establish an optimal signal-to-noise ratio, using a new optimization technique based on 2-dimensional spline functions.
Let's consider the holographic plate (see Figure 1) as a square, S, having the unit side, H=1, hence being defined by the Cartesian product $[0,1]\times[0,1]$.

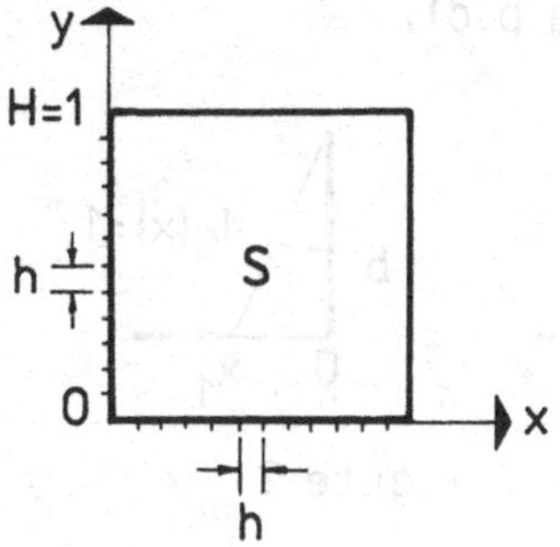

Figure 1

The sides of the square S will be divided, along the x and y axes, into the smallest possible equidistant discretization steps, h.
Let's choose the following approximation spline functions on the x axis:

$$l_i(x) = \begin{cases} \dfrac{x-x_{i-1}}{h}, & \text{for } x \in [x_{i-1}, x_i] \\ \dfrac{x_{i+1}-x}{h}, & \text{for } x \in [x_i, x_{i+1}] \\ 0, & \text{for } 0 \leq x \leq x_{i-1}, x_{i+1} \leq x \leq 1 \end{cases} \tag{1}$$

$$l_0(x) = \begin{cases} \dfrac{x_1-x}{h}, & \text{for } 0 \leq x \leq x_1 \\ 0, & \text{for } x_1 \leq x \leq 1 \end{cases} \tag{2}$$

$$l_{n+1}(x) = \begin{cases} \dfrac{x-x_n}{h}, & \text{for } x_n \leq x \leq 1 \\ 0, & \text{for } 0 \leq x \leq x_n \end{cases} \tag{3}$$

The functions expressed by equations (1), (2) and (3) are graphically represented in Figure 2 (a,b,c).

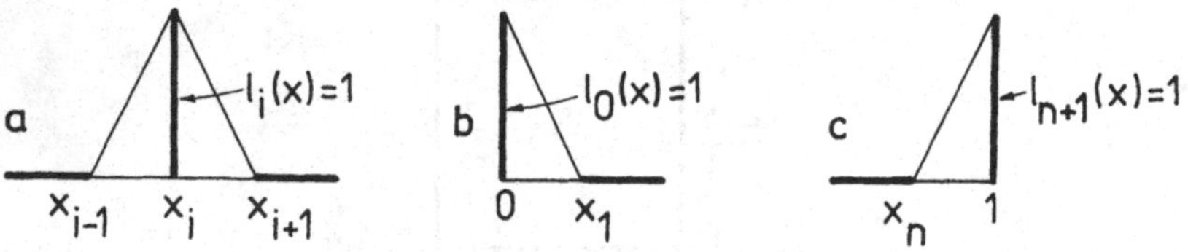

Figure 2

Let's choose now the following approximation spline function, on the y axis, having the same step h:

$$l_j(y) = \begin{cases} \dfrac{y-y_{j-1}}{h}, & \text{for } y \in [y_{j-1}, y_j] \\ \dfrac{y_{j+1}-y}{h}, & \text{for } y \in [y_j, y_{j+1}] \\ 0, & \text{otherwise} \end{cases} \tag{4}$$

$$l_0(y) = \begin{cases} \dfrac{y_1-y}{h}, & \text{for } 0\leqslant y\leqslant y_1 \\ 0, & \text{for } y_1\leqslant y\leqslant 1 \end{cases} \qquad (5)$$

$$l_{n+1}(y)= \begin{cases} \dfrac{y-y_n}{h}, & \text{for } y_n\leqslant y\leqslant 1 \\ 0, & \text{for } 0\leqslant y\leqslant y_n \end{cases} \qquad (6)$$

For these equations, the graphical representations are similar to those in Figure 2.

Let's choose now the basis spline functions on the plate S as $\left\{l_i(x)\, l_j(y)\right\}$, i.e.all the possible products, where $i=\overline{0,n+1}$, and $j=\overline{0,n+1}$. These basis spline functions will optimally approximate the transmittance T(E), where E is the exposure. Then, we will have the optimal approximated transmittance, T*(E), as:

$$T^*(E)=\sum_{k=1}^{(n+2)^2}\beta_k\, \mathcal{B}_k(x,y) \qquad (7)$$

where $E = E(x,y)$ and $\mathcal{B}_k(x,y) = \left\{l_i(x)\ l_j(y)\right\}$.

It is worth mentioning that:

$$\mathcal{B}_1 = \left\{l_o(x)\ l_o(y)\right\}$$

In order to optimally approximate the transmittance, we need to determine the β_k values (where $k=\overline{1,(n+2)^2}$), which can be obtained from the following linear system:

$$A\beta = K, \qquad (8)$$

where A is a matrix, and β is:

$$\beta = \begin{bmatrix} \beta_1 \\ \beta_2 \\ \vdots \\ \beta_{(n+2)^2} \end{bmatrix}$$

The right hand term K from equation (8) will be calculated as:

$$K = \left[k_i\right] = \left[\int_0^1\!\!\int_0^1 T(E)\mathcal{B}_i(x,y)\,dx\,dy\right]_{i=\overline{1,(n+2)^2}} \qquad (9)$$

The integral in the equation (9) is called by us, the <u>averaged transmittance.</u> It represents the value of T(E) in the points (x_i, y_i) (having $i=\overline{1,(n+2)^2}$), and it is obtained by means of system (8), in which the transmittance T(E) is included (inside the coefficients of the right hand term). To get the optimal approximated transmittance, T*(E), it is necessary to point out the value of the β's coefficients from the system (8).

In all, there are $(n+2)^2$ coefficients, where the averaged transmittance is included.

Let's consider now the way in which the matrix A is going to be obtained. For this purpose, we have the following matrix from the linear approximation[2,3]:

$$B = \frac{h}{6}\begin{bmatrix} 2 & 1 & & & 0 \\ 1 & 4 & \ddots & & \\ & \ddots & \ddots & \ddots & \\ & & \ddots & 4 & 1 \\ 0 & & & 1 & 2 \end{bmatrix} \qquad (10)$$

having (n+1) rows and (n+1) columns.

The matrix A is expressed by the tensorial product $B \otimes B$, thus having the following structure:

A = 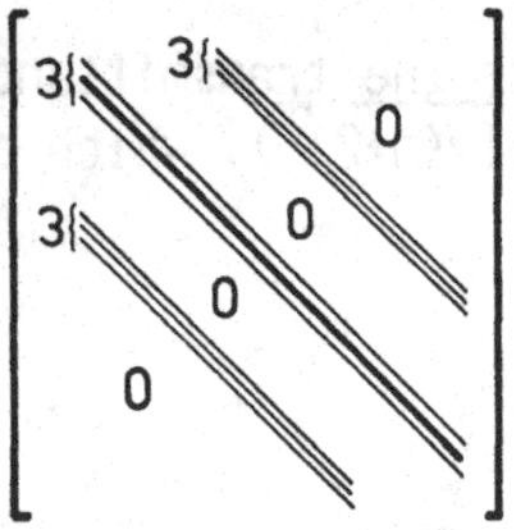 (11)

where the thicker lines indicate the main diagonal, and the thinner ones, the 8 codiagonals (nonzero); for the rest, the matrix A has null elements.

Analysing equation (10), we notice that B can be written as:

$$B = \begin{bmatrix} \frac{h}{3} & \frac{h}{6} & & & 0 \\ \frac{h}{6} & \frac{2h}{3} & \ddots & & \\ & \ddots & \ddots & \ddots & \\ & & \ddots & \frac{2h}{3} & \frac{h}{6} \\ 0 & & & \frac{h}{6} & \frac{h}{3} \end{bmatrix} \qquad (12)$$

It is obvious that the matrix A is made up of blocks which are obtained by multiplying each element of the matrix B with the whole matrix B. Consequently, the first block of the matrix A will be:

$$C = \begin{bmatrix} \frac{h^2}{9} & \frac{h^2}{18} & & & 0 \\ \frac{h^2}{18} & \frac{2h^2}{9} & \ddots & & \\ & \ddots & \ddots & \ddots & \\ & & \ddots & \frac{2h^2}{9} & \frac{h^2}{18} \\ 0 & & & \frac{h^2}{18} & \frac{h^2}{9} \end{bmatrix} \qquad (13)$$

In order to get the rest of the blocks, it is necessary to continue the same procedure. Analysing matrix C, we can notice that it is obtained by multiplying the matrix B with h/3.

Let's write now the matrix C from equation (13) as:

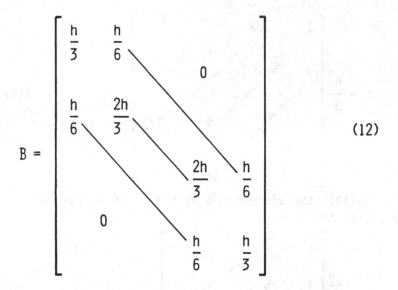

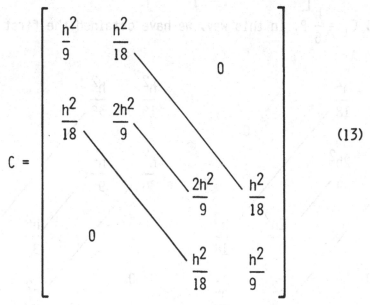

$$C = \frac{h^2}{9}\begin{bmatrix} 1 & 1/2 & & & 0 \\ 1/2 & 2 & \ddots & & \\ & \ddots & \ddots & \ddots & \\ & & \ddots & 2 & 1/2 \\ 0 & & & 1/2 & 1 \end{bmatrix} \tag{14}$$

and let's multiply the element h/6 by the whole matrix B.
The result will be:

$$C_1 = \frac{h^2}{36}\begin{bmatrix} 2 & 1 & & & 0 \\ 1 & 4 & \ddots & & \\ & \ddots & \ddots & \ddots & \\ & & \ddots & 4 & 1 \\ 0 & & & 1 & 2 \end{bmatrix} \tag{15}$$

Therefore, $C_1 = \frac{h}{6}$ B. In this way, we have obtained the first line of the matrix A:

$$\left[\begin{array}{ccccc|ccccc} \frac{h^2}{9} & \frac{h^2}{18} & & & 0 & \frac{h^2}{18} & \frac{h^2}{36} & & & 0 \\ \frac{h^2}{18} & \frac{2h^2}{9} & \ddots & & & \frac{h^2}{36} & \frac{h^2}{9} & \ddots & & \\ & \ddots & \ddots & \ddots & & & \ddots & \ddots & \ddots & \\ & & \ddots & \frac{2h^2}{9} & \frac{h^2}{18} & & & \ddots & \frac{h^2}{9} & \frac{h^2}{36} \\ 0 & & & \frac{h^2}{18} & \frac{h^2}{9} & 0 & & & \frac{h^2}{36} & \frac{h^2}{18} \end{array}\right] \tag{16}$$

Continuing the procedure, we can notice that in the matrix A there are 9 nonzero diagonals (one of them being the main diagonal), and zeroes for the remaining.
Now we can define the <u>transmittance density</u> as:

$$\iint_S \mathcal{B}_i(x,y)dxdy, \qquad (17)$$

where $\mathcal{B}_i(x,y)=\left\{l_k(x)\, l_j(y)\right\}$

The calculation of the vector K (see equation (8)), which has k_i components, is the following:

$$k_i = \int_0^1\int_0^1 T(E)\, \mathcal{B}_i(x,y)dxdy = T(E(x_i,y_i)), \qquad (18)$$

where (x_i,y_i) are network knots of the square S.
Accordingly, the linear system expressed by equation (8) which is $(n+2)^2$-dimensional, has to be solved. The vector K is $(n+2)^2$ - dimensional too, having the elements $T(E(x_i,y_i))$, that is, the values of the transmittance in all the network knots considered.
To solve the linear system (8), we have to use Gauss-Seidel's method[4].
In order to optimize the signal-to-noise ratio (S/N), we will use the optimal transmittance expressed by equation (7). The expression of the transmittance taken into account[5], is:

$$T \simeq \exp\left\{-bM\left[1 - \exp\left(-\varrho(E)\right)\right]\right\}, \qquad (19)$$

where $b = \frac{1}{2}\ln 10$, M is the total number of grains in the cell, i.e. in the minimum resolvable image area[6], and $\varrho(E)$ is:

$$\varrho(E) = \frac{1}{2}\left(\frac{E}{\alpha}\right)^2$$

In the latest equation, α is an appropriate constant corresponding to the maximum derivative of the Hurter-Driffield curve. That means that the parameter α corresponds to the exposure time when dD/dt is maximum (D is the photographic density). On the other hand, the signal-to-noise ratio[5] is:

$$\frac{S}{N} = \frac{T^2}{\sigma_T^2}$$

where the variance, σ_T^2, is given by Francis Yu[5] as:

$$\sigma_T^2 \simeq Mb^2\exp[-\varrho(E)]\exp\left\{-2bM\left[1-\exp(-\varrho(E))\right]\right\}\left\{1-\exp[-\varrho(E)]\right\} \qquad (20)$$

The optimal signal-to-noise ratio will then be:

$$\left(\frac{S}{N}\right)^* = \frac{T^{*2}}{\sigma_T^{*2}} \qquad (21)$$

To evaluate the equation (21), we must calculate the approximated variance, i.e. the σ_T^{*2}, by using the same procedure as in the case of the optimal transmittance. Thus:

$$\sigma_T^* = \sum_{p=1}^{(n+2)^2} \gamma_p \cdot \mathcal{B}_p(x,y) \tag{22}$$

where γ_p is calculated from the following system:

$$A\ \gamma = P \tag{23}$$

The matrix A is the same expressed by equation (11), and P will be:

$$P = \begin{bmatrix} \int_0^1\int_0^1 \sigma_T(x,y)\mathcal{B}_1(x,y)\,dxdy \\ \int_0^1\int_0^1 \sigma_T(x,y)\mathcal{B}_2(x,y)\,dxdy \\ \vdots \\ \int_0^1\int_0^1 \sigma_T(x,y)\mathcal{B}_{(n+2)^2}(x,y)\,dxdy \end{bmatrix} \tag{24}$$

In this way, two linear polynomials depending on x and y, will be obtained for T^* and σ_T^*. Thus, the optimal signal-to-noise ratio expressed by equation(21)will represent a ratio between two quadratic polynomials.

ACKNOWLEDGEMENTS

We express our gratitude to Dr.Alexandru I.Şchiop, professor at the Politechnical Institute of Bucharest, Department of Mathematics, for his competent suggestions and mathematical assistance.

REFERENCES

1. Vlad,V.I., et al.:
   Prelucrarea optică a informaţiei (Optical Processing of Information), Editura Academiei R.S.Romania, Bucharest (1976)
2. Schultz, M.H.:
   Spline Analysis, Prentice-Hall (1973)
3. Homescu,R.:
   On Linear Optimization in Optical Data Processing, in Economic Computation and Economic Cybernetics Studies and Research, 1 (1979)
4. Schiop, A.I.:
   Analiza unor metode de discretizare (Analysis of Some Discretization Methods), Editura Academiei R.S.Romania, Bucharest (1978)
5. Yu,F.T.S.:
   Introduction to Diffraction, Information Processing, and Holography, The MIT Press, Cambridge, Massachusetts (1973)
6. Altman,J.H. and Zweig, H.J.:
   Effect of Spread Function on the Storage of Information on Photographic Emulsion, in Phot.Sci.Eng., 7 (1963)

AN ASYMPTOTIC APPROACH TO THE DYNAMIC OPTIMIZATION OF COMPLEX CYCLIC PROCESS

W. Jankowski
Development Center for Automation
of Chemical and Oil Industry "Chemoautomatyka"
Rydygiera 8, Warsaw / Poland

A b s t r a c t. The optimization of process which consists of slow, noncyclic and fast, cyclic elements is considered. The asymptotic problem was found to be easier to solve and it allows to construct suboptimal control. Convergence properties of suboptimal control are presented. Two-layer control structure results consistently from asymptotic approach. The operation of chemical reactor is exemplified. Catalyst deactivation, regeneration and exchange are considered.

1. INTRODUCTION

The process cycles can be connected with:

- batch processing,
- periodic control,
- external disturbances.

The existence of slowly varying elements causes that the optimization of control is often stated in exploitation time being much longer then the process cycle interval.

For the optimization of catalyst deactivation Ermini [1] proposed three-level hierarchy: number of regeneration cycles, time and activity scheduling, cycle trajectory determined optimally at successive levels. Optimization of process with large number of cycles makes problems similar to those which appear when stiff system of differential equations has to be solved.

The singular perturbation was used to decompose fast and slow problems in linear state regulator [5] and in optimization of nonlinear process with fast measured disturbances [3]. In batch cycles states of fast variables at the beginning and at the end of each cycle are forced and equal - this case is considered in detail.

2. PROBLEM STATEMENT

The state of considered system is described by vector $\underline{y}$ which is composed of vectors:

$\underline{w}$ - slow, noncyclic coordinates,

$\underline{x}$ - fast, cyclic coordinates.

The state equations take the following form:

$$\frac{d\underline{w}}{dt} = \underline{g}(\underline{w},\underline{x},\underline{u}) \qquad \underline{w}(0) = \underline{w}_o \tag{1}$$

$$\frac{d\underline{x}}{dt} = \underline{f}(\underline{w},\underline{x},\underline{u}) \qquad \underline{x}(0) = \underline{x}_o \tag{2}$$

where $\underline{u}$ is the vector of control variables. The system is considered in exploitation interval $\Delta = [0, T_E]$. Let n denote a number of cycles in exploitation interval and vector $\underline{T} = [T_1, T_2, \ldots, T_n]$ denotes moments when the end of cycle occurs. In batch processing the cycle condition is:

$$\underline{x}(T_i) = \underline{x}_o \qquad i \in \overline{1,n} \tag{3}$$

It is clear that $T_n = T_E$. The performance function:

$$J(\underline{w}_\Delta, \underline{x}_\Delta, \underline{u}_\Delta) = K(\underline{w}(T_E)) + \int_0^{T_E} f_o(\underline{w},\underline{x},\underline{u})\, dt \tag{4}$$

where $\underline{w}_\Delta$, $\underline{x}_\Delta$, $\underline{u}_\Delta$ are trajectories over interval Δ, means the profit over exploitation time.

The trajectories $\underline{w}_\Delta$, $\underline{x}_\Delta$, $\underline{u}_\Delta$ that fulfil equations (1-3) and give maximum of performance (4) are the solution of necesary conditions that results from maximum principle. For each cycle they take form:

$$\frac{d\underline{x}_i}{dt} = \underline{f}(\underline{w}_i, \underline{x}_i, \hat{\underline{u}}_i) \qquad \underline{x}_i(T_{i-1}) = \underline{x}_i(T_i) = \underline{x}_o \tag{5a}$$

$$\frac{d\underline{\Psi}_i}{dt} = -\frac{\partial}{\partial \underline{x}} H(\underline{w}_i, \underline{x}_i, \underline{\lambda}_i, \underline{\Psi}_i, \hat{\underline{u}}_i) \tag{5b}$$

$$\frac{d\underline{w}_i}{dt} = \underline{g}(\underline{w}_i, \underline{x}_i, \hat{\underline{u}}_i) \qquad \underline{w}_i(T_{i-1}) = \underline{w}_{i-1}(T_{i-1}) \tag{5c}$$

$$\frac{d\underline{\lambda}_i}{dt} = -\frac{\partial}{\partial \underline{w}} H(\underline{w}_i, \underline{x}_i, \underline{\lambda}_i, \underline{\Psi}_i, \hat{\underline{u}}_i) \qquad \underline{\lambda}_i(T_{i-1}) = \underline{\lambda}_{i-1}(T_{i-1}) \tag{5d}$$

$$\hat{\underline{u}}_i = \arg\max_{\underline{u}} \left\{ H(\underline{w}_i, \underline{x}_i, \underline{\lambda}_i, \underline{\Psi}_i, \underline{u}) \right\} \tag{5e}$$

$$H(\underline{w}_i, \underline{x}_i, \underline{\lambda}_i, \underline{\Psi}_i, \underline{u}_i) = \mu \tag{5f}$$

where $H(\underline{w},\underline{x},\underline{\lambda},\underline{\Psi},\underline{u}) = f_o(\underline{w},\underline{x},\underline{u}) + \langle \underline{\lambda}, \underline{g}(\underline{w},\underline{x},\underline{u})\rangle + \langle \underline{\Psi}, \underline{f}(\underline{w},\underline{x},\underline{u})\rangle$. The initial values $T_{i-1}, \underline{w}_i(T_{i-1}), \underline{\lambda}_i(T_{i-1})$ are given recurrently, parameter μ - forced value of hamiltonian determines the moment T_i of the end of cycle. The coordination problem is to choose the initial value of slow costate $\underline{\lambda}(0)$ fulfilling the transversality condition:

$$\underline{\lambda}(T_E) = \left.\frac{dK}{d\underline{w}}\right|_{\underline{w}=\underline{w}(T_E)} \tag{6}$$

and such a value of parameter μ that gives optimal number of cycles.

The conditions (5),(6) are useful in solving the optimization problem for low values of n. The computational effort increases linearly with number of cycles, which makes problem difficult to solve for large number of cycles.

3. ASYMPTOTIC PROBLEM

Let us introduce a positive parameter $\varkappa$ to the fast state equation:

$$\frac{d\underline{x}}{dt} = \frac{1}{\varkappa}\,\underline{f}(\underline{w},\underline{x},\underline{u}) \qquad \underline{x}(0) = \underline{x}_0 \tag{7}$$

When $\varkappa \to 0_+$ the singularly perturbated process is considered, its cycle interval decreases with parameter $\varkappa$:

$$\lim_{\varkappa \to 0_+} \left| T_i(\varkappa) - T_{i-1}(\varkappa) \right| = 0 \qquad i \in \overline{1,n(\varkappa)} \tag{8}$$

Two variable expansion technique [5] is used. An artificial time variable $\tau = \frac{t}{\varkappa} \in \Delta_t = [0,\delta(t)]$ expresses the changes of fast variable along the cycle. The approximation of 0-order is denoted by subscript A, its slow variables do not change with τ and for each $t \in \Delta$ are the solution of the following system:

$$\frac{\partial \underline{x}_A(t,\tau)}{\partial \tau} = \underline{f}(\underline{w}_A,\underline{x}_A,\underline{u}_A) \qquad \underline{x}(t,0) = \underline{x}(t,\delta(t)) = \underline{x}_0 \tag{9a}$$

$$\frac{\partial \underline{\Psi}_A(t,\tau)}{\partial \tau} = -H_x(\underline{w}_A,\underline{x}_A,\underline{\lambda}_A,\underline{\Psi}_A,\underline{u}_A) \tag{9b}$$

$$\underline{u}_A(t,\tau) = \arg\max_{\underline{u}} \left\{ H(\underline{w}_A,\underline{x}_A,\underline{\lambda}_A,\underline{\Psi}_A,\underline{u}) \right\} \tag{9c}$$

$$H(\underline{w}_A,\underline{x}_A,\underline{\lambda}_A,\underline{\Psi}_A,\underline{u}_A) = \mu_A \tag{9d}$$

Slow trajectories $\underline{w}_{A\Delta},\underline{\lambda}_{A\Delta}$ are the solution of averaged differential system with two-boundary problem:

$$\frac{d\underline{w}_A}{dt} = \underset{\tau}{E}\{\underline{g}(\underline{w}_A,\underline{x}_A,\underline{u}_A)\} \qquad \underline{w}_A(0) = \underline{w}_0 \tag{10a}$$

$$\frac{d\underline{\lambda}_A}{dt} = -\underset{\tau}{E}\left\{H_w(\underline{w}_A,\underline{x}_A,\underline{\lambda}_A,\underline{\Psi}_A,\underline{u}_A)\right\} \qquad \underline{\lambda}_A(T_E) = \left.\frac{dK}{d\underline{w}}\right|_{\underline{w}=\underline{w}_A(T_E)} \tag{10b}$$

where

$$\underset{\tau}{\underline{E}}\{\underline{g}(\underline{w}_A,\underline{x}_A,\underline{u}_A)\} = \frac{1}{\delta(t)} \int_0^{\delta(t)} \underline{g}(\underline{w}(t),\underline{x}_A(t,\tau),\underline{u}_A(t,\tau))\,dt\,.$$

The numerical procedure of solving system (10) requires solving the system (9) at finite number of points $(\underline{w},\underline{\lambda})$. The number of this points does not depend on number of process cycles - this makes asymptotic problem attractive even when the substitution (5) by (9) does not give substantial computational advantage.

The cycle parameter μ is the solution of supervisory optimization problem:

$$\mu_A = \arg\max \left\{ K(\underline{w}_A(T_E)) + \int_0^{T_E} \underset{\tau}{E}\{f_0(\underline{w}_A, \underline{x}_A, \underline{u}_A)\}\, dt \right\} \tag{11}$$

Suboptimal control $\tilde{\underline{u}}_\Delta(\varkappa)$ can be derived directly from the asymptotic solution.

4. SUBOPTIMAL CONTROL - CONVERGENCE PROPERTIES

The control $\tilde{\underline{u}}_\Delta(\varkappa)$ is based on asymptotic solution:

$$\underline{u}(t,\tau) = \underline{u}_A(T_i, \frac{t-T_i}{\varkappa}) \qquad i = \max\{\, i:\ T_i < t \,\} \tag{12}$$

where the recursion for T_i takes the following form:

$$T_i(\varkappa) = T_{i-1}(\varkappa) + \varkappa \cdot \delta(T_{i-1}(\varkappa)) \qquad T_0(\varkappa) = 0 \tag{13}$$

The state trajectory and the performance factor of the system under the control $\tilde{\underline{u}}_\Delta(\varkappa)$ are denoted by $\tilde{\underline{w}}_\Delta(\varkappa), \tilde{\underline{x}}_\Delta(\varkappa), \tilde{J}(\varkappa)$. Generally $\tilde{\underline{x}}(T_i(\varkappa),\varkappa) = \underline{x}_0$ for i>0, which means that cycle condition (3) is not fulfilled, however $\tilde{\underline{u}}_\Delta(\varkappa)$ is useful for rough analysis of optimization problem.

The following sets are introduced to define convergence properties of suboptimal control:

- domain Γ of hamiltonian convexity:

$$\Gamma = \left\{ (\underline{y}, \underline{\eta}, \underline{u}):\ H_{uu} > 0,\ \ H_{yy} - H_{yu} H_{uu}^{-1} \cdot H_{uy} \geqslant 0 \right\}$$

where $\underline{\eta} = [\underline{\lambda}', \underline{\Psi}']'$ is the costate vector,

- domain Γ_K of the performance function final state component convexity

$$\Gamma_K = \left\{ \underline{w}:\ \ K_{ww} \geqslant 0 \right\},$$

- set of asymptotic solution

$$G = \left\{ (\underline{w}, \underline{x}, \underline{\lambda}, \underline{\Psi}, \underline{u}):\ \ \exists\, t \in \Delta,\ \tau \in [0, \delta(t)] \text{ that } \begin{array}{l} \underline{w} = \underline{w}_A(t),\ \underline{x} = \underline{x}_A(t,\tau),\ \underline{\lambda} = \underline{\lambda}_A(t) \\ \underline{\Psi} = \underline{\Psi}_A(t,\tau),\ \underline{u} = \underline{u}_A(t,\tau) \end{array} \right\}$$

T h e o r e m 1. Let $G \subset \Gamma$ and $\underline{w}_A(T_E) \in \Gamma_K$ then exists $\varkappa_0 > 0$ that for all $\varkappa \in (0, \varkappa_0]$ exists $\hat{\underline{u}}_\Delta(\varkappa), \hat{\underline{w}}_\Delta(\varkappa), \hat{\underline{x}}_\Delta(\varkappa)$, $\hat{J}(\varkappa) = J(\hat{\underline{w}}_\Delta, \hat{\underline{x}}_\Delta, \hat{\underline{u}}_\Delta)$, which is a local optimum of problem (1-4). ∎

T h e o r e m 2. Let $G \subset \Gamma$ and $\underline{w}_A(T_E) \in \Gamma_K$ then exists $\varkappa_0 > 0$ and $L < \infty$ that for all $\varkappa \in (0, \varkappa_0]$:

$$\max \left\{ \|\hat{\underline{w}}_\Delta(\varkappa) - \tilde{\underline{w}}_\Delta(\varkappa)\|,\ \|\hat{\underline{x}}_\Delta(\varkappa) - \tilde{\underline{x}}_\Delta(\varkappa)\|,\ \|\hat{\underline{u}}_\Delta(\varkappa) - \tilde{\underline{u}}_\Delta(\varkappa)\| \right\} < \varkappa \cdot L$$

$$|\hat{J}(\varkappa) - \tilde{J}(\varkappa)| < \varkappa^2 \cdot L\,.$$

∎

The proof results from the parametric sensitivity of the two-boundary differential problem [4].

5. CHEMICAL REACTOR EXAMPLE

The exploitation of a simple catalytic system, Fig. 1, over time T_E= 50 month is considered. Cycle is connected with catalyst coking in production phase and decoking, during regeneration. The main reaction is A→B; due to low conversion the output concentraction takes the form:

$$C_B = a \exp\left(-\frac{E_k}{R\,T}\right) \tag{14}$$

where: a= w·x - catalyst activity,
w - parameter of catalyst structure,
x - parameter of catalyst coking,
E_k- activation energy,
R - gas constant.

The catalyst coking is side reaction A→C:

$$\frac{dx}{dt} = -\,a \exp\left(\frac{-E_c}{R\,T}\right) \qquad x(0) = 1\,, \tag{15a}$$

the rate of decoking is described by linear relation:

$$\frac{dx}{dt} = (1-x) + u_2\,. \tag{15b}$$

The catalyst structure changes due to thermal sintering during production phase:

$$\frac{dw}{dt} = -\,a \exp\left(\frac{-E_s}{R\,T}\right) \qquad w(0) = 100 \tag{16}$$

The profit results from the production of B-component, and costs of regeneration:

$$f_o = \begin{cases} C_B & \text{for production phase} \\ -0.5\cdot(u_2)^2 & \text{for regeneration phase} \end{cases} \tag{17}$$

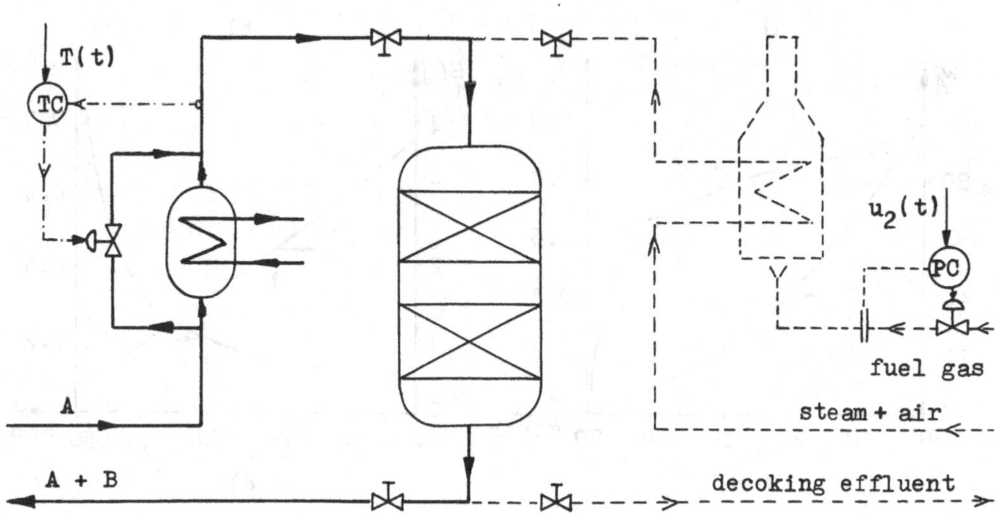

Fig. 1. Catalytic reactor system.

The fast subproblem (9) was solved for certain combination of activation energies: $E_k = E_s = E$, $E_c = 2 \cdot E$, the auxiliary control variable $u_1 = \exp(-E/RT)$ replaced temperature. The slow trajectories $w_{A\Delta}, \lambda_{A\Delta}$ were computed numerically. The set of points $(\mu, \lambda_A(0))$ for which transversality condition $\underline{\lambda}_A(T_E)=0$ was fulfilled is presented at Fig.2a. The point $\mu_A=0.273$, $\lambda_A=0.9$ was chosen to maximize the performance function. For the first cycle production phase is longer then regeneration, the rate of coking increases; the control increases during both production and regeneration phase - Fig.2b.

Catalyst sintering is rather fast at the beginning of exploitation time, λ_A - the penalty coefficient for sintering rate decreases at the end - Fig.2c. Medium profit in cycle decreases with time in spite of rapid increase of temperature at the end of exploitation time - Fig.2d.

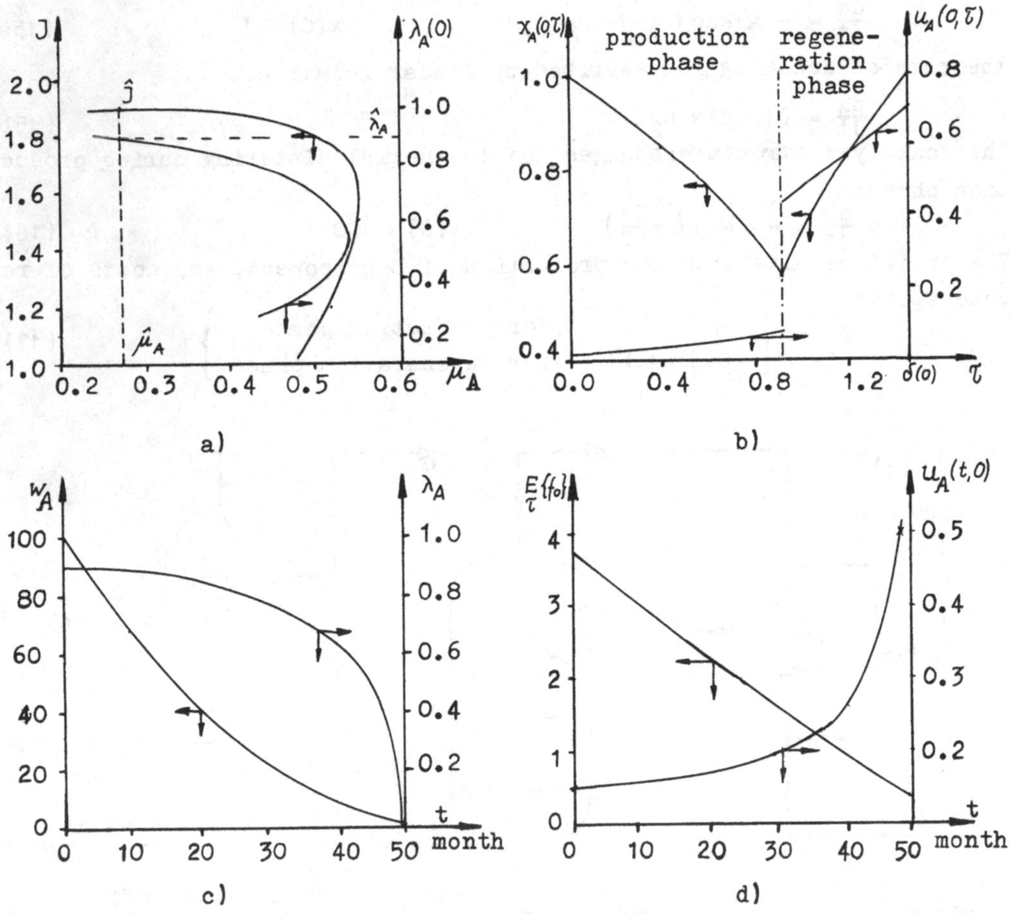

Fig.2. Asymptotic trajectories for catalytic system.

6. CONTROL STRUCTURE

In real process control uncertainties ought to be considered. The repetition control structure is the most common approach to the dynamic system. Let T_R denotes repetition interval - at $t_o=kT_R$ state and parameters of state equations are estimated using new measurements from interval $[0,t_o]$ and new optimization problem for interval $[t_o,T_E]$ is stated.

For model of batch processing the parameters $\underline{\alpha}, \underline{\beta}$ were distinguished and:

- $\underline{\alpha}$ denotes parameters constant over exploitation interval Δ,
- $\underline{\beta}$ denotes parameters constant over cycle interval Δ_i.

The suitable repetition interval is: $T_R'\approx 0.05\,T_E$ for slow state $\underline{w}$ and parameters $\underline{\alpha}$, $T_R''\approx 0.5\cdot T_i$ for fast state $\underline{x}$ and parameters $\underline{\beta}$. Consistently layer control structure was derived - Fig.3. At higher optimization layer more complex asymptotic problem over interval $[t_o,T_E]$ is solved.

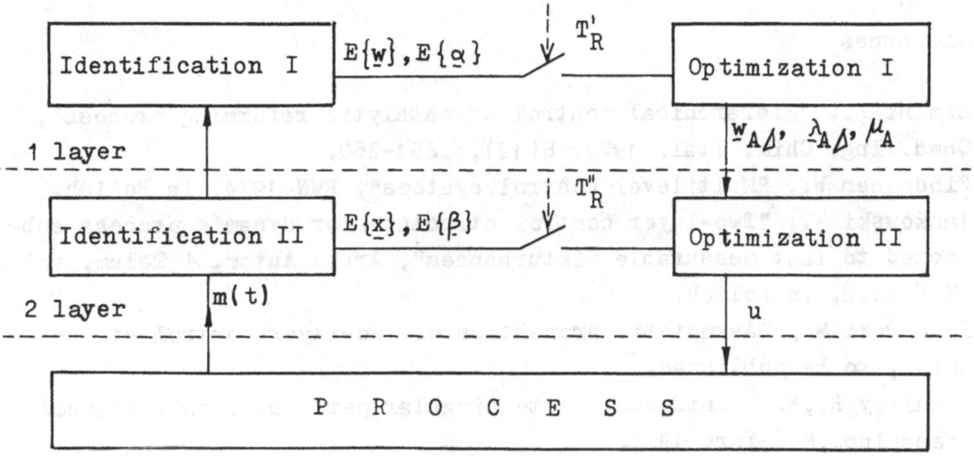

Fig.3. Two-layer control structure.

Slow variables are used to formulate second layer optimization problem as follows:

$$\max_{\underline{x},\underline{u}} \left\{ \int_{t_o}^{T_i} f_o(\underline{w}_A,\underline{x},\underline{u}) + \langle \underline{\lambda}_A, \underline{g}(\underline{w}_A,\underline{x},\underline{u})\rangle \, dt \right\} = J_i(t_o,T_i) \qquad (18a)$$

$$\frac{dx}{dt} = \underline{f}(\underline{w}_A,\underline{x},\underline{u}) \qquad \underline{x}(t_o) = E_{t_o}\{\underline{x}\} \quad \underline{x}(T_i) = \underline{x}_o \qquad (18b)$$

The end of current cycle T_i is determined at lower optimization layer to assure constant performance function improvement:

$$\frac{\partial J(t_o,T_i)}{\partial T_i} = \mu_A \qquad (18c)$$

For large number of cycles the performance function of this structure converges to the expected value in the single layer structure that works with repetition interval $T_R = T''_R$. The computational savings in two -layer structure are obvious. Due to repetition in fast layer cycle condition (3) is fulfilled.

7. CONCLUSIONS

Some advantages are connected with asymptotic approach to optimization of cyclic process over exploitation time:

- problem is decomposed and more analytical,
- computational effort does not depend on number of cycles,
- family of problems with parameter $\varkappa$ can be investigated.

The paper was motivated by problem of catalyst deactivation control, nontrivial example shows the advanteges of presented approach which may be useful for optimization of other cyclic systems, too.

References

[1] Ermini L., "Hierarchical control of catalytic reforming process", Quad. Ing. Chim. Ital. 1972, 8(12), p.251-260.

[2] Findeisen W., "Multilevel control systems", PWN 1974, in Polish.

[3] Jankowski W., "Two-layer control structure for dynamic process subjected to fast measurable disturbances", Arch. Autom. i Telem. vol XXIV No.2, in Polish.

[4] Jankowski W., "Asymptotic properties of two-layer control structure", to be published.

[5] O'Malley R.,E., "Introduction to singular perturbations", Academic Press Inc., New York 1974.

METHODS OF PERIODIC OPTIMIZATION IN STABILIZATION PROBLEMS OF BIPED APPARATUS

V.B.Larin

Academy of Sciences of Ukr.SSR Institute of Mathematics, Kiev, USSR

Consideration of legged vehicles (LV) as locomotion robots /I/ reflects the point of view, existing for present time, about hierarchical (multilevel) structure of control system of these vehicles and makes possible both the global treatment of multilevel system of human- or locomotion robot gait control and the investigations of more simple systems (containing the less number of levels). To the second direction it may be added the results of the investigations of the optimization linear periodic systems connected with the problem of LV stabilization systems synthesis. In spite of the existed opinion (/2/ p.20, /3/) about useless of application of optimization methods to the solution of the artificial gait problems, in the report it is shown that due to the use of optimization methods solving in linear model the LV stabilization problem, it is succeeded to weaken essentialy the acuteness of main problem, which hampered mathematical analysis of LV - the problem of control of great number of degrees of freedom of the system with varying constraints.
Let us proceed to the statement of the problem. In /4/ it is shown, that one may synthesize a satisfactory control system of simple biped LV (which is idelizely treated as a inverted simple pendulum) deviding the general problem into two ones: vertical vehicle stabilization can be realized by changing the force in a leg and the control problem of horizontal motion may be solved by suitable choice of coordinates of leg point of support for every step. Further LV scheme complication is connected with the addition of new elements in control system of horizontal motion (foot, compensation mass and so on). In connection with this the horizontal motion stabilization of each systems can be realized both in way of impulse control, i.e. by the choice of coordinates of leg point of support for every step (by step length control /2/), and also by continuous control (by compensation mass motion control, by posture change of zero moment point in foot and so on). Therefore for more complicated LV dynamic scheme it is necessary to combine both stabilization methods. In other words the specificity of synthesis problems of LV stabilization system (with weighted and weightless legs) is caused by the fact, that in conse-

quence of varying constraints (changing supporting legs), LV as control object in different phases of its motion is described either by differential equations or by difference equations /5,6/[x]. Thus, if it is considered the one-supported gait of biped LV (leg change is accured in time-interval τ), then, as it is shown in /6/, the varying of the error vector $\varepsilon = x - \eta$ of vehicle program motion (which is characterized by vector η) reproduction, during k-th step $((K-1)\tau < t < K\tau)$, is described by differential equation

$$\dot{\varepsilon} = F\varepsilon + Gu \tag{I}$$

and at the moments of supporting leg change $(t = K\tau)$ - by difference equation. In the case of inertiallessness of the carried leg it is described by the relation

$$\varepsilon(K\tau+0) = \varepsilon(K\tau-0) + MV(K)$$

In the case when inertion of the leg is taken into account, then/5,6/

$$\varepsilon(K\tau+0) = N\varepsilon(K\tau-0)$$

Assume that for $t=K\tau$ the changing of the vector ε is determined by the correlation

$$\varepsilon(K\tau+0) = N\varepsilon(K\tau-0) + MV(K) \tag{2}$$

In these equations x is a vehicle phase coordinates vector, u, V are the control vectors.

Let us formulate the problem. Let in time intervals $(K-1)\tau < t < K\tau$, $K=1,2,\ldots$ the object motion be described by the system of differential equations (I) and in the moments $t = K\tau$ the change of vector ε be submited to difference correlation (2). It is necessary to find such a strategy (regulator equation) of continuous and impulse controls $(u(t) = f(\varepsilon(t)),\ V(K) = \varphi(\varepsilon(K\tau-0)))$, that "object+regulator" closed system should be asymptotically stable and that this strategy should minimize the following quadratic functional (performance criteria):

$$J(t_0) = \int_{t_0}^{\infty} (\varepsilon^T Q\varepsilon + u^T Bu)\,dt + \sum_{K=1}^{\infty} V^T(K)\,C\,V(K) \tag{3}$$

In (I)-(3) matrices $F, G, B=B^T, Q=Q^T$ are periodic over t with period τ, matrices $N, M, C=C^T$ are constant. Using the usual procedure for tasks of linear-quadratic-gaussian problem of finding the minimum of the functional (3) in the quadratic form:

---

[x] As examples of such objects besides LV can serve also modern transportation facilities /7,8/.

$$\min_{u,v} \mathcal{J}(t_0) = \varepsilon^T(t_0)\, S(t_0)\, \varepsilon(t_0)$$

we shall find, that for $t \neq k\tau$

$$u = -B^{-1} G^T S \varepsilon \tag{4}$$

and for $t = k\tau$

$$v(k) = -(M^T S(k\tau+0) M + C)^{-1} M^T S(k\tau+0) N \varepsilon(k\tau-0) \tag{5}$$

Matrix S for $t \neq k\tau$ satisfies differential Riccati equation

$$-\dot{S} = SF + F^T S - SGB^{-1}G^T S + Q \tag{6}$$

and the jumps of this matrix at $t = k\tau$ are decribed by following correlation

$$S(k\tau-0) = N^T[S(k\tau+0) - S(k\tau+0) M (C + M^T S(k\tau+0) M)^{-1} M^T S(k\tau+0)] N \tag{7}$$

Periodicity of the problem in consideration (strategy does not change with changes of t_0 in (3) on the whole number of periods τ or, in other words, regulator parameters should not depend on the step number) impose the periodicity condition on the matrix S

$$S(k\tau+0) = S((k-1)\tau+0) \tag{8}$$

which together with (6), (7) and the requirement of asymptotical stability of systems (I), (2), (4), (5) completely determines the periodic matrix S. Concretization of the condition (8) leads to the discrete Riccati equation for the matrix $S(+0)$

$$S(+0) = \Phi(\tau) N^T [S(+0) - S(+0) D (\Pi^{-1} + D^T S(+0) D)^{-1} D^T S(+0)] N \Phi^T(\tau) - R(\tau). \tag{9}$$

The matrices containing in this equation (except the matrices mentioned in statement of the problem) are determined by following manner (E is the unit matrix):

$$\dot{\Phi} = \Phi(QW + F^T), \quad \Phi(0) = E,$$
$$\dot{R} = -\Phi Q \Phi^T, \quad R(0) = 0,$$
$$\dot{W} = FW + WF^T + WQW - GB^{-1}G^T, \quad W(0) = 0$$

$$MC^{-1}M^T - NW(\tau)N^T = D\Pi D^T$$

Factorization of the latest matrix is carried out in such a way,that Π^{-1} exists. The searched solution is the solution of equation (9), for which the matrix

$$(E + D\Pi D^T)^{-1} N\Phi^T(\tau) \tag{IO}$$

has eigenvalues inside of the unit circle. Thus the solution of $S(+0)$ can be found by usual methods (see, for example /9/). In special case, if in (2) $M=0$, $N=E$, then $S(+0)$ found in such a way, determines the solution of Riccati equation, that arises in problems of periodic optimization /IO/.

Let us concretize the obtained correlations, if in (3) $Q=0$. In this case $R(\tau)=0$ and the solution of Riccati equation (9) is replaced by more simple procedure - by the solution of Ljapunov equation.

Let the eigenvalues of matrix $N\Phi^T(\tau)$ be not lying on the unit circle, i.e. there exists such a matrix T, that

$$T^{-1}N\Phi^T(\tau)T = \begin{Vmatrix} \Lambda_+ & 0 \\ 0 & \Lambda_- \end{Vmatrix}$$

eigenvalues of quadratic matrix Λ_+ lying outside of the unit circle and those of Λ_- inside of the unit circle. In this case matrix $S(+0)$ can be presented in the form:

$$S(+0) = (T^T)^{-1} \begin{Vmatrix} Y & 0 \\ 0 & 0 \end{Vmatrix} T^{-1}$$

Symmetric matrix Y is determined by Ljapunov equation

$$Y^{-1} - \Lambda_+^{-1} Y^{-1} (\Lambda_+^T)^{-1} = \Lambda_+^{-1} g_{11} (\Lambda_+^T)^{-1}$$

The matrix g_{11} contained in this equation is obtained by breaking the matrix in blocks

$$T^{-1}D\Pi D^T (T^T)^{-1} = \begin{Vmatrix} g_{11} & g_{12} \\ g_{21} & g_{22} \end{Vmatrix}$$

(The dimensions of matrices g_{11}, Y and Λ_+ are coincide).

By such a choice of matrix $S(+0)$ the eigenvalues of matrix (IO) coincide with the eigenvalues of matrices Λ_+^{-1} and Λ_-. The searched periodic solution of equations (6), (7) for $0 < t < \tau$ has a form:

$$S = (T^T \Phi)^{-1} \left\| \begin{matrix} (Y^{-1} + \gamma_{11})^{-1} & 0 \\ 0 & 0 \end{matrix} \right\| (\Phi^T T)^{-1}$$

where the matrix γ_{11} is determined in the following way:

$$\left\| \begin{matrix} \gamma_{11} & \gamma_{12} \\ \gamma_{21} & \gamma_{22} \end{matrix} \right\| = \Gamma, \ \dot{\Gamma} = -(\Phi^T)^{-1} G B^{-1} G^T \Phi^{-1}, \ \Gamma(0) = 0$$

The given expression of matrix S to points that optimization of system of stabilization in accordance with criterien (3), generally speaking, leads to the non-stationary feed-back coefficients matrix, determined by (4), even in the case of stationary object (matrices F, G do not depend on time). It is interesting to note that the results of some researches (see /2/,p.422) are pointing to expediency of the use of variable coefficients of feed-back circuit in the tracking systems of anthropomorphous mechanisms. One of the possible ways of overcoming the difficulties connected with the realization of non-stationary coefficients matrix of feed-back loop is to realize the LV control in discrete time moments by means of the digital computer. Let us restrict ourselves only by the description of discrete version of synthesis problem of LV stabilization system. Let the moments $t_{i,K}$ break the time of k-th step into ℓ equal intervals, on everyone of which $(t_{i-1,K} < t < t_{i,K})$ the controling influence u, contained in (I), is constant (components of vector u are step functions of time). Assumption about the piecewise constancy of the vector u permits to describe vehicle motion during the step not by differential equation (I), but by suitable difference correlation

$$\varepsilon(i+1) = \Psi(i)\,\varepsilon(i) + \Theta(i)\,u(i) \qquad \text{(II)}$$

the transition to which from equation (I) is carried out in the usual manner (see, for example, /II/). As it is followed from the periodicity of system (I), the difference equations (2), (II) are the periodic finite-difference system, the optimization procedure of which by the quadratic performance criterion is similar to the above considered and is reduced to the solution of equation of type (9) /I2/. Due to the use of this procedure, finally, the stabilization algorithm forming controling influences in discrete time moments is obtained. Such a way of control can be comparatively easily realized on the

base of technical means of modern digital techniques.
In frame of the linear-quadratic-gaussian problem basing on the algorithm of construction of periodic solution of Riccati matrix equation described above, it is possible to consider more complicated statements of synthesis problem. Thus, for example, one may to assume that only a part of phase coordinates of object is measured and the results of measurements are distorted by the additive random noise /I3/, to take into account the delay, which is caused by the time of navigation data processing and by the time, that is necessary for the controling signal forming by electronic computer /I4/, to synthesize stabilization system for jumping vehicle /I5/, to use the visual information for dynamic loads descent by movement on the uneven surface /I6/ and so on. It should be pointed out that it is the described optimization approach to the problem of LV stabilization system which gives the possibility to solve the problems of such kind, using the unified mathematical apparatus. However, in spite of this it should not be forgotten that considered algorithmes of LV regulators synthesis guarrantee the asymptotical stability of "object+regulator" closed systems only in linear approximation. Strictly speaking, it should be asserted that the stabilization system assures going out of LV on program trajectory only under the small enough influences. Therefore, under finite influences in view of non-linearity of LV mathematical model, the question about estimation influences area under which the regulator realizes going on program motion, requires the special consideration. Furthermore, in working out the control systems of comparatively complicated (anthropomorphous) LV the supplementary questions are arising: is it possible to obtain the effective control system by separately synthesizing both the vertical stabilization and horizontal stabilization systems in more complicated (in comparison with the described in /4/) LV; is it possible to use the available arbitrariness in statement of synthesis stabilization system problem (for example, the choice of elements of matrices B, Q, C in functional (3) or in its analogue in case of synthesis of descrete stabilization system for suppression of non-desirable non-linear effects and so on. These questions were investigated in /I7/ by the mathematical modelling of plane motion of LV, which is idealized as a three-link (weighted body and two weighted legs) but, in contrast to /5/, in/I7/ it was assumed the telescopic structure of legs (it was necessary for solution of vertical motion control problem) and was postulated that every leg is supplied with the foot. The given in /I7/ results of mathematical modelling of LV plane motion demonstrate effectiveness of

synthesized linear stabilization algorithmes. Thus, under the fixed parameters of stabilization system the vehicle can stand on the spot, set out, rise on inclination in 23°, stop approximately at a distance of I.5 steps by gait speed 3 km/hour.

REFERENCES

[I] Popov, E.P., A.F. Vereshchagin, S.L. Zenkevich: Manipulating Robots: Dynamics and Algorithms. /in Russian/, Nauka, Moscow, 1978.

[2] Vukobratovich, M.: Legged Robots and Antropomorphous Mechanisms. /in Russian/, Mir, Moscow, 1976

[3] Vukobratovich, M, D. Stokich: A Simplified Control Procedure for Strongly Linked Nonlinear Large-Scale Mechanical Systems. /in Russian/, Avtomatika i Telemekhanika, No. 2, 1978, pp. 12-25

[4] Larin, V.B.: Stabilization of Biped Apparatus. /in Russian/, Izv. A.N. SSSR, Mekhanika Tverdogo Tela, No. 5, 1976, pp. 4-13

[5] Golliday, C.L., Jr, H. Hemami: An Approach to Analyzing Biped Locomotion Dinamics and Designing Robot Locomotion Controls. IEEE Trans. Auto. Control, AC-22, 6, 1977, pp. 963-972

[6] Larin, V.B.: Stabilization of Horizontal Motion of Biped Apparatus. /in Russian/, Izv. A.N. SSSR, Mekhanika Tverdogo Tela, No. 5, 1978, pp. 35-44

[7] Meisinger, R.: Optimale Regelung periodischer System mit sprungformiger Zu-stansanderung. ZAMM 57, 1977, T.79-T.81

[8] Popp, K.: Stabilitätsuntersuchung für das System Magnetschwebefahrzug-Fahrweg. ZAMM 58, 1978, T.165-T.168

[9] Vaughan, D.R.: A nonrecursive Algebraic Solutution for the Discrete Riccati Equation. IEE Trans. Auto. Control, AC-15, 5, 1970, pp. 597-599

[10] Bittanti, S., A. Locatelli, C. Maffezzoni: Second-Variation Methods in Periodic Optimization. J. Optimizat. Theory and Appl., 14, No. 1, 1974, pp. 31-49

[11] Astrom, K.J.: Introduction to Stochastic control Theory. Academ. Press, New York, 1970

[12] Larin, V.B.: Optimization of Periodical Systems. /in Russian/, Doklady A.N. SSSR, Vol. 239, No. 1, 1978, pp. 67-70

[13] Larin, V.B.: Stabilization of Biped Apparatus without Full Information about its Phase Coordinates. /in Russian/, Matematicheskaya Fizika, Vypusk 25, Naukova Dumka, Kiev, 1979, pp. 38-49

[14] Naumenko, K.I.: Stabilization of Horizontal Motion of Biped Apparatus without full Information. /in Russian/, Preprint No.78.31, Inst. of. Mathematics of the USSR Acad. of Sci., Kiev, 1978

[15] Larin, V.B.: Control of Locomotion Systems. /in Russian/, Preprint No. 78.15, Inst. of Mathematics of the USSR Acad. of Sci. Kiev, 1978

[16] Bordiug, B.A., V.B. Larin: Utilization of Visual Information in Control of Legged Apparatus. /in Russian/, Preprint No. 79.9, Inst. of Mathematics of the USSR Acad. of Sci., Kiev, 1979

[17] Karpinskii, F.G.: A Model of Biped Apparatus /description and motion control/. /in Russian/, Preprint No. 78.23, Inst. of Mathematics of the USSR Acad. of Sci., Kiev, 1978.

COMPARISON OF OPTIMAL AND SUBOPTIMAL METHODS FOR PULP MILL PRODUCTION CONTROL

K. Leiviskä
University of Oulu
Department of Process Engineering
Division of Control Engineering
Oulu, Finland

INTRODUCTION

In this paper the application of two algorithms to the calculation of the production schedules of the sulphate pulp mill consisting of fibre lines and a chemical recovery cycle is discussed. The first one is based on the time delay algorithm originally developed by Tamura and the second one is the suboptimal algorithm developed by Singh and Coales. These algorithms were modified so that the specific features of the problem considered can be taken into account. These include the compensation of planned shut-downs, identification of infeasible situations, consideration of the steam balance etc.

In the modelling of the pulp mill six different processes must be considered, namely: a drying plant, a bleach plant, a digester house, an evaporation plant, a recovery furnace and a causticization plant. For energy balance calculations also an auxiliary boiler, a bark boiler, must be considered.

The performance of these algorithms was compared using simulations with UNIVAC 1100/20 computer of the University of Oulu. The comparison was carried out using a simplified form of the original problem, namely the optimization of the pulp mill fibre line, only. This is much simpler problem, really, than the original one, but all the essential features can be included, except the balancing of the generation and consumption of the steam.

MATHEMATICAL MODEL

Figure 1 shows the simplified flow diagram of the sulphate mill fibre line. In the modelling of this system the state vector consists of the amounts of material in each intermediary storage, $x_1 \ldots x_3$, and the control vector correspondingly of the production rates $u_1 \ldots u_3$. The given pulp production is considered as a deterministic disturbance. In addition to that also following assumptions must be made:

1. The consistencies of the fibre flows are constants during the scheduling period.
2. The dynamics of each process can be neglected.

Now we can write the model equations as

$$\frac{dx_1}{dt} = b_1 u_1(t-\theta_1) - v_1(t)$$

$$\frac{dx_2}{dt} = b_2 u_2(t-\theta_2) - u_1(t) \qquad (1)$$

$$\frac{dx_3}{dt} = u_3(t) - u_2(t),$$

where θ_1 and θ_2 are the delays characteristic to the processes u_1 and u_2. The coefficients b_1 and b_2 can be determined, if the following parameters are given:

1. The consistency of the fibre flow from the bleach plant, 12 %
2. The consistency of the fibre flow to the bleach plant, 10 %
3. The consistency of the fibre flow from the washing plant, 10 %
4. The consistency of the fibre flow to the washing plant, 10 %

No fibre losses are considered, but, of course, constant losses can be included. Now we have

$$\frac{dx_1}{dt} = 0.8056\ u_1(t-\theta_1) - v_1(t)$$

$$\frac{dx_2}{dt} = u_2(t-\theta_2) - u_1(t) \qquad (2)$$

$$\frac{dx_3}{dt} = u_3(t) - u_2(t).$$

The discrete time presentation is the most natural way to formulate the production control problems. Therefore the model equations must also be discretized. Of course the variables are also constrained as

$$\bar{x}^{Min}(k) \le \bar{x}(k) \le \bar{x}^{Max}(k) \qquad (3)$$

$$\bar{u}^{Min}(k) \le \bar{u}(k) \le \bar{u}^{Max}(k)$$

PROBLEM FORMULATION

The solution of the production control problem must fulfil following requirements /4/:

1. The given production schedule of the dried pulp must be realized.
2. The number of production rate changes must be minimized
3. The intermediary storages must not be empty or flow over.
4. The given target levels of the intermediary storages at the end of the scheduling period must be reached.
5. The generation and consumption of the steam must be balanced.
6. The indirect storaging of the steam must be possible.

Let us consider a general objective function

$$J = \frac{1}{2}\sum_{i=1}^{N} (\pi_i(x_i(K)) + \sum_{k=0}^{K-1} f_i(x_i(k),u_i(k),k)), \tag{4}$$

where N is the number of the subprocesses and K is the number of the scheduling intervals.

For the state variables the reference trajectories $\bar{x}^o(k)$ can be determined which provide the most advantageous situation as for the unplanned shut-downs. For instance, the production of a bottleneck process can be maximized by using a high target level of the storage before this process and, of course, a low one of the storage after this process. Also in some other situations the reference trajectories are practical. Now we can write

$$\begin{aligned} \pi_i(x_i(K)) &= \pi_i(x_i(K) - x_i^o(K)) \\ f_{ix}(x_i(k)) &= f_{ix}(x_i(k) - x_i^o(k)) \end{aligned} \tag{5}$$

Because the production rate changes cause disturbances, they must be avoided. This is possible by denoting

$$f_{iu}(u_i(k)) = f_{iu}(u_i(k) - u_i^o). \tag{6}$$

The reference production can be determined when the required pulp production and the planned shut-downs of the processes are given. In the following a linear-quadratic objective function is used

$$J = \frac{1}{2}\Big(\sum_{i=1}^{N} ||(x_i(K) - x_i^o(K)||^2_{\Omega_i(K)} + \sum_{k=0}^{K-1} (||x_i(k) - x_i^o(k)||^2_{\Omega_i(k)}$$

$$+ ||u_i(k) - u_i^o||^2_{R_i(k)} \Big)\Big). \quad (7)$$

The problem is to

$$\min_{x,u} \quad J \quad \text{when } \bar{x}(0) = x_o \quad (8)$$

and taking the system constraints into account.

COMPARISON OF THE ALGORITHMS

The methods

Figure 2 shows a hierarchical structuring of the procedure using Tamura's algorithm /1,2,3/. This algorithm uses, in principle, the decomposition of the Lagrangian according to the discrete time index k. If the weighting matrices in the linear-quadratic objective function are diagonal, only parametric one variable optimization problems on the lowest level exist. These are solved using the values of the costate variables calculated on the upper level. This can be done using a standard conjugate gradient algorithm. Tamura's algorithm has shown out to be very efficient, because it is capable to deal with the time delays and constraints in a very easy way /4/.

The idea of suboptimal control proposed by Singh and Coales /5/ was applied using the following procedure (Fig. 3):

1. The schedule for the fibre flow from the bleach plant is calculated so that the general requirements, that were presented before, are fulfilled. This is done on the lower level using Tamura's algorithm. On the upper level the flow to the bleach plant is calculated using the process model of the bleach plant.
2. The same procedure is repeated to calculate the production rates of the washing plant and the digester house.

Examples

Here the scheduling period is 48 hours and the scheduling interval is 4 hours.

Example 1: The shut-down of the drying plant during intervals 5 and 6.

Example 2: Example 1 together with the filling of the washed pulp storage.

Example 3: Example 1 together with the shut-down of the digester house during interval 3.

Table 1 shows the computing time, the number of iterations and the relative value of the objective function for both algorithms together with the computer memory requirements.

CONCLUSIONS

According to Table 1 the suboptimal approach has some advantages

1. Faster computing
2. Smaller use of computer memory
3. Routines are very simple and the functioning of them can be understood very easily.

It has also some disadvantages

1. The suboptimal performance. It calculates much more production rate changes than Tamura's algorithm, also in simple cases. The smoothing of these changes is, of course, no problem.
2. It can be applied only for serial systems. Therefore the chemical recovery cycle must be 'cut', in the whole mill case.
3. There are problems, if we want to include the steam balance in the objective function.
4. In the bottleneck cases some a priori information may be required. This is shown in Fig. 4, which shows graphically the production rates and the storage situations for both approaches in the case of Example 3. Now the small size of storage x_3 limits the solution very strongly and because of the shut-down of the

digester house (u_3) also the production rate of the washing plant (u_2) must be diminished. This is calculated by Tamura's algorithm (Fig. 4a). The suboptimal approach calculates the production rates starting from u_1 and finishing to u_3. Therefore the required restriction of the production rate of the washing plant during the 3rd interval must be given as a priori information.

As a conclusion it can be said that Tamura's algorithm can be directly applied to all the Examples given before. It has been shown to be applicable to the whole mill optimization, too /6/. The suboptimal approach can be applied to serial systems and in some cases, it required additional information.

LITERATURE

1. Singh M.G., Drew A.W., Coales J.F., (1975). Comparisons of practical hierarchical control methods for interconnected dynamical systems. Automatica, 11, 331-350.

2. Tamura H. (1973). Application of duality and decomposition in high order multistage decision processes. Cambridge University Engineering Dept., Report CUED/B-Control TR 49.

3. Tamura H. (1973). A discrete dynamical model with distributed transport delays and its hierarchical optimization for preserving stream quality. IEEE Trans. Systems, Man and Cybernetics SMC-4, 424-431.

4. Leiviskä K., Uronen P. (1979). Dynamic optimization of a sulphate mill pulp line. Preprints of IFAC/IFORS Symposium, Toulouse, France, 6-8 March.

5. Singh M., Coales J. (1975). A heuristic approach to the hierarchical control of multivariable serially connected dynamical systems. Int. J. Control, 21, 4, p. 575-586.

6. Leiviskä K., Uronen P., (1979). Hierarchical control of an integrated pulp and paper mill - Principles and examples. Report No. 113. Purdue Laboratory for Applied Industrial Control, West Lafayette, Indiana.

Table 1. The computing time, the number of iterations, the relative value of the objective function (that for Tamura's algorithm denoted by 1) for each example. Also the computer memory use for both algorithms is presented. Here the main programs that read the necessary data for calculations and a subroutine including the algorithm are separately considered

	Example	Tamura's algorithm	The suboptimal algorithm
Computing time (s)	1	3.2	1.2
	2	3.2	1.2
	3	4.3	1.3
Number of iterations	1	44	75
	2	44	75
	3	68	78
Relative value of the objective function.	1	1	3.4
	2	1	3.5
	3	1	1.3
Memory used (Main program)		15.8	5.5
Memory used (Algorithm)		5.7	5.0

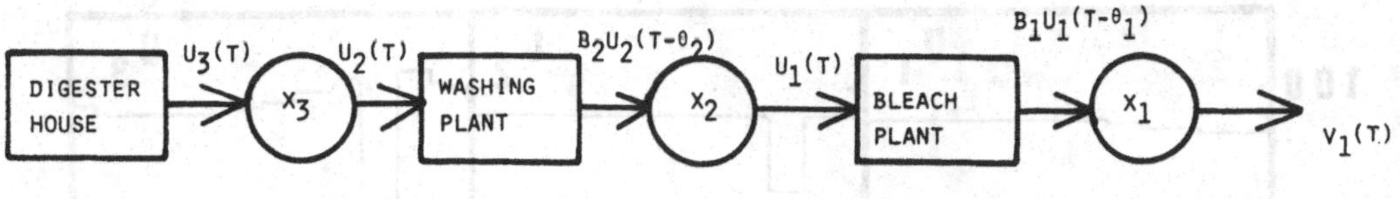

Figure 1. The simplified flow diagram of a pulp mill fibre line.

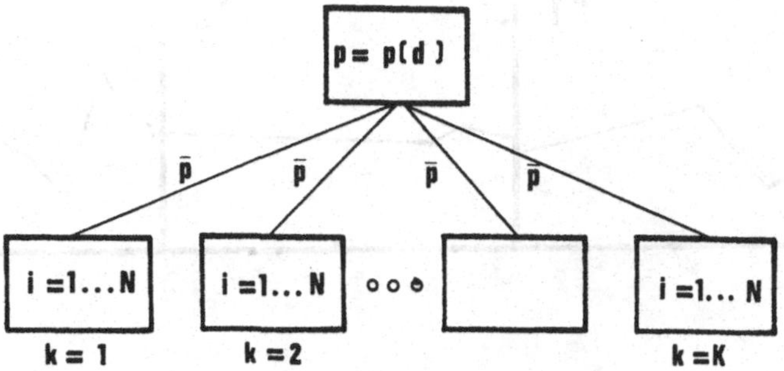

Figure 2. The hierarchical structure of Tamura's algorithm.

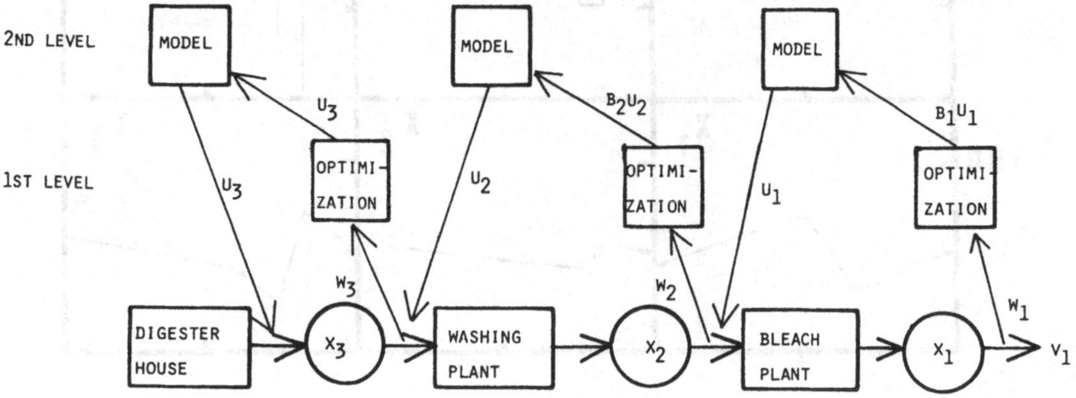

Figure 3. The suboptimal approach, principle of application.

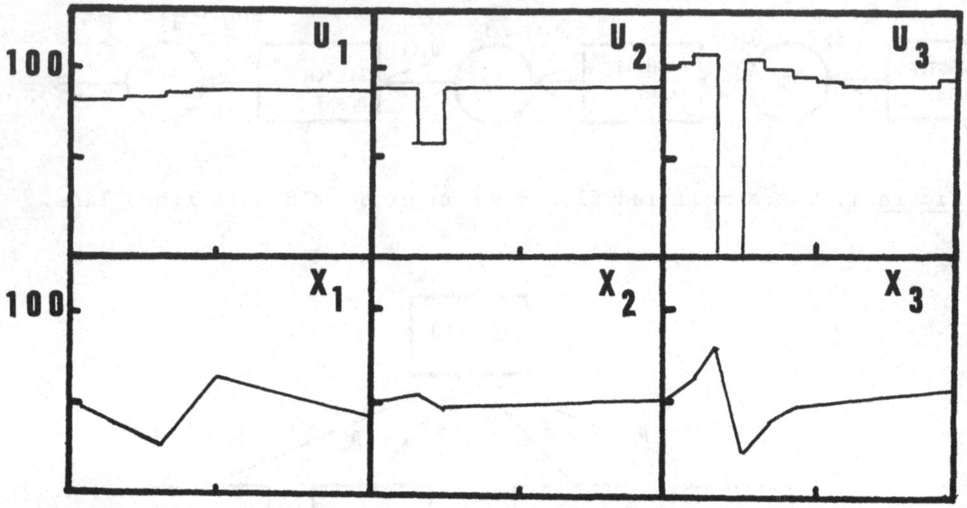

A.

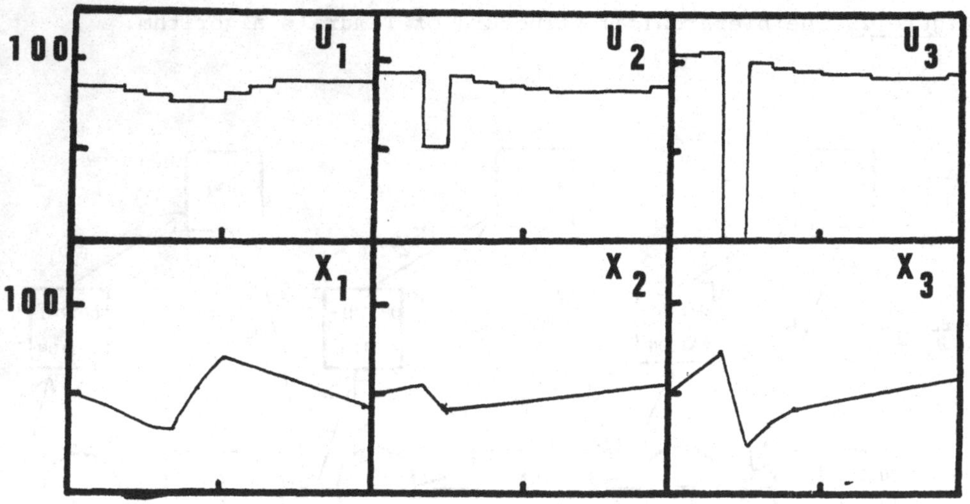

B.

Figure 4. The sample of results. The production rates and state trajectories for Example 3. A. Tamura's algorithm. B. The suboptimal approach. All the symbols as denoted in Fig. 1.

STREAMS OF INFORMATION IN THE PROCESS OF SYSTEMATIC MODELLING OF COMPLEX TECHNICAL OBJECTS ON THE EXAMPLE OF VESSEL ENGINES

Antoni Podsiadło, Jacek Sobociński
The Institute of Basic Engineering Sciences
Merchant Navy Academy
ul.Czerwonych Kosynierów
Gdynia, Poland

1. Introduction

The attempt to increase the efficiency of working of technical objects is accompanied by the increased demand for additional information. This justifies current tendencies leading to the intensification and development of information systems.

The actual designing of a given information system should be preceeded by the working out of certain models which could be treated as a simplification of reality. That simplification results from the elimination of any relations and elements that contain information which is useless and unimportant as far as the aim of a research is concerned. Therefore, the stage of constructing a given model must neccessarily include the estimation of the value of information involved. Information representing maximum profitability is regarded as a difference between the value of information and the cost of this obtainment. If seems only right to accept the value of information as the criterion of the similarity of a given model to an object being modelled.

Since 1970 - Merchant Navy Academy in Gdynia has been exploring the problem "Optimization of utilization of vessels, engines and other appliances in the sea economy". The purpose of a part of that research is to define a possibility and method of constructing a model of a technical objects, taking into account the criterion of the information value of its elements. Within the same research an attempt has been made to analyse the information concerning technical efficiency of working of a given object. That analysis consists in quantitative evaluation of the influence exercised by defects of all elements of a given object upon the realization of exploitation tasks. It is suggested that this evaluation should be based on a factor; called - the information value of an element. This information value of an element in turn, is based on a particular analysis and synthesis of information concerning the structure and function of an object. All the elements of a given object are arranged according to their diminishing information value and thus they create a hierarchic model of an object. The possibility of constructing such a model is specially important for testing of complex technical objects. The knowledge of the hierarchic model makes it possible, for example, to indicate, among thousands of elements, those which determine the efficiency of working of an object. As it was proved, by various research, those elements constituate only a small per cent of the total number of all elements.

2. The process of modelling of a given object

Modelling of an object consists in a gradual, proporcional to the collected information eliminating of those elements of a given object which are unimportant from the point of view of the technical efficiency of working.

Three stages can be distinquish in the process of modelling [fig.1]:

- modelling of the structural content of the information,
- modelling of the functional content of the information,
- information analysis of the system.

2.1. Modelling of the structural content of the information

The main aim of this stage is the identification of a modelled object as a system. A model of such a system should make it possible to evaluate the influence exercised by defects of each element upon the possibility of realization of a technical task by the system. The starting-point is to define a relation creating the system while taking into consideration the aim of research and a task of the system [2]. That relation makes possible to include the particular elements into the system as well as to distinquisch those qualities of the elements which are important from the point of view of the realized research. The relation creating the system introduces the division of the set containing elements of the system into subsets containing basic elements, reserve elements and passive elements. The system $\sigma'_0 = \langle E', S' \rangle$ as an isomorphic representation of the object described in the technical documentation, is obtained as a result of the structural analysis based on the relation creating the

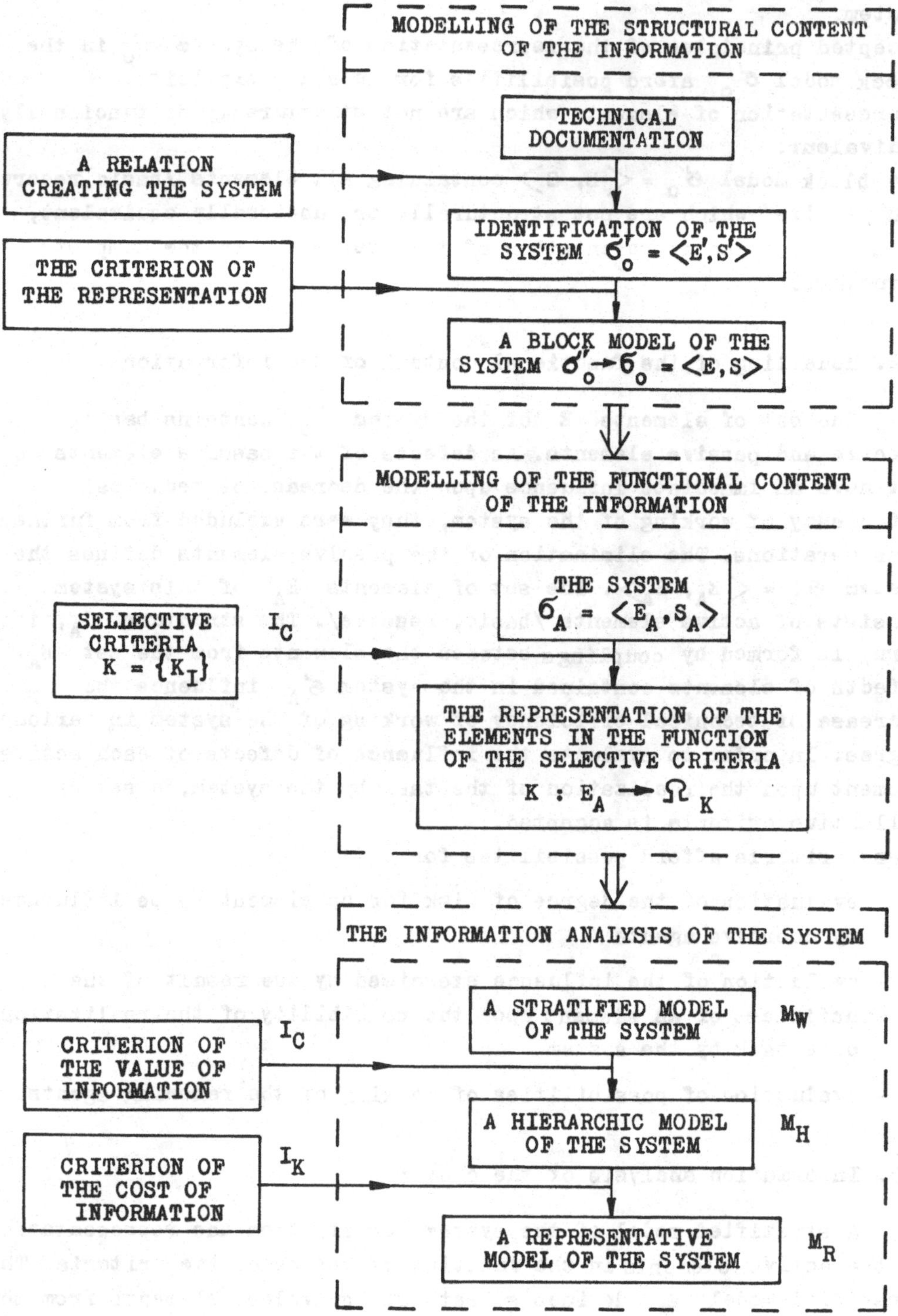

FIG. 1. The process of modelling of a given object

system.
Accepted principles of the representation of the system σ'_0 in the block model σ_0 aford posibilities for mutually explicit representation of elements which are not structurally or funcionally equivalent.
The block model $\sigma_0 = \langle E, S \rangle$ containing all elements /basic, reserve and passive/ which are not structurally or funcionally equivalent, constituates the representation of the system σ'_0 in the complete structure.

2.2. Modelling of the functional content of the information

The set of elements E of the system σ_0 contains basic, reserve and passive elements. As defects of the passive elements do not have an immediate influence upon the decrease of technical efficiency of working of the system, they were excluded from further considerations. The elimination of the passive elements defines the system $\sigma_A = \langle E_A, S_A \rangle$. The set of elements E_A of this system consists of active elements /basic, reserve/. The structure S_A, in turn, is formed by couplings between the elements from the set E_A. Defects of elements contained in the system σ_A influence the decrease of technical efficiency of working of the system in various degree. In order to evaluate the influence of defects of each active element upon the realization of the task by the system, a set of sellective criteria is accepted.
These criteria afford possibilites for

- evaluation of the degree of risk for on element to be influenced by coercive agents
- evaluation of the influence exercised by the result of the unfithess of an element upon the possibility of the realization of a task by the system
- evaluation of possibilities of working of the renewing agents.

2.3. Information analysis of the system

A stratified model of the system results from the representation of the active elements in the function of the selective criteria. The stratified model M_W defines subsets of equivalent elements from the point of view of accepted criteria of evaluation. The employment of these criteria leads to the division of the set of elements into several subsets which differ in number /strata/ and kind /substrata/ of the qualities considered as important ones.

The purpose of the information analysis is a quantitative evaluation of the information value of elements. This creates the necessity for the hierarchism of selective criteria, determined by their contribution to the information value. In this way, all the sellective criteria are reduced to one synthetic estimation.

The method of evaluating the information value of elements, based on the results of their sellection according to accepted criteria has been presented in [1].
Arranged in accordance with the information value, the subsets of elements /substrata distinquished in the model M_W/ constituate the hierarchic model M_H of the system. These subsets form the suceeding hierarchic levels /h = 1,2 .../ of the model.

When the evaluation of the cost of information is possible, the optimal choice of a model /a representative model M_R/ is defined by the maximum profitability of information refarded as a difference between the value of information and the cost of its obtainment.

3. The empiric verification of the process of modelling

In order to check the usefullness of the presented process of modelling for practical applications, hierarchic models of vessel engines of main propulsion /constructed by Sulzer firm - RD and RND/ have been worked out.

Calculated value of the factor $I_{C(h)}$ /the information value/ as a function of a number of elements forming the succeeding selections of a model of the tested engines is presented in the figure no 2.

A number of 85 choices /h = 1 ÷ 85/ has been distinquished. Consequently; the first choice incluedes $N_{e(1)}$= 17 elements /these are elements which satisfy all criteria/ while the fifth choice includes $N_{e(50)}$= 323 elements. The last - eighty fifth choice comprises $N_{e(85)}$= 1022 elements; that is to say, all those active elements of the system which satisfy at least one criterion.

The increase in number of elements of the hierarchic model results in the diminishing increase in the value of information. For example /fig.2/ for the first hundred elements of the model, the value of information is I_C = 0,53, and for the next hundred and twenty the increase in value is the same as for all the rest 802 elements.

Knowing the value of information $I_{C(h)}$ for suceeding choices the model, it is possible to define the representative model /i-e.

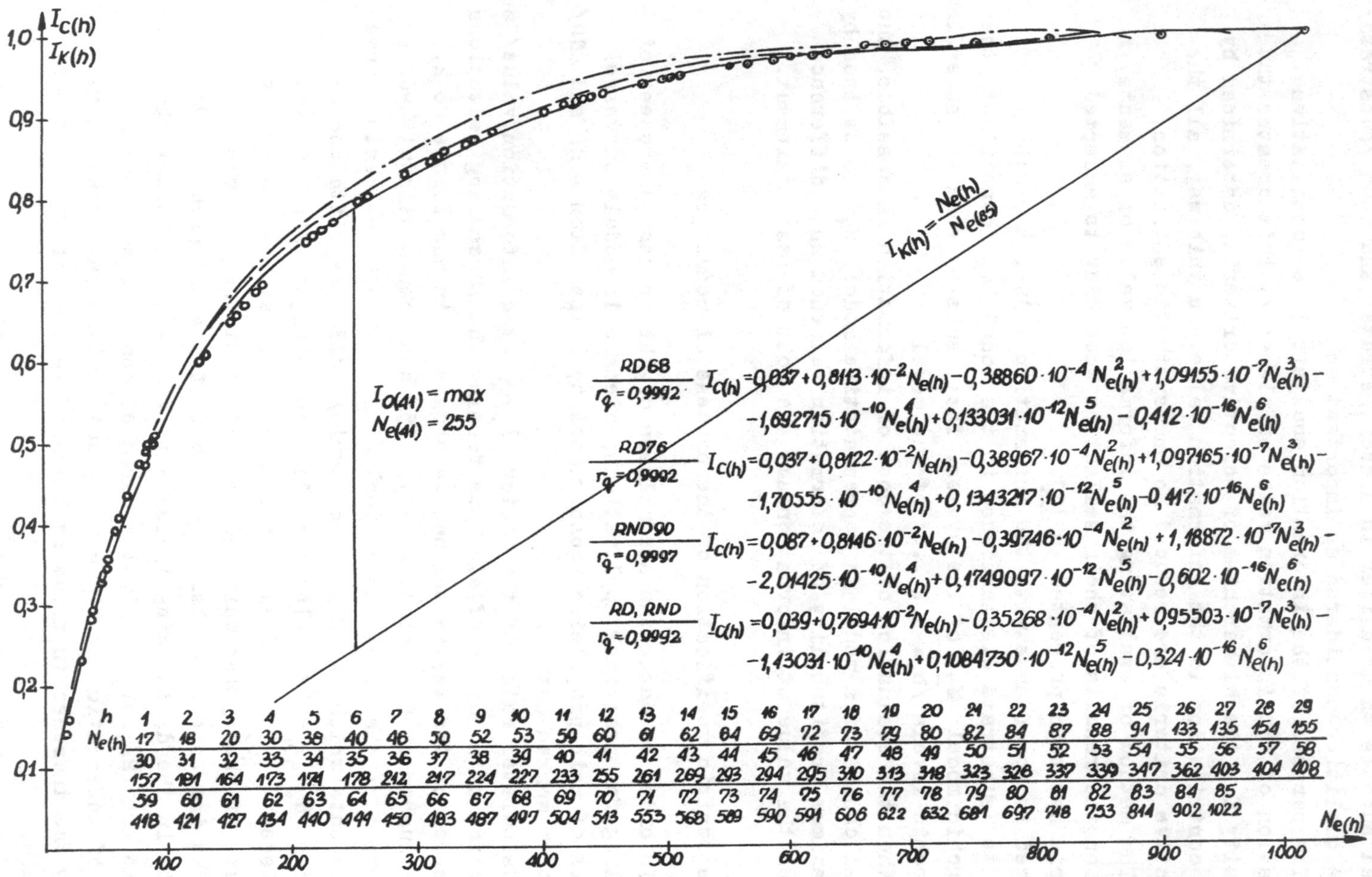

FIG.2. The information value as a function of a number of elements forming the suceeding selections of a model of a vessel engine of main propulsion represented by Sulzer's RD and RND.

the one consisting of important elements/ while taking into consideration the cost of information.
The choice of the representative model is determined by the maximum profitability of information, /i-e the maximum difference between its value $I_{C(h)}$ and the relative cost $I_{K(h)}$ of its obtainment/.

In the particular case of the tested types of engines, the representative model is the forty first choice, which comprises 255 elements characterised by the value $I_{C(41)} = 0{,}797$.

The checking of the adequacy of this model requires the comparaison of the calculated "a priori" value information to the data drown from the eyploitation of 67 vessels representing a total time of exploitation of 305 years were used.

The possibility to define a representation model is a potential source of valuable information, because

- representative model includes only 8% /per cent/ of all kinds of elements
- defects of these elements are the cause of up to 84,5% of the lost time of exploitation as it was recorded in the set of tested objects.

4. References

[1] Podsiadło A.: Sposób wstępnej oceny wpływu niezdatności elementów na efektywność działania systemu /rozprawa doktorska/ IBMER 1978.

[2] Ujemow A.I.: Metody budowy i rozwoju ogólnej teorii systemów. Prakseologia Nr 2/46/1973.

CONTENTS

LIST OF AUTHORS

---

[x] paper not received

Lecture Notes in Control and Information Sciences

Edited by A. V. Balakrishnan and M. Thoma

Vol. 22: Optimization Techniques
Proceedings of the 9th IFIP Conference on
Optimization Techniques,
Warsaw, September 4–8, 1979
Part 1
Edited by K. Iracki, K. Malanowski, S. Walukiewicz
XVI, 569 pages. 1980

Vol. 23: Optimization Techniques
Proceedings of the 9th IFIP Conference on
Optimization Techniques,
Warsaw, September 4-8, 1979
Part 2
Edited by K. Iracki, K. Malanowski, S. Walukiewicz
XV, 621 pages. 1980

Vol. 26: D. L. Iglehart, G. S. Shedler
Regenerative Simulation of Response
Times in Networks of Queues
XII, 204 pages. 1980

Vol. 27: D. H. Jacobson, D. H. Martin, M. Pachter, T. Geveci
Extensions of Linear-Quadratic Control Theory
XI, 288 pages. 1980